NUMBER WORLDS

Accelerate Math Success

Teacher Edition

Sharon Griffin

Douglas H. Clements

Julie Sarama

McGraw Hill

mheonline.com

Copyright © 2015 McGraw-Hill Education

All rights reserved. No part of this publication may be reproduced or distributed in any form or by any means, or stored in a database or retrieval system, without the prior written consent of McGraw-Hill Education, including, but not limited to, network storage or transmission, or broadcast for distance learning.

Permission is granted to reproduce the material contained on pages A1–A58 on the condition that such material be reproduced only for classroom use; be provided to students, teachers, or families without charge; and be used solely in conjunction with *Number Worlds, Intervention Level D Teacher Edition*.

Send all inquiries to:
McGraw-Hill Education
8787 Orion Place
Columbus, OH 43240

ISBN: 978-0-02-129477-0
MHID: 0-02-129477-1

Printed in the United States of America.

10 11 12 13 LKV 24 23 22 21 20

Common Core State Standards© Copyright 2010. National Governors Association Center for Best Practices and Council of Chief State School Officers. All rights reserved.

Contents

Getting Started T6

Unit 1 Number Sense within 100 — 16

Week 1 Constructing Whole Numbers
Overview 18
Lesson 20
Project-Based Learning 30

Week 2 Numbers on a Line
Overview 32
Lesson 34
Project-Based Learning 44

Week 3 Tens and Ones
Overview 46
Lesson 48
Project-Based Learning 58

Week 4 Visualizing and Constructing Whole Numbers
Overview 60
Lesson 62
Project-Based Learning 72

Week 5 Number Patterns
Overview 74
Lesson 76
Project-Based Learning 86

Week 6 Whole Number Relationships
Overview 88
Lesson 90
Project-Based Learning 100

Unit 2 Number Sense to 1,000 — 102

Week 1 Understanding the Base-Ten Number System
Overview 104
Lesson 106
Project-Based Learning 116

Week 2 Constructing Whole Numbers to 999
Overview 118
Lesson 120
Project-Based Learning 130

Week 3 Representing Number Systems
Overview 132
Lesson 134
Project-Based Learning 144

Week 4 Place Value to 1,000
Overview 146
Lesson 148
Project-Based Learning 158

Week 5 Skip Counting within 1,000
Overview 160
Lesson 162
Project-Based Learning 172

Week 6 Comparing Whole Numbers
Overview 174
Lesson 176
Project-Based Learning 186

Contents

Unit 3 Addition — 188

Week 1 Addition Fundamentals
- Overview 190
- Lesson 192
- Project-Based Learning 202

Week 2 Mastering the Basic Facts
- Overview 204
- Lesson 206
- Project-Based Learning 216

Week 3 Solving Addition Problems
- Overview 218
- Lesson 220
- Project-Based Learning 230

Week 4 Addition Tools and Strategies
- Overview 232
- Lesson 234
- Project-Based Learning 244

Week 5 Addition Word Problems within 100
- Overview 246
- Lesson 248
- Project-Based Learning 258

Week 6 Solving Addition Word Problems within 1,000
- Overview 260
- Lesson 262
- Project-Based Learning 272

Unit 4 Subtraction — 274

Week 1 Subtraction Fundamentals
- Overview 276
- Lesson 278
- Project-Based Learning 288

Week 2 Mastering Basic Subtraction Facts
- Overview 290
- Lesson 292
- Project-Based Learning 302

Week 3 Solving Subtraction Problems
- Overview 304
- Lesson 306
- Project-Based Learning 316

Week 4 Subtraction Tools and Strategies
- Overview 318
- Lesson 320
- Project-Based Learning 330

Week 5 Subtraction Word Problems within 100
- Overview 332
- Lesson 334
- Project-Based Learning 344

Week 6 Solving Subtraction Word Problems within 1,000
- Overview 346
- Lesson 348
- Project-Based Learning 358

Contents

Unit 5 Geometry and Measurement 360

Week 1 Linear Measurement
Overview 362
Lesson 364
Project-Based Learning 374

Week 2 Measurement Tools
Overview 376
Lesson 378
Project-Based Learning 388

Week 3 Time Measurement to the Half Hour
Overview 390
Lesson 392
Project-Based Learning 402

Week 4 Time Measurement to the Nearest Five Minutes
Overview 404
Lesson 406
Project-Based Learning 416

Week 5 Attributes of Shapes
Overview 418
Lesson 420
Project-Based Learning 430

Week 6 Graphs
Overview 432
Lesson 434
Project-Based Learning 444

Appendix

Blackline Masters A1
About Math Intervention B1
Program Research B2
Content Strands B6
Scope and Sequence B17
Glossary C1

Math Intervention for Grades Pre-K to 8

Help struggling students accelerate math success with a proven approach.

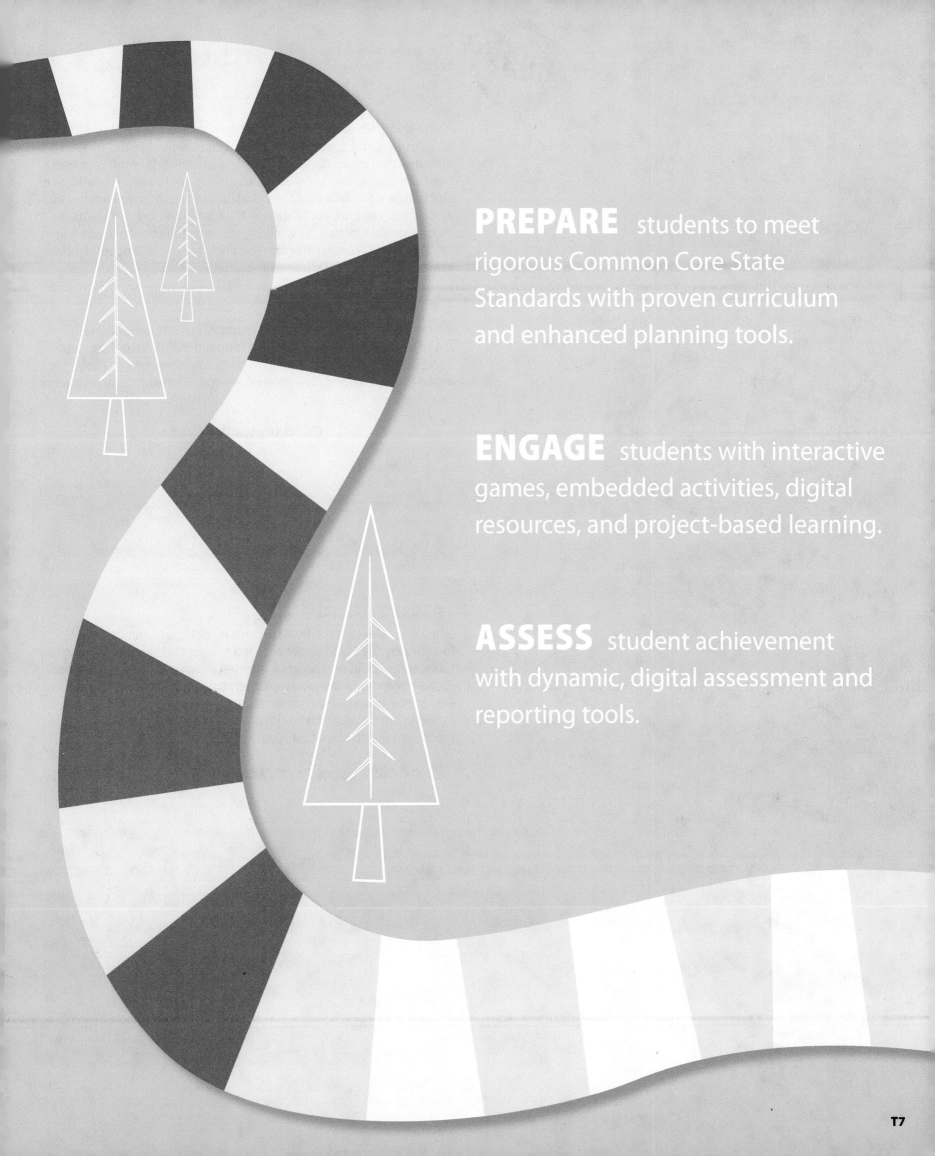

PREPARE students to meet rigorous Common Core State Standards with proven curriculum and enhanced planning tools.

ENGAGE students with interactive games, embedded activities, digital resources, and project-based learning.

ASSESS student achievement with dynamic, digital assessment and reporting tools.

Meet the Common Core State Standards

Number Worlds is now built to the Common Core State Standards, concentrating on specific standards to effectively accelerate mathematical understanding. Daily lesson activities emphasize using communication, logic, reasoning, modeling, tools, precision, structure, and patterns to solve problems. All student activities, reflections, and assessments require application of the Common Core Standards for Mathematical Practice. By allowing students to focus on mastering key standards, **Number Worlds** effectively connects concepts to close achievement gaps for students in grades Pre-K to 8.

Prevention

Prevention Levels (A–C) focus on foundations in number sense. These levels prepare students with foundational skills and concepts necessary to succeed with math in the future. Prevention Levels include 32 weeks of daily instruction—now with additional instruction on money and time!

CCSS topics Per Grade

Level A Grade Pre-K	Level B Grade K CCSS Key Standards	Level C Grade 1 CCSS Key Standards
Students acquire well-developed counting and quantity schemas.	Students develop a well-consolidated central conceptual structure for single-digit numbers.	Students link their central conceptual structure of numbers to the formal number system.

Intervention

Intervention levels (D–J) help students unlock the Common Core by focusing on Key Standards at each grade level (2–8) and supporting the foundational skills and concepts needed to achieve these objectives. Each level contains five, six-week intensive units that focus on concepts aligned to the mathematical domains of the CCSS. Designed for flexibility, units can be taught in any order or in isolation with a placement test to identify student needs.

CCSS Topics Per Grade and Unit

	Level D Grade 2	Level E Grade 3	Level F Grade 4	Level G Grade 5	Level H Grade 6	Level I Grade 7	Level J Grade 8
Unit 1	Number Sense within 100	Number Sense	Number Sense	Number Sense	Number Sense	Number Sense	Number Sense
Unit 2	Number Sense to 1,000	Addition	Addition & Subtraction	Multiplication & Division	Operations Sense	Operations Sense	Operations Sense
Unit 3	Addition	Subtraction	Multiplication	Operations with Decimals	Algebra	Algebra	Algebra
Unit 4	Subtraction	Multiplication & Division	Division	Operations with Fractions	Statistical Analysis	Statistical Analysis	Statistical Analysis
Unit 5	Geometry & Measurement	Geometry & Measurement	Geometry & Measurement	Geometry & Measurement	Geometry & Measurement	Geometry & Measurement	Geometry & Measurement

Accelerate success with a research-proven approach

Results with Number Worlds

Rigorous field testing shows that students who began at a disadvantage surpassed the performance of students who began on level with their peers simply with the help of the Number Worlds program. A longitudinal study, measuring the progress of three groups of children from the beginning of Kindergarten to the end of Grade 2, demonstrates the program's efficacy.

Both the treatment and control groups tested one to two years behind normative measures in mathematical knowledge, while the normative group was on track. The treatment group used Number Worlds, while the other two groups used a variety of other mathematical programs during the course of the study. The treatment group using Number Worlds met and exceeded mean developmental scores by the end of Grade 2. The control group continued to fall behind its peers.

Results with Building Blocks

Building Blocks software, the result of research funded by the National Science Foundation, features online activities and an adaptive assignment engine that guides children through research-based learning trajectories. This software is utilized throughout the Number Worlds program.

In research studies, Building Blocks software was shown to increase children's knowledge of essential mathematical concepts and skills. One study tested Building Blocks against a comparable math program and a no-treatment control group. All classrooms were randomly assigned the "gold standard" of scientific evaluation. Building Blocks children outperformed both the comparison group and the control group. Results indicate strong positive effects with achievement gains near or exceeding those recorded for individual tutoring.

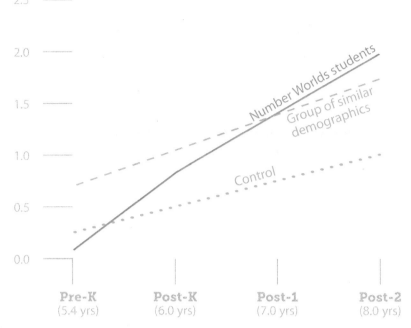

Longitudinal study showing mean developmental scores in mathematical knowledge during Grades K-2

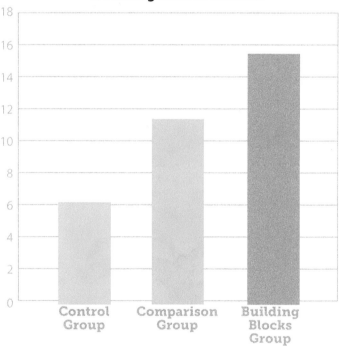

For additional program research visit *SRANumberWorlds.com*

Your roadmap to mastering the CCSS

The Number Worlds Weekly Planner provides helpful information before lessons begin, making teacher preparation simple and effective. Weekly Planners map out the entire week of lessons, complete with Learning Objectives and all of the resources needed to maximize instructional time. Teachers can access all digital planning tools and resources online, making Number Worlds easy-to-use in a variety of educational settings.

Skills Focus gives teachers a quick overview of how the week progresses across learning objectives to effectively meet weekly Key Standards.

Skills Focus
- Use objects and pictures to create equal groups and to identify a product.
- Relate repeated addition to pictures and groups to identify a product.
- Develop a conceptual knowledge of multiplication through the use of models.

All instruction and activities are purposefully tied to the Key Standard for the Week.

Math at Home extends learning and encourages support from home.

Daily lessons build math connections

Every lesson includes features designed to easily build connections between daily instruction and the Key Standard for the week. Tools to help students build math vocabulary skills and relate concepts to real-world situations are embedded in each lesson.

Daily lesson activities emphasize using communication, logic, reasoning, modeling, tools, precision, structure, and patterns to solve problems. Student activities, reflections, and assessments require application of the Common Core Standards for Mathematical Practice.

Key Common Core State Standards are listed and defined in each lesson, reinforcing the focus for the week.

Standard CCSS
3.OA.1 Interpret products of whole numbers, (e.g., interpret 5 × 7 as the total number of objects in 5 groups of 7 objects each). *For example, describe a context in which a total number of objects can be expressed as 5 × 7.*

Find the Math helps students relate math concepts to real-world situations with a weekly discussion. *

Find the Math
Encourage students to identify objects in groups.
Use the following to begin the discussion.

- If you own a pet, what are some additional items to you need to purchase for it? Possible answers: food; treats; bedding; vitamins

Have students complete *Student Workbook*, page 5.

Vocabulary builds mathematical language and understanding throughout the lesson.

* Featured in the Intervention levels (D–J).

Suggestions for Implementation

If You Have 30 Minutes

		Levels A-C		Levels D-J	
		TIME	ACTIVITY	TIME	ACTIVITY
Days 1-4	ENGAGE	30	Activity Card	15	Activity Card (Develop)
				15	Student Workbook (Practice)
Day 5	ENGAGE	15	Free-Choice Activity Card	15	Practice
	ASSESS	15	Formal Assessment	15	Formal Assessment

If You Have 45 Minutes

		Levels A-C		Levels D-J	
		TIME	ACTIVITY	TIME	ACTIVITY
Days 1-4	ENGAGE	30	Activity Card	15	Activity Card (Develop)
				15	Student Workbook (Practice)
		15	Interactive Differentiation	15	Interactive Differentiation
Day 5	ENGAGE	15	Free-Choice Activity Card	15	Student Workbook (Practice)
	ASSESS	15	Formal Assessment	15	Formal Assessment
		15-20	Project-Based Learning	15-20	Project-Based Learning

If You Have 60 Minutes

		Levels A-C		Levels D-J	
		TIME	ACTIVITY	TIME	ACTIVITY
Days 1-4	WARM UP	5	Warm Up Card	5	Prepare
	ENGAGE	30	Activity Card	15	Activity Card (Develop)
				15	Student Workbook (Practice)
		15	Interactive Differentiation	15	Interactive Differentiation
	REFLECT	5	Extended Response	5	Think Critically
	ASSESS	5	Informal Assessment	5	Informal Assessment
Day 5	WARM UP	5	Warm Up Card	5	Prepare
	ENGAGE	15	Free-Choice Activity Card	15	Student Workbook (Practice)
	REFLECT	5-10	Extended Response	5-10	Think Critically
	ASSESS	15	Formal Assessment	15	Formal Assessment
		15-20	Project-Based Learning	15-20	Project-Based Learning

Provide an effective RtI solution

With a research-proven curriculum and extensive field testing, Number Worlds supports RtI and helps schools meet their academic objectives. As RtI encourages working with at-risk students early on, Number Worlds is the only math intervention curriculum with a built-in prevention program for grades Pre-K–1.

Number Worlds gives students the confidence and skills to excel in math. Best of all, it allows teachers to make a positive difference in their students' lives.

Meet student needs with flexible lesson structures

Daily routines drive each Number Worlds lesson to simplify classroom instruction. Adaptable lesson structures are designed to be customized for various educational settings, making Number Worlds an effective solution for RtI Tiers 2 and 3.

WARM UP: Prepare students for each lesson with a quick warm-up activity related to a real-world topic.

ENGAGE: Student Engagement is the heart of the Number Worlds program. Students develop and practice key math concepts with engaging games and activities.

REFLECT: Critical thinking reinforces students' understanding of math concepts through writing and discussion activities.

ASSESS: Ongoing progress monitoring with informal and formal assessments allow teachers to quickly track progress, accelerate learning, and provide additional practice to struggling students.

Let us help you get started!

McGraw-Hill Education offers a variety of customized Professional Development solutions.

For additional information and PD resources visit *SRANumberWorlds.com*

Engage all learners with built-in differentiation

Number Worlds is designed to be effective for a variety of learners. Each lesson includes embedded resources to engage struggling students, English Language Learners, special-needs students, and other diverse student populations.

Support English Language Learners by building academic language and creating context before the lessons begin.

Creating Context
Math activities are an excellent way to give English Learners practice listening to and speaking English. The natural repetition of procedural and counting language replaces tedious drill with authentic, active experience. Build wait time into the process, and provide a low-stakes environment that makes the activities enjoyable.

Interactive Differentiation provides opportunities for students to spend more time learning critical math concepts with hands-on and digital interactive games.

Interactive Differentiation
Consult the **Teacher Dashboard** for grouping suggestions. You can also use performance on the Engage activity to guide students.

Independent Practice

For additional practice with repeated addition use the Set Tool. Have students select the Stamp tool and the number of marbles in each group and place groups on the mat.

Supported Practice

For additional support, use the 100 Table with students.

- Tell students that you will use the 100 Table to show repeated addition.
- In the Highlight menu, choose either the top or bottom "count by" drop-down menu.
- Pick a number from 1–12, then choose a "start on" value of 0.
- A pattern will highlight on the table in yellow or blue, depending on which menu you chose.
- Have students identify the sums, such as the sums for 9 + 9 + 9 + 9 + 9 + 9 + 9 + 9 + 9 + 9. 9, 18, 27, 36, 45, 54, 63, 72, 81, 90, 99
- Repeat until students can relate addition sentences, such as 9 + 9 + 9 + 9 + 9 + 9 + 9 + 9 + 9 = 81, along with their corresponding multiplication facts, such as 9 × 9.

Alternative Grouping Suggestions help teachers adapt instructions to fit various learning environments such as one-on-one tutoring, summer school, and after-school settings. *

Alternative Groupings
Pair: Lay out Picture Cards for students; four for each number. Ask students to use two Picture Cards with the same number to form a Club with 4. Then ask them to use three cards with the same number to form a Club with 15. Repeat with different numbers of cards and different Clubs.

Daily Progress Monitoring allows teachers to swiftly adapt instructions to provide immediate feedback and targeted remediation for struggling students.

Progress Monitoring
If... students have difficulty remembering the repeated addition sentence in **Counting Clubs—2**,
Then... distribute paper clips to students so that they can model the amount on their cards along the edges.

T11-A

* Featured in the Intervention levels (D–J).

Real-World Applications allow students to think critically about how the concepts covered in each lesson relate to everyday scenarios and demonstrate higher levels of mastery. *

Real-World Application
Suppose that your school had a fund-raiser by selling bumper stickers for $3 each.

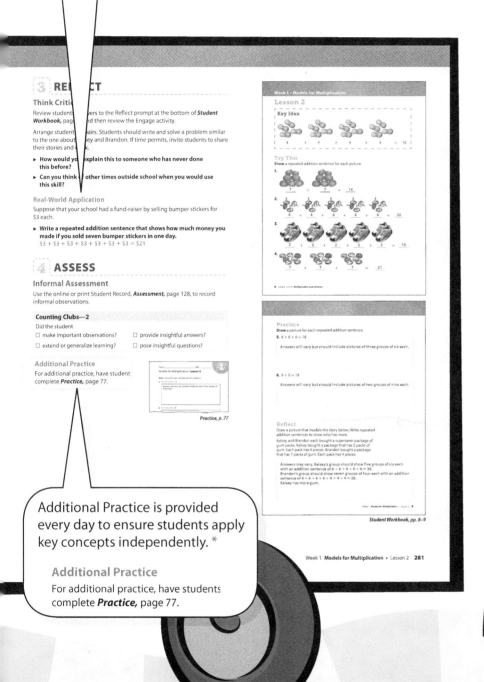

Additional Practice is provided every day to ensure students apply key concepts independently. *

Additional Practice
For additional practice, have students complete *Practice,* page 77.

* Featured in the Intervention levels (D–J).

Boost student engagement with Project-Based Learning

NEW! Project-Based Learning

Weekly, standards-driven Project-Based Learning increases long-term retention of concepts and has been shown to be more effective than traditional instruction. Students collaborate on projects to answer an essential question and each week builds on the next. Students are challenged to apply and demonstrate mastery of concepts and skills by expressing understanding through discussion, research, and presentation.

Each project includes an easy-to-follow routine. Real-world scenarios are used to help build college and career readiness for all students. Each project wraps up with a discussion, presentation, or reflection of the project.

Project-evaluation-criteria rubrics allow teachers and students to discuss aspects of the project and expectations for completing the project.

Project Evaluation Criteria
Review project evaluation criteria with students prior to beginning the project.

Exceeds Expectations
- ☐ Project result is explained and can be extended.
- ☐ Project result is explained in context and can be applied to other situations.
- ☐ Project result is explained using advanced mathematical vocabulary.
- ☐ Project result is explained and extended, and shows advanced knowledge of mathematical concepts and skills.

Meets Expectations
- ☐ Project result is explained.
- ☐ Project result is explained in context.
- ☐ Project result is explained using mathematical vocabulary.
- ☐ Project result is described, and mathematics are used correctly.
- ☐ Project result is explained, and shows satisfactory knowledge of mathematical concepts and skills.

Does Not Meet Expectations
- ☐ Project result is not explained.
- ☐ Project result is explained, but out of context.
- ☐ Project result is explained, but mathematical vocabulary is oversimplified.
- ☐ Project result is described, but mathematics are not used correctly.
- ☐ Project result is not explained and or extended, or shows less than satisfactory knowledge of mathematical concepts and skills.

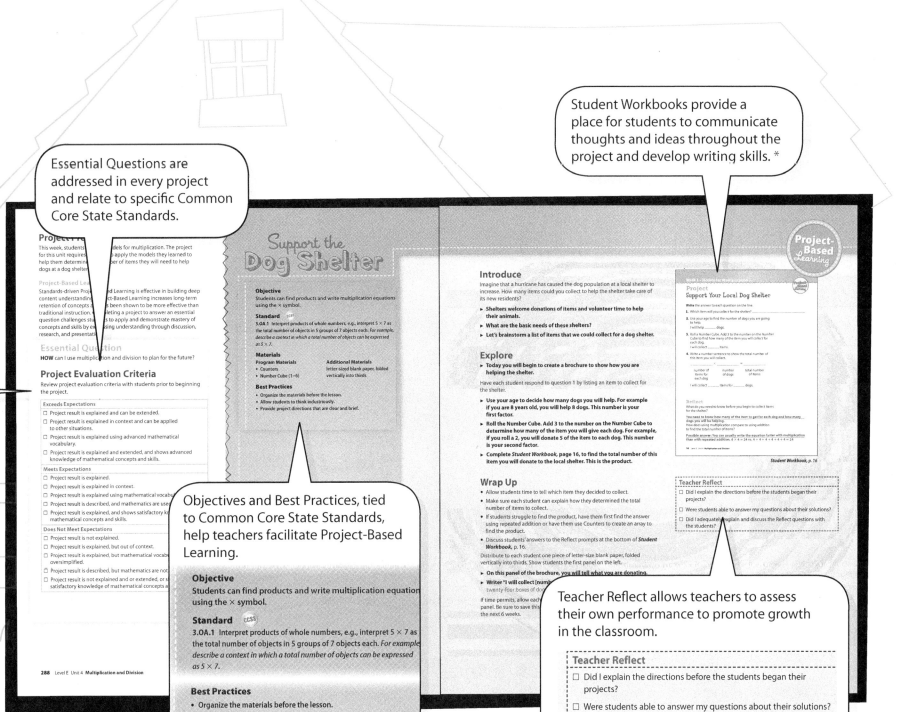

Accelerate student learning with interactive games

In **Number Worlds**, math is designed to be fun for all students.

- An effective balance between hands-on and digital games helps students apply concepts in a variety of settings.
- Every Number Worlds activity is tied to a Key Common Core State Standard, providing the instructional support necessary to get struggling students back on track.
- Through purposeful activities, students develop conceptual understanding and computational fluency as they actively apply concepts to realistic scenarios.
- Both group and individual activities are purposefully embedded in the curriculum to provide many opportunities for collaboration and teaming as well as personalized, individual learning.

Building Blocks Activities

Support and motivate students with an adaptive, personalized learning system that is proven to work. With **Building Blocks**, students are actively engaged in their learning as they progress through adaptive math activities.

Building Blocks, the result of NSF-funded research, develops students' mathematical thinking through interactive, web-based practice activities. Students progress through research-proven learning trajectories, making connections and effectively building mathematical understanding.

Strategic Digital Modeling

Students and teachers have access to a variety of web-based digital math tools. These powerful tools support reasoning skills and problem solving by allowing students to virtually explore and model mathematical concepts.

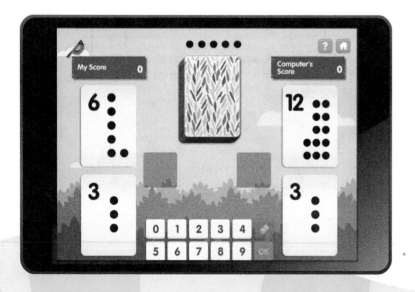

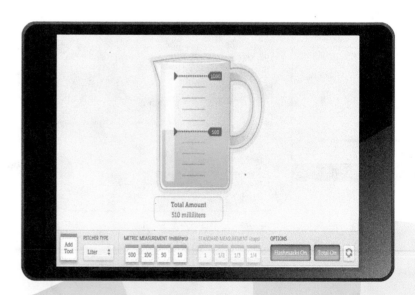

Digital Game Boards for Group Interaction

New digital game boards and number lines support group instruction. Game Boards can be accessed online and projected, allowing students to model their mathematical understanding.

Hands-On games and activities for every lesson

Every Number Worlds lesson utilizes hands-on activities that incorporate manipulatives. This allows students to explore and demonstrate abstract concepts in a concrete way.

Manipulatives are provided in every teacher materials kit to effectively support student learning. Digital manipulatives can be accessed by students and teachers through the Digital Math Tools resource.

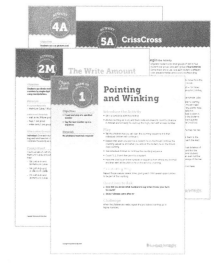

PHOTO: McGraw-Hill Education

Activity Cards are now embedded in every lesson to engage students in math learning.

Game Boards are provided in the teacher materials kits and are accessible online to support student engagement in a variety of settings.

Achieve success with dynamic assessments and reporting

Assessment tools allow teachers and administrators to correctly place and monitor at-risk students. A variety of assessment options evaluate student proficiency, inform instruction, and visually track progress.

Placement Tests

Students in grades Pre-K–5 (levels A–G) can be placed into Number Worlds using the Number Knowledge Test, which measures students' ability to make sense of quantitative problems. The Number Knowledge test can be found in the Placement Test Guide.

Students can be placed into Number Worlds using the Placement Test. Based on the results, students are placed at their appropriate level.

Informal Assessments

Each lesson includes an Informal Assessment to record informal observations. Teachers can use the online assessment tool or the Student Assessment Record for recording observations.

Formal Assessments

Weekly Assessments measure the concepts covered in the week's daily lessons and provide actionable items for the teacher.

Cumulative Assessments evaluate how well students mastered and retained mathematical concepts covered every 4 weeks. *

Unit Assessments are available for levels D–J. Pre-tests and Post-tests can be used as a placement tool if units are being taught in isolation. Unit Assessments can help support grouping and differentiated instruction for students in tiers 2 and 3 of the RtI model. Assessments can be taken online or with the Assessment Book.

Benchmark Level Assessments determine areas of potential remediation and mastery of critical math concepts. Level pre-tests and post-tests can be taken online. Online reporting allows teachers to collect and share student data. Teachers use student and class reports to effectively tailor instruction to the needs of the students.

* Featured in the Prevention levels (A–C).

District-Level Reports allow administrators and district leaders to easily make informed decisions by comparing student proficiency across schools. Actionable progress reports allow trends to be effectively communicated and problem areas to be quickly and accurately identified.

Building Blocks Reports Real-time Building Blocks reports provide formative assessment of student progress. Building Blocks reports are available at the student, class, and school levels, allowing teachers and administrators to easily and effectively monitor student progress and pinpoint problem areas as they arise.

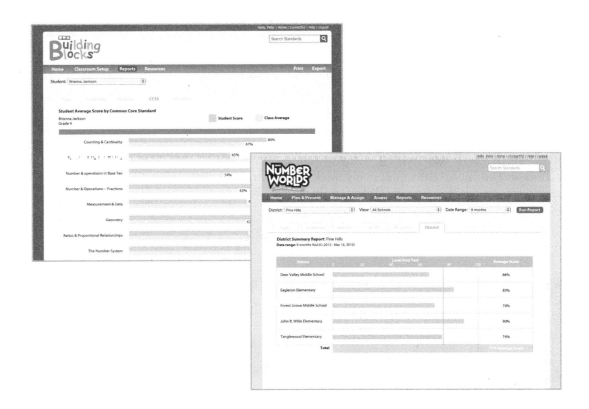

Online Progress Reports

Online reporting allows teachers to collect and share student data. Teachers can see student and class reports to effectively tailor instruction to the needs of students.

Student reports allow teachers to see which skills an individual student has mastered and where students can improve.

Class reports allow teachers and administrators to see how an entire class is performing on specific skills and topics.

District level reporting allows administrators and leaders to make critical decisions and compare student proficiency across schools. With these reports, trends can be effectively communicated, and issues can be quickly and accurately identified.

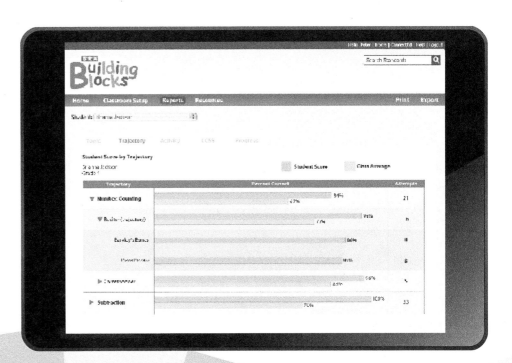

T14-B

Comprehensive packages for flexible implementation

Teacher Materials Package, Levels A–J

Teacher Dashboard is the control center for all ongoing instruction and allows the teacher to quickly plan, teach, and assess student progress.

Digital Game Boards enhance motivation for students and connect key skills in weekly lessons.

Digital Math Tools help students solve problems, explore concepts, and demonstrate understanding.

Assessments and Reporting are available online to help teachers and administrators effectively track, record, and monitor student progress at the individual student level, the class level, and the school level.

Activity Cards engage students and develop math inquiry through games and activities. Challenge Activities differentiate and extend learning for students who demonstrate mastery-level comprehension.

Playing Cards allow students to test hypotheses and ideas as they work through the skills in each lesson.

Vocabulary Cards help students understand mathematical language and help students communicate their understanding. *

Assessment Book allows students to take any in-program assessment offline. *

Implementation Guide provides flexible planning options.

Placement Test Guide allows teachers to easily and effectively place students into the appropriate Number Worlds level. The Placement Test Guide includes a bonus Number Knowledge Test for students in levels A–G.

English Learner Support Guide provides lessons, strategies, and resources to effectively support English Language Learners.

Practice eBooks help students reinforce key math concepts and sharpen problem-solving skills.

Teacher Editions and eBooks allow teachers to develop critical math concepts with routine lesson plans, integrated project-based learning, and a guide for facilitating differentiated learning in the classroom.

* Items available as separate purchase.

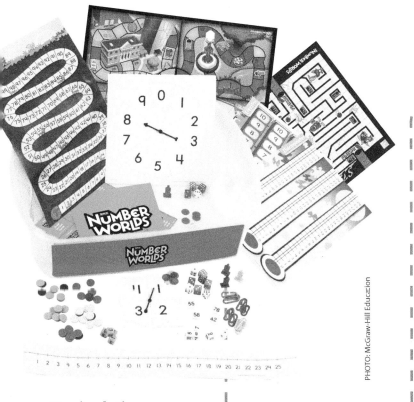

Manipulatives are designed to support conceptual development and student engagement. They are included with the Teacher Materials Kit.

Student Materials Package, Levels A–J

Student Dashboard provides online resources for each lesson in one place to support conceptual development and at-home connections. Device-friendly resources maximize accessibility. A to-do list provides quick access to assignments and activities. Online student assessments reflect CCSS objectives and goals.

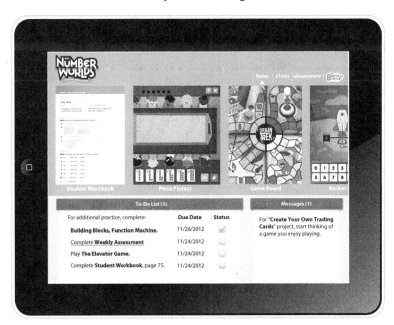

Building Blocks Activities are adaptive, research-proven online activities designed to engage students and reinforce levels of mathematical development. Now available for grades Pre-K to 8.

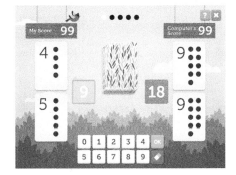

Student Workbooks and eBooks include developmental activities to help students develop higher-order thinking skills and practice basic skills.

For pricing and purchasing information visit
MHEonline.com/NumberWorlds

T15-B

UNIT 1: Number Sense within 100

Unit at a Glance
This **Number Worlds** unit builds on a prior knowledge of whole numbers. Students will explore different ways to visualize and represent numbers and become familiar with equivalent representations of the same number. Students will study the importance of 10 and compare the values of whole numbers to 100.

Skills Trace

Before Level D	Level D	After Level D
Level C Students can count, read, and write numbers to 100, understand number lines and other representations of numbers, and begin to understand the base-ten system.	By the end of this unit, students should be able to count, read, write, compare and order whole numbers to 100. They have learned the fundamentals of the base-ten system and can count and group objects in ones and tens. Students can identify sets as greater than, less than, or equal to, and can count by 2s, 5s, and 10s to 100.	**Moving on to Level E** Students will learn strategies for visualizing and constructing whole numbers to 1,000, explore the place value and monetary systems, and examine relationships among numbers.

Learning Technology
The following activities are available online to support the learning goals in this unit.

Building Blocks
- Before and After Math
- Bright Idea
- Build Stairs 1
- Build Stairs 2
- Dinosaur Shop 1
- Dinosaur Shop 2
- Easy as Pie: Add Numbers
- Number Compare 1
- Number Compare 3
- Number Compare 4
- Number Compare 5
- Number Snapshots 3
- Number Snapshots 8
- Rocket Blast 2
- School Supply Shop

Digital Tools
- Base 10 Blocks Tool
- Number Line

Unit Overview

Week	Focus
1	**Constructing Whole Numbers** • *Teacher Edition,* pp. 16–31 • *Activity Cards,* 1A, 1B, 1C, 1D • *Student Workbook,* pp. 5–16 • *English Learner Support Guide,* pp. 22–23 • *Assessment,* pp. 13–14
2	**Numbers on a Line** • *Teacher Edition,* pp. 32–45 • *Activity Cards,* 1E, 1F, 1G, 1H • *Student Workbook,* pp. 17–28 • *English Learner Support Guide,* pp. 24–25 • *Assessment,* pp. 15–16
3	**Tens and Ones** • *Teacher Edition,* pp. 46–59 • *Activity Cards,* 1I, 1J, 1K, 1L • *Student Workbook,* pp. 29–40 • *English Learner Support Guide,* pp. 26–27 • *Assessment,* pp. 17–18
4	**Visualizing and Constructing Whole Numbers** • *Teacher Edition,* pp. 60–73 • *Activity Cards,* 1M, 1N, 1O, 1P • *Student Workbook,* pp. 41–52 • *English Learner Support Guide,* pp. 28–29 • *Assessment,* pp. 19–20
5	**Number Patterns** • *Teacher Edition,* pp. 74–87 • *Activity Cards,* 1Q, 1R • *Student Workbook,* pp. 53–64 • *English Learner Support Guide,* pp. 30–31 • *Assessment,* pp. 21–22
6	**Whole Number Relationships** • *Teacher Edition,* pp. 88–101 • *Activity Cards,* 1S, 1T, 1U, 1V • *Student Workbook,* pp. 65–76 • *English Learner Support Guide,* pp. 32–33 • *Assessment,* pp. 23 24

Essential Question

HOW can I use comparing and ordering numbers outside the classroom?

In this unit, students will explore how comparing and ordering can be used to solve real-world problems by counting and combining numbers to create a picture of items in a park.

Learning Goals	CCSS Key Standards
Students can visualize and represent numbers up to 20 on an array, make target numbers by adding, removing, or combining numbers, count on and count back from a given number, and recognize that 10 ones equal 1 ten.	**Domain:** Number and Operations in Base Ten **Cluster:** Understand place value. **1.NBT.2a:** 10 can be thought of as a bundle of ten ones—called a "ten."
Students can visualize numbers up to 100 on a number line, skip count by tens, and add and subtract small quantities using a number line.	**Domain:** Number and Operations in Base Ten **Cluster:** Understand place value. **2.NBT.1:** Understand that the three digits of a three-digit number represent amounts of hundreds, tens, and ones; e.g., 706 equals 7 hundreds, 0 tens, and 6 ones.
Students can skip count by tens, make groups of ten, and think of two-digit numbers in terms of tens and ones.	**Domain:** Number and Operations in Base Ten **Cluster:** Understand place value. **1.NBT.2:** Understand that the two digits of a two-digit number represent amounts of tens and ones.
Students can make 10 by combining two and three numbers, visualize and represent two-digit numbers several different ways. This week lays a foundation for regrouping.	**Domain:** Number and Operations in Base Ten **Cluster:** Understand place value. **2.NBT.1:** Understand that the three digits of a three-digit number represent amounts of hundreds, tens, and ones; e.g., 706 equals 7 hundreds, 0 tens, and 6 ones.
Students can skip count by twos, fives, and tens, detect errors in skip-counting sequences, and use number patterns to solve problems.	**Domain:** Number and Operations in Base Ten **Cluster:** Understand place value. **2.NBT.2:** Count within 1000; skip-count by 5s, 10s, and 100s.
Students can visualize amounts of 10, 25, and 50, compare and order whole numbers to 100, and make comparisons between numbers with "greater than" or "less than" and with the symbols > and <.	**Domain:** Number and Operations in Base Ten **Cluster:** Understand place value. **2.NBT.4:** Compare two three-digit numbers based on meanings of the hundreds, tens, and ones digits, using >, =, and < symbols to record the results of comparisons.

Daily lesson activities emphasize using communication, logic, reasoning, modeling, tools, precision, structure, and patterns to solve problems. All student activities, reflections, and assessments require application of the **Common Core Standards for Mathematical Practice.**

WEEK 1: Constructing Whole Numbers

Week at a Glance

This week, students begin **Number Worlds,** Level D, Number Sense within 100. Students will visualize and construct numbers to 20.

Skills Focus

- Visualize and represent numbers up to 20 on an array.
- Make target numbers by adding, removing, or combining numbers.
- Count on and count back from a given number.
- Lay the foundation for regrouping by recognizing that 10 ones equal 1 ten.

How Students Learn

Many instructional tools and activities can be used to support students in developing part-whole thinking. This week, students work with ten frames to help them organize visual patterns in terms of 5 and 10. The ten frame will help them see that a quantity such as 8 can be thought of as a group of 5 and a group of 3, or that the quantity 10 can be thought of as 8 and 2 and as 5 and 3 and 2.

English Learners ELL

For language support, use the **English Learner Support Guide,** pages 22–23, to preview lesson concepts and teach academic vocabulary. **Number Words** Vocabulary Cards are listed as additional materials in many lessons and can be used to preteach and reinforce academic vocabulary.

Math at Home

Give one copy of the Letter to Home, page 1, to each student. Encourage students to share and complete the activity with their caregivers.

Weekly Planner

Lesson	Learning Objectives
1 pages 20–21	Students can identify numerals and amounts from 1 to 10.
2 pages 22–23	Students can use a ten frame to visualize and construct numbers 1–10.
3 pages 24–25	Students can quickly visualize the number shown on a ten frame and how many more are needed to make ten.
4 pages 26–27	Students can count on from ten to construct numbers 11–20.
5 pages 28–29	**Review and Assess** Students review skills learned this week and complete the weekly assessment.
Project pages 30–31	Students can identify numerals and amounts from 1–20.

18 Level D Unit 1 **Number Sense within 100**

Key Standard for the Week

Domain: Number and Operations in Base Ten
Cluster: Understand place value.
1.NBT.2a 10 can be thought of as a bundle of ten ones—called a "ten."

Materials		Technology
Program Materials • *Student Workbook,* pp. 5–7 • *Practice,* p. 5 • Activity Card 1A, **Concentration** • Dot Set Cards • Number Cards (1–10)	**Additional Materials** • Vocabulary Card 44, *ten*	**Teacher Dashboard** Building Blocks Dinosaur Shop 1
Program Materials • *Student Workbook,* pp. 8–9 • *Practice,* p. 6 • Activity Card 1B, **How Many?** • Ten Frame • Ten Frame Flash Cards • Counters		**Teacher Dashboard** Building Blocks Dinosaur Shop 2; Build Stairs 1
Program Materials • *Student Workbook,* pp. 10–11 • *Practice,* p. 7 • Activity Card 1C, **Five and Ten** • Ten Frame • Counters • Dot Set Cards	**Additional Materials** math-link cubes*	**Teacher Dashboard** Building Blocks Number Snapshots 8
Program Materials • *Student Workbook,* pp. 12–13 • *Practice,* p. 8 • Activity Card 1D, **Concentration to 20** • Dot Set Cards • Dot Set 10 Cards • Number Cards (1–20)		**Teacher Dashboard** Building Blocks Number Snapshots 3
Program Materials • *Student Workbook,* pp. 14–15 • Weekly Test, *Assessment,* pp. 13–14		Review previous activities.
Program Materials • *Student Workbook,* p. 16 • Dot Set Cards (1–10)	**Additional Materials** • crayons • paper, 11 × 17 inches, for student drawing	

*Available from McGraw-Hill Education

WEEK 1
Constructing Whole Numbers

Find the Math

In this week, students will identify numerals and amounts from 1–20.

Use the following to begin a guided discussion:

▶ **How can we tell how many ducklings there are in this picture?**
I can start at one and count on from there.

Have students complete **Student Workbook,** page 5.

Student Workbook, p. 5

Lesson 1

Objective
Students can identify numerals and amounts from 1 to 10.

Standard
1.NBT.2a 10 can be thought of as a bundle of ten ones—called a "ten."

Vocabulary
ten

Creating Context
Review the basic routines for playing games. Make sure students understand the rules for taking turns, deciding who goes first, and what to do in the event of a tie.

Materials

Program Materials
- Dot Set Cards, 1 set per student group
- Number Cards (1–10), 1 set per student group

Additional Materials
Vocabulary Card 44, *ten*

1 WARM UP

Prepare
Make two columns on the board—one labeled *Students* and one labeled *Teacher*.

▶ **Listen carefully to hear whether or not I make a mistake when I count. If you catch my mistake, you get a point. If no one catches my mistake, I get a point.**

▶ **1, 2, 3, 4, 5, 6, 8, 9, 10. Did I make a mistake?**

- Repeat the activity several times.

2 ENGAGE

Develop: Concentration

"Today we are going to match numerals to numbers of dots." Follow the instructions on the Activity Card **Concentration.** As students complete the activity, be sure to use the Questions to Ask.

Activity Card 1A

Alternative Groupings

Small Group: Use two sets of Dot Set Cards and Number Cards for each group of three to four students.

Individual: Partner with the student, and complete the activity as written.

Progress Monitoring

| **If...** students are struggling to catch your mistakes, | **Then...** write the numbers on the board as you say them aloud. |

Practice

Have students complete **Student Workbook,** pages 6–7. Guide students through the Key Idea example and the Try This exercises.

Interactive Differentiation

Consult the **Teacher Dashboard** for grouping suggestions. You can also use performance on the Engage activity to guide students.

Independent Practice

For additional practice understanding the numbers 1–10, have students use Dinosaur Shop 1: Label Boxes. Students will identify the numerals that represent a target number of dinosaurs in a number frame.

Supported Practice

For additional support, use Dot Set Cards and Number Cards (1–10).

- Give pairs of students Dot Set Cards and Number Cards (1–10), one set of each per student pair.
- Have students turn the number cards facedown and spread out the dot set cards faceup.
- Tell students to turn over a number card and match it to a dot set card. For example, if a student turns over a 5, she should find a dot card with five dots.
- Help students match the correct dot card to the number card.
- Students in each pair should take turns until all the number cards are matched to a dot set card.

REFLECT

Think Critically

Review students' answers to the Reflect prompt at the bottom of **Student Workbook,** page 7, and then review the Engage activity.

Discuss to reinforce the concept that patterns can help students quickly recognize the number of objects in a set.

▶ **How did you figure out how many the card shows?**

▶ **Do the dot-set patterns remind you of anything?**

ASSESS

Informal Assessment

Use the online or print Student Record, **Assessment,** page 128, to record informal observations.

What's Missing?
Did the student
☐ respond accurately?   ☐ respond with confidence?
☐ respond quickly?   ☐ self-correct?

Additional Practice

For additional practice, have students complete **Practice,** p. 4.

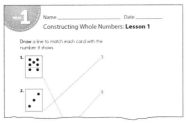

Practice, p. 4

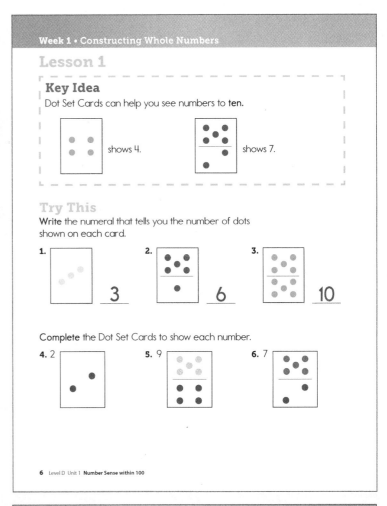

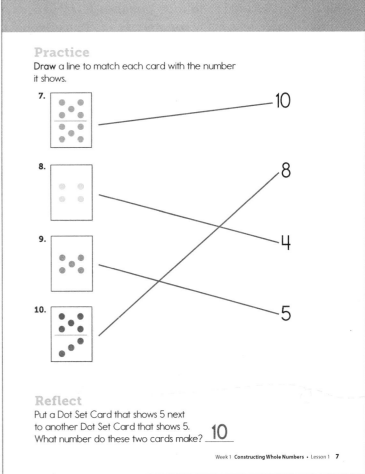

Student Workbook, pp. 6–7

Week 1 **Constructing Whole Numbers** • Lesson 1   **21**

WEEK 1
Constructing Whole Numbers

Lesson 2

Objective
Students can use a ten frame to visualize and construct numbers 1–10.

Standard
1.NBT.2a 10 can be thought of as a bundle of ten ones—called a "ten."

Creating Context
To give students practice listening and counting, have them listen to a recording of a song or a story and tally the number of times they hear a particular word, such as *the*.

Materials
Program Materials
- Ten Frame, 1 per student
- Ten Frame Flash Cards, 1 set per student
- Counters, 2 different colors, 10 per color per student

1 WARM UP

Prepare
Show students an empty ten frame.

▶ **This is a tool we will use to show numbers. If I filled each space with a Counter, how many counters do you think I would use?** Answers will vary, but they should be close to 10.

▶ **Help me find out how many Counters I need. Count with me.**

Show students the correct way to fill a ten frame while counting, first filling the top row left to right, and then the bottom row left to right.

▶ **This is called a *ten frame* because it shows the number 10.**

Distribute a ten frame and Counters to students, and ask them to show 4. Then have them show 7 on the ten frame.

Just the Facts
Display a ten frame with the number *6* shown in it. Tell students that after they say a number they should clap that number of times. Use questions such as the following:

▶ **How many counters are there?** Students say *six* and clap six times.

▶ **How many empty spaces are there?** Students say *four* and clap four times.

▶ **What is the greatest number of counters you can fit on a ten frame?** Students say *ten* and clap ten times.

2 ENGAGE

Develop: How Many?
"Today we are going to use the ten frame to count numbers quickly." Follow the instructions on the Activity Card **How Many?** As students complete the activity, be sure to use the Questions to Ask.

Activity Card 1B

Alternative Groupings
Pair: Give each student a set of Ten Frame Flash Cards, and have the students take turns quizzing one another.

Progress Monitoring

| If... students have trouble recognizing the number needed to make 10, | Then... have them use Counters in one color to show the number you call out and Counters in another color to fill the spaces as they count. |

Practice
Have students complete **Student Workbook,** pages 8–9. Guide students through the Key Idea example and the Try This exercises.

Interactive Differentiation
Consult the **Teacher Dashboard** for grouping options. You can also use performance on the Engage activity to guide students.

Independent Practice
For additional practice understanding how to make numbers to 10, use Dinosaur Shop 2: Fill Orders. Students add dinosaurs to a box to match target numerals.

Supported Practice
For additional support, use Build Stairs 1. In this activity students add stairs to a stair frame outline to reach a target height.

22 Level D Unit 1 **Number Sense within 100**

REFLECT

Think Critically

Review students' answers to the Reflect prompt at the bottom of **Student Workbook,** page 9, and then review the Engage activity.

Discuss all strategies students used to find their answers.

- How did you figure out the answer?
- Is there another way to do it?

Real-World Application

- People often use tally marks when recording counts of items. Tally marks are grouped in sets of five. To count a set of tally marks, skip count by fives until you do not have a set of five. Then count by ones for each mark.

Write sets of tally marks to 10 on the board, and ask students to count aloud to find how many marks there are.

ASSESS

Informal Assessment

Use the online or print Student Record, **Assessment,** page 128, to record informal observations.

How Many?
Did the student
- ☐ respond accurately?
- ☐ respond quickly?
- ☐ respond with confidence?
- ☐ self-correct?

Additional Practice

For additional practice, have students complete **Practice,** p. 5.

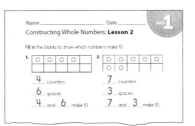

Practice, p. 5

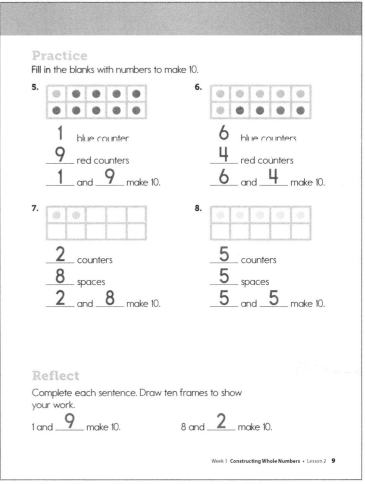

Student Workbook, pp. 8–9

Week 1 **Constructing Whole Numbers** • Lesson 2   23

WEEK 1
Constructing Whole Numbers

Lesson 3

Objective
Students can quickly visualize the number shown on a ten frame and how many more are needed to make ten.

Standard
1.NBT.2a 10 can be thought of as a bundle of ten ones—called a "ten."

Creating Context
Teach students the names and values of the U.S. penny, nickel, and dime. Make a chart that shows each coin, its name, and its value.

Materials
Program Materials
- Ten Frame, 1 per student
- Counters, 2 different colors, 10 per color per student
- Dot Set Cards

Additional Materials
math-link cubes

1 WARM UP

Prepare
Distribute a Ten Frame worksheet and 10 Counters in each of two different colors to each student.

▶ **Use your Counters to fill the ten frame. Make sure all the spaces are filled.**

Fill a ten frame with Counters, and then write on the board the number of Counters of each color you used. Have students tell you the two numbers they have and record each new set on the board.

▶ **Look at your ten frame to see whether there is any combination of numbers that we have not listed.**

After providing time for students to look over the list, ask the following question:

▶ **What do these pairs of numbers have in common?** They fill a ten frame. Each pair of numbers makes 10.

Just the Facts
Show seven counters in a ten frame. Have students hold up the number of fingers that matches the number of counters in the ten frame and say the number. Use questions such as the following:

▶ **How many counters are there?** Students say *seven* and show seven fingers.

▶ **How many spaces are empty?** Students say *three* and show three fingers.

▶ **If I fill in the three spaces, how many counters are there now?** Students say *ten* and show ten fingers.

2 ENGAGE

Develop: Five and Ten
"Today we are going to use Counters and ten frames to show different numbers." Follow the instructions on the Activity Card **Five and Ten** As students complete the activity, be sure to use the Questions to Ask.

Activity Card 1C

Alternative Groupings
Pair: Have students take turns saying the numbers to be made on the ten frame. After one partner places the Counters on the frame, have both partners verify that the Counters are correct.

Progress Monitoring
| If... students struggle with the idea that ten can be broken into parts, | Then... give them a stack of ten math-link cubes. Have them split the stack into two sections and count how many are in each part. |

Practice
Have students complete **Student Workbook,** pages 10–11. Guide students through the Key Idea example and the Try This exercises.

Interactive Differentiation
Consult the **Teacher Dashboard** for grouping suggestions. You can also use performance on the Engage activity to guide students.

Independent Practice

For additional practice, use Number Snapshots 8: Dots to Numerals up to 10. Students will combine collections up to 10 and match them to a corresponding numeral.

Supported Practice

For additional support, use Dot Set Cards. Show students a card that has three dots.

▶ **How many dots are showing?** 3

▶ **If I want 5 dots, do I need more dots or fewer dots?** more

- Replace the 3-dot card with a 5-dot card now that students have identified a need for 2 more dots.

▶ **If I want 3 dots, do I need more dots or fewer dots?** fewer

- Replace the 5-dot card with a 3-dot card now that students have identified a need for 2 fewer dots.

- Repeat with other numbers 1–10 to reinforce the concepts of addition and subtraction.

REFLECT

Think Critically

Review students' answers to the Reflect prompt at the bottom of *Student Workbook,* page 11, and then review the Engage activity.

Explain that 5 and 10 are important numbers in math, and that grouping numbers or objects in sets of 5 or 10 is a valuable skill.

▶ If a ten frame has 3 Counters and you need 8, do you need more or fewer Counters? more

▶ How do you know that? 8 is more than 3.

Real-World Application

▶ The value of ten pennies equals the value of 1 dime.

▶ If I have 6 pennies, how many more do I need to make the value of one dime? 4

▶ If I have 2 pennies, how many more do I need to make the value of one dime? 8

▶ If I have 5 pennies, how many more do I need to make the value of one dime? 5

ASSESS

Informal Assessment

Use the online or print Student Record, *Assessment,* page 128, to record informal observations.

Five and Ten

Did the student

☐ respond accurately?   ☐ respond with confidence?

☐ respond quickly?   ☐ self-correct?

Additional Practice

For additional practice, have students complete *Practice,* p. 6.

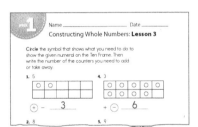

Practice, p. 6

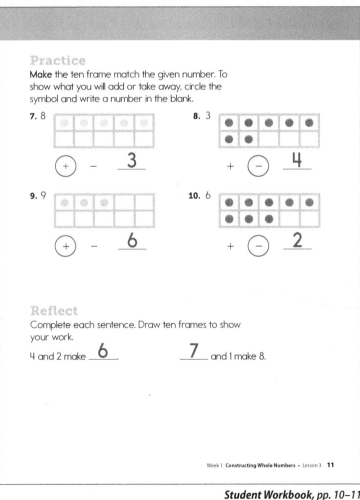

Student Workbook, pp. 10–11

Week 1 Constructing Whole Numbers • Lesson 3   **25**

WEEK 1
Constructing Whole Numbers

Lesson 4

Objective
Students can count on from ten to construct numbers 11–20.

Standard
1.NBT.2a 10 can be thought of as a bundle of ten ones—called a "ten."

Creating Context
Review the terms *greater than* and *less than*. Students may know *great* as a synonym for *fantastic*. Explain that in this context it means "larger than" or "worth more than" and that *less than* means "smaller than."

Materials
Program Materials
- Dot Set Cards, 2 sets per group
- Dot Set 10 Cards, 1 set per group
- Number Cards (1–20), 1 set per group

1 WARM UP

Prepare

▶ **You can use two Dot Set Cards to show numbers that are higher than ten.**

Display a Dot Set 10 Card and a Dot Set 3 Card.

▶ **What two numbers are shown here?**

▶ **We already know this is a set of ten, so let's count the dots on the other card from ten—*11, 12, 13*. What number is shown by the 10 and 3 together?** 13

Repeat the activity with a Dot Set 10 Card and a Dot Set 6 Card.

Just the Facts

Present students with a number and have them chorally call out the number that comes next. Use questions such as the following:

▶ **Which number comes after 10?** 11
▶ **Which number comes after 14?** 15
▶ **Which two numbers come after 15?** 16, 17

2 ENGAGE

Develop: Concentration to 20

"Today we are going to match Dot Set Cards with Number Cards that show numbers from 1 to 20." Follow the instructions on the Activity Card **Concentration to 20**. As students complete the activity, be sure to use the Questions to Ask.

Activity Card 1D

Alternative Groupings

Whole Class: Organize the class into groups of four. Monitor game play using the Questions to Ask.

Individual: Partner with the student to complete the activity.

Progress Monitoring

If… students have difficulty seeing the set of 10 in numbers 11–20,

Then… allow them to use ten frames to construct numbers.

Practice

Have students complete **Student Workbook,** pages 12–13. Guide students through the Key Idea example and the Try This exercises.

Interactive Differentiation

Consult the **Teacher Dashboard** for grouping suggestions. You can also use performance on the Engage activity to guide students.

Independent Practice

For additional practice with number recognition and counting up to 20, use Number Snapshots 3. Students identify an image that correctly matches a target image from four multiple-choice selections.

Supported Practice

For additional support, use Dot Set Cards. Place two sets of dot cards in front of a student. Place the 10-dot card in front of the student along with a 4-dot card.

▶ **What number do the two cards make?** 14
▶ **What number comes after 14?** 15
▶ **How can you make 15 using the 10-dot card and another card?**
  Answers may vary; possible answer: I can use the 5-dot card.

Continue until you have covered all numbers from 11 through 20.

REFLECT

Think Critically

Review students' answers to the Reflect prompt at the bottom of **Student Workbook,** page 13, and then review the Engage activity.

Discuss to reinforce that quantities can be combined to show numbers.

▶ **How do you show numbers higher than 10?** Possible answers: I put a 10 card next to the card I flipped over; I put the 10 card next to the 5 card to make 15, and I counted from 10.

▶ **How would you show numbers higher than 10 on a ten frame?**

Real-World Application

A ten-dollar bill is like a ten frame full of one-dollar bills because one ten-dollar bill is equal to ten one-dollar bills.

▶ **How many bills and what types are needed to make $12?** 3; 1 ten, 2 ones

▶ **How many bills and what types are needed to make $14?** 5; 1 ten, 4 ones

▶ **How many bills and what types are needed to make $20?** 2; 2 tens

ASSESS

Informal Assessment

Use the online or print Student Record, **Assessment,** page 128, to record informal observations.

Concentration to 20

Did the student
- ☐ make important observations?
- ☐ extend or generalize learning?
- ☐ provide insightful answers?
- ☐ pose insightful questions?

Additional Practice

For additional practice, have students complete **Practice,** page 7.

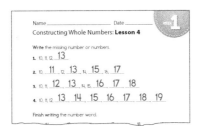
Practice, p. 7

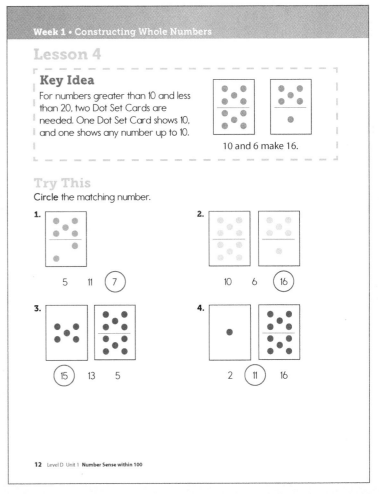

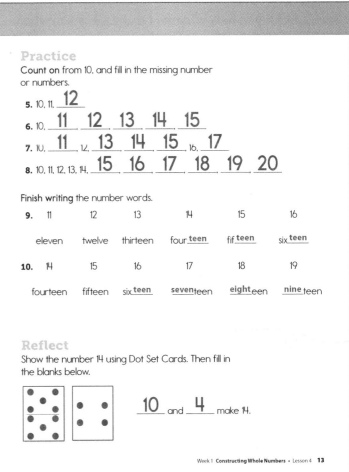

Student Workbook, pp. 12–13

Week 1 **Constructing Whole Numbers** • Lesson 4   **27**

WEEK 1
Constructing Whole Numbers

Lesson 5 *Review*

Objective
Students review skills learned this week and complete the weekly assessment.

Standard
1.NBT.2a 10 can be thought of as a bundle of ten ones—called a "ten."

Vocabulary
Review vocabulary introduced during the week.

Creating Context
Understanding place value is necessary for success in mathematics. This week uses ten frams designed to clearly separate the amounts in each ten. This is an excellent strategy for English Learners, especially those at early English proficiency levels who may not be able to verbally explain their understanding. On their ten frames, students can manipulate objects and point to answers.

1 WARM UP

Prepare

Make two columns on the board—one labeled *Students* and one labeled *Teacher*.

▶ **Listen carefully to hear whether I make a mistake when I count backward. If you catch my mistake, you get a point. If no one catches my mistake, I get a point.**

▶ **10, 9, 8, 7, 5, 4, 3, 2, 1. Did I make a mistake?**

• Repeat the activity several times.

2 ENGAGE

Practice

Have students complete **Student Workbook,** pages 14–15. Guide students through the Key Idea example.

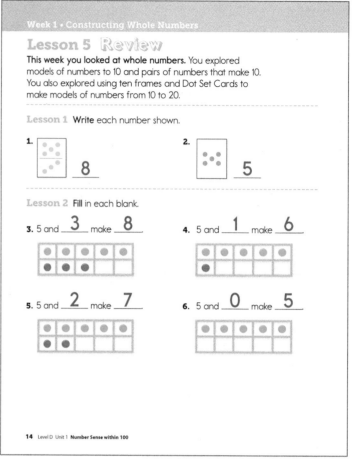

Student Workbook, pp. 14–15

28 Level D Unit 1 **Number Sense within 100**

3 REFLECT

Think Critically

Review students' answers to the Reflect prompt at the bottom of **Student Workbook,** page 15.

Discuss the answer with the group to reinforce Week 1 concepts.

4 ASSESS

Formal Assessment

Students may take the weekly assessment online.

As an alternative, students may complete the weekly test on **Assessment,** pages 13–14. Record progress using the Student Assessment Record, **Assessment,** page 128.

Going Forward

Use the **Teacher Dashboard** to view results of the online assessments, to input the results of print student assessments, and to review progress before making decisions about next steps. Use the weekly test results and observations to determine the next steps for each student.

Retention	
Student displays good grasp of this week's concepts and skills.	Have students catch your mistakes as you incorrectly count forward to twenty and backward from twenty.

Remediation	
Student is still struggling with the week's concepts and skills.	Build tens in sequence. First, use a Dot Set 1 card and have students count 1 dot. Repeat with Dot Set cards 2–5, having students count the dots on each card.
	Then repeat this activity, using a Dot Set 5 card as a base card. Use a Dot Set 1 card and have students count dots to 6. Repeat with Dot Set 2–5 cards as students count numbers 7 to 10.

Suggestions for Re-Evaluation: If a student has struggled without success for several weeks, use observations and test results to place the student at a level in which he or she can find success and build confidence to move forward.

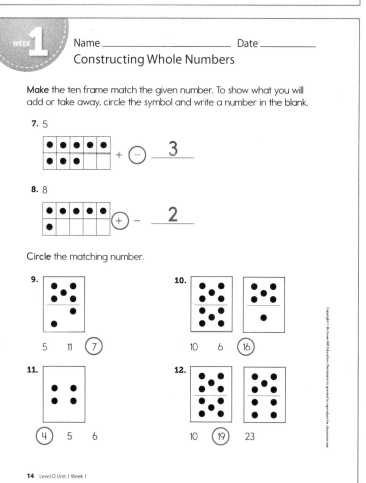

Assessment, pp. 13–14

Project Preview

This week, students learned how to make numbers up to 20 and model the numbers using Dot Set Cards. The project for this unit requires students to extend the knowledge they gained in Find the Math and what they have learned this week. They will use counting and combining numbers to determine how many items are in a park.

Project-Based Learning

Standards-driven Project-Based Learning is effective in building deep content understanding. Project-Based Learning increases long-term retention of concepts and has been shown to be more effective than traditional instruction. By completing a project to answer an essential question, students are challenged to apply and demonstrate mastery of concepts and skills by expressing understanding through discussion, research, and presentation.

Essential Question

HOW can I use comparing and ordering numbers outside the classroom?

Project Evaluation Criteria

Review project evaluation criteria with students prior to beginning the project.

Exceeds Expectations
☐ Project result is explained and can be extended.
☐ Project result is explained in context and can be applied to other situations.
☐ Project result is explained using advanced mathematical vocabulary.
☐ Project result is described, and mathematics are used correctly and can be extended.
☐ Project result is explained and extended, and shows advanced knowledge of mathematical concepts and skills.

Meets Expectations
☐ Project result is explained.
☐ Project result is explained in context.
☐ Project result is explained using mathematical vocabulary.
☐ Project result is described, and mathematics are used correctly.
☐ Project result is explained, and shows satisfactory knowledge of mathematical concepts and skills.

Does Not Meet Expectations
☐ Project result is not explained.
☐ Project result is explained, but out of context.
☐ Project result is explained, but mathematical vocabulary is oversimplified.
☐ Project result is described, but mathematics are not used correctly.
☐ Project result is not explained and or extended, or shows less than satisfactory knowledge of mathematical concepts and skills.

Picturing the Park

Objective
Students can identify numerals and amounts from 1–20.

Standard
1.NBT.2a 10 can be thought of as a bundle of ones—called a "ten."

Materials

Program Materials
Dot Set Cards (1–10)

Additional Materials
- crayons
- paper, 11 × 17 inches, for student drawing

Best Practices
- Clearly enunciate instructions.
- Provide project directions that are clear and brief.
- Provide meaning and organization to the lessons and concepts.

Introduce

Pretend that you are going to the park. What are some of the things you would see?

▶ **What types of people would you see?** Answers may vary. Possible answers: children, runners, people picnicking

▶ **Name some other things you would see at the park.** Answers may vary. Possible answers: animals, swings, walking trails

▶ **What do you like to play at the park?** Answers may vary. Possible answers: soccer, hopscotch, tennis

Explore

▶ **Today you will begin to create a picture of a park.**

▶ **Complete *Student Workbook*, page 16, to find the number of items you will draw in your picture of the park.**

▶ **Write down two items you might see in the park.** Answers may vary. Possible answers: animals, swings, walking trails

▶ **You will use a Dot Set Card to determine how many of each item you will add to your drawing. Turn over a Dot Set Card and write down the number that matches the card next to the name of your item.** for example, *4 swings*

▶ **For the next item, take out the 10 card and turn it face up on your desk. Now pull any card from the deck of Dot Set Cards and place it next to the 10 card. This will be the amount used for your next item.** for example, *12 children*

▶ **Draw those items on your large sheet of paper.**

Wrap Up

- Allow students time to count the dots on the Dot Set Card(s), write down the number, and draw the corresponding quantity of the item on their papers.
- Make sure each student can correctly match the number of items on the Dot Set Card to the correct numeral. Have students recount each item they drew to make sure it matches the number they wrote down.
- If students struggle with finding the right number, have them count aloud as they tap each dot on the Dot Set Card(s).
- Discuss students' answers to the Reflect prompt at the bottom of ***Student Workbook*,** page 16.
- Save students' pictures for use in Week 2.

If time permits, allow each student to look at another student's picture and match the correct Dot Set Card to each item the other student drew. For example, if another student drew seven people, the student should match the Dot Set 7 Card to the seven people. The other student verifies this by looking at his sheet and saying, "Yes, I have the number 7 written for people."

Week 1 · Constructing Whole Numbers

Project
Picturing the Park

Name, count, and draw a picture of items found in a park.

1. What item found in a park will you draw?
   Answers may vary. Possible answers: tree, child, dog

2. Turn over a Dot Set Card and write the number you see.
   Answers may vary. Possible answers: 1–10

3. Draw that many of the item on your picture.

4. What other item found in a park will you draw?
   Answers may vary. Possible answers: tree, child, dog

5. Place a Dot Set 10 card and a Dot Set Card face up on your desk. Write the number that they make altogether.
   Answers may vary. Possible answers: 11–20

6. Draw that many of the item on your picture.

Reflect
How did you know how many items to draw?
I had to count the number of dots on all of the Dot Set Cards.

16 Level D Unit 1 **Number Sense within 100**

Student Workbook, p. 16

Teacher Reflect

☐ Did I explain the directions before students began their project?

☐ Did I supply the necessary materials?

☐ Did students correctly use art, objects, graphs, or posters to explain their solutions?

WEEK 2: Numbers on a Line

Week at a Glance
This week, students continue with **Number Worlds,** Level D, Number Sense within 100, by exploring number lines.

Skills Focus
- Visualize numbers up to 100 on a number line.
- Skip count by tens.
- Add and subtract small quantities using a number line.

How Students Learn
Helping students to develop interconnections between different representations of numbers helps to improve their number sense. Students with good number sense have a rich variety of ways to represent any quantitative situation, and to be able to move back and forth among those representations with ease, picking the form of representation that makes it easier for them to understand or work on the particular problem with which they are faced.

English Learners ELL
For language support, use the **English Learner Support Guide,** pages 24–25, to preview lesson concepts and teach academic vocabulary. **Number Words** Vocabulary Cards are listed as additional materials in many lessons and can be used to preteach and reinforce academic vocabulary.

Math at Home
Give one copy of the Letter to Home, page 2, to each student. Encourage students to share and complete the activity with their caregivers.

Weekly Planner

Lesson	Learning Objectives
1 pages 34–35	Students can count and identify numbers from 1 to 100.
2 pages 36–37	Students can count by tens.
3 pages 38–39	Students can add small quantities to progress through a number line to 100.
4 pages 40–41	Students can use plus and minus symbols to determine in which direction to move on a number line.
5 pages 42–43	**Review and Assess** Students review skills learned this week and complete the weekly assessment.
Project pages 44–45	Students can count, recognize tens and ones, and add small quantities.

Key Standard for the Week

Domain: Number and Operations in Base Ten
Cluster: Understand place value.
2.NBT.1 Understand that the three digits of a three-digit number represent amounts of hundreds, tens, and ones; e.g., 706 equals 7 hundreds, 0 tens, and 6 ones.

Materials		Technology
Program Materials • **Student Workbook,** pp. 17–19 • **Practice,** p. 8 • Activity Card 1E, **Steve's New Bike** • The Neighborhood • Delivery Building Picture • Neighborhood Number Line • Number Cards (1–100) • Steve's Bike Pawn	**Additional Materials** • container • math-link cubes*, 30 • small stickers • Vocabulary Card 14, *even number* • Vocabulary Card 30, *odd number*	*Teacher Dashboard* Building Blocks Bright Idea; Before and After Math
Program Materials • **Student Workbook,** pp. 20–21 • **Practice,** p. 9 • Activity Card 1F, **One Hundred Houses** • The Neighborhood • Neighborhood Number Line • Number Cards 1–100	**Additional Materials** • self-sticking notes, 100 • Vocabulary Card 45, *tens*	*Teacher Dashboard* Building Blocks School Supply Shop Number Line
Program Materials • **Student Workbook,** pp. 22–23 • **Practice,** p. 10 • Activity Card 1G, **Number Line to 100 Game**	• Neighborhood Number Line • Number Line to 100 Game Board • Number 1–6 Cubes • Pawns	*Teacher Dashboard* Building Blocks Rocket Blast 2
Program Materials • **Student Workbook,** pp. 24–25 • **Practice,** p. 11 • Activity Card 1H, **Plus-Minus Race to 100** • Number Line to 100 Game Board	• Number 1–6 Cubes • Pawns • Plus-Minus Cards	*Teacher Dashboard* Building Blocks Easy as Pie: Add Numbers
Program Materials • **Student Workbook,** pp. 26–27 • **Assessment,** pp. 15–16		Review previous activities.
Program Materials • **Student Workbook,** p. 28 • Counters	• crayons • dot cube • drawing of park from Week 1	

*Available from McGraw-Hill Education

WEEK 2
Numbers on a Line

Find the Math

In this week, students will identify numerals that come before and after numbers on a number line.

Use the following to begin a guided discussion:

▶ **How can you tell what number you are on a line? How do you know how many people are BEFORE you?** *I can start at the first person and count up to find out how many people are before me. If I know the number of the people before me, I know I am the number after.*

Have students complete **Student Workbook,** page 17.

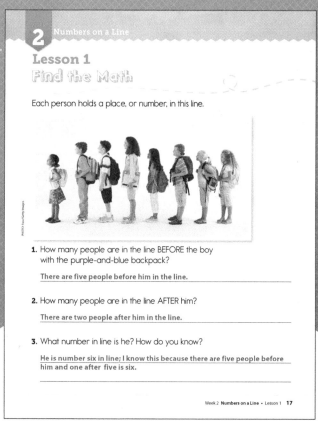

Student Workbook, p. 17

Lesson 1

Objective
Students can count and identify numbers from 1 to 100.

Standard
2.NBT.1 Understand that the three digits of a three-digit number represent amounts of hundreds, tens, and ones; e.g., 706 equals 7 hundreds, 0 tens, and 6 ones.

Vocabulary
- even number
- odd number

Creating Context
Practice the concepts of *nearest* and *farthest* with classroom objects. Have students walk to objects that are nearest and farthest from them.

Materials

Program Materials
- The Neighborhood
- Delivery Building Picture
- Neighborhood Number Line
- Number Cards (1–100)
- Steve's Bike Pawn

Additional Materials
- container
- math-link cubes, 30
- small stickers
- Vocabulary Card 14, *even number*
- Vocabulary Card 30, *odd number*

Prepare Ahead
Place the Delivery Building before the first house on the Neighborhood Number Line. Place the Number Cards in the container.

1 WARM UP

Prepare
Have students help you assemble the Neighborhood Number Line. Put up the first section and have students count the houses with you. Then have students predict what number comes next. Repeat the procedure until all sections are displayed. For additional instructions and variations, see **The Neighborhood**.

2 ENGAGE

Develop: Steve's New Bike

"Today we are going to explore numbers to 100." Follow the instructions on the Activity Card **Steve's New Bike.** As students complete the activity, be sure to use the Questions to Ask.

Activity Card 1E

Alternative Groupings

Individual: Partner with the student and complete the activity as written.

> **Progress Monitoring**
>
> **If…** students need practice with specific portions of the 1–100 sequence,
>
> **Then…** have them focus on a specific section of the Neighborhood Number Line.

Practice

Have students complete **Student Workbook,** pages 18–19. Guide students through the Key Idea example and the Try This exercises.

34 Level D Unit 1 **Number Sense within 100**

Interactive Differentiation

Consult the **Number Worlds Teacher Dashboard** for differentiated instruction groupings. You can also use performance on the Engage activity to guide students.

Independent Practice

For additional practice understanding numbers on a line, have students use Bright Idea: Counting On Game. Students count on from a numeral to identify number amounts. Student then move forward the corresponding number of spaces on the digital game board (up to 100).

Supported Practice

For additional support, use Before and After Math.

- Tell students they will identify and select numbers that come either right before or right after a target number.
- Students will click on the correct number.
- After students have finished the activity, ask them to give the number before and after a specific number to test understanding.

REFLECT

Think Critically

Review students' answers to the Reflect prompt at the bottom of **Student Workbook,** page 19, and then review the Engage activity.

Verify that students have labeled the number lines correctly.

▶ **What number comes after 57?** 58

▶ **What number comes before 45?** 44

▶ **What two numbers are on either side of 59?** 58 and 60

ASSESS

Informal Assessment

Use the online or print Student Record, **Assessment,** page 128, to record informal observations.

Steve's New Bike
Did the student
☐ make important observations?   ☐ provide insightful answers?
☐ extend or generalize learning?   ☐ pose insightful questions?

Additional Practice

For additional practice, have students complete **Practice,** page 8.

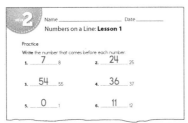

Practice, p. 8

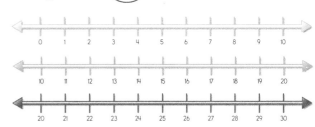

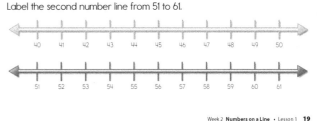

Student Workbook, pp. 18–19

WEEK 2
Numbers on a Line

Lesson 2

Objective
Students can count by tens.

Standard
2.NBT.1 Understand that the three digits of a three-digit number represent amounts of hundreds, tens, and ones; e.g., 706 equals 7 hundreds, 0 tens, and 6 ones.

Vocabulary
tens

Creating Context
Explain to students that *next-door neighbors* are people who live beside your home. Next-door neighbors live in the house, apartment, or home next to you.

Materials
Program Materials
- The Neighborhood
- Neighborhood Number Line
- Number Cards (1–100)

Additional Materials
- self-sticking notes, 100
- Vocabulary Card 45, *tens*

Prepare Ahead
Place the self-sticking notes over the numbers on the Neighborhood Number Line.

1 WARM UP

Prepare
Continue using The Neighborhood to explore the number line with students. Focus on skip counting by tens, starting at the decade markers (0, 10, 20, and so on). If students are successful at this, try skip counting by tens from other numbers in the sequence, such as 5.

Just the Facts
Have students chorally call out the number that you are asking for.

Use questions such as the following:

▶ **Skip count by 10. What number comes after 20?** 30

▶ **Start at 100 and skip count backward by ten once. What number do you get?** 90

▶ **Start at 70 and skip count backward by ten two times. What number do you get?** 50

2 ENGAGE

Develop: One Hundred Houses
"Today we are going to learn about multiples of ten." Follow the Instructions on the Activity Card **One Hundred Houses**. As students complete the activity, be sure to use the Questions to Ask.

Activity Card 1F

Alternative Groupings
Small Group: Have students work together to number the houses.

Pair: Have partners work together to number the houses 1–9, 11–19, and so on.

Progress Monitoring
If... students are not identifying multiples of ten correctly,

Then... practice skip counting by tens, and discuss the patterns they see on the Neighborhood Number Line.

Practice
Have students complete **Student Workbook,** pages 20–21. Guide students through the Key Idea example and the Try This exercises.

Interactive Differentiation
Consult the **Teacher Dashboard** for grouping suggestions. You can also use performance on the Engage activity to guide students.

Independent Practice
For additional practice understanding how to skip count, use School Supply Shop. Students count school supplies, bundled in groups of ten. Students will count the bundled supplies until they reach a target number (up to 100).

Supported Practice
For additional support understanding how to skip count, use the Number Line tool.

Select a number format from the palette to use for counting. Select a starting point for counting in the Number Line Begin area of the palette. You can choose a positive or negative number by selecting "+" or "−" next to the number.

- Tell students they will count forward and backward by using the Number Line tool.
- As they count to the right, values become larger, and they will use the "+" sign in the Skip Count area of the palette. Each "jump" will be drawn from left to right to show skip-counting.
- To count backward select the negative ("−") sign in the Skip Count area. Each "jump" will be drawn from right to left to show skip-counting.
- Give students numbers, such as 15 or 20, and ask them to skip count until they reach 25 or 50.
- Ask students where they think the jump will land before each click or how many clicks it will take to reach the targeted number.

REFLECT

Think Critically

Review students' answers to the Reflect prompt at the bottom of **Student Workbook,** page 21, and then review the Engage activity.

Have students discuss the strategies they used to create 100s.

Real-World Application

Ask students to think about standing in a lunch line.

- **Who will be served before you? Who will be served after you?**
- **How is this similar to a number line?**
- **How many houses are in each block of the Neighborhood Number Line?** 10
- **How many houses are in 10 blocks of the neighborhood—the whole neighborhood?** 100
- **If I start at House 50 and I want to get to House 90, how many houses do I have to walk past?** 40
- **Can you write a number sentence to show what we just did?**
  $50 + 40 = 90$

ASSESS

Informal Assessment

Use the online or print Student Record, **Assessment,** page 128, to record informal observations.

One Hundred Houses
Did the student
- ☐ make important observations?
- ☐ provide insightful answers?
- ☐ extend or generalize learning?
- ☐ pose insightful questions?

Additional Practice

For additional practice, have students complete **Practice,** p. 9.

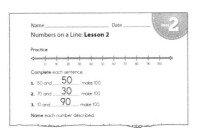

Practice, p. 9

Week 2 • Numbers on a Line

Lesson 2

Key Idea
You can count larger numbers by skip counting.

(number line 0 to 100 by tens)

Try This
Use the number line to skip count by **tens**.

1. 10, 20, 30, __40__ __50__ __60__
2. 50, 60, 70, __80__ __90__ __100__
3. __30__ __40__ 50, 60, 70, __80__ __90__ __100__

Write the number you are at after you skip count.

4. Begin at 20, and skip count by tens three times. __50__
5. Begin at 0, and skip count by tens six times. __60__
6. Begin at 50, and skip count by tens five times. __100__
7. Begin at 10, and skip count by tens two times. __30__

20 Level D Unit 1 Number Sense within 100

Practice
Complete each sentence.

8. 80 and __20__ make 100.
9. 90 and __10__ make 100.

Write each number described.

10. I am the number you land on when you start at 30 and skip count by tens three times. __60__
11. I am the number you land on when you start at 50 and skip count backward by tens two times. __30__
12. I am the number you land on when you start at 90 and skip count by tens once. __100__
13. I am the number you land on when you start at 80 and skip count backward by tens four times. __40__

Reflect
Complete the sentence in two different ways. Use the Neighborhood Number Line if you need help. *Sample answers shown*

__20__ and __60__ and __20__ make 100.
__10__ and __50__ and __40__ make 100.

Week 2 **Numbers on a Line** • Lesson 2 21

Student Workbook, pp. 20–21

WEEK 2
Numbers on a Line

Lesson 3

Objective
Students can add small quantities to progress through a number line to 100.

Standard
2.NBT.1 Understand that the three digits of a three-digit number represent amounts of hundreds, tens, and ones; e.g., 706 equals 7 hundreds, 0 tens, and 6 ones.

Vocabulary
- greater than
- less than

Creating Context
Practice the concepts of *before* and *after* with a calendar. Discuss the facts that yesterday was before today, 10 comes before 11, and so on.

Materials

Program Materials
- Neighborhood Number Line
- Number Line to 100 Game Board, 1 per group
- Number 1–6 Cubes, 1 per group
- Pawns, 1 per student

1 WARM UP

Prepare

Distribute the ten sections of the Neighborhood Number Line to students, and tell them that it is their job to put the blocks in order without speaking. When they have completed the number line, have students confirm that the sections are in order by counting up to 100 and back from 100, pointing at each house to make sure the numbers match.

Just the Facts

Have students point to the correct number on a number line. Use questions such as the following:

▶ **I am the number after 7. What number am I?** 8

▶ **I am the number before 70. What number am I?** 69

▶ **I am the number in between 54 and 52. What number am I?** 53

2 ENGAGE

Develop: Number Line to 100 Game

"Today we are going to play a game in which you will try to be the first person to reach 100." Follow the instructions on the Activity Card **Number Line to 100 Game.** As students complete the activity, be sure to use the Questions to Ask.

Activity Card 1G

Alternative Groupings

Whole Class: Organize students into groups of three or four, and have them complete the activity as written.

Progress Monitoring

If... students need help keeping track of their positions on the Number Line to 100 Game Board,

Then... have them use markers to write their initials on each number section they land on to avoid confusion.

Practice

Have students complete **Student Workbook,** pages 22–23. Guide students through the Key Idea example and the Try This exercises.

Interactive Differentiation

Consult the **Teacher Dashboard** for grouping suggestions. You can also use performance on the Engage activity to guide students.

Independent Practice

For additional practice, have students use Rocket Blast 2. Students will be given a number line with only the initial and final endpoints labeled. Students will be given a location on that line and asked to determine the correct number label for that point.

Supported Practice

For additional support, have students use the Number Line to 100 Game Board activity cards.

- Tell students that they will play a game in which they will try to get to 100.
- Begin the activity by having students take turns rolling the Number 1–6 Cube and moving their Pawns forward that number of spaces.
- After each turn, have each student enter on the score sheet the number he or she landed on during that turn.
- The first student to reach 100 is the winner. The remaining students should continue playing until all players have reached 100.

REFLECT

Think Critically

Review students' answers to the Reflect prompt at the bottom of **Student Workbook,** page 23, and then review the Engage activity.

Have students discuss the strategies they used to distinguish right from left.

▶ **What number is one greater than 57?** 58

▶ **What number is one less than 34?** 33

Real-World Application

▶ **In real neighborhoods, do house numbers increase by ten as you go up or down the street?**

▶ **What is your address?**

▶ **What do you think your neighbors' addresses are? Why?**

ASSESS

Informal Assessment

Use the online or print Student Record, **Assessment,** page 128, to record informal observations.

Number Line to 100 Game

Did the student

☐ respond accurately?   ☐ respond with confidence?

☐ respond quickly?   ☐ self-correct?

Additional Practice

For additional practice, have students complete **Practice,** p. 10.

Practice, p. 10

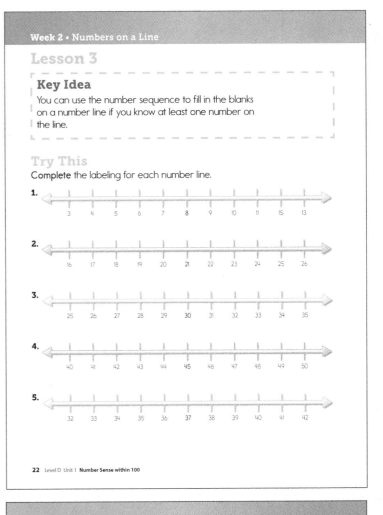

Student Workbook, pp. 22–23

Week 2 **Numbers on a Line** • Lesson 3   **39**

WEEK 2
Numbers on a Line

Lesson 4

Objective
Students can use plus and minus symbols to determine in which direction to move on a number line.

Standard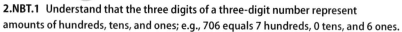
2.NBT.1 Understand that the three digits of a three-digit number represent amounts of hundreds, tens, and ones; e.g., 706 equals 7 hundreds, 0 tens, and 6 ones.

Creating Context
Continue to reinforce the concepts of *before* and *after* with a calendar. Discuss the facts that tomorrow is after today, 15 comes after 14, and so on.

Materials
Program Materials
- Number Line to 100 Game Board, 1 per group
- Number 1–6 Cubes, 1 per group
- Pawns, 1 per student
- Plus-Minus Cards, 1 set per group

Prepare Ahead
Shuffle the Plus-Minus Cards, and place them facedown. Make sure several Plus Cards are at the top of the stack.

1 WARM UP

Prepare
Present students with "What Number Am I?" mystery problems. Allow volunteers to guess the answers, and ask them to explain their reasoning. Use questions such as the following:

▶ I am three more than 32. What number am I? 35
▶ I am one fewer than 29. What number am I? 28

Just the Facts
Have students write down the number they think is correct on a piece of paper and hold it up.

▶ What number is 10 more than 11? 21
▶ What number is 10 less than 12? 2
▶ What number is 10 more than 15? 25

2 ENGAGE

Develop: Plus-Minus Race to 100
"Today we are going to play the Number Line to 100 Game, but we are adding rules." Follow the instructions on the Activity Card **Plus-Minus Race to 100**. As students complete the activity, be sure to use the Questions to Ask.

Activity Card 1H

Alternative Groupings
Individual: Partner with the student, and complete the activity as written.

Progress Monitoring

If... students are comfortable with the game procedures and want to add a challenge,

Then... tell them that they must land on 100 to win. Ask them to predict what number they need to roll to reach 100 without going over.

Practice
Have students complete **Student Workbook,** pages 24–25. Guide students through the Key Idea example and the Try This exercises.

Interactive Differentiation
Consult the **Teacher Dashboard** for suggested groupings. You can also use performance on the Engage activity to guide students.

Independent Practice
For additional practice, have students use Easy as Pie: Add Numbers. Students will identify numerals (zero through eight) and total number amounts (one through ten) and then move forward a corresponding number of spaces on a game board (up to 100).

Supported Practice
For additional support, have students use the **Plus-Minus Race to 100** Plus-Minus Cards.

- Shuffle the Plus-Minus Cards, and place them facedown. Make sure several Plus Cards are at the top of the stack so students will have a chance to move along the number line before picking a Minus Card.
- Organize students into groups of three or four. Distribute a Number Line to 100 Game Board, a set of Pawns, a Number 1–6 Cube, and a set of Plus-Minus Cards to each group.
- Tell students they will play a game in which they will try to be the first player to reach 100. They will use cards that speed them up or slow them down as they try to reach 100.

REFLECT

Think Critically

Review students' answers to the Reflect prompt at the bottom of *Student Workbook,* page 25, and then review the Engage activity.

Discuss to reinforce the concept of the number sequence.

▶ **Write a description for 15 using either *more than* or *less than*.**

▶ **Write a description for 79 using either *more than* or *less than*.**

Real-World Application

▶ **What other times do you use *more than* and *less than*?**

ASSESS

Informal Assessment

Use the online or print Student Record, *Assessment,* page 128, to record informal observations.

Plus-Minus Race to 100

Did the student

☐ pay attention to the contributions of others? ☐ improve on a strategy?

☐ contribute information and ideas? ☐ reflect on and check accuracy of work?

Additional Practice

For additional practice, have students complete *Practice,* page 11.

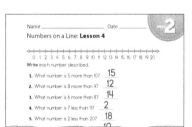

Practice, p. 11

Week 2 • Numbers on a Line

Lesson 4

Key Idea
Move right along the number line toward the bigger numbers for *more than.*
Move left along the number line toward the smaller numbers for *less than.*

Try This
Write each number.

1. What number is 1 more than 4? __5__
2. What number is 1 more than 11? __12__
3. What number is 2 more than 20? __22__
4. What number is 2 more than 12? __14__
5. What number is 1 less than 3? __2__
6. What number is 1 less than 18? __17__
7. What number is 2 less than 28? __26__
8. What number is 2 less than 10? __8__
9. What number is 2 more than 9? __11__

24 Level D Unit 1 **Number Sense within 100**

Practice
Use the Neighborhood Number Line to answer each question.

10. What number is 3 more than 8? __11__
11. What number is 10 more than 6? __16__
12. What number is 8 less than 20? __12__
13. What number is 2 less than 21? __19__

Complete each sentence.

14. 4 less than __17__ is 13.
15. 3 more than __30__ is 33.
16. 10 less than __55__ is 45.

Reflect
Write a description for the number 25 using the words *more than.*

Possible answer: 25 is 4 more than 21.

Write a description for the number 16 using the words *less than.*

Possible answer: 16 is 4 less than 20.

Week 2 **Numbers on a Line** • Lesson 4 25

Student Workbook, pp. 24–25

WEEK 2
Numbers on a Line

Lesson 5 Review

Objective
Students review skills learned this week and complete the weekly assessment.

Standard
2.NBT.1 Understand that the three digits of a three-digit number represent amounts of hundreds, tens, and ones; e.g., 706 equals 7 hundreds, 0 tens, and 6 ones.

Vocabulary
Review vocabulary introduced during the week.

Creating Context
Continue to reinforce the concepts of *forward* and *backward* with a number line. Discuss the fact that if you move forward from 16, the next number is 17, and so on.

1 WARM UP

Prepare

▶ **Which numbers come immediately before and after 36?**
35 and 37

▶ **Which numbers come immediately before and after 35?**
34 and 36

▶ **Which numbers come immediately before and after 37?**
36 and 38

2 ENGAGE

Practice
Have students complete **Student Workbook,** pages 26–27.

Week 2 • Numbers on a Line

Lesson 5 Review
This week you looked at our number system. You explored numbers that are *before* and *after*. You also practiced skip counting by tens.

Lesson 1 Write the number that is one count before each number.

1. _11_ 12
2. _75_ 76

Write the number that is one count after each number.

3. 67 _68_
4. 29 _30_
5. 44 _45_
6. 50 _51_

Lesson 2 Write the number you are at after each count.

7. Begin at 40, and skip count by tens five times. _90_
8. Begin at 20, and skip count by tens three times. _50_
9. Begin at 30, and skip count by tens four times. _70_

Lesson 3 Name each number. Use a number line if you need help.

10. I am the number before 24. _23_
11. I am the number after 17. _18_
12. I am the number before 50. _49_

Lesson 4 Name each number. Use a number line if you need help.

13. What number is 3 more than 46? _49_
14. What number is 10 more than 52? _62_
15. What number is 5 less than 65? _60_

Reflect
Write a description for the number 32 using the words *more than*. Write another description for the number 32 using the words *less than*.

Possible answer: 32 is 2 more than 30. Possible answer: 32 is 3 less than 35.

Student Workbook, pp. 26–27

42 Level D Unit 1 **Number Sense within 100**

3 REFLECT

Think Critically

Review students' answers to the Reflect prompt at the bottom of *Student Workbook,* page 27.

Discuss the answer with the group to reinforce Week 2 concepts

4 ASSESS

Formal Assessment

Students may take the weekly assessment online.

As an alternative, students may complete the weekly test on *Assessment,* pages 15–16. Record progress using the Student Assessment Record, *Assessment,* page 128.

Going Forward

Use the **Teacher Dashboard** to view results of the online assessments, to input the results of print student assessments, and to review progress before making decisions about next steps. Use the weekly test results and observations to determine the next steps for each student.

Retention	
Student displays good grasp of this week's concepts and skills.	Have students identify numerals (zero through nine) and total number amounts (one through ten) and then move backward a corresponding number of spaces on a game board (from 100).
Remediation	
Student is still struggling with the week's concepts and skills.	Have students use the Number Line to 100 Game Board and a Number 1–6 Cube.
	Students should take turns rolling the Number 1–6 Cube and moving a Pawn forward the corresponding number of spaces. Have each student record each number on which he or she lands on a score sheet. Continue the activity until students reach a number above 90.

Suggestions for Re-Evaluation: If a student has struggled without success for several weeks, use observations and test results to place the student at a level in which he or she can find success and build confidence to move forward.

Name _____ Date _____

Numbers on a Line — WEEK 2

Write the number that comes just before each number.

1. __18__ 19
2. __41__ 42
3. __51__ 52
4. __33__ 34

Write the number that comes just after each number.

5. 23 __24__
6. 90 __91__
7. 41 __42__
8. 19 __20__

Fill in the blanks.

9. The number 14 is located between __13__ and __15__
10. The number 31 is located between __30__ and __32__

Follow the directions to find the answers.

11. Begin at 0, and skip count by tens four times. __40__
12. Begin at 30, and skip count by tens three times. __60__
13. Begin at 20, and skip count by tens five times. __70__
14. Begin at 50, and skip count by tens four times. __90__

Level D Unit 1 Week 2 15

WEEK 2 Name _____ Date _____

Numbers on a Line

Complete the labeling for each number line.

15. ← 7 8 9 10 11 12 13 14 15 16 17 →

16. ← 20 21 22 23 24 25 26 27 28 29 30 →

Name each number described.

17. I am the number before 13. __12__
18. I am the number after 24. __25__
19. I am the number before 50. __49__
20. I am the number between 16 and 18. __17__

Answer each question.

21. What number is 1 more than 6? __7__
22. What number is 1 more than 21? __22__
23. What number is 2 more than 30? __32__
24. What number is 2 more than 14? __16__

16 Level D Unit 1 Week 2

Assessment, pp. 15–16

Project Preview

This week, students explored the number system, learning which numbers come before and after a particular number on a number line. The project for this unit requires students to extend the knowledge they gained in Find the Math and what they have learned this week. They will determine how many items they need to pack for a lunch in the park.

Project-Based Learning

Standards-driven Project-Based Learning is effective in building deep content understanding. Project-Based Learning increases long-term retention of concepts and has been shown to be more effective than traditional instruction. By completing a project to answer an essential question, students are challenged to apply and demonstrate mastery of concepts and skills by expressing understanding through discussion, research, and presentation.

Essential Question

HOW can I use comparing and ordering numbers outside the classroom?

Project Evaluation Criteria

Review project evaluation criteria with students prior to beginning the project.

Exceeds Expectations
☐ Project result is explained and can be extended.
☐ Project result is explained in context and can be applied to other situations.
☐ Project result is explained using advanced mathematical vocabulary.
☐ Project result is described, and mathematics are used correctly and can be extended.
☐ Project result is explained and extended, and shows advanced knowledge of mathematical concepts and skills.

Meets Expectations
☐ Project result is explained.
☐ Project result is explained in context.
☐ Project result is explained using mathematical vocabulary.
☐ Project result is described, and mathematics are used correctly.
☐ Project result is explained, and shows satisfactory knowledge of mathematical concepts and skills.

Does Not Meet Expectations
☐ Project result is not explained.
☐ Project result is explained, but out of context.
☐ Project result is explained, but mathematical vocabulary is oversimplified.
☐ Project result is described, but mathematics are not used correctly.
☐ Project result is not explained and or extended, or shows less than satisfactory knowledge of mathematical concepts and skills.

Picnicking in the Park

Objective
Students can count, recognize tens and ones, and add small quantities.

Standard
2.NBT.1 Understand that the three digits of a three-digit number represent amounts of hundreds, tens, and ones; e.g., 706 equals 7 hundreds, 0 tens, and 6 ones.

Materials
Program Materials
Counters

Additional Materials
- crayons
- dot cube
- drawing of park from Week 1

Best Practices
- Coach, demonstrate, and model.
- Create adequate time lines for each project.
- Focus students on their work to maintain engagement.

Introduce

Imagine that your parents are taking you and three of your friends to the park.

- **What will you take with you for lunch?** Answers may vary. Possible answer: sandwiches and apples
- **How many of each item will you need?** Answers may vary. Possible answer: I will bring 5 sandwiches and 10 apples.

Explore

- **Today you will add to the picture you began drawing last week. Today you will draw the food you will take with you on your trip to the park.**
- **Complete *Student Workbook*, page 28 to find the number of food items you will bring.**
- Have students identify two items they want to bring for lunch.
- **Roll a dot cube. The number you roll tells you how many of the first food item to bring. Write down that number.**
- **Next, follow the instructions your friend gives you on how many items to bring. For example, you roll 4 on the dot cube. You will bring four sandwiches, but your friend wants you to bring 3 more than the number you rolled. What is 3 more than 4? You will bring 7 sandwiches.**
- **Draw a sandwich on your picture of the park and place the number 7 inside it.**
- **Repeat for the second item.**
- **Complete *Student Workbook*, page 28.**

Wrap Up

- Allow students time to decide which items they want to bring.
- Make sure all students can explain how they determined the total number of items they brought.
- If students struggle to find the number of items to bring, have them use Counters to represent the items.
- Discuss students' answers to the Reflect prompts at the bottom of **Student Workbook,** page 28.
- Save students' drawings for use in Week 3.

If time permits, allow each student to check another student's work.

Week 2 • Numbers on a Line

Project
Picnicking in the Park

Name and draw a picture of food you will pack for a picnic.

1. What food will you pack?

   Answers may vary. Possible answers: sandwiches, carrots

2. Roll a dot cube to find how many servings you will pack.

   Answers may vary. Possible answer: I rolled a 4 on the dot cube. I will pack 4 sandwiches.

3. Your friend asks you to pack 5 more. How many do you pack?

   Answers may vary. Possible answer: I rolled a 4. I have to take 5 more, so I will take 9.

4. Draw a picnic table and the food you will pack on your picture.

Reflect

How do you know how many items to start with?

I counted the number of dots on the dot cube.

How did you know how many items to draw?

Answers will vary. Possible answer: I counted the dots and then counted on five more.

28 Level D Unit 1 **Number Sense within 100**

Student Workbook, p. 28

Teacher Reflect

- ☐ Did I explain the directions before students began their project?
- ☐ Did I supply the necessary materials?
- ☐ Did students correctly use art, objects, graphs, or posters to explain their solutions?

Week 2 **Numbers on a Line** • Lesson 5

WEEK 3: Tens and Ones

Week at a Glance

This week, students continue **Number Worlds,** Level D, Number Sense within 100, by exploring the fundamentals of the base-ten number system. Students will begin to group ones into tens and understand the value of ten.

Skills Focus

- Skip count by tens.
- Make groups of ten.
- Think of two-digit numbers in terms of tens and ones.

How Students Learn

As students move through the activities, provide multiple opportunities for them to encounter interesting features of the number system and to make interesting discoveries about properties of this system. Helping students to build up an extensive network of associations in this context can serve as an intuitive basis for higher-order learning and discovery in later years.

English Learners ELL

For language support, use the **English Learner Support Guide,** pages 26–27, to preview lesson concepts and teach academic vocabulary.

Math at Home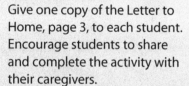

Give one copy of the Letter to Home, page 3, to each student. Encourage students to share and complete the activity with their caregivers.

Weekly Planner

Lesson	Learning Objectives
1 pages 48–49	Students can coordinate movement along tens and ones on a number line to arrive at a target number.
2 pages 50–51	Students can determine the value of digits in double-digit numbers.
3 pages 52–53	Students can model double-digit numbers as tens and ones.
4 pages 54–55	Students can trade ones for tens.
5 pages 56–57	**Review and Assess** Students review skills learned this week and complete the weekly assessment.
Project pages 58–59	Students can trade ones for tens.

46 Level D Unit 1 **Number Sense within 100**

Key Standard for the Week

Domain: Number and Operations in Base Ten
Cluster: Understand place value.
1.NBT.2 Understand that the two digits of a two-digit number represent amounts of tens and ones.

Materials		Technology
Program Materials • *Student Workbook,* pp. 29–31 • *Practice,* p. 12 • Activity Card 1I, **Rosemary's Super Shoes** • Neighborhood Number Line • Number Cards (1–100) • Rosemary's Shoes Pawns • Delivery Building Picture	**Additional Materials** • small stickers • Vocabulary Card 32, *ones* • Vocabulary Card 45, *tens*	**Teacher Dashboard** Number Line Building Blocks Clean the Plates
Program Materials • *Student Workbook,* pp. 32–33 • *Practice,* p. 13 • Activity Card 1J, **Hotel Room Service** • Introduction to the Hotel • Room Service Delivery Slips • Hotel Game Board	• Number Cards (1–100) • Pawns • Counters **Additional Materials** • Vocabulary Card 32, *ones* • Vocabulary Card 45, *tens*	**Teacher Dashboard** Base 10 Blocks Tool
Program Materials • *Student Workbook,* pp. 34–35 • *Practice,* p. 14 • Activity Card 1K, **Double-Digit Number Comparison** • Number Cards (0–9) • Number Construction Mat	**Additional Materials** • Vocabulary Card 32, *ones* • Vocabulary Card 45, *tens*	**Teacher Dashboard** Building Blocks Number Compare 5
Program Materials • *Student Workbook,* pp. 36–37 • *Practice,* p. 15 • Activity Card 1L, **Building and Trading Numbers** • Number Cards (0–9) • Number Construction Mat	• Place-Value Mat • Counters **Additional Materials** • base-ten blocks • Vocabulary Card 32, *ones* • Vocabulary Card 45, *tens*	**Teacher Dashboard** Building Blocks Number Snapshots 10
Program Materials • *Student Workbook,* pp. 38–39 • Weekly Test, *Assessment,* pp. 17–18	**Additional Materials** small stickers	Review previous activities.
Program Materials • *Student Workbook,* p. 40 • Counters	**Additional Materials** • crayons • drawing of park from week 2	

*Available from McGraw-Hill Education

WEEK 3
Tens and Ones

Find the Math

In this week, students will identify how tens and ones make up numbers.

Use the following to begin a guided discussion:

▶ **What's the difference between giving a friend the bowl of peaches and giving him or her one peach?** If I give a friend the bowl of peaches, I would give my friend ten peaches. If I give my friend one peach, I would only give him or her one peach, not ten.

Have students complete **Student Workbook,** page 29.

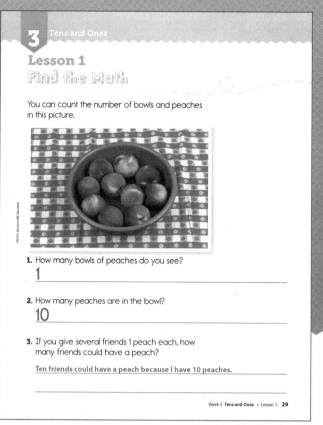

Student Workbook, p. 29

Lesson 1

Objective
Students can coordinate movement along tens and ones on a number line to arrive at a target number.

Standard
1.NBT.2 Understand that the two digits of a two-digit number represent amounts of tens and ones.

Vocabulary
tens

Creating Context
Playing games that encourage competition, such as **Rosemary's Super Shoes**, helps students acquire math strategies and concepts. Make sure students practice talking about math through each step of the game while they play.

Materials

Program Materials
- Neighborhood Number Line
- Number Cards (1-100)
- Rosemary's Shoes Pawns
- Delivery Building Picture

Additional Materials
- small stickers
- Vocabulary Card 32, *ones*
- Vocabulary Card 45, *tens*

1 WARM UP

Prepare

Play a game of What Number Am I? For this lesson, use several mystery questions that require students to think of numbers as tens and ones and to identify numbers when they are described in expanded form. For example:

▶ **I am 1 group of tens and 4 ones. What number am I?**

2 ENGAGE

Develop: Rosemary's Super Shoes

"Today we are going to help Rosemary make deliveries by using tens and ones." Follow the instructions on the Activity Card **Rosemary's Super Shoes.** As students complete the activity, be sure to use the Questions to Ask.

Activity Card 1I

Alternative Groupings

Individual: Partner with the student, and complete the activity as written.

Progress Monitoring

| **If…** students struggle in a specific portion of the 1–100 sequence | **Then…** have them focus on the section of the Neighborhood Number line where they began to struggle. |

Practice

Have students complete **Student Workbook,** pages 30–31. Guide students through the Key Idea example and the Try This exercises.

Interactive Differentiation

Consult the **Teacher Dashboard** for grouping suggestions. You can also use performance on the Engage activity to guide students.

Independent Practice

For additional practice understanding skip counting, have students use the Number Line Tool. Select a Number Format from the palette to use for counting. Select a starting point for counting in the Number Line Begin area of the palette. You can choose a positive or negative number by selecting "+" or "−" next to the number.

- Tell students they will count forward or backward by 10 using the Number Line Tool.
- As they count to the right and values become larger, they will use the "+" sign in the Skip Count area of the palette. Each "jump" will be drawn from left to right to show skip-counting.
- To count backward select the negative ("−") sign in the Skip Count area. Each "jump" will be drawn from right to left to show skip-counting.
- Give students numbers, such as 10 or 20, and tell them to skip count until they reach 50.
- Give students a number, such as 40, and tell them to skip count 3 times. Ask students where they think the "jump" will end.

Supported Practice

For additional support, use Clean the Plates. Students use skip counting to produce products that are multiples of 10s, 5s, 2s, and 3s. The task is to identify how many bundles you had to add to make the product.

REFLECT

Think Critically

Review students' answers to the Reflect prompt at the bottom of *Student Workbook,* page 31, and then review the Engage activity.

Discuss to reinforce the concept of place value.

▶ **How many tens and ones does 54 have?** 5 tens, 4 ones

▶ **If we grouped 54 into sets of 10, have many groups would we have and how many would be left over?** 5 groups with 4 left over

ASSESS

Informal Assessment

Use the online or print Student Record, **Assessment,** page 128, to record informal observations.

Rosemary's Super Shoes

Did the student
- ☐ make important observations?
- ☐ provide insightful answers?
- ☐ extend or generalize learning?
- ☐ pose insightful questions?

Additional Practice

For additional practice, have students complete **Practice,** page 12.

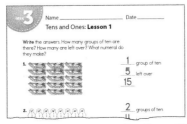

Practice, p. 12

Week 3 • Tens and Ones

Lesson 1

Key Idea
To count a group of objects quickly, sort the objects into groups of ten and skip count by tens.

Try This
Skip count by tens. Fill in the missing numbers.

1. 10, 20, __30__
2. 10, __20__, 30, __40__
3. 20, 30, __40__, __50__
4. __50__, 60, 70, __80__

Circle sets of ten. Count how many groups of ten there are. Then write the numeral that tells how many there are.

5. __3__ groups of ten
   __30__

6. __2__ groups of ten
   __20__

30 Level D Unit 1 Number Sense within 100

Practice
Tell how many groups of ten there are. How many are left over? What numeral do they make?

7. __1__ groups of ten
   __2__ left over
   __12__

8. __3__ groups of ten
   __3__ left over
   __33__

Reflect
Draw 23 circles. Group them into tens and ones.

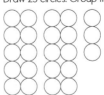

Week 3 Tens and Ones • Lesson 1 31

Student Workbook, pp. 30–31

Week 3 **Tens and Ones** • Lesson 1 49

WEEK 3
Tens and Ones

Lesson 2

Objective
Students can determine the value of digits in double-digit numbers.

Standard
1.NBT.2 Understand that the two digits of a two-digit number represent amounts of tens and ones.

Vocabulary
- ones
- tens

Creating Context
In English, some words end with a silent *e*. Help English Learners find examples, and make sure they know that when the silent *e* appears at the end of a word, the other vowel in the word is pronounced as a long vowel. Some examples in this lesson include the following: *combine, place, value, write, made, one, make,* and *trade*.

Materials
Program Materials
- Introduction to the Hotel
- Room Service Delivery Slips, 1 per student
- Hotel Game Board, 1 per group
- Number Cards (1-100)
- Pawns, 1 per student
- Counters, 100 per group

Additional Materials
- Vocabulary Card 32, *ones*
- Vocabulary Card 45, *tens*

1 WARM UP

Prepare
Show students the Hotel Game Board. Ask whether anyone has ever stayed at a hotel and knows what room service is. If necessary, explain that at some hotels, guests can call the kitchen to order food and a server delivers it to their room. Tell students that they are going to pretend to be the waiters at a room-service hotel, and their job will be to deliver room-service orders. For additional directions, see Introduction to the Hotel.

Just the Facts
Have students chorally call out the number that you are asking for. Use questions such as the following:

▶ **Twenty ones equals how many sets of ten?** 2
▶ **One ten equals how many ones?** 10
▶ **Thirteen ones equals how many tens and ones?** 1 ten and 3 ones

2 ENGAGE

Develop: Hotel Room Service
"Today we are going to make deliveries to rooms on the Hotel Game Board." Follow the instructions on the Activity Card **Hotel Room Service**. As students complete the activity, be sure to use the Questions to Ask.

Activity Card 1J

Alternative Groupings
Pair: Allow additional time to complete the activity.

Progress Monitoring

If... students need additional practice examining number patterns, **Then...** stack the Delivery Slips to create situations resulting in patterns. For example, you might stack them so the players reach rooms 39, 49, 59, and 69. Then have students discuss differences and similarities in their number positions.

Practice
Have students complete **Student Workbook,** pages 32–33. Guide students through the Key Idea example and the Try This exercises.

Interactive Differentiation
Consult the **Teacher Dashboard** for grouping suggestions. You can also use performance on the Engage activity to guide students.

Independent Practice

For additional practice, use the Base 10 Blocks Tool. Students will break down larger base-ten blocks into smaller base-ten blocks. Start students with blocks of ten.

- Tell students after placing blocks on a counting mat they will use the Cursor Tool to select a block.
- Click the Break-Down button in the Options area of the palette. Your block will be broken down into ten of the next smallest base-ten blocks.
- After placing blocks on a counting mat, use the Cursor Tool to select groups of ten blocks.
- Click the Combine button in the Options area of the palette. Your blocks will be combined into the next largest base-ten unit.
- Give students numbers, such as "7 and 4," "9 and 3," "27 and 6," and tell them to highlight ten ones, and make them into tens. Then students can see how many ones are left over.
- Give students numbers, such as "4 tens and 3 ones," "6 tens and 11 ones," and see how many tens and ones there are.

Supported Practice

For additional support, have students use Counters. Give students thirty Counters in sets of ten to build different numbers using tens and ones.

▶ **Have students show 15. How many sets of ten do you have and how many ones do you have?** I have one set of 10 and 5 ones.
▶ **Have students show 17. How many sets of ten do you have and how many ones?** I have one set of 10 and 7 ones.
▶ **Have students show 25. How many sets of ten do you have and how many ones?** I have two sets of 10 and 5 ones.

REFLECT

Think Critically

Review students' answers to the Reflect prompt at the bottom of *Student Workbook,* page 33, and then review the Engage activity.

Have students discuss the strategies they used to make trades.

- **What number is represented by 2 tens and 11 ones?** 31
- **What number is represented by 6 tens and 15 ones?** 75

Real-World Application

- **In the metric system, ten of one unit can be traded for one of the next larger-sized unit. These trades are similar to the trades made for groups of ten. For example, there are 10 millimeters in 1 centimeter.**
- **How many centimeters can you make with 20 millimeters?** 2 centimeters
- **Explain the trade.** I traded 10 millimeters for 1 centimeter and another 10 millimeters for another centimeter.

ASSESS

Informal Assessment

Use the online or print Student Record, *Assessment,* page 128, to record informal observations.

Hotel Room Service

Did the student

☐ respond accurately?   ☐ respond with confidence?

☐ respond quickly?   ☐ self-correct?

Additional Practice

For additional practice, have students complete *Practice,* page 13.

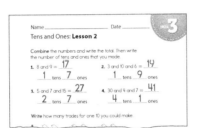

Practice, p. 13

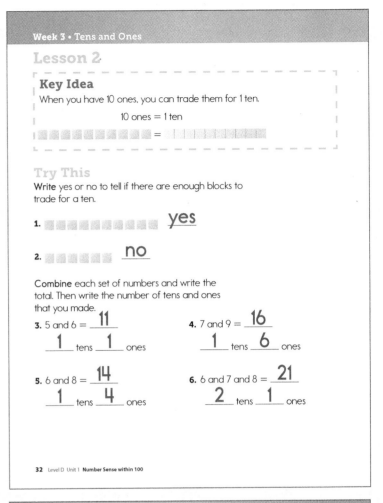

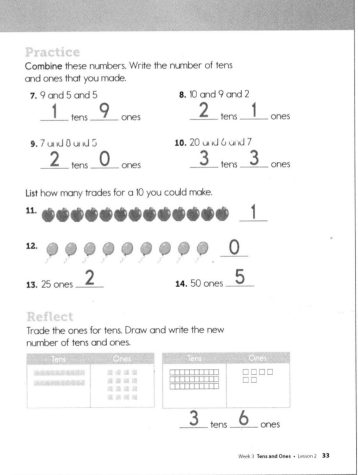

Student Workbook, pp. 32–33

Week 3 **Tens and Ones** • Lesson 2   **51**

WEEK 3
Tens and Ones

Lesson 3

Objective
Students can model double-digit numbers as tens and ones.

Standard
1.NBT.2 Understand that the two digits of a two-digit number represent amounts of tens and ones.

Vocabulary
- ones
- tens

Creating Context
Students often benefit from seeing new skills modeled. As you demonstrate and explain skills and concepts, be sure to pause between each step to check for understanding.

Materials
Program Materials
- Number Cards (0–9), 2 sets per student
- Number Construction Mat, 1 per student

Additional Materials
- Vocabulary Card 32, *ones*
- Vocabulary Card 45, *tens*

1 WARM UP

Prepare
▶ **Can anyone tell me what a single-digit number is?** Possible answer: a number with only one numeral in it
▶ **What are the single-digit numbers?** 1, 2, 3, 4, 5, 6, 7, 8, 9, 0
▶ **What is a double-digit number?** Possible answer: a number with two numerals in it
▶ **What is the smallest double-digit number?** 10 **What is the largest double-digit number?** 99

Just the Facts
Have students chorally call out the correct number shown by base-ten blocks.
- Display one rod and one unit.
▶ **What number is shown?** 11
- Display two rods and two units.
▶ **What number is shown?** 22
- Display three rods and four units.
▶ **What number is shown?** 34

2 ENGAGE

Develop: Double-Digit Number Composition
"Today we are going to identify and compose two-digit numerals." Follow the instructions on the Activity Card **Double-Digit Number Composition.** As students complete the activity, be sure to use the Questions to Ask.

Activity Card 1K

Alternative Groupings
Whole Class: Display the Number Construction Mat in front of the class. Invite volunteers to place the numbers to create the smallest and largest numbers and discuss their answers as a class.

Pair: Distribute a set of cards and a mat to the pair. Have one student create the largest possible number, and have the other student create the smallest possible number. Have students alternate roles.

Progress Monitoring
If… students have trouble remembering the definitions of *single-digit* and *double-digit*,

Then… explain that *single* means "one" and *double* means "two."

Practice
Have students complete **Student Workbook,** pages 34–35. Guide students through the Key Idea example and the Try This exercises.

Interactive Differentiation
Consult the **Teacher Dashboard** for grouping suggestions. You can also use performance on the Engage activity to guide students.

Independent Practice

For additional practice, have students use Number Compare 5: Dot Arrays to 100. Students will compare two cards and choose the one with the greater value, up to 100.

Supported Practice

For additional support, have students use the Double-Digit Number Composition activity card and the Number Construction Mat. Distribute one Number Construction Mat to each student. Tell students that they will identify and compose two-digit numerals. Tell students that they are going to be using two columns on this form: the one labeled *Ones* and the one labeled *Tens*. Explain that the form can be used to determine how much any number is worth.

▶ **If a number, such as 6, is in the ones column, how much do you think it is worth?** 6 ones, or 6
▶ **If the same number, 6, is in the tens column, how much do you think it is worth?** 6 tens, or 60

52 Level D Unit 1 **Number Sense within 100**

REFLECT

Think Critically

Review students' answers to the Reflect prompt at the bottom of *Student Workbook,* page 35, and then review the Engage activity.

Discuss the value of a rod and of a unit block.

▶ Can you look at a rod and immediately know that it is worth ten?

▶ What is the value of 5 rods and 3 unit blocks? 53

Real-World Application

▶ A decade is a length of time of 10 years, so three decades is the same as 30 years. Sometimes it helps to picture a number in your head. You can think of decades in terms of tens. One decade equals one ten, and single years are similar to ones.

▶ Show how 4 decades and 7 years, or 47 years, would be built on a **Number Construction Mat.** 4 rods and 7 unit blocks

ASSESS

Informal Assessment

Use the online or print Student Record, *Assessment,* page 128, to record informal observations.

Double-Digit Number Composition

Did the student

☐ respond accurately? ☐ respond with confidence?

☐ respond quickly? ☐ self-correct?

Additional Practice

For additional practice, have students complete *Practice,* page 14.

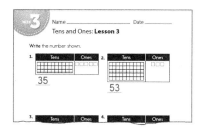

Practice, p. 14

Student Workbook, pp. 34–35

WEEK 3
Tens and Ones

Lesson 4

Objective
Students can trade ones for tens.

Standard
1.NBT.2 Understand that the two digits of a two-digit number represent amounts of tens and ones.

Vocabulary
- ones
- tens

Creating Context
Look through each lesson to find idioms and synonyms. Have English Learners keep a list of these expressions and words and their definitions.

Materials
Program Materials
- Number Cards (0–9), 1 set per student
- Number Construction Mat, 1 per student
- Place-Value Mat, 1 per student
- Counters, 26 per student

Additional Materials
- base-ten blocks, 4 rods and 20 blocks per student
- Vocabulary Card 32, *ones*
- Vocabulary Card 45, *tens*

1 WARM UP

Prepare
▶ **Use Counters to show the number 26. Organize the Counters so you can see the number easily.**

▶ **How did you organize your Counters?**

As a class, look at all the models. If some students have modeled 26 as 26 ones, or as 5 groups of five and 1 one, or as 2 groups of ten and 6 ones, point out the differences and remind students that all models show 26. If no one has created the last model mentioned, show students how 26 can be distributed into tens and ones. Tell students that this is the model that is used most often.

▶ **The number names you say when you are counting are the names for tens and ones. What number do you say after 19?** 20; Twenty is 2 tens. As soon as you count up to 9 in any decade, the next number name you say describes the number of groups of 10 you have counted.

▶ **What number do you say after 29?** 30; Thirty is 3 tens.

Just the Facts
Have students chorally call out how many tens and how many ones are shown by base-ten blocks.

- Display two rods and eight unit blocks.
▶ **How many tens and how many ones do you see?** 2 tens; 8 ones
- Repeat the activity for different numbers as time permits.

2 ENGAGE

Develop: Building and Trading Numbers
"Today we are going to use base-ten blocks to regroup numbers into tens and ones." Follow the instructions on the Activity Card **Building and Trading Numbers.** As students complete the activity, be sure to use the Questions to Ask.

Activity Card 1L

Alternative Groupings
Pair: Have students work together to complete the Conclude the Activity section.

Progress Monitoring
If... students are successful with Building and Trading Numbers

Then... allow them to use pennies and dimes with the Number Construction Mat

Practice
Have students complete **Student Workbook,** pages 36–37. Guide students through the Key Idea example and the Try This exercises.

Interactive Differentiation
Consult the **Teacher Dashboard** for grouping suggestions. You can also use performance on the Engage activity to guide students.

Independent Practice

For additional practice, have students use Number Snapshots 10. Students will combine collections of dots up to 50 and match them to a corresponding numeral.

Supported Practice

For additional support, have students use Counters.

▶ **Give students a number, such as 27. How can you use Counters to show 27?** Answers may vary. Possible answer: 2 sets of ten and 7 ones

▶ **What is another way to show 27?** Answers may vary. Possible answer: 1 set of ten and 17 ones

▶ **How can you show 32 in two different ways?** Answers may vary. Possible answer: 3 sets of ten and two ones

REFLECT

Think Critically

Review students' answers to the Reflect prompt at the bottom of *Student Workbook*, page 37, and then review the Engage activity.

Have a volunteer draw a third way to model 52 on the board.

▸ **Is this model correct? Are there other ways to show 52?**

Discuss to reinforce the concept that a number can be represented many different ways.

Real-World Application

▸ **In some board games you can buy houses to put on your property. After you have a certain number of houses, you can trade them for a hotel. Imagine that a hotel is worth ten houses.**

▸ **Would you rather have 1 hotel and 2 houses or 11 houses on your property? Explain.** I would rather have 1 hotel and 2 houses because the hotel represents 10 houses, and 10 and 2 is more than 11.

▸ **If a player landed on your property, would he or she owe you the same amount if you had 13 houses or 1 hotel and 3 houses?** yes

ASSESS

Informal Assessment

Use the online or print Student Record, *Assessment,* page 128, to record informal observations.

Building and Trading Numbers

Did the student
- ☐ make important observations?
- ☐ extend or generalize learning?
- ☐ provide insightful answers?
- ☐ pose insightful questions?

Additional Practice

For additional practice, have students complete *Practice,* page 15.

Practice, p. 15

Week 3 • Tens and Ones

Lesson 4

Key Idea
You can use blocks to make models of numbers in different ways.

Try This

Write the number of tens and ones. Then write the total number.

1. __3__ tens __5__ ones
   35

2. __2__ tens __15__ ones
   35

3. __4__ tens __11__ ones
   51

4. __5__ tens __1__ ones
   51

Practice

Make a model of each number in two different ways, using groups of blocks.

5. 21
   - Sample answer: 1 tens block and 11 ones blocks
   - Sample answer: 2 tens blocks and 1 ones block

6. 54
   - Sample answer: 5 tens blocks and 4 ones blocks
   - Sample answer: 4 tens blocks and 14 ones blocks

Reflect

Draw another way to model 52.

Sample answer: Students could draw 5 tens blocks and 2 ones blocks.

Student Workbook, pp. 36–37

WEEK 3
Tens and Ones

Lesson 5 Review

Objective
Students review skills learned this week and complete the weekly assessment.

Standard
1.NBT.2 Understand that the two digits of a two-digit number represent amounts of tens and ones.

Vocabulary
Review vocabulary introduced during the week.

Creating Context
English Learners may find it difficult to know how the letter *-s* is pronounced in the words *tens* and *ones*. Explain that after the letter *n* or the sound of /n/, the letter *s* is pronounced /z/.

1 WARM UP

Prepare

▶ What is the value of a single rod? 10
▶ What is the value of a single unit block? 1
▶ What is the value of 4 rods and 8 unit blocks? 48

2 ENGAGE

Practice
Have students complete **Student Workbook,** pages 38–39.

Week 3 • Tens and Ones

Lesson 5 Review

This week you explored groups of tens and ones. You discovered that you can trade 10 ones for 1 ten.

Lesson 1 Skip count by tens to find the number of buttons. Each row has ten buttons.

1. 54 buttons

Lesson 2 Combine these numbers. Write the number of tens and ones made.

2. 6 and 8 and 4
tens 1 ones 8

3. 4 and 9 and 2
tens 1 ones 5

4. 9 and 7 and 5
tens 2 ones 1

38 Level D Unit 1 Number Sense within 100

Lesson 3 Write each number that is shown by the groups of blocks.

5. 17
6. 43

Lesson 4 Draw a model of each number.

7. 42
8. 14

Reflect

Write the number that is shown. Show that number in another way.

51

Week 3 Tens and Ones • Lesson 5 39

Student Workbook, pp. 38–39

56 Level D Unit 1 **Number Sense within 100**

3 REFLECT

Think Critically

Review students' answers to the Reflect prompt at the bottom of **Student Workbook,** page 39.

Discuss the answer with the group to reinforce Week 3 concepts

4 ASSESS

Formal Assessment

Students may take the weekly assessment online.

As an alternative, students may complete the weekly test on **Assessment,** pages 17–18. Record progress using the Student Assessment Record, **Assessment,** page 128.

Going Forward

Use the **Teacher Dashboard** to view results of the online assessments, to input the results of print student assessments, and to review progress before making decisions about next steps. Use the weekly test results and observations to determine the next steps for each student.

Retention	
Student displays good grasp of this week's concepts and skills.	Have students continue to use the Activity Rosemary's Super Shoes as time permits.
Remediation	
Student is still struggling with the week's concepts and skills.	Use the Double-Digit Number Composition activity card and the Number Construction Mat. Ask questions such as the following ▶ **If a number, such as 5, is in the ones column, how much is it worth?** 5 ones, or 5 ▶ **If the same number, 5, is in the tens column, how much is it worth?** 5 tens and zero ones, or 50

Suggestions for Re-Evaluation: If a student has struggled without success for several weeks, use observations and test results to place the student at a level in which he or she can find success and build confidence to move forward.

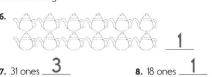

Assessment, pp. 17–18

Project Preview

This week, students explored groups of tens and ones. Students learned that you can trade 10 ones for 1 ten. The project for this unit requires students to extend the knowledge of what they have learned this week. They will change numbers from sets of ten to ones as they figure out how to share food to feed to the animals in the park.

Project-Based Learning

Standards-driven Project-Based Learning is effective in building deep content understanding. Project-Based Learning increases long-term retention of concepts and has been shown to be more effective than traditional instruction. By completing a project to answer an essential question, students are challenged to apply and demonstrate mastery of concepts and skills by expressing understanding through discussion, research, and presentation.

Essential Question

HOW can I use comparing and ordering numbers outside the classroom?

Project Evaluation Criteria

Review project evaluation criteria with students prior to beginning the project.

Exceeds Expectations
☐ Project result is explained and can be extended.
☐ Project result is explained in context and can be applied to other situations.
☐ Project result is explained using advanced mathematical vocabulary.
☐ Project result is described, and mathematics are used correctly and can be extended.
☐ Project result is explained and extended, and shows advanced knowledge of mathematical concepts and skills.

Meets Expectations
☐ Project result is explained.
☐ Project result is explained in context.
☐ Project result is explained using mathematical vocabulary.
☐ Project result is described, and mathematics are used correctly.
☐ Project result is explained, and shows satisfactory knowledge of mathematical concepts and skills.

Does Not Meet Expectations
☐ Project result is not explained.
☐ Project result is explained, but out of context.
☐ Project result is explained, but mathematical vocabulary is oversimplified.
☐ Project result is described, but mathematics are not used correctly.
☐ Project result is not explained and or extended, or shows less than satisfactory knowledge of mathematical concepts and skills.

Feeding Animals in the Park

Objective
Students can trade ones for tens.

Standard
1.NBT.2 Understand that the two digits of a two-digit number represent amounts of tens and ones.

Materials

Program Materials
Counters

Additional Materials
- crayons
- drawing of park from Week 2

Best Practices
- Organize materials ahead of time.
- Provide project directions that are clear and brief.
- Check for student understanding frequently.

Introduce

Imagine that you and your two friends are going to the park. You will buy animal food that you will all share.

- **What types of animal food can you buy?** Answers may vary. Possible answers: sunflower seeds, fish pellets, peanuts
- **How many pieces of animal food will each of you get?** Answers may vary. Possible answers: Each person will get the same amount.

Explore

- Today you will decide what animal food you will share among your friends.
- Look at *Student Workbook* page 40 to find how much animal food you and your friends will get.
- You can choose from three types of food: sunflower seeds, fish pellets, or peanuts.
- If each friend gets a set of ten, how many pieces of food will be left over? Students can use counters to represent the number of each animal food. Answers may vary based on food chosen.
- Do you have any sets of ten left? If so, how many? Answers may vary based on food chosen.
- Do you have any ones left? If so, how many? Answers may vary based on food chosen.
- Complete *Student Workbook*, page 40.

Wrap Up

- Allow students time to decide which food they want to choose.
- Make sure each student can explain how they separated the numbers (pieces of food) into groups of ten and can explain how many were left over.
- If students struggle to share the items, remind them that each set of 10 is 10 separate items.
- Discuss students' answers to the Reflect prompt at the bottom of **Student Workbook,** page 40.
- Save students' drawings for use in Week 4.

If time permits, allow each student to share another type of food.

Week 3 • Tens and Ones

Project
Feeding Animals in the Park

Circle which food you will give to your friends to feed to the animals. Then answer the questions.

44 sunflower seeds   34 fish pellets   37 peanuts

1. Share the animal food you pick equally with two friends. How many pieces do each of you have?

   10 pieces each

2. Do you have any sets of ten left over?

   Answers may vary. Possible answer: Yes, I have one set of ten left over.

3. Do you have any ones left over?

   Possible answer: Yes, I have 4 ones left over.

4. In your picture of the park, draw the animal food you chose. Add a drawing of the animal that will eat it.

Reflect

How do you know if you have ones or a set of ten left over?

Answers may vary. Possible answer: I know I have a set of ten if I have ten or more snacks left over. If I have fewer than ten, then I have only ones.

40 Level D Unit 1 **Number Sense within 100**

Student Workbook, p. 40

Teacher Reflect

☐ Did I explain the directions before students began their project?

☐ Were students able to answer my questions about their solutions?

☐ Did I adequately explain and discuss the Reflect questions with students?

WEEK 4
Visualizing and Constructing Whole Numbers

Week at a Glance
This week, students continue **Number Worlds,** Level D, Number Sense within 100. Students will explore different ways to visualize and represent numbers and become familiar with equivalent representations of the same number. Students will study the importance of 10 and the fundamentals of regrouping.

Skills Focus
- Make 10 by combining two and three numbers.
- Visualize and represent two-digit numbers several different ways.
- Lay the foundation for regrouping.

How Students Learn
Students' development and understanding of the part-whole concept this week lays the foundation for work with place value throughout the rest of the unit. Students begin by working with the benchmark number 10. Relationships for numbers through 10 are then used to develop number concepts. Encourage students to think flexibly about numbers and investigate multiple ways of composing and decomposing numbers.

English Learners ELL
For language support, use the **English Learner Support Guide,** pages 28–29, to preview lesson concepts and teach academic vocabulary.

Math at Home
Give one copy of the Letter to Home, page 4, to each student. Encourage students to share and complete the activity with their caregivers.

Weekly Planner

Lesson	Learning Objectives
1 pages 62–63	Students can identify combinations of numbers that make 10.
2 pages 64–65	Students can trade 10 ones for 1 ten and can understand their equivalence.
3 pages 66–67	Students can identify the number of tens and ones in a two-digit number.
4 pages 68–69	Students can make models of two-digit numbers as tens and ones and trade ones for tens.
5 pages 70–71	**Review and Assess** Students review skills learned this week and complete the weekly assessment.
Project pages 72–73	Students can represent two-digit numbers.

60   Level D Unit 1 **Number Sense within 100**

Key Standard for the Week

Domain: Number and Operations in Base Ten

Cluster: Understand place value.

2.NBT.1 Understand that the three digits of a three-digit number represent amounts of hundreds, tens, and ones; e.g., 706 equals 7 hundreds, 0 tens, and 6 ones.

Materials		Technology
Program Materials • *Student Workbook*, pp. 41–43 • *Practice*, p. 16 • Activity Card 1M, **Groups of 10**	**Additional Materials** • math-link cubes* • Vocabulary Card 32, *ones* • Vocabulary Card 45, *tens*	*Teacher Dashboard* Building Blocks Build Stairs 2
Program Materials • *Student Workbook*, pp. 44–45 • *Practice*, p. 17 • Activity Card 1N, **Let's Make a Trade** • Dot Set Cards • Place-Value Mat • Counters	**Additional Materials** • math-link cubes* • Vocabulary Card 10, *digit* • Vocabulary Card 32, *ones* • Vocabulary Card 45, *tens*	*Teacher Dashboard* Base 10 Blocks Tool
Program Materials • *Student Workbook*, pp. 46–47 • *Practice*, p. 18 • Activity Card 1O, **Regrouping**	**Additional Materials** • classroom objects for counting • math-link cubes* • Vocabulary Card 32, *ones* • Vocabulary Card 45, *tens*	*Teacher Dashboard* Base 10 Blocks Tool
Program Materials • *Student Workbook*, pp. 48–49 • *Practice*, p. 19 • Activity Card 1P, **More Building and Trading Numbers** • Number Cards (0–9) • Number Construction Mat • Place-Value Mat	**Additional Materials** • base-ten blocks* • Vocabulary Card 23, *hundreds* • Vocabulary Card 32, *ones* • Vocabulary Card 45, *tens*	*Teacher Dashboard* Base 10 Blocks Tool
Program Materials • *Student Workbook*, pp. 50–51 • Weekly Test, *Assessment*, pp. 19–20		Review previous activities.
Program Materials • *Student Workbook*, p. 52 • Counters	**Additional Materials** • crayons • drawing of park from Week 3	

*Available from McGraw-Hill Education

WEEK 4

Visualizing and Constructing Whole Numbers

Find the Math

In this week, students will construct whole numbers in different ways.

Use the following to begin a guided discussion:

▶ **Can you think of another way to show 10 objects?** Answers may vary. Possible answer: 1 and 9

Have students complete **Student Workbook,** page 41.

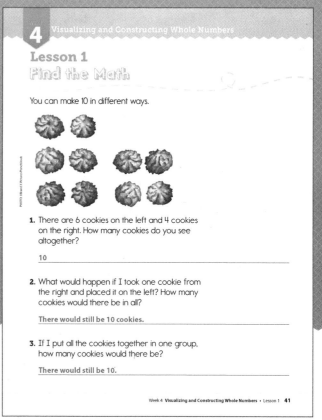

Student Workbook, p. 41

Lesson 1

Objective
Students can identify combinations of numbers that make 10.

Standard
2.NBT.1 Understand that the three digits of a three-digit number represent amounts of hundreds, tens, and ones; e.g., 706 equals 7 hundreds, 0 tens, and 6 ones.

Vocabulary
- ones
- tens

Creating Context
In English, the word *make* is very useful. *Make* is part of many descriptive phrases we use in this lesson. It usually means "to do" or "to construct something," but it can be used in many ways. Find phrases using the word make to explain to English Learners.

Materials
Additional Materials
- math-link cubes, 20 per student
- Vocabulary Card 32, *ones*
- Vocabulary Card 45, *tens*

1 WARM UP

Prepare

▶ **Your hands can serve as a good model for a group of 10. Hold your hands out in front of you.**

▶ **How many fingers (including the thumb) are on each hand?** 5

▶ **Make a math statement about how many fingers are on each hand and how many you have altogether.** 5 and 5 make 10.

▶ **Tuck down three fingers. Make a statement that tells how many straight fingers you have, how many tucked fingers you have, and how many fingers you have altogether.** 7 and 3 make 10.

▶ **Today, you are going to find other combinations of numbers that make ten.**

2 ENGAGE

Develop: Groups of Ten

"Today we are going to see what numbers make ten." Follow the instructions on the Activity Card **Groups of Ten.** As students complete the activity, be sure to use the Questions to Ask.

Activity Card 1M

Alternative Groupings

Pair: Complete the activity as written, making sure that the pair has eighty cubes for the second part of the activity.

Individual: Work with the student to find the possible combinations that equal ten, and then discuss the Questions to Ask.

Progress Monitoring

| **If...** students struggle with groups of ten, | **Then...** have students build groups of five with two colors of cubes before they build groups of ten. |

Practice

Have students complete **Student Workbook,** pages 42–43. Guide students through the Key Idea example and the Try This exercises.

62 Level D Unit 1 **Number Sense within 100**

Interactive Differentiation

Consult the **Teacher Dashboard** for grouping suggestions. You can also use performance on the Engage activity to guide students.

Independent Practice

For additional practice understanding different ways to make 10, have students use Build Stairs 2: Order Steps. Students will identify the appropriate stacks of unit cubes needed to fill in a series of staircase steps.

Supported Practice

Give each student 10 math-link cubes. Show students a way to make 10 using math-link cubes. Show students that 2 math-link cubes and 8 math-link cubes make 10 math-link cubes.

▸ **Can you show me another way to make 10 using two sets of math-link cubes?** Answers may vary; possible answers: 7 and 3; 6 and 4

▸ **Can you show me another way to make 10 using three sets of math-link cubes?** Answers may vary; possible answers: 7, 2, 1; 2, 5, 3

Keep going until students understand that there are many different ways to show 10.

REFLECT

Think Critically

Review students' answers to the Reflect prompt at the bottom of **Student Workbook,** page 43, and then review the Engage activity.

Discuss the different ways the people could be arranged in the vehicles. Ask students to share their combinations and describe how they found their answers.

▸ **What did you learn about the number 10?** Answers may vary. Possible answer: I learned you can make a ten using different combinations of numbers.

▸ **If I gave you three numbers, how would you know if they make ten?** Possible answer: I would combine them and see if I get ten.

ASSESS

Informal Assessment

Use the online or print Student Record, **Assessment,** page 128, to record informal observations.

Groups of Ten

Did the student
- ☐ make important observations?
- ☐ provide insightful answers?
- ☐ extend or generalize learning?
- ☐ pose insightful questions?

Additional Practice

For additional practice, have students complete **Practice,** page 16.

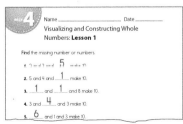

Practice, p. 16

Week 4 • Visualizing and Constructing Whole Numbers

Lesson 1

Key Idea
There are several ways to make 10 with three numbers.

3 and 5 and 2 make 10.   6 and 2 and 2 make 10.

Try This
Fill in the blanks to show your answer.

1. __1__ and __2__ and __7__ make 10.
2. __4__ and __5__ and __1__ make 10.

Practice
Add enough math-link cubes to make 10. Color each new cube with a second or third color. Then fill in the blanks.

3. 5 and ___ and ___ make 10.
   Answers will vary. Possible answers are 1 and 4 or 2 and 3.

4. 3 and ___ and ___ make 10.
   Answers will vary. Possible answers are 1 and 6, 2 and 5, or 3 and 4.

42 Level D Unit 1 Number Sense within 100

Find the missing number or numbers.

5. 1 and 6 and __3__ make 10.
6. 2 and 2 and __6__ make 10.
7. 1 and 4 and __5__ make 10.
8. 4 and __4__ and 2 make 10.
9. 8 and __1__ and 1 make 10.
10. __2__ and 7 and 1 make 10.
11. 7 and 1 and __2__ make 10.
12. __5__ and 3 and 2 make 10.

Use math-link cubes to decide if these numbers together make 10.

13. 1, 2, 7   **yes**
14. 7, 3, 2   **no**
15. 7, 2, 1   **yes**
16. 3, 3, 3   **no**
17. 4, 2, 5   **no**
18. 4, 4, 2   **yes**
19. 5, 5, 1   **no**
20. 6, 3, 2   **no**

Reflect
There are 10 people riding in 3 vehicles. One vehicle holds up to 8 people. The other vehicles hold up to 4 people, each. Write three possible ways the 10 people can be arranged in the vehicles.

Answers will vary. Possible answers:
(1, 1, 8) (1, 2, 7) (1, 3, 6) (1, 4, 5) (2, 2, 6) (2, 3, 5) (2, 4, 4) and (3, 3, 4)

Week 4 Visualizing and Constructing Whole Numbers • Lesson 1 43

Student Workbook, pp. 42–43

WEEK 4
Visualizing and Constructing Whole Numbers

Lesson 2

Objective
Students can trade 10 ones for 1 ten and can understand their equivalence.

Standard
2.NBT.1 Understand that the three digits of a three-digit number represent amounts of hundreds, tens, and ones; e.g., 706 equals 7 hundreds, 0 tens, and 6 ones.

Vocabulary
- digit
- ones
- tens

Creating Context
In English, there are many words that sound the same but are spelled differently and have different meanings. In this lesson, students are asked if the answer makes sense. The word *sense* sounds like *cents* and may cause some confusion. Write both words on the board, and explain the difference between these two homophones.

Materials

Program Materials
- Dot Set Cards, 1 set per student
- Place-Value Mat, 1 per student
- Counters, 20 per student

Additional Materials
- math-link cubes
- Vocabulary Card 10, *digit*
- Vocabulary Card 32, *ones*
- Vocabulary Card 45, *tens*

1 WARM UP

Prepare
▶ We have been making ten by combining two and three numbers together. Ten is an important number. How do we write this number?

Ask each student to write the number 10 on a piece of paper. Review their written work, correcting any errors, and then write *10* on the board. Remind students that a *digit* is *any number from 0 to 9*.

▶ **There are two digits in this number: a 1 and a 0. When there are two digits in any written number, we say the number has two place values. There is a place to write one of the digits** (underline the 1 in the 10 to demonstrate this) **and there is a place to write the second digit** (underline the 0 in the ten to demonstrate this).

▶ **What does the 1 stand for in this written number?** 10; 1 ten.

▶ **What does the 0 stand for in this written number?** 0; 0 ones.

Give several students an opportunity to answer and then provide the correct answers if students say the 1 in 10 is worth 1 or if they are unable to answer at all. Tell students the next activity will help them understand these place values a lot better.

Just the Facts
Have students chorally call out the answer for each question.

▶ **How many tens are there in 9 ones? In 11 ones?** 0; 1

▶ **How many tens are there in 23 ones? How many ones are left over?** 2 tens; three ones are left over.

2 ENGAGE

Develop: Let's Make a Trade
"Today we are going to trade Base-Ten Blocks to see how numbers can be shown different ways." Follow the instructions on the Activity Card **Let's Make a Trade**. As students complete the activity, be sure to use the Questions to Ask.

Activity Card 1N

Alternative Groupings
Individual: Play the game with the student after modeling how to use the mat.

Progress Monitoring
If... students do not understand that the Counters in the ones column have different values from the Counters in the tens column,

Then... substitute pennies for ones and dimes for tens.

Practice
Have students complete **Student Workbook,** pages 44–45. Guide students through the Key Idea example and the Try This exercises.

Interactive Differentiation
Consult the **Teacher Dashboard** for grouping suggestions. You can also use performance on the Engage activity to guide students.

Independent Practice
For additional practice, students will use the Base 10 Blocks Tool. Students will show numbers in different ways.

▶ **With the Base 10 Blocks Tool, how can you show 13 in different ways using ones or using tens and ones?** Answers may vary. Possible answers: 1 ten and 3 ones; 13 ones

▶ **With the Base 10 Blocks Tool, how can you show 20 in different ways using tens and ones?** Answers may vary. Possible answers: 2 tens; 20 ones

▶ **With the Base 10 Blocks Tool, how can you show 24 in different ways using tens and ones?** Answers may vary. Possible answers: 2 tens and 4 ones; 24 ones

Supported Practice
For additional support, use math-link cubes to show tens and ones.

Make math-link cube trains of six cubes and eight cubes each. Display the 6-cube train to the left of the 8-cube train.

▶ **How many ones are in the left cube train?** 6
▶ **How many ones are in the right cube train?** 8
▶ **How many ones are there altogether?** 14
▶ **How many sets of 10 are in 14?** 1
▶ **How many are left over?** 4
▶ **Show me what a set of 10 and 5 left over would look like.** Students show a group of 10 and 5 cubes separate from the 10.
▶ **How many ones is that?** 15

Have students continue to build models of additional numbers under 50.

64 Level D Unit 1 **Number Sense within 100**

REFLECT

Think Critically

Review students' answers to the Reflect prompt at the bottom of *Student Workbook,* page 45, and then review the Engage activity.

Invite students to share their answers to the prompt. Make sure that the idea that 1 ten holds a greater value than 1 one is addressed.

▶ **Why is 1 ten more than 1 one?** Possible answer: because it takes 10 ones to make 1 ten.

▶ **What would you do if you had 10 tens?** Possible answer: Trade them for 1 hundred.

Real-World Application

▶ You can trade when you are doing something as simple as counting. When you count past 9, you automatically group the ones into tens. After each ten we make, we move up one number in the tens column and then begin counting from 1 to 9 again ... *11, 12, 13, 14, and so on.* Count from 1 to 30. How many times did we move up a number in the tens place? three

ASSESS

Informal Assessment

Use the online or print Student Record, *Assessment,* page 128, to record informal observations.

Let's Make a Trade

Did the student

☐ pay attention to the contributions of others?

☐ contribute information and ideas?

☐ improve on a strategy?

☐ reflect on and check accuracy of work?

Additional Practice

For additional practice, have students complete *Practice,* p. 17.

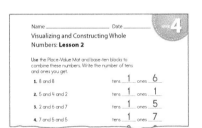

Practice, p. 17

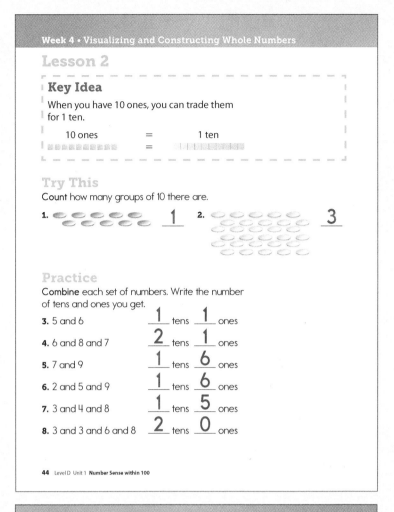

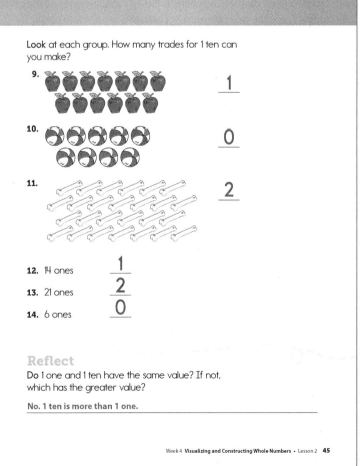

***Student Workbook,* pp. 44–45**

WEEK 4
Visualizing and Constructing Whole Numbers

Lesson 3

Objective
Students can identify the number of tens and ones in a two-digit number.

Standard
2.NBT.1 Understand that the three digits of a three-digit number represent amounts of hundreds, tens, and ones; e.g., 706 equals 7 hundreds, 0 tens, and 6 ones.

Vocabulary
- ones
- tens

Creating Context
In this lesson, we ask students about birthday candles. Remember that not all families have the same cultural traditions surrounding holidays, birthdays, and other celebrations.

Materials
Additional Materials
- classroom objects, such as pennies, marbles, or Counters
- math-link cubes, 50 per pair
- Vocabulary Card 32, *ones*
- Vocabulary Card 45, *tens*

Prepare Ahead
Create a two-column chart on the board. Label one column *Tens* and the other *Ones*. Have students copy the chart onto a sheet of paper.

1 WARM UP

Prepare
As a class, count objects you have on hand (pennies, marbles, or Counters). When you reach 10, point to a student to stand up to represent the 10, then push the objects aside, go back to 1, and count up to 10 again. When you have counted all the objects, skip count by tens for each student who is standing up.

▶ **How many objects do these students represent?** Allow discussion to make sure students understand that each student stands for ten objects.

Just the Facts
Have students take turns answering questions aloud.

▶ **How many sets of tens and ones equal 42 peanuts?** 4 sets of ten and 2 ones

▶ **How many sets of tens and ones equal 59 peanuts?** 5 sets of ten and 9 ones

▶ **How many sets of tens and ones equal 72 peanuts?** 7 sets of ten and 2 ones

2 ENGAGE

Develop: Regrouping
"Today we are going to regroup cubes into tens and ones so you can see how much a two-digit number is worth." Follow the instructions on the Activity Card **Regrouping**. As students complete the activity, be sure to use the Questions to Ask.

Activity Card 10

Alternative Groupings
Small Group: Organize students into pairs or groups of three, and complete the activity as written.

Progress Monitoring

| **If...** students are struggling to display tens and ones visually, | **Then...** give them a two-digit number under twenty, and ask them to make a model of the tens and ones in the number using math-link cubes. |

Practice
Have students complete **Student Workbook,** pages 46–47. Guide students through the Key Idea example and the Try This exercises.

Interactive Differentiation
Consult the **Teacher Dashboard** for grouping suggestions. You can also use performance on the Engage activity to guide students.

Independent Practice

For additional practice, students will use the Base 10 Blocks Tool to count and model two-digit numbers.

▶ **Show students 5 tens and 3 ones with the Base 10 Blocks Tool and ask: What number is shown?** 53

▶ **How can you model 43 with base-ten blocks?** Answers may vary. Possible answers: 4 tens and 3 ones, 3 tens and 13 ones

▶ **How can you model 61 with base-ten blocks?** Answers may vary. Possible answers: 6 tens and 1 one, 5 tens and 11 ones

▶ **How can you model 97 with base-ten blocks?** Answers may vary. Possible answers: 9 tens and 7 ones, 8 tens and 17 ones

Supported Practice

For additional support, use math-link cubes to show two-digit numbers. Show students 3 sets of ten and 2 single math-link cubes. Point out that each set is ten separate cubes and that all the cubes together add up to 32 cubes. Give students fifty separate cubes.

▶ **How can you show 23 by using as many tens as you can?** Students show 2 tens and 3 ones.

▶ **How can you show 39 by using as many tens as you can?** Students show 3 tens and 9 ones.

▶ **I am showing you 4 tens and 5 ones. What number did I make?** 45

REFLECT

Think Critically

Review students' answers to the Reflect prompt at the bottom of *Student Workbook,* page 47, and then review the Engage activity.

Make sure students identify that there is one ten in 14 ones. They should trade those 10 ones for 1 ten and have 4 ones remaining.

▶ **What is the next step when you have 10 ones?** Possible answer: Trade them for 1 ten.

▶ **How can you determine how many tens are in the number 52?** Possible answer: Make a model using base-ten rods and cubes. You would have five rods and two cubes, so 52 has 5 tens.

Real-World Application

▶ **Imagine you are sent to the cafeteria to get seventy-six apples for the third-grade picnic. There are plenty of bags to carry the apples. How could you use groups of ten to count the correct amount of apples?** Possible answer: I would count ten apples and put them in a bag until I have seven bags. Then I'd need six more apples.

ASSESS

Informal Assessment

Use the online or print Student Record, *Assessment,* page 128, to record informal observations.

Regrouping

Did the student

☐ make important observations?   ☐ provide insightful answers?

☐ extend or generalize learning?   ☐ pose insightful questions?

Additional Practice

For additional practice, have students complete *Practice,* p. 18.

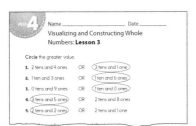

Practice, p. 18

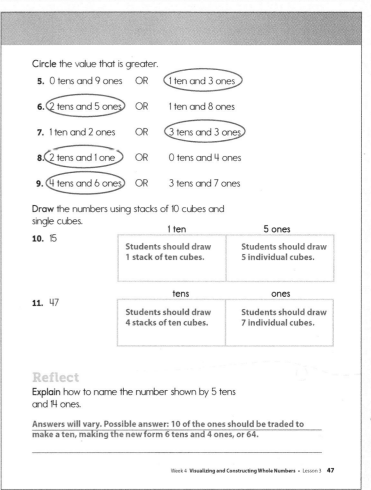

Student Workbook, pp. 46–47

WEEK 4
Visualizing and Constructing Whole Numbers

Lesson 4

Objective
Students can make models of two-digit numbers as tens and ones and trade ones for tens.

Standard
2.NBT.1 Understand that the three digits of a three-digit number represent amounts of hundreds, tens, and ones; e.g., 706 equals 7 hundreds, 0 tens, and 6 ones.

Vocabulary
- hundreds
- ones
- tens

Creating Context
Checking for understanding regularly is very important with English Learners. Sometimes they cannot produce a long speech sample but understand the concept and can show that they understand in other ways. Be sure to include some questions that ask for *yes/no* answers in addition to the open-ended questions in each lesson.

Materials
Program Materials
- Number Cards (0–9), 1 set per student
- Number Construction Mat, 1 per student
- Place-Value Mat, 1 per student

Additional Materials
- base-ten blocks, 4 rods and 20 units per student
- Vocabulary Card 23, *hundreds*
- Vocabulary Card 32, *ones*
- Vocabulary Card 45, *tens*

1 WARM UP

Prepare
Show students one base-ten rod and one base-ten unit block.

▶ **These are base-ten blocks. A rod shows a certain number of unit blocks. How many unit blocks do you think it shows?** ten

Check answers by lining up unit blocks next to the rod. Have students discover together that ten unit blocks equal one rod.

Just the Facts
Have students chorally call out the number that you are asking for.

Use questions such as the following:

▶ **What number is 3 sets of ten and 2 ones?** 32
▶ **What number is 1 set of ten and 5 ones?** 15
▶ **What number is 15 ones?** 15

2 ENGAGE

Develop: Building and Trading Numbers

"Today we are going to model numbers with Base-Ten Blocks." Follow the instructions on the Activity Card **More Building and Trading Numbers.** As students complete the activity, be sure to use the Questions to Ask.

Activity Card 1P

Alternative Groupings
Pair: Have students work together to complete the Conclude the Activity section.

Progress Monitoring

If... a student cannot identify a number represented with base-ten blocks,

Then... pair the student with another student who has grasped this skill to practice making and naming numbers using base-ten rods and unit blocks.

Practice
Have students complete **Student Workbook,** pages 48–49. Guide students through the Key Idea example and the Try This exercises.

Interactive Differentiation
Consult the **Teacher Dashboard** for grouping suggestions. You can also use performance on the Engage activity to guide students.

Independent Practice

For additional practice, students will use the Base 10 Blocks Tool to show two-digit numbers in different ways.

▶ **How can you show 45 as tens and ones in two different ways?**
Answers may vary. Possible answer: 3 tens and 15 ones, 4 tens and 5 ones

▶ **How can you show 32 as tens and ones in two different ways?**
Answers may vary. Possible answer: 2 tens and 12 ones, 3 tens and 2 ones

▶ **How can you show 19 as tens and ones in two different ways?**
Answers may vary. Possible answer: 19 ones, 1 ten and 9 ones

Supported Practice

For additional support, use base-ten blocks. Use two base-ten rods and two unit blocks. Point out that each rod equals ten units and that all the units together are 22 units. If you trade one set of ten for ones, you will have 1 set of ten and 12 ones, but you still have 22 units. Continue to use base-ten blocks to show two-digit numbers. Show students 2 base-ten rods and 12 base-ten unit cubes.

▶ **How can you show 32 in tens and ones?** I can show 3 base-ten rods and 2 unit cubes.

▶ **If I trade one base-ten rod for ten units, what number do I have?**
You still have 32.

▶ **How many sets of ten do I have now and how many ones do I have?**
There are now 2 sets of ten and 12 ones.

REFLECT

Think Critically

Review students' answers to the Reflect prompt at the bottom of *Student Workbook,* page 49, and then review the Engage activity.

Have students share how they found the value of the rods and unit blocks in the prompt.

- ▶ **Show me how you counted the cubes and the rods to determine the amount.** Students should count the cubes by ones and the rods by tens.
- ▶ **What is the value of a rod?** A rod is equal to ten ones.

Real-World Application

- ▶ A *decade* means *ten years*. So, three decades is the same as thirty years. Sometimes it helps to picture a number in your head. You can think of decades in terms of base-ten rods. One decade equals one base-ten rod, and single years are like base-ten unit blocks.

ASSESS

Informal Assessment

Use the online or print Student Record, *Assessment,* page 128, to record informal observations.

Building and Trading Numbers

Did the student

☐ respond accurately?   ☐ respond with confidence?

☐ respond quickly?   ☐ self-correct?

Additional Practice

For additional practice, have students complete *Practice,* page 19.

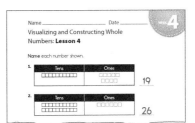

Practice, p. 19

Week 4 • Visualizing and Constructing Whole Numbers

Lesson 4

Key Idea

▪ = 1   ▬▬▬▬ = 10

To find the value of a group of unit blocks, count by ones.

To find the value of a group of rods, skip count by tens.

Try This

Name each number shown.

1. Tens | Ones — **49**
2. Tens | Ones — **13**
3. Tens | Ones — **25**
4. Tens | Ones — **48**

48 Level D · Unit 1 Number Sense within 100

Practice

Make a model of each number by drawing base-ten blocks.

5. 27 — Students should draw 2 rods in the tens place and 7 blocks in the ones place.

6. 32 — Students should draw 3 rods in the tens place and 2 blocks in the ones place.

Reflect

Fill in the blanks to name the number shown.

__7__ tens __6__ ones

Explain how to find the value of the rods and unit blocks.

Skip by tens to find the value of the group of 7 rods and count by ones to find the value of the group of six unit blocks.

What number do the blocks show? **76**

Week 4 Visualizing and Constructing Whole Numbers • Lesson 4   49

Student Workbook, pp. 48–49

WEEK 4
Visualizing and Constructing Whole Numbers

Lesson 5 Review

Objective
Students review skills learned this week and complete the weekly assessment.

Standard
2.NBT.1 Understand that the three digits of a three-digit number represent amounts of hundreds, tens, and ones; e.g., 706 equals 7 hundreds, 0 tens, and 6 ones.

Vocabulary
Review vocabulary introduced during the week.

Creating Context
Book titles, the names of shops such as hair salons, and the names of games often have titles that use a play on words or a short statement to tell what the book, shop, or game is about. In this week, students played **Groups of 10, Let's Make a Trade, Regrouping,** and **Building and Trading Numbers.** Select two of these activities, and have students tell how their titles describe something about the activities.

1 WARM UP

Prepare
Provide students with large handfuls of math-link cubes. Select a student to name a two-digit number that is less than fifty. All students will then make a model of that number by making stacks of ten for the digit in the tens place and leaving unconnected cubes to represent the digit in the ones place.

2 ENGAGE

Practice
Have students complete **Student Workbook,** pages 50–51.

Week 4 • Visualizing and Constructing Whole Numbers

Lesson 5 Review

This week you explored whole numbers. You learned how to represent whole numbers using base-ten blocks. You also looked at many ways to make 10 using two numbers and three numbers.

Lesson 1 Fill in the blanks to name the number that makes 10.

1. 6 and __2__ and __2__ make 10.
2. 4 and __4__ and 2 make 10.
3. 1 and 4 and __5__ make 10.
4. 3 and __1__ and 6 make 10.

Lesson 2 Combine the numbers. Then write how many tens and ones each total has.

5. 2 and 9 and 1 — __1__ ten(s) and __2__ one(s)
6. 7 and 2 and 1 and 6 — __1__ ten(s) and __6__ one(s)
7. 7 and 3 and 6 and 4 — __2__ ten(s) and __0__ one(s)
8. 4 and 4 and 2 and 3 — __1__ ten(s) and __3__ one(s)

Lesson 3 Write the number the blocks show.

9. __16__
10. __43__

Make trades and write the new number of tens and ones.

11. 3 tens and 21 ones — __5__ tens __1__ one
12. 5 tens and 30 ones — __8__ tens __0__ ones

Lesson 4 Name each number shown.

13. __17__
14. __43__

Reflect
Write two ways you could express the number 18.

1 ten and 8 ones; 18 ones

Student Workbook, pp. 50–51

3 REFLECT

Think Critically

Review students' answers to the Reflect prompt at the bottom of *Student Workbook,* page 51.

Discuss the answer with the group to reinforce Week 4 concepts.

4 ASSESS

Formal Assessment

Students may take the weekly assessment online.

As an alternative, students may complete the weekly test on *Assessment,* pages 19–20. Record progress using the Student Assessment Record, *Assessment,* page 128.

Going Forward

Use the **Teacher Dashboard** to view results of the online assessments, to input the results of print student assessments, and to review progress before making decisions about next steps. Use the weekly test results and observations to determine the next steps for each student.

Retention	
Student displays good grasp of this week's concepts and skills.	Have students use math-link cubes to show two-digit numbers in different ways and write the name of each number they build.
Remediation	
Student is still struggling with the week's concepts and skills.	Continue to have students use math-link cubes to show two-digit numbers. Show students 2 sets of ten and 6 math-link cubes. Point out that each set is ten separate cubes and that all the cubes together add up to 26 cubes.

Suggestions for Re-Evaluation: If a student has struggled without success for several weeks, use observations and test results to place the student at a level where they can find success and build confidence to move forward.

Assessment, pp. 19–20

Week 4 **Visualizing and Constructing Whole Numbers** • Lesson 5

Project Preview

This week, students learned how to represent whole numbers using Counters. Students also looked at many ways to make numbers using two numbers and three numbers. The project for this unit requires students to extend the knowledge they gained in Find the Math and what they have learned this week. They will use Counters to help them find tens and ones as they separate and share flowers.

Project-Based Learning

Standards-driven Project-Based Learning is effective in building deep content understanding. Project-Based Learning increases long-term retention of concepts and has been shown to be more effective than traditional instruction. By completing a project to answer an essential question, students are challenged to apply and demonstrate mastery of concepts and skills by expressing understanding through discussion, research, and presentation.

Essential Question

HOW can I use comparing and ordering numbers outside the classroom?

Project Evaluation Criteria

Review project evaluation criteria with students prior to beginning the project.

Exceeds Expectations
☐ Project result is explained and can be extended.
☐ Project result is explained in context and can be applied to other situations.
☐ Project result is explained using advanced mathematical vocabulary.
☐ Project result is described, and mathematics are used correctly and can be extended.
☐ Project result is explained and extended, and shows advanced knowledge of mathematical concepts and skills.

Meets Expectations
☐ Project result is explained.
☐ Project result is explained in context.
☐ Project result is explained using mathematical vocabulary.
☐ Project result is described, and mathematics are used correctly.
☐ Project result is explained, and shows satisfactory knowledge of mathematical concepts and skills.

Does Not Meet Expectations
☐ Project result is not explained.
☐ Project result is explained, but out of context.
☐ Project result is explained, but mathematical vocabulary is oversimplified.
☐ Project result is described, but mathematics are not used correctly.
☐ Project result is not explained and or extended, or shows less than satisfactory knowledge of mathematical concepts and skills.

Bundling Blooms in the Park

Objective
Students can represent two-digit numbers.

Standard
2.NBT.1 Understand that the three digits of a three-digit number represent amounts of hundreds, tens, and ones; e.g., 706 equals 7 hundreds, 0 tens, and 6 ones.

Materials

Program Materials	Additional Materials
Counters	• crayons
	• drawing of park from Week 3

Best Practices
- Set clear expectations, rules, and procedures.
- Create adequate time lines for each project.
- Coach, demonstrate, and model.

Introduce

Imagine that you are asked to place flowers on two different tables at the park. You will have 25 flowers.

▶ **Will you place one flower at a time on each table or is there a quicker way?** Answers may vary. Possible answer: I will place flowers on one table then the other until they are all on the tables.

▶ **Do you think you could give each table the same amount?** No, there will be some left over.

Explore

▶ **Today you will place flowers on two separate tables at the park in two different ways. Draw another table in your picture of the park.**

▶ **Complete *Student Workbook* page 52 to find how out how many flowers will be on each table.**

- Give each student 25 Counters.

▶ **Pretend each counter is a flower. The first way you will separate the flowers is by tens. Place as many sets of ten as you can on one table and give the other table the rest of the flowers.**

▶ **Next, find another way to separate the flowers. Each table must have at least one set of ten.**

▶ **Draw flowers on each table in your picture. Above each table, write the number of flowers that the table has. For example, if the table has 15 flowers on it, write the number 15 above the table.**

▶ **Take the counters off your pictures. Now draw the correct number of flowers on each table.**

Wrap Up

- Allow students time to tell how they decided to separate the flowers.
- Make sure each student can explain how many sets of ten are on each table.
- If students struggle to write how many sets of ten they used, remind them that 10 ones is equal to 1 set of 10.
- Discuss students' answers to the Reflect prompts at the bottom of **Student Workbook,** page 52.
- Save students' drawings for use in Week 5.

If time permits, allow each student to count how many flowers other students have placed on their tables.

Week 4 • Visualizing and Constructing Whole Numbers

Project
Bundling Blooms in the Park

Use 25 counters to show how many sets of ten flowers can be given to each table.

1. Draw another picnic table in your park picture. How many flowers did each table get?

   One table has 20 flowers, and the other table has 5 flowers.

2. Put all 25 flowers on the tables. Each table must get at least one set of ten. How many sets of ten did each table get?

   Each table received one set of ten.

3. How many ones does each table have?

   Answers may vary. Possible answer: One has 1 one and the other 4 ones.

4. Place a stack of counters on each table in your picture. Each counter equals one flower. Above each table, write the number of flowers on it.

5. Remove the counters from your picture. Draw the number of flowers shown on the tables in your picture.

Reflect

How did you know how many flowers to give each table?

Answers will vary. Possible answer: I knew I have to give each table 1 ten.

Student Workbook, p. 52

Teacher Reflect

☐ Did I explain the directions before students began their project?

☐ Did I supply the necessary materials?

☐ Did I adequately explain and discuss the Reflect questions with the students?

WEEK 5: Number Patterns

Week at a Glance
This week, students continue with **Number Worlds,** Level D, Number Sense within 100 by skip counting to explore patterns.

Skills Focus
- Skip count by twos, fives, and tens.
- Detect errors in skip-counting sequence.
- Use number patterns to solve problems.

How Students Learn
Exploring patterns helps students develop the ability to form generalizations. As students identify missing numbers in a pattern while skip counting, they should give justification for the numbers identified.

English Learners ELL
For language support, use the **English Learner Support Guide,** pages 30–31, to preview lesson concepts and teach academic vocabulary.

Math at Home
Give one copy of the Letter to Home, page 5, to each student. Encourage students to share and complete the activity with their caregivers.

Weekly Planner

Lesson	Learning Objectives
1 pages 76–77	Students can detect errors when skip counting by twos.
2 pages 78–79	Students can detect errors when skip counting by fives and tens.
3 pages 80–81	Students can use number patterns to predict which number will come before and after a specified number.
4 pages 82–83	Students can use number patterns to predict which number will come before and after a specified number.
5 pages 84–85	**Review and Assess** Students review skills learned this week and complete the weekly assessment.
Project pages 86–87	Students can skip count by 2s, 5s, and 10s.

74 Level D Unit 1 **Number Sense within 100**

Key Standard for the Week

Domain: Number and Operations in Base Ten
Cluster: Understand place value.
2.NBT.2 Count within 1000; skip-count by 5s, 10s, and 100s.

Materials	Technology
Program Materials • **Student Workbook,** pp. 53–55 • **Practice,** p. 20 • Activity Card 1Q, **Catch the Teacher**	*Teacher Dashboard* Building Blocks® Tire Recycling
Program Materials • **Student Workbook,** pp. 56–57 • **Practice,** p. 21 • Activity Card 1Q, **Catch the Teacher** • Neighborhood Number Line	*Teacher Dashboard* Building Blocks® Clean the Plates
Program Materials • **Student Workbook,** pp. 58–59 • **Practice,** p. 22 • Activity Card 1R, **What Number Am I?** • Neighborhood Number Line	*Teacher Dashboard* 100 Table
Program Materials • **Student Workbook,** pp. 60–61 • **Practice,** p. 23 • Activity Card 1R, **What Number Am I?** • Neighborhood Number Line	*Teacher Dashboard* 100 Table
Program Materials • **Student Workbook,** pp. 62–63 • Weekly Test, **Assessment,** pp. 21–22	Review previous activities.
Program Materials **Student Workbook,** p. 64 **Additional Materials** • crayons • drawing of park from Week 4 • number line *Available from McGraw-Hill Education	

WEEK 5
Number Patterns

Find the Math
In this week, students will learn number patterns by skip counting.

Use the following to begin a guided discussion:

▶ **What is something you might need to count quickly and you could count by skip counting?** Possible answer: I would use skip counting to count the pennies I have saved.

Have students complete *Student Workbook,* page 53.

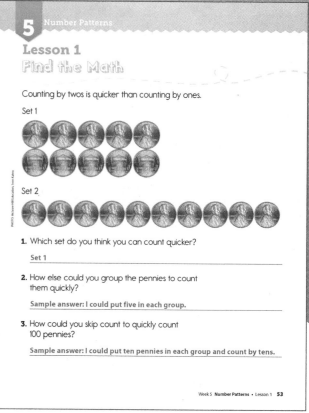

Student Workbook, p. 53

Lesson 1

Objective
Students can detect errors when skip counting by twos.

Standard
2.NBT.2 Count within 1000; skip-count by 5s, 10s, and 100s.

Creating Context
Some students who have attended school outside the United States may find it puzzling to hear the teacher make a mistake or ask questions when the teacher clearly knows the answer. Help students and their families realize that games where the teacher purposefully makes mistakes are designed to enhance learning.

Materials
No materials needed.

1 WARM UP

Prepare
▶ **Look around the room. What do you see that comes in sets of two?** Possible answers: shoes; eyes; feet; hands

Have students count the objects named, skip counting by twos.

2 ENGAGE

Develop: Catch the Teacher

"Today we are going to practice using number patterns by skip counting." Follow the instructions on the Activity Card **Catch the Teacher.** As students complete the activity, be sure to use the Questions to Ask.

Alternative Groupings
Individual: Complete the activity as written.

Activity Card 1Q

Progress Monitoring

| If... students have trouble identifying the missing number, | Then... practice skip counting with them throughout the day. |

Practice
Have students complete *Student Workbook,* pages 54–55. Guide students through the Key Idea example and the Try This exercises.

76 Level D Unit 1 **Number Sense within 100**

Interactive Differentiation

Consult the **Teacher Dashboard** for grouping suggestions. You can also use performance on the Engage activity to guide students.

Independent Practice

For additional practice understanding skip counting, have students use Tire Recycling. Students will skip count by twos to count tires as they are moved.

Supported Practice

For additional support, take number cards and place them in a line or write down the numbers on a piece of paper. Tell students they are skip counting by two and ask them to point out the mistake.

▶ 2, 4, 6, 7, 10   7
▶ 22, 24, 26, 28, 32   32
▶ 33, 34, 35, 37, 39   34
▶ 61, 63, 65, 66, 67   66

REFLECT

Think Critically

Review students' answers to the Reflect prompt at the bottom of **Student Workbook** page 55, and then review the Engage activity.

Discuss to reinforce the idea that the number sequences students know from skip counting are actually patterns.

▶ What do you notice about skip counting by twos?
▶ Does this remind you of anything else? Why?

ASSESS

Informal Assessment

Use the online or print Student Record, **Assessment,** page 128, to record informal observations.

Catch the Teacher

Did the student

☐ respond accurately?   ☐ respond with confidence?

☐ respond quickly?   ☐ self-correct?

Additional Practice

For additional practice, have students complete **Practice,** page 20.

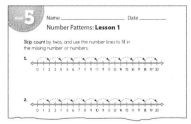

Practice, p. 20

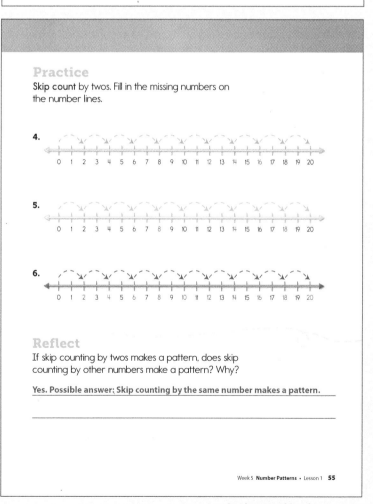

Student Workbook, pp. 54–55

Week 5 **Number Patterns** • Lesson 1   **77**

WEEK 5
Number Patterns

Lesson 2

Objective
Students can detect errors when skip counting by fives and tens.

Standard
2.NBT.2 Count within 1000; skip-count by 5s, 10s, and 100s.

Creating Context
To help English Learners at early proficiency levels, teach a sentence frame they can use when they catch the teacher in a counting mistake; for example, "The number _____ is missing." Some students may find it odd for the teacher to make such rudimentary mistakes. Make sure they understand that this is a practice exercise to make sure they are listening.

Materials
Program Materials
Neighborhood Number Line

1 WARM UP

Prepare

▶ **Look around the room. What do you see that comes in sets of five?** Answers may vary. Possible answer: fingers on a hand

Have students count objects you name, skip counting by fives. Repeat with tens.

Just the Facts

Have students call out the number that comes next in each set:

▶ 5, 10, 15, _____ 20
▶ 35, 40, 45, _____ 50
▶ 90, 80, 70, _____ 60

2 ENGAGE

Develop: Catch the Teacher

"Today you are going to continue to catch me making mistakes with number patterns." Follow the instructions on the Activity Card **Catch the Teacher** (Variation 1: Fives and Tens). As students complete the activity, be sure to use the Questions to Ask.

Activity Card 1Q

Alternative Groupings

Individual: Complete the activity as written.

Progress Monitoring

If... students have trouble remembering the number sequence for skip counting by fives or tens,

Then... have them refer to the Neighborhood Number Line.

Practice

Have students complete **Student Workbook,** pages 56–57. Guide students through the Key Idea example and the Try This exercises.

Interactive Differentiation

Consult the **Teacher Dashboard** for grouping suggestions. You can also use performance on the Engage activity to guide students.

Independent Practice

For additional practice understanding skip counting by fives and tens, have students use Clean the Plates.

Students use skip counting to produce products that are multiples of tens, fives, twos, and threes.

Supported Practice

For additional support, give each student a number line. Tell them they can use a number line as a reference if needed. Skip count by five and have them follow on the number line: 5, 10, 15, 20, 25.

Now tell students you will skip count by five again and they should raise their hands when they hear a mistake.

▶ **5, 10, 15, 16** Students should raise their hands when you get to 16.

▶ **55, 50, 45, 35, 30** Students should raise their hand when you get to 35.

• Now tell students you will skip count by ten, and they should raise their hands when they hear a mistake.

▶ **10, 20, 30, 50** Students should raise their hands when you get to 50.

• If students do not raise their hand at the correct time, slowly repeat the numbers and place a finger on the corresponding number on the number line as you count.

REFLECT

Think Critically
Review students' answers to the Reflect prompt at the bottom of **Student Workbook** page 57, and then review the Engage activity.

Real-World Application
▶ **People often use tally marks when recording counts of items. Tally marks are grouped in sets of five. To count a set of tally marks, skip count by fives until you do not have another set of five. Then count by ones for each mark.**

Write sets of tally marks on the board, and ask students to figure out how many marks there are by counting aloud.

ASSESS

Informal Assessment
Use the online or print Student Record, **Assessment,** page 128, to record informal observations.

Catch the Teacher
Did the student
- ☐ respond accurately?   ☐ respond with confidence?
- ☐ respond quickly?   ☐ self-correct?

Additional Practice
For additional practice, have students complete **Practice,** page 21.

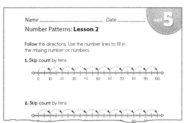
Practice, p. 21

Week 5 • Number Patterns

Lesson 2

Key Idea
Skip counting by fives and tens also makes number patterns.

Try This
Skip count, and fill in the missing numbers on the number lines.

1. Skip count by fives.

2. Skip count by tens.

3. Skip count by fives.

56 Level D Unit 1 Number Sense within 100

Practice
Skip count, and fill in the missing numbers on the number lines.

4. Skip count by tens.

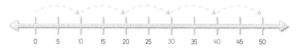

5. Skip count by fives.

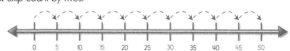

6. Skip count by tens.

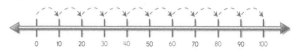

Reflect
Are number patterns or shape patterns easier to work with?

Answers will vary. Possible answer: Number patterns are easier to work with, because I can skip count by a number.

Week 5 Number Patterns • Lesson 2 57

Student Workbook, pp. 56–57

WEEK 5
Number Patterns

Lesson 3

Objective
Students can use number patterns to predict which number will come before and after a specified number.

Standard
2.NBT.2 Count within 1000; skip-count by 5s, 10s, and 100s.

Creating Context
Acquiring prepositions is an intermediate to advanced skill on the continuum of English proficiency, so some English Learners may find them difficult. Because not all languages share the same structure, the notion of prepositions may be unfamiliar to some students. Review prepositions used in this activity, such as *before, after,* and *between*.

Materials
Program Materials
Neighborhood Number Line

1 WARM UP

Prepare
Review the meaning of the terms *before, after,* and *between* by having students line up and say who comes before, between, or after them in line.

Just the Facts
Have students write the correct number on a piece of paper and hold it up.

▶ I am 2 less than 55. What number am I? 53
▶ I am 2 more than 68. What number am I? 70
▶ If you count by twos, I am the tenth number you say. What number am I? 20

2 ENGAGE

Develop: What Number Am I?
"Today we are going to use number patterns to solve puzzles." Follow the instructions on the Activity Card **What Number Am I?** (Variation: Number Pattern Mysteries). As students complete the activity, be sure to use the Questions to Ask.

Activity Card 1R

Alternative Groupings
Individual: Complete the activity as written.

Progress Monitoring
If... students have trouble remembering verbal clues,

Then... have them use a paper number line and a pencil to mark off eliminated numbers.

Practice
Have students complete **Student Workbook,** pages 58–59. Guide students through the Key Idea example and the Try This exercises.

Interactive Differentiation
Consult the **Teacher Dashboard** for grouping suggestions. You can also use performance on the Engage activity to guide students.

Independent Practice

For additional practice, have students explore "count-by" patterns on the 100 Table Tool. Guide them, as follows:

▶ **Adjust the settings below the grid:** Click the *Add* Table icon, and set *Grid Type* to "0–100." Then set *Skip Counts* to "2" and *Skip Count Speed* to "Medium."

▶ **At the left of the grid,** set up *Skip Count A* as follows: *Start* = "0," *End* = "100," and *Count By* = "2."

▶ **Predict what the pattern will look like.** Then click *Start* to check. Explain the answer.

▶ **Now set up** *Skip Count B* as follows: *Start* = "1," *End* = "100," and *Count By* = "2."

▶ **Describe the pattern you expect to see.** Then click *Start* to check and explain the answer.

• Allow students to explore and explain other "count-by" questions.

Supported Practice

For additional support, give students a number and ask them to figure out which number you are describing.

▶ I am the number that comes two after 15. 17
▶ I am less than 25. I am after 23. 24
▶ I am 2 groups of ten and 6 ones. 26

• Help students make up their own descriptions for numbers. Have students use the number line.

③ REFLECT

Think Critically

Review students' answers to the Reflect prompt at the bottom of **Student Workbook** page 59, and then review the Engage activity.

▸ **How did the Neighborhood Number Line make it easier to answer questions?**

▸ **What does this tell you about numbers?**

Explain that even though a student may be comfortable with their answers, it is always a good idea to double-check your work.

Real-World Application

▸ **If the houses on one side of the street are all even numbered, and the houses on the other side are odd numbered, and the house between house 28 and house 32 is missing, what would its number be?** 30

④ ASSESS

Informal Assessment

Use the online or print Student Record, **Assessment,** page 128, to record informal observations.

What Number Am I?

Did the student
- ☐ respond accurately?
- ☐ respond with confidence?
- ☐ respond quickly?
- ☐ self-correct?

Additional Practice

For additional practice, have students complete **Practice,** page 22.

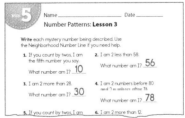

Practice, p. 22

Week 5 • Number Patterns

Lesson 3

> **Key Idea**
> You can use number patterns to solve problems.

Try This

Skip count by twos. Write the three numbers that come after the number shown.

1. 2, __4__, __6__, __8__
2. 16, __18__, __20__, __22__
3. 22, __24__, __26__, __28__
4. 8, __10__, __12__, __14__

Skip count by twos. Write the number that comes between the numbers shown.

5. 8, __10__, 12
6. 22, __24__, 26
7. 12, __14__, 16
8. 6, __8__, 10

Skip count by twos. Write the three numbers that come before the number shown.

9. __4__, __6__, __8__, 10
10. __24__, __26__, __28__, 30
11. __10__, __12__, __14__, 16
12. __12__, __14__, __16__, 18
13. __18__, __20__, __22__, 24
14. __6__, __8__, __10__, 12

58 Level D Unit 1 **Number Sense within 100**

Practice

Write the mystery number that is being described. Use the Neighborhood Number Line if you need help.

15. I am 2 numbers before 40 and 2 numbers after 36. What number am I? __38__

16. If you count by twos, I am the fourth number you say. What number am I? __8__

17. I am 2 more than 64. What number am I? __66__

18. I am 2 less than 84. What number am I? __82__

19. I am 2 numbers before 94 and 2 numbers after 90. What number am I? __92__

20. I am 2 numbers before 48 and 2 numbers after 44. What number am I? __46__

21. If you count by twos, I am the twelfth number you say. What number am I? __24__

22. I am 2 more than 98. What number am I? __100__

Reflect

If I count and say "2, 4, 6, 8…," what number am I counting by? How do you know?

Two. Possible answer: You are adding 2 to each number.

Week 5 **Number Patterns** • Lesson 3 59

Student Workbook, pp. 58–59

WEEK 5
Number Patterns

Lesson 4

Objective
Students can use number patterns to predict which number will come before and after a specified number.

Standard
2.NBT.2 Count within 1000; skip-count by 5s, 10s, and 100s.

Creating Context
Discuss with English Learners that riddles are sometimes written in the first person. Explain that riddles give clues and information a little bit at a time, and you are supposed to guess the answer to a question based on these clues.

Materials
Program Materials
Neighborhood Number Line

1 WARM UP

Prepare
Review the What Number Am I? activity by asking students questions such as those covered in Lesson 3.

Just the Facts
Have students write down each correct number on a piece of paper and hold it up.

▶ **If you start at 10 and count by tens, I am the sixth number. What number am I?** 60

▶ **I am 10 numbers before 88. What number am I?** 78

▶ **I am 5 more than 50 and 5 less than 60. What number am I?** 55

2 ENGAGE

Develop: What Number Am I?
"Today we are going to continue to use number patterns to solve puzzles." Follow the instructions on the Activity Card **What Number Am I?** (Variation: Number Pattern Mysteries). As students complete the activity, be sure to use the Questions to Ask.

Activity Card 1R

Alternative Groupings
Individual: Complete the activity as written.

Progress Monitoring
If... students have trouble remembering the sequence of numbers when skip counting,

Then... model skip counting for them throughout the day.

Practice
Have students complete **Student Workbook,** pages 60–61. Guide students through the Key Idea example and the Try This exercises.

Interactive Differentiation
Consult the **Teacher Dashboard** for grouping suggestions. You can also use performance on the Engage activity to guide students.

Independent Practice
For additional practice, have students use the 100 Table to view "count by" patterns.

Supported Practice
For additional support, show students a pattern that skips by 5.

▶ **20, 25, 30, 35. What pattern do you see?** The pattern goes up by 5.
▶ **What number do you think will come next?** 40
▶ **What number do you think will come after that?** 45
▶ **What number do you think would come before 20?** 15
▶ **If I want to start at 50 and skip count by 5, what three numbers would come next?** 55, 60, 65
- Tell students to make a 5 pattern of their own.
- Repeat questions, starting with the number 20 but using a 10 pattern.

82 Level D Unit 1 **Number Sense within 100**

REFLECT

Think Critically

Review students' answers to the Reflect prompt at the bottom of **Student Workbook** page 61, and then review the Engage activity.

Discuss the activity and student pages with students. Challenge them to create and solve mystery number problems and explain how they solve each problem.

Real-World Application

▶ **What are some examples of places you skip count?** Answers may vary. Possible answers: in gym class, when choosing teams

ASSESS

Informal Assessment

Use the online or print Student Record, **Assessment**, page 128, to record informal observations.

What Number Am I?
Did the student
- ☐ provide a clear explanation?
- ☐ choose appropriate strategies?
- ☐ communicate reasons and strategies?
- ☐ argue logically?

Additional Practice

For additional practice, have students complete **Practice,** page 23.

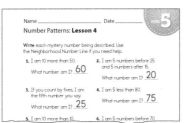

Practice, p. 23

Week 5 • Number Patterns

Lesson 4

Key Idea
You can use number patterns to solve problems.

Try This
Skip count by fives. Write the three numbers that come after the number shown.

1. 5, **10**, **15**, **20**
2. 25, **30**, **35**, **40**
3. 50, **55**, **60**, **65**
4. 85, **90**, **95**, **100**

Skip count by tens. Write the three numbers that come before the number shown.

5. **60**, **70**, **80**, 90
6. **10**, **20**, **30**, 40
7. **50**, **60**, **70**, 80
8. **30**, **40**, **50**, 60

Skip count by fives. Write the number that comes between the numbers shown.

9. 10, **15**, 20
10. 45, **50**, 55
11. 75, **80**, 85
12. 15, **20**, 25

60 Level D Unit 1 Number Sense within 100

Practice
Write the mystery number described. Use the Neighborhood Number Line if you need help.

13. I am 5 numbers before 30 and 5 numbers after 20. What number am I? **25**
14. If you count by tens, I am the fourth number you say. What number am I? **40**
15. I am 10 less than 100. What number am I? **90**
16. I am 5 more than 40. What number am I? **45**
17. If you count by fives, I am the eighth number you say. What number am I? **40**
18. I am 10 more than 90. What number am I? **100**
19. I am 5 numbers before 55 and 5 numbers after 45. What number am I? **50**
20. I am 5 less than 70. What number am I? **65**

Reflect
Create a mystery number problem. Have a classmate solve your problem.

Answers will vary. Student problems should match the format in Practice.

Week 5 Number Patterns • Lesson 4 61

Student Workbook, pp. 60–61

WEEK 5
Number Patterns

Lesson 5 Review

Objective
Students review skills learned this week and complete the weekly assessment.

Standard
2.NBT.2 Count within 1000; skip-count by 5s, 10s, and 100s.

Creating Context
Using graphic organizers helps English Learners access higher-level thinking skills without requiring complete proficiency in English. Using a Venn diagram, ask students to list the differences and similarities between a pattern and a rule.

1 WARM UP

Prepare
Set out a variety of objects that could be used to create patterns. Have a student volunteer arrange the objects in a pattern and display them so the rest of the class can see the pattern.

▸ Are these objects arranged in a pattern?

▸ How would you describe this pattern?

2 ENGAGE

Practice
Have students complete **Student Workbook,** pages 62–63. Guide students through the Key Idea example.

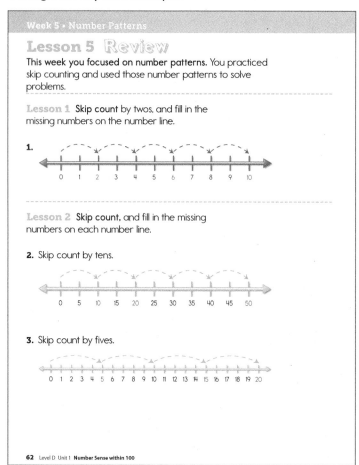

Student Workbook, pp. 62–63

3 REFLECT

Think Critically

Review students' answers to the Reflect prompt at the bottom of **Student Workbook** page 63.

Discuss the answer with the group to reinforce Week 5 concepts

4 ASSESS

Formal Assessment

Students may take the weekly assessment online.

As an alternative, students may complete the weekly test on **Assessment**, pages 21–22. Record progress using the Student Assessment Record, **Assessment**, page 128.

Going Forward

Use the **Teacher Dashboard** to view results of the online assessments, to input the results of print student assessments, and to review progress before making decisions about next steps. Use the weekly test results and observations to determine the next steps for each student.

Retention	
Student displays good grasp of this week's concepts and skills.	Have students skip count backwards by 5s from 35. Make sure students know that the number they will name after 5 is 0.
Remediation	
Student is still struggling with the week's concepts and skills.	Give each student a number line. Skip count by five and have them follow on the number line: 5, 10, 15, 20, 25. Have students skip count by 5s until they reach the number 95 without hesitation.

Suggestions for Re-Evaluation: If a student has struggled without success for several weeks, use observations and test results to place the student at a level in which he or she can find success and build confidence to move forward.

Name _____ Date _____   WEEK 5

Number Patterns

Skip count by twos, and fill in the missing number or numbers on the number lines.

1.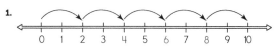

Follow the directions, and use the number lines to fill in the missing number or numbers.

2. Skip count by fives.

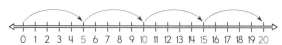

3. Skip count by tens.

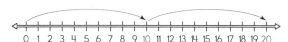

Level D Unit 1 Week 5   21

WEEK 5   Name _____ Date _____

Number Patterns

Write the mystery number being described.

4. I am 2 numbers before 58 and 2 numbers after 54. What number am I? **56**

5. If you count by twos, I am the third number you say. What number am I? **6**

6. I am 2 more than 42. What number am I? **44**

7. I am 5 less than 40. What number am I? **35**

8. I am 10 more than 80. What number am I? **90**

9. I am 5 numbers before 80 and 5 numbers after 70. What number am I? **75**

10. If you count by tens, I am the fourth number you say. What number am I? **40**

22   Level D Unit 1 Week 5

Assessment, pp. 21–22

Project Preview

This week, students learned to skip count and used skip counting to solve pattern problems. The project for this unit requires students to extend the knowledge they gained in Find the Math and what they have learned this week. They will use skip counting to create a hopscotch board.

Project-Based Learning

Standards-driven Project-Based Learning is effective in building deep content understanding. Project-Based Learning increases long-term retention of concepts and has been shown to be more effective than traditional instruction. By completing a project to answer an essential question, students are challenged to apply and demonstrate mastery of concepts and skills by expressing understanding through discussion, research, and presentation.

Essential Question

HOW can I use comparing and ordering numbers outside the classroom?

Project Evaluation Criteria

Review project evaluation criteria with students prior to beginning the project.

Exceeds Expectations
☐ Project result is explained and can be extended.
☐ Project result is explained in context and can be applied to other situations.
☐ Project result is explained using advanced mathematical vocabulary.
☐ Project result is described, and mathematics are used correctly and can be extended.
☐ Project result is explained and extended, and shows advanced knowledge of mathematical concepts and skills.

Meets Expectations
☐ Project result is explained.
☐ Project result is explained in context.
☐ Project result is explained using mathematical vocabulary.
☐ Project result is described, and mathematics are used correctly.
☐ Project result is explained, and shows satisfactory knowledge of mathematical concepts and skills.

Does Not Meet Expectations
☐ Project result is not explained.
☐ Project result is explained, but out of context.
☐ Project result is explained, but mathematical vocabulary is oversimplified.
☐ Project result is described, but mathematics are not used correctly.
☐ Project result is not explained and or extended, or shows less than satisfactory knowledge of mathematical concepts and skills.

Skipping in the Park

Objective
Students can skip count by 2s, 5s, and 10s.

Standard
2.NBT.2 Count within 1000; skip-count by 5s, 10s, and 100s.

Materials
Additional Materials
- crayons
- drawing of park from Week 4
- number line

Best Practices
- Provide project directions that are clear and brief.
- Allow active learning with noise and movement.
- Provide meaning and organization to the lessons and concepts.

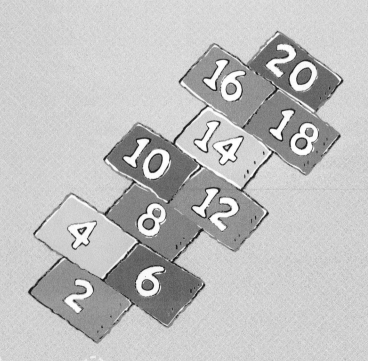

Introduce

Imagine that you are in the park and you are going to play a game of hopscotch with your friends. Instead of placing the numbers 1 to 10 on the board, you will use skip counting to make the hopscotch board.

- ▶ **How do you play hopscotch?** Answers may vary. Possible answer: You throw a beanbag and hop to the rectangle it lands on.
- ▶ **What number could you start with on the hopscotch board?** Answers may vary. Possible answer: I will start with 2.
- ▶ **How could you set up your hopscotch board?** Answers may vary. Possible answer: I will start with 5 and count by 5s.

Explore

- ▶ Today you will add to your picture by placing a hopscotch board on it.
- ▶ Complete *Student Workbook* page 64 to find out what kind of hopscotch board you will draw.
- ▶ First have each student draw a blank hopscotch board on his or her picture of the park.
- ▶ It's up to you to decide how many boxes you would like on your hopscotch board. You must have at least 10, but no more than 20.
- ▶ You may also choose if you would like to skip count by 2, 5, or 10. If you choose 10, you should have no more than 10 squares.

Wrap Up

- Allow students time to decide how many boxes they would like and what number they would like to skip count by.
- Make sure each student correctly draws his or her hopscotch board.
- If students struggle to find the correct numbers for their boards, have them use a number line and skip count to identify the numbers they will use.
- Discuss students' answers to the Reflect prompt at the bottom of **Student Workbook** page 64.
- Save students' drawings for use in Week 6.

If time permits, allow each student to make a second board using a different skip count pattern.

Week 5 • Number Patterns

Project

Sipping in the Park

Draw a hopscotch board in your picture of the park. Draw at least 10 boxes but no more than 20.

1. Choose a number for skip counting. Use either 2, 5, or 10. Which number did you choose?

   Answers may vary. Possible answer: I will skip count by 5.

2. Fill in the numbers you skip count by on your hopscotch board.

3. What number did your hopscotch game end with?

   Answers may vary. Possible answer: 50

4. If you added another box to your hopscotch game, what would that number be?

   Answers may vary. Possible answer: 55

Reflect

How did you know which number came next on your hopscotch board?

Answers may vary. Possible answer: I started at 5 and know 5 more is 10, so 10 comes next.

If you counted by a different number, would you take more hops or fewer hops?

Answers may vary. Possible answer: If I skip counted by 2, I would take more hops.

Student Workbook, p. 64

Teacher Reflect

☐ Did students correctly use art, objects, graphs, or posters to explain their solutions?

☐ Did I explain the directions before students began their project?

☐ Were students able to correctly answer the Reflect questions?

WEEK 6: Whole Number Relationships

Week at a Glance

This week students complete **Number Worlds,** Level D, Number Sense within 100. Students will compare and order the vales of one-and two-digit numbers and will estimate using the amounts of 10, 50, and 100 as visual guidelines.

Skills Focus

- Visualize amounts of 10, 25, and 50.
- Compare and order whole numbers to 100.
- Make comparisons between numbers with "greater than" or "less than" or symbols > and <.

How Students Learn

Encourage students to think flexibly about numbers and to investigate multiple ways of composing and decomposing numbers using manipulatives and reasoning strategies. Give them opportunities to extend their knowledge of the numbers from 1 to 10 as they begin to work with numbers from 11 to 20, and then with numbers from 21 to 100.

English Learners ELL

For language support, use the **English Learner Support Guide,** pages 32–33, to preview lesson concepts and teach academic vocabulary.

Math at Home

Give one copy of the Letter to Home, page 6, to each student. Encourage students to share and complete the activity with their caregivers.

Weekly Planner

Lesson	Learning Objectives
1 pages 90–91	Students can use quantities of 10 to estimate amounts up to 100.
2 pages 92–93	Students can order a series of amounts and compare numbers up to 20.
3 pages 94–95	Students can use numbers to compare groups and say whether there are enough, too many, or too few.
4 pages 96–97	Students can correctly use the symbols <, >, and = to relate two numbers.
5 pages 98–99	**Review and Assess** Students review skills learned this week and complete the weekly assessment.
Project pages 100–101	Students can compare numbers and write an expression using the "greater than" or "less than" symbol.

Key Standard for the Week

Domain: Number and Operations in Base Ten
Cluster: Understand place value.
2.NBT.4 Compare two three-digit numbers based on meanings of the hundreds, tens, and ones digits, using >, =, and < symbols to record the results of comparisons.

Materials		Technology
Program Materials • **Student Workbook,** pp. 65–67 • **Practice,** p. 24 • Activity Card 1S, **Estimation**	**Additional Materials** • paper plates • plastic sandwich bags • small objects, such as beans • Vocabulary Card 21, *greater than* • Vocabulary Card 24, *less than*	**Teacher Dashboard** Building Blocks Number Compare 1
Program Materials • **Student Workbook,** pp. 68–69 • **Practice,** p. 25 • Activity Card 1T, **Feed the Animals** • Number Cards (11–20) • Zoo Pictures • Counters	**Additional Materials** • plastic sandwich bags • Vocabulary Card 21, *greater than* • Vocabulary Card 24, *less than*	**Teacher Dashboard** Building Blocks Number Compare 3
Program Materials • **Student Workbook,** pp. 70–71 • **Practice,** p. 26 • Activity Card 1U, **Party!** • Party!	**Additional Materials** • books • Vocabulary Card 21, *greater than* • Vocabulary Card 24, *less than*	**Teacher Dashboard** Building Blocks Number Compare 4
Program Materials • **Student Workbook,** pp. 72–73 • **Practice,** p. 27 • Activity Card 1V, **Hungry Alligators**	**Additional Materials** • construction paper • markers • Vocabulary Card 21, *greater than* • Vocabulary Card 24, *less than*	**Teacher Dashboard** Building Blocks Number Compare 5
Program Materials • **Student Workbook,** pp. 74–75 • Weekly Test, **Assessment,** pp. 23–24		Review previous activities.
Program Materials • **Student Workbook,** p. 76 • Dot Cubes	**Additional Materials** • crayons • drawing of park from Week 5	

WEEK 6
Whole Number Relationships

Find the Math

In this week, students will learn how to compare numbers to figure out which is greater and which is less.

Use the following to begin a guided discussion:

▶ **Why would you need to compare two amounts?** Answers may vary. Possible answers: I may need to add to something to make the amounts the same; I may need to know who has more of something.

Have students complete **Student Workbook,** page 65.

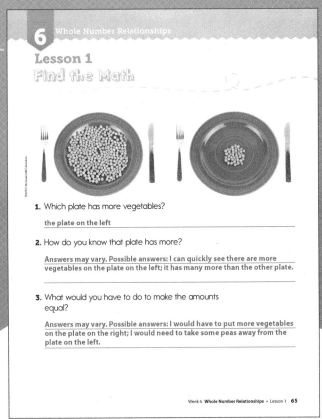

Student Workbook, p. 65

Lesson 1

Objective
Students can use quantities of 10 to estimate amounts up to 100.

Standard
2.NBT.4 Compare two three-digit numbers based on meanings of the hundreds, tens, and ones digits, using >, =, and < symbols to record the results of comparisons.

Vocabulary
- greater than
- less than

Creating Context
As English Learners work to understand adjectives, review with them the English pattern of comparatives and superlatives, such as *great, greater,* and *greatest*.

Materials
Additional Materials
- paper plates, 6
- plastic sandwich bags, 6
- small objects, such as beans, 6 sets of 15–99
- Vocabulary Card 21, *greater than*
- Vocabulary Card 24, *less than*

Prepare Ahead
Place between 15 and 99 small objects, such as beans, on six poster boards or paper plates. Make sure each plate has a different number of objects. Label each plate with a letter or number. Next to each display, place exactly 10 of the same object in a plastic sandwich bag.

1 WARM UP

Prepare
Set out two different amounts (up to 20) of a small object, such as seeds or beans, on two plates labeled *A* and *B*.

▶ **Look at these two groups of objects and make a prediction about which group has more (or which group has less). Record your prediction on a piece of paper by writing the letter that is on the plate:** *A* **or** *B*.

▶ **What information did you use to make your prediction?** Possible answer: the overall size of each group and how much space it took

▶ **How can you check to see if your prediction is correct?** Possible answer: by counting the number of objects on each plate

2 ENGAGE

Develop: Estimation

"Today we are going to learn how to estimate." Follow the instructions on the Activity Card **Estimation.** As students complete the activity, be sure to use the Questions to Ask.

Activity Card 1S

Alternative Groupings
Individual: Complete the activity as written.

Progress Monitoring

| **If...** students are having difficulty estimating, | **Then...** have them practice with smaller amounts. |

Practice
Have students complete **Student Workbook,** pages 66–67. Guide students through the Key Idea example and the Try This exercises.

Interactive Differentiation
Consult the **Teacher Dashboard** for grouping suggestions. You can also use performance on the Engage activity to guide students.

Independent Practice
For additional practice with comparing numbers, students will use Number Compare 1: Dots and Numerals. Students compare two cards and choose the one with the greater value. (Cards contain dots and numerals.)

Supported Practice

For additional support, use small objects, such as beans, to show greater than, less than or equal to. Give the beans to the students.

▶ **Place ten beans in a row.** Students should place ten beans in a row.

▶ **What if you wanted more than ten beans, what should you do?** I should add beans to the row.

▶ **Show me more than ten beans.** Answers may vary. Possible answer: Students should place more beans in the row.

▶ **If you wanted less than ten beans, show me what you should do.** Answers may vary. Possible answer: Students should take away enough beans to show less than ten.

• Place down sets of beans in two rows: one set with five beans and one set with eight beans.

▶ **Which set is greater? Which set is less?** The set with eight is greater; the set with five is less.

▶ **Place down two sets equal to the ones shown.** Students should place down five beans and eight beans.

REFLECT

Think Critically

Review students' answers to the Reflect prompt at the bottom of **Student Workbook** page 67, and then review the Engage activity.

Discuss to reinforce the concept of estimation and its use in comparing numbers.

▶ **When is estimating helpful for comparing numbers? Why?**
▶ **What does this remind you of?**

ASSESS

Informal Assessment

Use the online or print Student Record, **Assessment,** page 128, to record informal observations.

Estimation

Did the student
☐ respond accurately?   ☐ respond with confidence?
☐ respond quickly?   ☐ self-correct?

Additional Practice

For additional practice, have students complete **Practice,** page 24.

Practice, p. 24

Week 6 • Whole Number Relationships

Lesson 1

Key Idea
Use the words *more than* or *less than* to compare amounts to 10.

🥜🥜🥜🥜🥜🥜🥜🥜🥜🥜   equals ten
🥜🥜🥜🥜🥜🥜   less than ten
🥜🥜🥜🥜🥜🥜🥜🥜🥜🥜🥜🥜🥜   more than ten

Try This
Circle the group that has the most objects.

1. 🪵🪵🪵   (🪵🪵🪵🪵🪵🪵🪵🪵🪵🪵)

Circle the group that has the fewest objects.

2. (🐦🐦🐦🐦🐦)   🐦🐦🐦🐦🐦🐦🐦🐦🐦🐦

Choose the words that show if this group has more or less than 10 objects.

3. 👞👞👞👞👞👞👞👞👞👞👞👞👞👞👞   (more than 10)   less than 10

66 Level D Unit 1 Number Sense within 100

Practice
Use the 10 paper clips to estimate whether each group is more than or less than 10.

10 paper clips look like this: 📎📎📎📎📎📎📎📎📎📎

4. 📎📎📎📎📎   _less_ than 10
5. 📎📎📎📎📎📎📎 _less_ than 10
6. 📎📎📎📎📎📎📎📎📎📎📎📎📎 _more_ than 10

Use the 10 pushpins to estimate how many pushpins are in each group. Choose the closest estimate.

10 pushpins look like this: 📌📌📌📌📌📌📌📌📌📌

7. 📌📌📌📌📌📌📌📌📌📌📌📌📌📌📌📌📌📌📌📌
   (20)   30   60

8. 📌📌📌📌📌📌📌📌📌📌📌📌📌📌📌📌📌📌📌📌📌📌📌📌📌📌📌📌📌📌
   (30)   40   50

Reflect
The Midland Dodgers played the Fairfield Jets in baseball. The final score was Dodgers 8, Jets 11.

Which team lost the game? _Dodgers_

Which team scored more than 10 runs? _Jets_

Week 6 Whole Number Relationships • Lesson 1 67

Student Workbook, pp. 66–67

Week 6 **Whole Number Relationships** • Lesson 1 **91**

WEEK 6
Whole Number Relationships

Lesson 2

Objective
Students can order a series of amounts and compare numbers up to 20.

Standard
2.NBT.4 Compare two three-digit numbers based on meanings of the hundreds, tens, and ones digits, using >, =, and < symbols to record the results of comparisons.

Vocabulary
- greater than
- less than

Creating Context
Review with English Learners some of the commands that are used often in math to be sure they understand what to do. Commands are verbs, such as *find, reflect, read, practice,* and *complete.*

Materials

Program Materials
- Number Cards (11–20), 1 set per group
- Zoo Pictures, 1 set per group
- Counters, 50–74 in 4 different colors, 1 set per group

Additional Materials
- plastic sandwich bags, 4 per group
- Vocabulary Card 21, *greater than*
- Vocabulary Card 24, *less than*

1 WARM UP

Prepare
Play **What Number Am I?** For this lesson, use several mystery questions that require students to compare numbers up to 20. For example:

▶ I am less than 20. I am more than 18. What number am I?

▶ I am three numbers bigger than 8. What number am I?

Just the Facts
Have students review skip counting. Use the following questions:

▶ Which number comes next in the pattern? 2, 4, 6, 8, . . . 10

▶ Which number comes next in the pattern? 26, 28, 30, 32, . . . 34

▶ Which number comes next in the pattern? 40, 45, 50, 55, . . . 60

2 ENGAGE

Develop: Feed the Animals
"Today we are going to pretend to be zookeepers and figure out how much to feed the animals." Follow the instructions on the Activity Card **Feed the Animals**. As students complete the activity, be sure to use the Questions to Ask.

Activity Card 1T

Alternative Groupings
Individual: Assist the student as needed.

Progress Monitoring

If... a student makes a mistake identifying the number of Counters in his or her bag,

Then... help him or her find the correct number, say its name, and then recount the Counters to confirm that the number matches the set size.

Practice
Have students complete **Student Workbook,** pages 68–69. Guide students through the Key Idea example and the Try This exercises.

Interactive Differentiation
Consult the **Teacher Dashboard** for grouping suggestions. You can also use performance on the Engage activity to guide students.

Independent Practice

For additional practice with comparing numbers, students will use Number Compare 3: Dots to 10. Students compare two cards and choose the one with the greater value. (Cards contain dot arrangements to ten.)

Supported Practice

For additional support, use small objects (such as beans). Take 5 beans and 2 beans and explain to students that 5 is "more than" 2. Next, arrange two sets of beans. Place 20 beans in a row and 15 beans in a row below the 20.

▶ **Which row has more?** the top row with 20 beans

- Place 10 beans on the table. Place 20 beans in two rows below the 10 beans and compare.

▶ **Ten beans looks like this.** (Point to the row of 10 beans.)

▶ **How many sets of ten do you see here?** (Point to the row of 20 beans.) I see 2 sets of 10.

▶ **Which row is greater?** the row with two sets of ten

▶ **Which row is less?** the row with one set of ten

92 Level D Unit 1 **Number Sense within 100**

REFLECT

Think Critically

Review students' answers to the Reflect prompt at the bottom of **Student Workbook** page 69, and then review the Engage activity.

Discuss to reinforce the concept that estimation can help students solve problems more quickly.

- Why is estimating useful when comparing numbers?
- What other situations could you use estimates for?
- When would you want an exact answer instead of an estimate?

Real-World Application

- **A quarter is twenty-five cents. People often use this as a reference point for estimating change. Would you need more or less than a quarter if something costs 27 cents?** more **18 cents?** less **35 cents?** more

ASSESS

Informal Assessment

Use the online or print Student Record, **Assessment,** page 128, to record informal observations.

Feed the Animals

Did the student
- ☐ respond accurately?
- ☐ respond quickly?
- ☐ respond with confidence?
- ☐ self-correct?

Additional Practice

For additional practice, have students complete **Practice,** page 25.

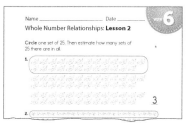

Practice, p. 25

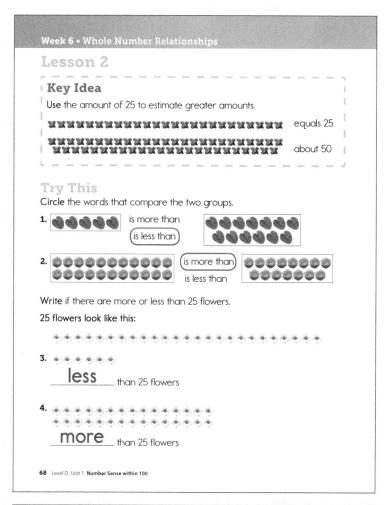

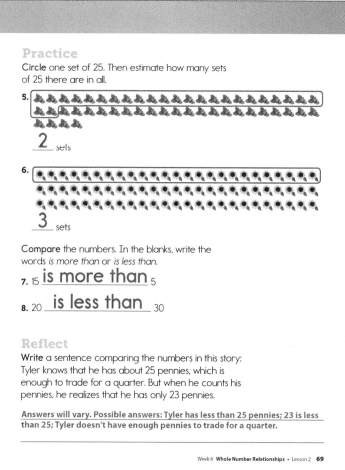

Student Workbook, pp. 68–69

Week 6 **Whole Number Relationships** • Lesson 2 **93**

WEEK 6
Whole Number Relationships

Lesson 3

Objective
Students can use numbers to compare groups and say whether there are enough, too many, or too few.

Standard
2.NBT.4 Compare two three-digit numbers based on meanings of the hundreds, tens, and ones digits, using >, =, and < symbols to record the results of comparisons.

Vocabulary
- greater than
- less than

Creating Context
In Problems 3 and 4, the instructions tell students to write *L* next to the smallest group. The *L* represents the word *least*. The instructions say to write *G* next to the largest group. The *G* stands for the word *greatest*.

Materials
Program Materials
- Party!, 1 per student

Additional Materials
- books, 3 or 4 for display to the class
- Vocabulary Card 21, *greater than*
- Vocabulary Card 24, *less than*

1 WARM UP

Prepare
Hold up a book so students can see its thickness.

For each book, ask the following questions:

▶ **Do you think this book has more than 50 pages or fewer than 50 pages?**

▶ **What information did you use to make your prediction?** Answers may vary. Possible answer: I looked at the size of the book. It looks really big (or not so big).

▶ **How can we check your predictions? Show me.** See how many pages the book has by looking at the number on the last page.

▶ **Now that we know how many pages the book has, do you want to change your prediction?**

Just the Facts
Have students nod "yes" if the numbers are ordered correctly and shake their heads "no" if the numbers are in an incorrect order. Use questions such as the following: **Are these numbers ordered from greatest to least?**

▶ **40, 30, 20** nod "yes"
▶ **10, 15, 17** shake head "no"
▶ **11, 12, 22** shake head "no"

2 ENGAGE

Develop: Party!
"Today we are going to make sure that everyone gets the same number of items as we plan a party." Follow the instructions on the Activity Card **Party!**. As students complete the activity, be sure to use the Questions to Ask.

Activity Card 1U

Alternative Groupings
Individual: Assist the student as needed.

Progress Monitoring
If... students have trouble counting the party items,

Then... make sure they are counting only one item at a time.

Practice
Have students complete *Student Workbook*, pages 70–71. Guide students through the Key Idea example and the Try This exercises.

Interactive Differentiation
Consult the **Teacher Dashboard** for grouping suggestions. You can also use performance on the Engage activity to guide students.

Independent Practice

For additional practice with comparing numbers, students should use Number Compare 4: Numerals to 100. Students will compare two cards and choose the one with the greater value. (Cards contain numbers to 100.)

Supported Practice

For additional support, use pictures and Counters to help students understand ordering.

- Show students three pictures with different amounts of an item. Have students place them in the correct order from greatest to least.
- Place Counters in three groups. Tell the students to place the groups in the correct order from least to greatest. Place Counters in the following amounts in groups:

▶ **7, 9, 4** 4, 7, 9
▶ **4, 8, 12** already in the correct order
▶ **20, 15, 12** 12, 15, 20

REFLECT

Think Critically

Review students' answers to the Reflect prompt at the bottom of **Student Workbook** page 71, and then review the Engage activity.

Discuss to reinforce the concept of ordering numbers.

▶ **How do you know if numbers are ordered correctly?**

▶ **What does this remind you of?**

Real-World Application

Explain to students that when you compare measurable characteristics about a person or thing, they are usually ordered. Give the example that in a family, the dad may be the *tallest,* the mom the *next tallest,* and the child the *shortest.*

▶ **A cheetah can run 70 miles per hour, a giraffe 32 miles per hour, and a greyhound 40 miles per hour. Order the animals from fastest to slowest.**
cheetah, greyhound, giraffe

ASSESS

Informal Assessment

Use the online or print Student Record, **Assessment,** page 128, to record informal observations.

Party!
Did the student
- ☐ respond accurately?
- ☐ respond quickly?
- ☐ respond with confidence?
- ☐ self-correct?

Additional Practice

For additional practice, have students complete **Practice,** page 26.

Practice, p. 26

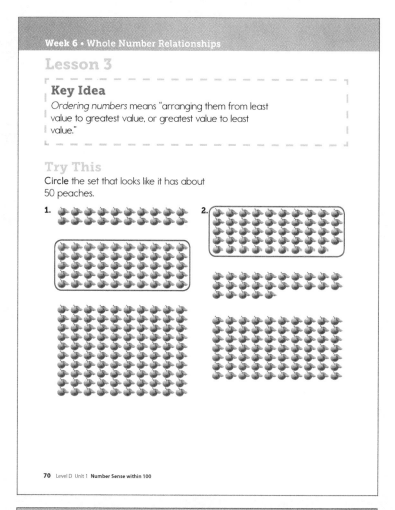

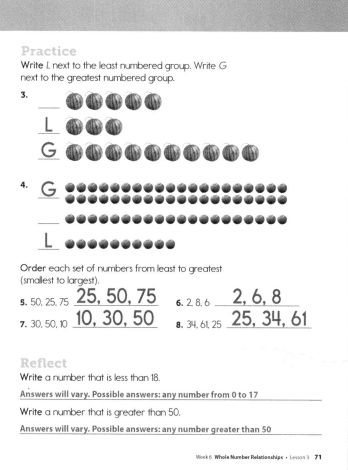

Student Workbook, pp. 70–71

WEEK 6
Whole Number Relationships

Lesson 4

Objective
Students can correctly use the symbols <, >, and = to relate two numbers.

Standard
2.NBT.4 Compare two three-digit numbers based on meanings of the hundreds, tens, and ones digits, using >, =, and < symbols to record the results of comparisons.

Vocabulary
- greater than
- less than

Creating Context
In English when we are talking about events that have already taken place, we usually speak in the past tense. We usually add *–ed* to the end of the verb or action word in the sentence. Some past-tense verbs are irregular and have a different pattern. The word *compare* becomes *compared,* but the word *eat* becomes *ate*. Make a two-column chart to show regular and irregular past-tense verbs.

Materials
Additional Materials
- construction paper, 3 sheets per student
- markers
- Vocabulary Card 21, *greater than*
- Vocabulary Card 24, *less than*

1 WARM UP

Prepare
As a class, brainstorm signs or symbols you see in math or in the real world. Write them on the board. Explain that these symbols usually stand for a word or words. Some examples are:

- $, *dollar*
- &, *and*
- +, *plus*
- a red octagon, *stop*

Just the Facts
Ask students to tell which symbol belongs in the blank spot.

▶ 12 _____ 22   < less than
▶ 70 _____ 50   > greater than
▶ 17 _____ 17   = equal to

2 ENGAGE

Develop: Hungry Alligators
"Today we are going to use the symbols for greater than and less than to show number comparisons." Follow the instructions on the Activity Card **Hungry Alligators**. As students complete the activity, be sure to use the Questions to Ask.

Activity Card 1V

Alternative Groupings
Individual: Have the student write the numbers being compared and fill in the symbol to make the correct comparison.

Progress Monitoring
If... students continue to struggle with choosing the correct symbol,

Then... have them form a greater than symbol with their arms and turn and walk toward items for which they wish they had more.

Practice
Have students complete **Student Workbook,** pages 72–73. Guide students through the Key Idea example and the Try This exercises.

Interactive Differentiation
Consult the **Teacher Dashboard** for grouping suggestions. You can also use performance on the Engage activity to guide students.

Independent Practice
For additional practice with comparing numbers, students will use Number Compare 5: Dot Arrays to 100. Students compare two cards and choose the one with the greater value. (Cards contain dot arrays to 100.)

Supported Practice
For additional support, draw pictures (as mnemonic devices) under the symbols. Display a picture of a fish with an open mouth. Tell students the fish always "eats" the bigger number. Write the numbers on the board and turn the fish's mouth towards the correct answer as students name it.

▶ **Which number is greater, 15 or 10?** 15
▶ **Which way do you place the fish to make this true? 15 _____ 10**
  mouth toward 15
▶ **Which way does the fish go to make this true? 9 _____ 7**
  mouth toward 9
▶ **Which way does the fish go to make this true? 17 _____ 25**
  mouth toward 25

96 Level D Unit 1 **Number Sense within 100**

REFLECT

Think Critically

Review students' answers to the Reflect prompt at the bottom of *Student Workbook* page 73, and then review the Engage activity.

Have students share their drawings or explanations with the class so that they can be exposed to a variety of ways to visualize or think about number comparisons.

▶ **Did you like someone else's strategy better than the one you used? Why?**

Real-World Application

The hungry alligator analogy is used to help students remember the meaning of the inequality symbols. Many students still struggle with using the symbols correctly. Discuss other analogies that work for the symbols. See whether your class can think of one that they will better remember and use.

ASSESS

Informal Assessment

Use the online or print Student Record, *Assessment,* page 128, to record informal observations.

Hungry Alligators

Did the student

☐ respond accurately?   ☐ respond with confidence?

☐ respond quickly?   ☐ self-correct?

Additional Practice

For additional practice, have students complete *Practice,* page 27.

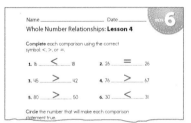

Practice, p. 27

Week 6 • Whole Number Relationships

Lesson 4

Key Idea
< means "less than." 13 < 49
> means "more than." 5 > 2
= means "equal to." 50 = 50

Try This

Complete each comparison using the correct symbol: <, >, or =.

1. 🌷🌷🌷🌷🌷🌷🌷🌷🌷 < 🌷🌷🌷🌷🌷🌷🌷🌷🌷🌷

2. 🥕🥕🥕🥕 = 🥕🥕🥕🥕

3. 1 < 4   4. 45 > 44
5. 42 = 42   6. 20 < 30
7. 39 > 33   8. 7 = 7

Write whether each comparison is *true* or *false*.

9. 72 > 62 **true**   10. 100 < 10 **false**
11. 8 < 13 **true**   12. 42 > 39 **true**

72 Level D Unit 1 Number Sense within 100

Practice

Complete each comparison using the correct symbol: <, >, or =.

13. 50 < 51
14. 98 > 89
15. 1 < 11
16. 71 > 63
17. 25 = 25
18. 20 < 40

Circle the number that will make each comparison statement true.

19. 5 <   4   1   (7)
20. 74 >   75 74 (73)

Reflect

Draw a picture or use words to explain why 3 < 13.

> Answers will vary. Possible answer: Because 13 is 1 ten and 3 ones but 3 is only 3 ones, 3 is less than 13.

Week 6 Whole Number Relationships • Lesson 4 73

Student Workbook, pp. 72–73

WEEK 6
Whole Number Relationships

Lesson 5 Review

Objective
Students review skills learned this week and complete the weekly assessment.

Standard
2.NBT.4 Compare two three-digit numbers based on meanings of the hundreds, tens, and ones digits, using >, =, and < symbols to record the results of comparisons.

Vocabulary
Review vocabulary introduced during the week.

Creating Context
Explain to students that when answering true/false questions, they should first identify what the question asks, solve the problem, and then mark whether it is true or false.

1 WARM UP

Prepare
Review comparing numbers and the symbols <, >, and =.

▶ **What are some numbers we use as reference points when estimating?** 10, 25, 50

▶ **Which symbols did we study this week?** <, >, =

2 ENGAGE

Practice
Have students complete *Student Workbook,* pages 74–75.

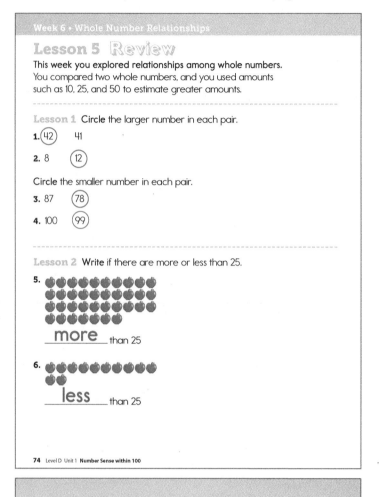

Student Workbook, pp. 74–75

98 Level D Unit 1 **Number Sense within 100**

3 REFLECT

Think Critically

Review students' answers to the Reflect prompt at the bottom of **Student Workbook** page 75.

Discuss the answer with the group to reinforce Week 6 concepts.

4 ASSESS

Formal Assessment

Students may take the weekly assessment online.

As an alternative, students may complete the weekly test on **Assessment,** pages 23–24. Record progress using the Student Assessment Record, **Assessment,** page 128.

Going Forward

Use the **Teacher Dashboard** to view results of the online assessments, to input the results of print student assessments, and to review progress before making decisions about next steps. Use the weekly test results and observations to determine the next steps for each student.

Retention	
Student displays good grasp of this week's concepts and skills.	Have students use the Building Blocks Activity, Number Compare 5: Dot Arrays to 100.
Remediation	
Student is still struggling with the week's concepts and skills.	Place Counters in three groups. Tell students to place the groups in the correct order from least to greatest. Begin with groups to 10, then to 20, then to 30.

Suggestions for Re-Evaluation: If a student has struggled without success for several weeks, use observations and test results to place the student at a level in which he or she can find success and build confidence to move forward.

Assessment, pp. 23–24

Week 6 **Whole Number Relationships** • Lesson 5   99

Project Preview

This week, students learned you could compare numbers to find out if someone has more or less of something than someone else. The project for this unit requires students to extend the knowledge they gained in Find the Math and that they have learned this week. They will compare and order numbers to determine who kicked a soccer ball farthest.

Project-Based Learning

Standards-driven Project-Based Learning is effective in building deep content understanding. Project-Based Learning increases long-term retention of concepts and has been shown to be more effective than traditional instruction. By completing a project to answer an essential question, students are challenged to apply and demonstrate mastery of concepts and skills by expressing understanding through discussion, research, and presentation.

Essential Question

HOW can I use comparing and ordering numbers outside the classroom?

Project Evaluation Criteria

Review project evaluation criteria with students prior to beginning the project.

Exceeds Expectations
☐ Project result is explained and can be extended.
☐ Project result is explained in context and can be applied to other situations.
☐ Project result is explained using advanced mathematical vocabulary.
☐ Project result is described, and mathematics are used correctly and can be extended.
☐ Project result is explained and extended, and shows advanced knowledge of mathematical concepts and skills.

Meets Expectations
☐ Project result is explained.
☐ Project result is explained in context.
☐ Project result is explained using mathematical vocabulary.
☐ Project result is described, and mathematics are used correctly.
☐ Project result is explained, and shows satisfactory knowledge of mathematical concepts and skills.

Does Not Meet Expectations
☐ Project result is not explained.
☐ Project result is explained, but out of context.
☐ Project result is explained, but mathematical vocabulary is oversimplified.
☐ Project result is described, but mathematics are not used correctly.
☐ Project result is not explained and or extended, or shows less than satisfactory knowledge of mathematical concepts and skills.

Playing in the Park

Objective
Students can compare numbers and write an expression using the "greater than" or "less than" symbol.

Standard
2.NBT.4 Compare two three-digit numbers based on meanings of the hundreds, tens, and ones digits, using >, =, and < symbols to record the results of comparisons.

Materials
Program Materials
Dot Cubes
Additional Materials
- crayons
- drawing of park from Week 5

Best Practices
- Provide project directions that are clear and brief.
- Select and provide the appropriate materials.
- Create adequate time lines for each project.

Introduce

Imagine that you are at the park kicking a soccer ball with your friends. You are trying to figure out who can kick the ball farthest.

▶ **How far do you think you can kick the ball?** Answers may vary. Possible answer: I can kick a soccer ball 30 feet.

▶ **How can you record how far you kick the ball?** I can write it down.

Explore

▶ **Today you will finish your picture by drawing soccer balls and labeling how far each soccer ball traveled.**
- Have each student sit with a partner.
▶ Complete *Student Workbook* page 76 to find out how far your soccer balls will go.
▶ Roll the dot cubes to determine how far the soccer ball will go. For example, if you roll a 9 the soccer ball travels 9 feet.
▶ Have the next person at your table roll the dot cubes. If he or she rolls a 12, his or her soccer ball will go 12 feet.
▶ Write an expression to show who kicked the ball farther. (Example: 12 > 9.)
▶ Ask another pair of students how far their soccer balls traveled. Write all four numbers in order from least to greatest in your *Student Workbook*.
▶ Draw four soccer balls on your picture of the park. Write the number showing how many feet each soccer ball traveled above each ball. Circle the soccer ball that went farthest.

Wrap Up

- Make sure each student can explain how they know which number is more.
- If students struggle to figure out which number is more, have them count aloud the dots on the dot cubes.
- Discuss students' answers to the Reflect prompts at the bottom of *Student Workbook,* page 76.
- Have students present their pictures to the class. Have each student talk about their drawings from the previous weeks. The pictures of the park can then be displayed.

If time permits, allow students to roll the dot cubes with their partners again to determine which of four soccer balls traveled the farthest.

Week 6 • Whole Number Relationships

Project
Playing in the Park

Roll the dot cubes to determine how far the soccer balls will travel.

1. How far did your soccer ball go? How far did your partner's soccer ball go?

   Answers may vary. Possible answer: any distances from 2 to 12 feet

2. Whose soccer ball went farther? Write a comparison showing whose went farther.

   Answers may vary. Possible answer: My soccer ball went farther, 12 > 9.

3. Ask another pair of students how far their soccer balls traveled. Write these distances along with yours, and then place all four distances in order from greatest to least.

   Answers may vary. Possible answer: 12 feet, 11 feet, 7 feet, 5 feet

4. Draw all four soccer balls in your picture. Above each ball, write the number of feet the ball traveled. Circle the ball that went farthest.
   Students will circle the ball with the highest number label.

Reflect
How do you know whose soccer ball went farthest?
I know which soccer ball went the farthest because it had the highest number.

76 Level D Unit 1 **Number Sense within 100**

Student Workbook, p. 76

Teacher Reflect

☐ Did I supply the necessary materials?

☐ Did I adequately explain and discuss the Reflect questions with students?

☐ Were students able to answer my questions about their solutions?

UNIT 2 Number Sense to 1,000

Unit at a Glance

This **Number Worlds** unit builds on a prior knowledge of whole numbers. Students will explore different ways to visualize and represent numbers and become familiar with equivalent representations of the same number. Students will read and write numerals to 1,000 and skip count by 5s, 10s, and 100s. They will also compare three-digit numbers.

Skills Trace

Before Level D	Level D	After Level D
Level C Students can count read and write numbers to 100, understand number lines and other representations of numbers, and begin to understand the base-ten system.	By the end of this unit, students should be able to count, read, write, compare and order whole numbers to 1,000. They have learned the fundamentals of the base-ten system and can count and group objects in ones and tens. Students can identify sets as greater than, less than or equal to, and can count by 2s, 5s, 10s, and 100s to 1,000.	**Moving on to Level E** Students will learn strategies for visualizing and constructing whole numbers to 1,000, explore the place value and monetary systems, and examine relationships among numbers.

Learning Technology

The following activities are available online to support the learning goals in this unit.

Building Blocks
- Before and After Math
- Book Stacks
- Build Stairs 1
- Build Stairs 2
- Bright Idea
- Count and Race
- Clean the Plates
- Dinosaur Shop 1
- Dinosaur Shop 2
- Easy as Pie: Add Numbers
- Number Compare 1
- Number Compare 3
- Number Compare 4
- Number Compare 5
- Number Snapshots 3
- Number Snapshots 8
- Number Snapshots 10
- Rocket Blast 2
- School Supply Shop
- Tire Recycling

Digital Tools
- Base 10 Blocks Tool
- Coins and Money
- Number Line Tool
- 100 Table Tool

Unit Overview

Week	Focus
1	**Understanding the Base-Ten Number System** • *Teacher Edition,* pp. 104–117 • *Activity Cards,* 1J, 2A, 2B, 2C • *Student Workbook,* pp. 5–16 • *English Learner Support Guide,* pp. 40–41 • *Assessment,* pp. 25–26
2	**Constructing Whole Numbers to 999** • *Teacher Edition,* pp. 118–131 • *Activity Cards,* 2D, 2E, 2F, 2G • *Student Workbook,* pp. 17–28 • *English Learner Support Guide,* pp. 42–43 • *Assessment,* pp. 27–28
3	**Representing Number Systems** • *Teacher Edition,* pp. 132–145 • *Activity Cards,* 2H, 2I, 2J, 2K • *Student Workbook,* pp. 29–40 • *English Learner Support Guide,* pp. 44–45 • *Assessment,* pp. 29–30
4	**Place Value to 1,000** • *Teacher Edition,* pp. 146–159 • *Activity Cards,* 2L, 2M, 2N, 2O • *Student Workbook,* pp. 41–52 • *English Learner Support Guide,* pp. 46–47 • *Assessment,* pp. 31–32
5	**Skip Counting within 1,000** • *Teacher Edition,* pp. 160–173 • *Activity Cards,* 2P, 2Q, 2R • *Student Workbook,* pp. 53–64 • *English Learner Support Guide,* pp. 48–49 • *Assessment,* pp. 33–34
6	**Comparing Whole Numbers** • *Teacher Edition,* pp. 174–187 • *Activity Cards,* 2N, 2S, 2T, 2U • *Student Workbook,* pp. 65–76 • *English Learner Support Guide,* pp. 50–51 • *Assessment,* pp. 35–36

Essential Question

WHY do I need to understand place value to use money?

In this unit, students will explore how comparing and ordering can be used to solve real-world problems by applying their knowledge of the base-ten number system to count and use money.

Learning Goals	CCSS Key Standards
Students can skip count by tens to 100, gain experience with numbers in a 10 × 10 arrangement, and see patterns in movements within the grid. **Project:** Students can use what they know about place value to count and exchange coins.	**Domain:** Number and Operations in Base Ten **Cluster:** Understand place value **2.NBT.2:** Count within 1000; skip-count by 5s, 10s, and 100s.
Students can recognize equivalent representations for the same number and separate and regroup two- and three- digit numbers into hundreds, tens, and ones. **Project:** Students can use place value to make trades between pennies and dimes and dimes and dollars.	**Domain:** Number and Operations in Base Ten **Cluster:** Understand place value. **2.NBT.3:** Read and write numbers to 1000 using base-ten numerals, number names, and expanded form.
Students can identify the value of a digit based on its position and apply knowledge of the base-ten system to money by making trades with pennies, dimes, and dollars. **Project:** Students can use place value to model a given value of money using the fewest number of coins possible.	**Domain:** Number and Operations in Base Ten **Cluster:** Understand place value. **2.NBT.1:** Understand that the three digits of a three-digit number represent amounts of hundreds, tens, and ones; e.g., 706 equals 7 hundreds, 0 tens, and 6 ones.
Students can trade and regroup numbers through the hundreds place and make a model of a number in two or more ways using base-ten blocks. **Project:** Students can use place value to count and compare two sets of coins to find the greater value.	**Domain:** Number and Operations in Base Ten **Cluster:** Understand place value. **2.NBT.3:** Read and write numbers to 1000 using base-ten numerals, number names, and expanded form.
Students can skip count by fives, tens, and 100s within 1,000 and complete counting sequences by counting by twos, fives, tens, and hundreds. **Project:** Students can use skip counting by fives and tens to count sets of nickels and dimes.	**Domain:** Number and Operations in Base Ten **Cluster:** Understand place value. **2.NBT.2:** Count within 1000; skip-count by 5s, 10s, and 100s.
Students can compare and order whole numbers to 999, and make comparisons between numbers with "greater than" or "less than" or symbols > and <. **Project:** Students can use the Number Sense skills they have gained to act out what it is like to work at a bank.	**Domain:** Number and Operations in Base Ten **Cluster:** Understand place value. **2.NBT.4:** Compare two three-digit numbers based on meanings of the hundreds, tens, and ones digits, using >, =, and < symbols to record the results of comparisons.

 Daily lesson activities emphasize using communication, logic, reasoning, modeling, tools, precision, structure, and patterns to solve problems. All student activities, reflections, and assessments require application of the **Common Core Standards for Mathematical Practice.**

WEEK 1: Understanding the Base-Ten Number System

Week at a Glance

This week, students begin **Number Worlds,** Level D, Number Sense to 1,000, by developing proficiency with the base-ten number system. Students will use number lines and charts to count and skip count by tens.

Skills Focus

- Use a number line to count to 100 and skip count by tens.
- Gain experience with numbers 1–100 in a ten-by-ten arrangement.
- Understand that in a ten-by-ten arrangement, horizontal movement affects the digit in the ones place, and vertical movement affects the digit in the tens place.

How Students Learn

Students should use models and other strategies to gain a level of familiarity and understanding about our numeration system. Make a variety of models available to students. When students can move from one type of model to another, it signals that they are internalizing the content and are ready to begin working without the use of models.

English Learners ELL

For language support, use the **English Learner Support Guide,** pages 40–41, to preview lesson concepts and teach academic vocabulary. **Number Words** Vocabulary Cards are listed as additional materials in many lessons and can be used to preteach and reinforce academic vocabulary.

Math at Home

Give one copy of the Letter to Home, page 7, to each student. Encourage students to share and complete the activity with their caregivers.

Weekly Planner

Lesson	Learning Objectives
1 pages 106–107	Students can count by ones to 100 and skip count by tens.
2 pages 108–109	Students can count by ones to 100 and skip count by tens forward or backward.
3 pages 110–111	Students can use a ten-by ten number grid to count to 100 and describe number patterns.
4 pages 112–113	Students can determine the value of digits in double-digit numbers.
5 pages 114–115	**Review and Assess** Students review skills learned this week and complete the weekly assessment.
Project pages 116–117	Students can use what they know about place value to count and exchange coins.

Key Standard for the Week

Domain: Number and Operations in Base Ten
Cluster: Understand place value.
2.NBT.2 Count within 1000; skip-count by 5s, 10s, and 100s.

Materials		Technology
Program Materials • **Student Workbook,** pp. 5–7 • **Practice,** p. 28 • Activity Card 2A, **Mystery Numbers in the Neighborhood** • What Number Am I? • Neighborhood Number Line	**Additional Materials** • math-link cubes* • Vocabulary Card 31, *ones* • Vocabulary Card 45, *tens*	*Teacher Dashboard*
Program Materials • **Student Workbook,** pp. 8–9 • **Practice,** p. 29 • Activity Card 2B, **Monsters in the Neighborhood** • Monster Cards • Neighborhood Number Line	• Number Cards 1–100 • Pawns, 2 **Additional Materials** • math-link cubes* • self-sticking notes, 54 • Vocabulary Cards 31 and 45	*Teacher Dashboard* Building Blocks Count and Race
Program Materials • **Student Workbook,** pp. 10–11 • **Practice,** p. 30 • Activity Card 2C, **Monsters in the Hotel** • Monster Cards • Hotel Game Board • Number 1–6 Cube	• Pawns, 1 per student • 10-and-1 Cube **Additional Materials** • beach ball • crayons • self-sticking notes, 100 • Vocabulary Cards 31 and 45	*Teacher Dashboard*
Program Materials • **Student Workbook,** pp. 12–13 • **Practice,** p. 31 • Activity Card 1J, **Hotel Room Service** • Room Service Delivery Slips • Hotel Game Board • Counters	• Number Cards (1–100) • Number 1–6 cubes • Pawns • 10-and-1 Cube **Additional Materials** • math-link cubes* • Vocabulary Cards 31 and 45	*Teacher Dashboard*
Program Materials • **Student Workbook,** pp. 14–15 • **Assessment,** pp. 25–26		Review previous activities.
Program Materials **Student Workbook,** p. 16	**Additional Materials** • deposit/withdrawal slip examples • math-link cubes • pictures of the inside of a bank • pictures of the outside of a bank • play money* *Available from McGraw-Hill Education	

WEEK 1
Understanding the Base-Ten Number System

Find the Math

In this week, students will use place value to skip count to 100 and describe number patterns.

Use the following to begin a guided discussion:

▶ **Why might you want to know the value of the coins?** Answers may vary. Possible answer: I might want to buy trading cards, and I want to know if I have enough money.

Have students complete **Student Workbook**, page 5.

Student Workbook, p. 5

Lesson 1

Objective
Students can count by ones to 100 and skip count by tens.

Standard
2.NBT.2 Count within 1000; skip-count by 5s, 10s, and 100s.

Vocabulary
- ones
- tens

Creating Context
In English, there are many words that have more than one meaning, such as *block*. This can confuse some English Learners when using math terms. An excellent strategy to use with English Learners is to preview the lesson to identify any potentially confusing words or phrases. Then you can introduce those words or explain them while teaching the lesson.

Materials

Program Materials
- What Number Am I?
- Neighborhood Number Line

Additional Materials
- math-link cubes
- Vocabulary Card 31, *ones*
- Vocabulary Card 45, *tens*

1 WARM UP

Prepare

With the whole class, practice skip counting by tens. For each example, stop at 100. Use the following starting numbers:

10, ___, ___, ___, ... 100   80, ___, 100

30, ___, ___, ___, ... 100   50, ___, ___, ___, ... 100

60, ___, ___, 100   40, ___, ___, ___, ... 100

0, ___, ___, ___, ... 100

20, ___, ___, ___, ... 100

2 ENGAGE

Develop: Mystery Numbers in the Neighborhood

"Today we are going to count by ones to 100 and skip count by tens." Follow the instructions on the Activity Card **Mystery Numbers in the Neighborhood.** As students complete the activity, be sure to use the Questions to Ask.

Activity Card 2A

Alternative Groupings

Whole Class: Lead the class in counting and encourage the class to discuss predictions throughout the activity.

Individual: Take turns counting and skip counting with the student.

Progress Monitoring

| **If...** students are at differing levels after the first game, | **Then...** assign students who are ready to work with greater numbers to one team and have that team start on 50. |

Practice

Have students complete **Student Workbook,** pages 6–7. Guide students through the Key Idea example and the Try This exercises.

Interactive Differentiation

Consult the **Teacher Dashboard** for grouping suggestions. You can also use performance on the Engage activity to guide students.

106 Level D Unit 2 **Number Sense to 1,000**

Independent Practice

For additional practice, have students model a number line by standing in a straight line. For the first round, guide students to start with the first person in line and count by ones. Then have the students start with the first person in line and count by tens. Then have the first student in line start at a number different than 1 or 10, such as 27 or 40.

Supported Practice

For additional support, use math-link cubes to model counting by ones and grouping by tens.

- Place fifteen math-link cubes in front of students. Have students count by ones as you point to each cube.
- Repeat with greater numbers of cubes, up to twenty-five.

Use the math-link cubes to illustrate how to group items into groups of ten and then skip count by tens.

▸ **How many cubes are in one group of ten?** 10
▸ **How many cubes are in two groups of ten?** 20
▸ **How many cubes are in three groups of ten?** 30

REFLECT

Think Critically

Review students' answers to the Reflect prompt at the bottom of **Student Workbook,** page 7, and then review the Engage activity.

▸ **Why are number lines helpful when you are finding a mystery number using clues?**

Guide students to understand how number lines can be helpful for visualization when given a range of numbers, such as 10–30. Number lines are also useful for counting from a given number to a mystery number, or finding a mystery number when given its neighbors. Using clues to eliminate numbers can be easier to see on a number line as well.

ASSESS

Informal Assessment

Use the online or print Student Record, **Assessment,** page 128, to record informal observations.

Mystery Numbers in the Neighborhood

Did the student
☐ make important observations?   ☐ provide insightful answers?
☐ extend or generalize learning?   ☐ pose insightful questions?

Additional Practice

For additional practice, have students complete **Practice,** page 28.

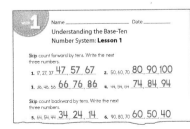

Practice, p. 28

Week 1 • Understanding the Base-Ten Number System

Lesson 1

Key Idea
When you count by **ones,** each number is 1 more than the number before it.
1, 2, 3, . . .
34, 35, 36, . . .

When you skip count by **tens,** each number is 10 more than the number before it.
10, 20, 30, 40, 50, . . .
12, 22, 32, 42, 52, . . .

Try This
Fill in the missing numbers on the number lines.

1. 1 2 3 4 5 6 7 8 9 10 11 12 13 14 15 16 17 18 19 20
2. 21 22 23 24 25 26 27 28 29 30 31 32 33 34 35 36 37 38 39 40
3. 41 42 43 44 45 46 47 48 49 50 51 52 53 54 55 56 57 58 59 60
4. 61 62 63 64 65 66 67 68 69 70 71 72 73 74 75 76 77 78 79 80

6   Level D Unit 2 Number Sense to 1,000

Practice
Skip count by tens. Write the next three numbers.

5. 20, 30, 40, __50 60 70__   6. 45, 55, 65, __75 85 95__
7. 33, 43, 53, __63 73 83__   8. 18, 28, 38, __48 58 68__
9. 100, 90, 80, __70 60 50__   10. 95, 85, 75, __65 55 45__
11. 81, 71, 61, __51 41 31__   12. 64, 54, 44, __34 24 14__

Count by ones. Write the next three numbers.

13. 23, 24, 25, __26 27 28__   14. 60, 61, 62, __63 64 65__
15. 78, 79, 80, __81 82 83__   16. 17, 18, 19, __20 21 22__
17. 100, 99, 98, __97 96 95__ 18. 45, 44, 43, __42 41 40__
19. 22, 21, 20, __19 18 17__   20. 63, 62, 61, __60 59 58__

Reflect
I am a number somewhere in the middle of the Neighborhood Number Line.
I am somewhere between 40 and 60.
One of my neighbors is 55.
If you start at 50 and move forward 4 numbers, you will land on me.
If you start at 60 and move backward 6 numbers, you will land on me.
What number am I? __54__

Week 1 Understanding the Base-Ten Number System • Lesson 1   7

Student Workbook, pp. 6–7

Week 1 **Understanding the Base-Ten Number System** • Lesson 1   **107**

WEEK 1
Understanding the Base-Ten Number System

Lesson 2

Objective
Students can count by ones to 100 and skip count by tens forward or backward.

Standard
2.NBT.2 Count within 1000; skip-count by 5s, 10s, and 100s.

Vocabulary
- ones
- tens

Creating Context
An excellent strategy to use with English Learners is to preview the lesson to identify any potentially confusing words or phrases. Help English Learners find and list the words *counting on, skip, difference, find,* and *steps.* Discuss their academic meanings and everyday meanings.

Materials

Program Materials
- Monster Cards
- Neighborhood Number Line
- Number Cards 1–100
- Pawns, 2

Additional Materials
- math-link cubes
- self-sticking notes, 54 (each note should cover only one house)
- Vocabulary Card 31, *ones*
- Vocabulary Card 45, *tens*

Prepare Ahead
Label each Monster Card a number from 1 to 100, put 55 Number Cards in a paper bag, and make sure to include the numbers that are on the Monster Cards.

1 WARM UP

Prepare
With the whole class, practice counting on.

Use a variety of starting numbers. As students count up by ones, point to each number on the Neighborhood Number Line.

Next have the whole class practice skip counting by tens. Use starting numbers that are not multiples of 10.

8, ___, ___, ___, ... 98

Just the Facts
Tell students to shake their heads "no" when they hear an incorrect number.

Use statements such as the following:

▶ **I'm going to count by tens. 10, 20, 30, 35, 40, 45, 50** Students should shake their heads "no" when they hear the number 35.

▶ **I'm going to count by tens. 33, 43, 53, 63, 70, 80, 90** Students should shake their heads "no" when they hear the number 70.

▶ **I'm going to count by tens backwards. 87, 77, 67, 57, 50, 40, 30** Students should shake their heads "no" when they hear the number 50.

2 ENGAGE

Develop: Monsters in the Neighborhood

"Today we are going to count and skip count by tens forward and backward." Follow the instructions on the Activity Card **Monsters in the Neighborhood.** As students complete the activity, be sure to use the Questions to Ask.

Activity Card 2B

Alternative Groupings
Pair: Provide assistance to each student on his or her turn as necessary.

Individual: Play the game with the student.

Progress Monitoring
If... some students are progressing rapidly,

Then... put those students on one team and have them start at 50.

Practice
Have students complete **Student Workbook,** pages 8–9. Guide students through the Key Idea example and the Try This exercises.

Interactive Differentiation
Consult the **Teacher Dashboard** for grouping suggestions. You can also use performance on the Engage activity to guide students.

Independent Practice

Students should use Count and Race to practice counting up to 50.

Supported Practice

For additional support, use math-link cubes to model the number 28. Guide students as they count and connect groups of ten cubes. Have students count by tens and then count on by ones to find the total number of cubes.

▶ **We have 28 cubes. How can we find a number that is 10 more than 28?** Add one more group of ten.

Have one student count out and make one more group of ten with additional math-link cubes.

▶ **When we add one more group of ten, which number increases, the tens or the ones?** the tens

▶ **If we add one group of ten to the 28 cubes, how many cubes will we have?** 38 cubes

Continue this activity by having students each make a group of ten cubes. Have students say the next number in the series as they take turns adding a group of ten. Then have students take turns taking away groups of ten cubes to practice counting by tens backward.

108 Level D Unit 2 **Number Sense to 1,000**

REFLECT

Think Critically

Review students' answers to the Reflect prompt at the bottom of *Student Workbook,* page 9, and then review the Engage activity.

Discuss to reinforce that the numbers in successive operations problems with addition and subtraction can be reordered to make solving easier.

▶ **Would the answer be the same if you changed the order of the moves?**

If the students are having trouble seeing that the order of the moves will not affect the answer, use the following example, using manipulatives if necessary, to help them:

▶ **Start on 10. Move back 1, then forward 10, then forward 3. What number are you on?** 22

▶ **Start on 10. Move forward 3, then forward 10, then back 1. What number are you on?** 22

▶ **Explain your answers by drawing a number line and then drawing your moves.**

Real-World Application

Have students create their own game using the Neighborhood Number Line. Have students write down the rules and play their game with other students.

ASSESS

Informal Assessment

Use the online or print Student Record, *Assessment,* page 128, to record informal observations.

Monsters in the Neighborhood

Did the student

☐ respond accurately?   ☐ respond with confidence?

☐ respond quickly?   ☐ self-correct?

Additional Practice

For additional practice, have students complete *Practice,* page 29.

Practice, p. 29

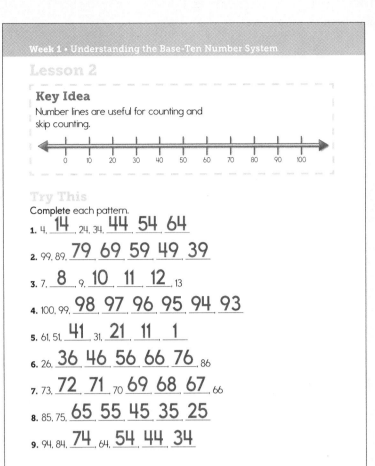

Student Workbook, pp. 8–9

Week 1 **Understanding the Base-Ten Number System** • Lesson 2   109

WEEK 1
Understanding the Base-Ten Number System

Lesson 3

Objective
Students can use a ten-by-ten number grid to count to 100 and describe number patterns.

Standard
2.NBT.2 Count within 1000; skip-count by 5s, 10s, and 100s.

Vocabulary
- ones
- tens

Creating Context
Suggest that English Learners work with a partner to solve Problem 1 in the *Student Workbook* for this lesson. Have them discuss in their primary language or in English which strategy they chose and why.

Materials

Program Materials
- Monster Cards
- Hotel Game Board
- Number 1–6 Cube
- Pawns, 1 per student
- 10-and-1 Cube

Additional Materials
- beach ball
- crayon, 1 per student
- self-sticking notes, 100
- Vocabulary Card 31, *ones*
- Vocabulary Card 45, *tens*

Prepare Ahead
Copy and cut out 25 monster cards and label each with numbers between 1–100. Cover the Hotel Game Board with paper.

1 WARM UP

Prepare
Play "What Number Am I?" (See Activity Card 1R).

If you have clue sets written by students, use those. Otherwise, create clue sets similar to those on the What Number Am I? worksheet.

Just the Facts
Have students use mental math to find numbers that are one more, one less, ten more, and ten less than a given number. Students may write their answers on individual dry-erase boards or on a blank piece of paper. Use questions such as the following:

▶ What is 1 more than 79? 80
▶ What is 10 less than 54? 44
▶ What is 10 more than 73? 83

2 ENGAGE

Develop: Monsters in the Hotel
"Today we are going to use a 10-by-10 number grid to count and describe patterns." Follow the instructions on the Activity Card **Monsters in the Hotel**. As students complete the activity, be sure to use the Questions to Ask.

Activity Card 2C

Alternative Groupings

Whole Class: Skip count by tens using the vertical columns of the Hotel Game Board.

Individual: Play the game with the student.

Progress Monitoring

If… students cannot decide whether to move up or down,

Then… suggest that they use the opposite move than the one used by the previous player.

Practice
Have students complete **Student Workbook,** pages 10–11. Guide students through the Key Idea example and the Try This exercises.

Interactive Differentiation

Consult the **Teacher Dashboard** for grouping suggestions. You can also use performance on the Engage activity to guide students.

Independent Practice

For additional practice play a game of "What's Next?" Assign one student to call out any number less than 90 and say, "What's next?" Then that student should throw a beach ball or other soft ball to another student in the classroom. The student who catches the ball should say the number that is 10 more or 10 less than the number the first student called out. The second student should think of a new number, call it out, say "What's next?" and toss the ball to a third student.

Supported Practice

For additional support, provide each student with several blank 100 Charts and a crayon.

- Say the number 20. Instruct students to color the number 20 on their charts, and then color in each number as they continue counting by tens to 100.
- Say the number 14. Instruct students to color in the number 14 on their charts, and then color in each number as they continue counting by tens to 94.

Repeat the activity as time allows.

▶ **When you count by tens, which digit in each number changes?** the number in the tens place

▶ **Look at the pattern on one of your charts. What do you notice?** Answers may vary. Possible answer: All the numbers that are colored are in the same column.

▶ **Why are the numbers you colored all in the same column?** There are ten numbers in each row, so when we count by tens, the number we color in is in the same position on each row.

REFLECT

Think Critically

Review students' answers to the Reflect prompt at the bottom of *Student Workbook,* page 11, and then review the Engage activity.

▶ **Is it easier to use a 100 chart or a number line to count or skip count?**

▶ **Which is easier to use to find patterns in numbers?**

Real-World Application

A 100 chart looks similar to some game boards. Students can create a game using a 100 chart and a Number 1–6 Cube. Students should write the basic rules for the game and then play the game with two classmates. Instruct students to write a sentence that describes the math skills they can practice by playing the game.

ASSESS

Informal Assessment

Use the online or print Student Record, *Assessment,* page 128, to record informal observations.

Monsters in the Hotel

Did the student

☐ respond accurately?   ☐ respond with confidence?

☐ respond quickly?   ☐ self-correct?

Additional Practice

For additional practice, have students complete *Practice,* page 30.

Practice, p. 30

Week 1 • Understanding the Base-Ten Number System

Lesson 3

Key Idea
The chart below is useful for counting by ones and skip counting by tens.

90	91	92	93	94	95	96	97	98	99
80	81	82	83	84	85	86	87	88	89
70	71	72	73	74	75	76	77	78	79
60	61	62	63	64	65	66	67	68	69
50	51	52	53	54	55	56	57	58	59
40	41	42	43	44	45	46	47	48	49
30	31	32	33	34	35	36	37	38	39
20	21	22	23	24	25	26	27	28	29
10	11	12	13	14	15	16	17	18	19
0	1	2	3	4	5	6	7	8	9

Try This
Complete each pattern

1. 32, 42, __52__ __62__ __72__ __82__
2. 32, 31, __30__ __29__ __28__ __27__
3. 58, 48, __38__ __28__ __18__ __8__
4. 58, 59, __60__ __61__ __62__ __63__
5. 7, 8, __9__ __10__ __11__ __12__
6. 7, 17, __27__ __37__ __47__ __57__
7. 60, 50, __40__ __30__ __20__ __10__
8. 60, 59, __58__ __57__ __56__ __55__

Practice
Complete the table.

9.

Starting Number	Move	Ending Number
12	forward 30	42
45	backward 40	5
22	forward 60	82
83	backward 70	13
3	forward 90	93
17	forward 80	97
59	forward 40	99

10. When you started at 3 and moved forward 90, did you trace your finger up or down the 99 chart to find the ending number? __up__

Reflect

Start at 7 on the number chart. Move up the column to 17, to 27, and then to 37. Describe what happens to the numbers as you move up any column on the chart.

The numbers increase by 10 each time you move up one row.

Start at 93. Move down the column to 83, to 73 and then to 63. Describe what happens to the numbers as you move down any column on the chart.

The numbers decrease by 10 each time you move down one row.

Student Workbook, pp. 10–11

Week 1 **Understanding the Base-Ten Number System** • Lesson 3

WEEK 1
Understanding the Base-Ten Number System

Lesson 4

Objective
Students can determine the value of each digit in a two-digit number.

Standard
2.NBT.2 Count within 1000; skip-count by 5s, 10s, and 100s.

Vocabulary
- ones
- tens

Creating Context
The Hotel Game gives an excellent visual cue for English Learners to support their understanding. They may already have a conceptual understanding of counting but may not have the English fluency to say the number names quickly.

Materials

Program Materials
- Room Service Delivery Slips
- Hotel Game Board, 1 per group
- Counters, 100 per group
- Number Cards (1–100)
- Number 1–6 Cube, 1 per pair
- Pawns, 1 per student

Additional Materials
- math-link cubes
- Vocabulary Card 31, *ones*
- Vocabulary Card 45, *tens*

1 WARM UP

Prepare
Have students practice moving around the Hotel Game Board. Give the class a starting number. Then have a student volunteer roll the 10-and-1 Cube. Ask the student to choose a move (forward or backward) and name the ending number.

Choose a new starting number, and continue until all students have had a turn. Keep a record of each turn where students can see it.

Just the Facts
Play a game of "Simon Says" using statements such as the following:

▶ **If 4 groups of 10 are equal to 40, touch your head.** Students should touch their heads.

▶ **If 9 groups of 9 are equal to 9, hop on one foot.** Students should not hop on one foot.

▶ **If 2 groups of 1 are equal to 20, spin in a circle.** Students should not spin in a circle.

After each incorrect statement, have students point out the error and correct answer.

2 ENGAGE

Develop: Hotel Room Service
"Today we are going to find out how much each digit in a two-digit number is worth." Follow the instructions on the Activity Card **Hotel Room Service**. As students complete the activity, be sure to use the Questions to Ask.

Activity Card 1J

Alternative Groupings
Pair: Allow additional time to complete the activity.

Progress Monitoring

If... students determine the equations easily,

Then... select four room numbers and instruct students to make a plan for making the deliveries to all four rooms.

Practice
Have students complete **Student Workbook,** pages 12–13. Guide students through the Key Idea example and the Try This exercises.

Interactive Differentiation
Consult the **Teacher Dashboard** for grouping suggestions. You can also use performance on the Engage activity to guide students.

Independent Practice
Have pairs of students roll two Number 1–6 Cubes. The first cube will represent the first digit in a two-digit number, and the second cube will represent the second digit. Instruct students to model each of the two digits using math-link cubes.

Supported Practice
For additional support, use math-link cubes to help students make the connection between digits and the values they represent. Write a number, such as 23, on the board. Guide students as they count out 23 math-link cubes and make groups of ten.

▶ **How many groups of ten do you have? How many ones do you have?** 2 groups of ten and 3 ones

▶ **Look at the two groups of ten. How many cubes are in the two groups altogether?** 20 cubes

Lead students to understand that because two groups of ten cubes are equal to 20, the digit 2 in the number 23 represents 20. There are 3 individual cubes, so the digit 3 in the number 23 represents 3.

Write other two-digit numbers on the board. Have students say the number of tens, the number of ones, and the value each digit represents.

Help struggling students build the numbers with math-link cubes.

REFLECT

Think Critically

Review students' answers to the Reflect prompt at the bottom of *Student Workbook,* page 13, and then review the Engage activity.

▶ **Why does the ten-by-ten arrangement work best for skip counting?**
You can skip count by tens simply by moving your finger up or down the chart.

Discuss why nine columns or rows, or eleven columns or rows, is not the best arrangement.

Real-World Application

▶ **You and your friends are staying at a hotel that is set up exactly like the Hotel Game Board. You are in Room 64. One of your friends is staying exactly 2 floors above you. How can you find what room your friend is in? What room is it?** Room 84

▶ **Your other friend is staying in the room next to yours, on the right. What room is your friend in?** Room 65

ASSESS

Informal Assessment

Use the online or print Student Record, *Assessment,* page 128, to record informal observations.

Hotel Room Service

Did the student
- ☐ make important observations?
- ☐ provide insightful answers?
- ☐ extend or generalize learning?
- ☐ pose insightful questions?

Additional Practice

For additional practice, have students complete *Practice,* page 31.

Practice, p. 31

Week 1 • Understanding the Base-Ten Number System

Lesson 4

Key Idea
The chart below is useful for counting by ones and skip counting by tens.

90	91	92	93	94	95	96	97	98	99
80	81	82	83	84	85	86	87	88	89
70	71	72	73	74	75	76	77	78	79
60	61	62	63	64	65	66	67	68	69
50	51	52	53	54	55	56	57	58	59
40	41	42	43	44	45	46	47	48	49
30	31	32	33	34	35	36	37	38	39
20	21	22	23	24	25	26	27	28	29
10	11	12	13	14	15	16	17	18	19
0	1	2	3	4	5	6	7	8	9

Try This

Complete each pattern.

1. 38, **48**, **58**, 68, **78**
2. **17**, 18, **19**, 20, **21**
3. 95, **94**, **93**, **92**, 91
4. **100**, 90, **80**, 70, **60**
5. 2, **12**, 22, **32**, **42**
6. **11**, 10, **9**, **8**, 7

Describe the move you would make to go from A to B

7. A. 43 B. 33 **backward 10**
8. A. 83 B. 84 **forward 1**
9. A. 67 B. 77 **forward 10**
10. A. 95 B. 94 **backward 1**

Practice

Complete the table.

11.

Starting Number	Move	Ending Number
28	forward 10	**38**
62	backward 10	**52**
94	backward 1	**93**
1	forward 10	**11**
56	forward 1	**57**
13	forward 10	**23**
77	backward 10	**67**

12. I am in the same column as 38.
    I am in the same row as 15.
    I am 4 less than 22.
    What number am I? **18**

Explain how you found the mystery number.
Did you need all three clues?

Answers may vary. Possible answer: I figured out the answer after two clues.

Reflect

Explain how you would move from 15 to 67 on a 99 chart.

Answers may vary. Possible answer: Move up five rows from 15 to 65. Then move right two columns from 65 to 67.

Student Workbook, pp. 12–13

WEEK 1
Understanding the Base-Ten Number System

Lesson 5 Review

Objective
Students review skills learned this week and complete the weekly assessment.

Standard
2.NBT.2 Count within 1000; skip-count by 5s, 10s, and 100s.

Vocabulary
Review vocabulary introduced during the week.

Creating Context
Be sure that English Learners understand that *use mental math* means "to figure out the answer without paper and pencil or calculator."

An excellent mental strategy is to think aloud or verbalize each step to figure out an answer.

1 WARM UP

Prepare

Display this table. Have students fill in the blanks as a group. Ask for student volunteers to give the answer belonging in each blank and fill in the chart after a student has answered.

Starting Number	Move	Ending Number
12	forward 1	13
29	forward 60	89
80	backward 40	40
74	forward 13	87
55	backward 50	5

2 ENGAGE

Practice

Have students complete **Student Workbook,** pages 14–15.

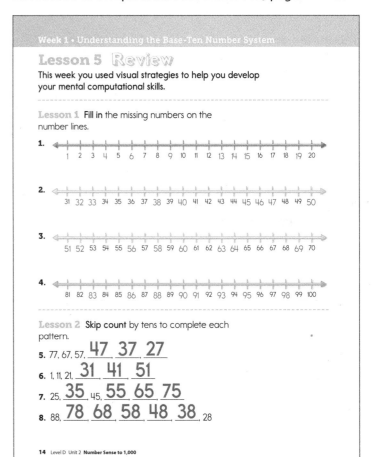

Student Workbook, pp. 14–15

114 Level D Unit 2 **Number Sense to 1,000**

3 REFLECT

Think Critically
Review students' answers to the Reflect prompt at the bottom of **Student Workbook,** page 15.

Discuss the answer with the group to reinforce Week 1 concepts.

4 ASSESS

Formal Assessment
Students may take the weekly assessment online.

As an alternative, students may complete the weekly test on **Assessment,** pages 25–26. Record progress using the Student Assessment Record, **Assessment,** page 128.

Going Forward
Use the **Teacher Dashboard** to view results of the online assessments, to input the results of print student assessments, and to review progress before making decisions about next steps. Use the weekly test results and observations to determine the next steps for each student.

Retention	
Student displays good grasp of this week's concepts and skills.	Have pairs of students roll two Number 1–6 Cubes. The first cube will represent the first digit in a two-digit number, and the second cube will represent the second digit. Instruct students to move the spaces representing each of the two digits on the Hotel Game Board.

Remediation	
Student is still struggling with the week's concepts and skills.	For additional support, provide each student with several blank 100 Charts and a crayon. • Say the number 37. • Instruct students to color in the number 37 on their charts, and then color in each number as they continue counting by tens to 97. • Repeat the activity as time allows.

Suggestions for Re-Evaluation: If a student has struggled without success for several weeks, use observations and test results to place the student at a level in which he or she can find success and build confidence to move forward.

Name _____ Date _____

WEEK 1

Understanding the Base-Ten Number System

Fill in the missing numbers in each pattern.

1. 28, __29__, 30, __31__, 32, 33

2. 16, __26__, 36, __46__, 56, 66

3. 38, __48__, 58, __68__, 78, 88, __98__

4. 63, 62, __61__, 60, __59__, __58__, 57

5. 92, __82__, 72, 62, __52__, 42, 32, __22__

Level D Unit 2 Week 1   25

WEEK 1

Name _____ Date _____

Understanding the Base-Ten Number System

6. If you start with 15 and move forward 10, what number do you land on? __25__

7. If you start with 57 and move backward 10, what number do you land on? __47__

8. Start on 21. Move forward 10. Then move back 3. What number are you on? __28__

9. Start on 33. Move forward 50. Then move back 10. What number are you on? __73__

10. Start on 67. Move backward 20. Then move forward 10. What number are you on? __57__

26   Level D Unit 2 Week 1

Assessment, pp. 25–26

Week 1 **Understanding the Base-Ten Number System** • Lesson 5   **115**

Project Preview

This week, students learned how to use the base-ten number system to count by ones and tens and to identify the value of each digit within a number. The project that students will complete by the end of this unit will require them to use what they have learned to count and use money outside the classroom. Students will use place value; count by ones, fives, and tens; and exchange groups of ones, fives, and tens to set up and manage a bank.

Project-Based Learning

Standards-driven Project-Based Learning is effective in building deep content understanding. Project-Based Learning increases long-term retention of concepts and has been shown to be more effective than traditional instruction. By completing a project to answer an essential question, students are challenged to apply and demonstrate mastery of concepts and skills by expressing understanding through discussion, research, and presentation.

Essential Question

WHY do I need to understand place value to use money?

Project Evaluation Criteria

Review project evaluation criteria with students prior to beginning the project.

Exceeds Expectations
☐ Project result is explained and can be extended.
☐ Project result is explained in context and can be applied to other situations.
☐ Project result is explained using advanced mathematical vocabulary.
☐ Project result is described, and mathematics are used correctly and can be extended.
☐ Project result is explained and extended, and shows advanced knowledge of mathematical concepts and skills.

Meets Expectations
☐ Project result is explained.
☐ Project result is explained in context.
☐ Project result is explained using mathematical vocabulary.
☐ Project result is described, and mathematics are used correctly.
☐ Project result is explained, and shows satisfactory knowledge of mathematical concepts and skills.

Does Not Meet Expectations
☐ Project result is not explained.
☐ Project result is explained, but out of context.
☐ Project result is explained, but mathematical vocabulary is oversimplified.
☐ Project result is described, but mathematics are not used correctly.
☐ Project result is not explained and/or extended, or shows less than satisfactory knowledge of mathematical concepts and skills.

What Is a Bank?

Objective
Students can use what they know about place value to count and exchange coins.

Standard
2.NBT.2 Count within 1000; skip-count by 5s, 10s, and 100s.

Materials
Additional Materials
- deposit/withdrawal slip examples
- math-link cubes
- pictures of the outside of a bank
- pictures of the inside of a bank
- play money

Prepare Ahead
Students will need to have a general understanding of the activities that go on inside a bank. If possible, encourage students to visit a bank with an adult prior to beginning the project. Begin collecting pictures of banks and bank logos, people working inside a bank, safe-deposit boxes, and so on. You can also collect examples of deposit and withdrawal slips to show to students.

Best Practices
- Make efficient use of cooperative learning groups.
- Allow active learning with noise and movement.
- Make decisions and contingency plans ahead of time.

Introduce

▶ **Pretend that you are in charge of a bank. You are going to collect information about what a banker does and how the people who work at a bank use place value to help people with their money.**

▶ **What is a bank?** Answers may vary. Possible answer: a place where people keep their money.

▶ **What do customers do at a bank?** Answers may vary. Possible answer: They give the bank their money to keep it safe or take their money out so they can spend it on things they need or want.

- Show students the pictures you have collected of banks, bank logos, and so on.

Explore

▶ **Today you will work in pairs to think of a name for your bank and design a logo.**

- Place students in groups of two. Allow each pair time to brainstorm a name for their bank and sketch a design for their logo.

▶ **If you were working at a bank, how would you count the value of a group of dimes?** by tens

▶ **How would you count a group of dimes and pennies?** Answers may vary. Possible answer: I would count the dimes first by tens. Then I would count on by ones to count the pennies.

▶ **Complete Student Workbook, page 16, to practice using place value to count sets of coins.**

- As students work through exercises 3 and 4 on **Student Workbook, page 16,** provide them with play money so that they can model counting the coins. Encourage students to count aloud as they skip count by tens and count on by ones to find the value of each set of coins.

Wrap Up

- Make sure students understand how to use skip counting by tens to count groups of dimes.
- Make sure students use a combination of skip counting by tens and counting on by ones to count sets of dimes and pennies.
- If students struggle to count the groups of coins, have them model the value of each coin with math-link cubes. For example, if there are 2 dimes and 3 pennies, have students connect two groups of 10 cubes together and count out 3 additional cubes. Model how to count by tens and then count on by ones.
- Discuss students' answers to the Reflect prompt at the bottom of **Student Workbook,** page 16.
- Provide each pair of students with a project folder. They should put their logo sketches and the name of their bank in the folder.
- If time permits, allow each student to practice counting groups of dimes and pennies with their partners.

Week 1 • Understanding the Base-Ten Number System

Project
What Is a Bank?

Answer the following questions to help you set up your bank.

1. What is the name of your bank?

   Answers will vary. Possible answer: Bank on It.

2. How can you use skip counting while you are working in your bank?

   Answers may vary. Possible answer: I can count a group of dimes by tens.

3. A customer at your bank wants to give you 8 dimes. How will you count the dimes? What is the value of 8 dimes?

   Answers may vary. Possible answer: I can count by tens: 10, 20, 30, 40, 50, 60, 70, 80. The value of eight dimes is 80¢.

4. A customer at your bank wants to give you 5 dimes and 6 pennies. How will you count the coins? What is their value?

   Answers may vary. Possible answer: I will count the dimes by tens and the pennies by ones: 10, 20, 30, 40, 50, 51, 52, 53, 54, 55, 56. The value of the coins is 56¢.

Reflect
How do people at a bank use what they know about place value to count money?

Answers may vary. Possible answer: The people at the bank can count dimes and pennies separately, because dimes are tens and pennies are ones.

16 Level D Unit 2 Number Sense to 1,000

Student Workbook, p. 16

Teacher Reflect

☐ Did I clearly explain how to organize the activity?

☐ Did students use brainstorming methods to create and organize their ideas?

☐ Was I able to answer questions when students did not understand?

WEEK 2
Constructing Whole Numbers to 999

Week at a Glance
This week, students continue with **Number Worlds,** Level D, Number Sense to 1,000. Students will explore different ways to visualize and represent quantities and understand that these are equivalent representations for the same number. Students will also build competence with regrouping.

Skills Focus
- Recognize equivalent representations for the same number and generate them by composing and decomposing quantities.
- Separate and regroup double-digit and triple-digit numbers into hundreds, tens, and ones.
- Gain experience with regrouping.

How Students Learn
Students' knowledge of numbers and quantity becomes more integrated as their experiences in mathematics continue. Students begin to link numbers to quantities and realize that questions about numbers can be answered with or without the use of concrete objects. Teach the number sense units with a focus on *quantity,* not *numbers.*

English Learners ELL
For language support, use the **English Learner Support Guide,** pages 42–43, to preview lesson concepts and teach academic vocabulary. **Number Words** Vocabulary Cards are listed as additional materials in many lessons and can be used to preteach and reinforce academic vocabulary.

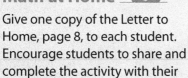

Math at Home
Give one copy of the Letter to Home, page 8, to each student. Encourage students to share and complete the activity with their caregivers.

Weekly Planner

Lesson	Learning Objectives
1 pages 120–121	Students can identify the number of tens and ones in a two-digit number.
2 pages 122–123	Students can identify the number of hundreds, tens, and ones in a three-digit number.
3 pages 124–125	Students can make trades between ones, tens, and hundreds.
4 pages 126–127	Students can make connections between the base-ten place-value system and our monetary system.
5 pages 128–129	**Review and Assess** Students review skills learned this week and complete the weekly assessment.
Project pages 130–131	Students can use place value to make trades between pennies and dimes and dimes and dollars.

Key Standard for the Week

Domain: Number and Operations in Base Ten
Cluster: Understand place value.
2.NBT.3 Read and write numbers to 1000 using base-ten numerals, number names, and expanded form.

Materials		Technology
Program Materials • *Student Workbook,* pp. 17–19 • *Practice,* p. 32 • Activity Card 2D, **Block Trading** • Number Construction Mat	**Additional Materials** • base-ten blocks* • math-link cubes* • Vocabulary Card 10, *digit* • Vocabulary Card 23, *hundreds* • Vocabulary Card 32, *ones* • Vocabulary Card 45, *tens*	*Teacher Dashboard* Base 10 Blocks Tool
Program Materials • *Student Workbook,* pp. 20–21 • *Practice,* p. 33 • Activity Card 2E, **The 500 Game** • Number Construction Mat • Number 1–6 Cubes	**Additional Materials** • base-ten blocks* • Vocabulary Card 10, *digit* • Vocabulary Card 23, *hundreds* • Vocabulary Card 32, *ones* • Vocabulary Card 45, *tens*	*Teacher Dashboard* Building Blocks Book Stacks Base 10 Blocks Tool
Program Materials • *Student Workbook,* pp. 22–23 • *Practice,* p. 34 • Activity Card 2F, **The 100 Game** • Number Construction Mat • Number 1–6 Cubes	**Additional Materials** • base-ten blocks* • Vocabulary Card 10, *digit* • Vocabulary Card 23, *hundreds* • Vocabulary Card 32, *ones* • Vocabulary Card 45, *tens*	*Teacher Dashboard* Base 10 Blocks Tool
Program Materials • *Student Workbook,* pp. 24–25 • *Practice,* p. 35 • Activity Card 2G, **The One-Dollar Game** • Money Construction Mat • Number 1–6 Cubes	**Additional Materials** • math-link cubes • play money* • Vocabulary Card 10, *digit* • Vocabulary Card 23, *hundreds* • Vocabulary Card 32, *ones* • Vocabulary Card 45, *tens*	*Teacher Dashboard* Coins and Money Tool
Program Materials • *Student Workbook,* pp. 26–27 • Weekly Test, *Assessment,* pp. 27–28		Review previous activities.
Program Materials *Student Workbook,* p. 28	**Additional Materials** • large pieces of construction paper or poster board • markers or crayons • play money*	

*Available from McGraw-Hill Education

WEEK 2
Constructing Whole Numbers to 999

Find the Math

In this week, students continue to examine place value.

Use the following to begin a guided discussion:

▶ **Why do some of the page numbers in the table of contents have more than one digit?** Only pages one through nine would have one digit. Any page number greater than 9 will have more than one digit.

Have students complete **Student Workbook,** page 17.

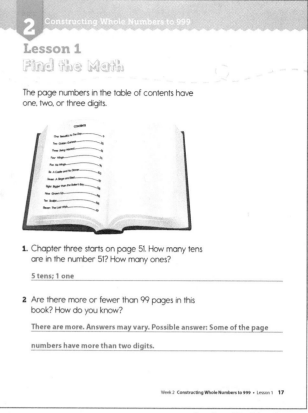

Student Workbook, p. 17

Lesson 1

Objective
Students can identify the number of tens and ones in a two-digit number.

Standard
2.NBT.3 Read and write numbers to 1000 using base-ten numerals, number names, and expanded form.

Vocabulary
- digit
- hundreds
- ones
- tens

Creating Context
Acquiring the academic language necessary to be successful in mathematics is challenging for English Learners. Often, the instructions given in a textbook are abbreviated and do not give sufficient direction. For example, the instructions for Problems 4 through 7 say to, "Tell how many tens and ones are in each number." Some English Learners may interpret this as a "repeat after me" instruction. Provide examples to clarify the instructions.

Materials
Program Materials
- Number Construction Mat, 1 per student

Additional Materials
- base-ten blocks, ones blocks and tens rods
- math-link cubes
- Vocabulary Card 10, *digit*
- Vocabulary Card 23, *hundreds*
- Vocabulary Card 32, *ones*
- Vocabulary Card 45, *tens*

1 WARM UP

Prepare
Instruct students to take a handful of unit blocks and to group the blocks into sets of 10. Students should name the number their blocks represent.

Have students join their blocks with those of another student and regroup the blocks into sets of 10. They should skip count by tens and count on to name the number represented by both students' blocks.

2 ENGAGE

Develop: Block Trading

"Today, we are going to identify the number of tens and ones in a two-digit number." Follow the instructions on the Activity Card **Block Trading**. As students complete the activity, be sure to use the Questions to Ask.

Activity Card 2D

Alternative Groupings

Whole Class: Display a Number Construction Mat and have one student demonstrate each example for the class.

Pair: Have both students work together on one mat to model problems.

Progress Monitoring

| If... students are making trades too early, | Then... have them write the numerals 1–10 in the ones section of the mat, placing units blocks on the numbers in order. When they place a unit block on 10, they make the trade. |

Practice
Have students complete **Student Workbook,** pages 18–19. Guide students through the Key Idea example and the Try This exercises.

120 Level D Unit 2 **Number Sense to 1,000**

Interactive Differentiation

Consult the **Teacher Dashboard** for grouping suggestions. You can also use performance on the Engage activity to guide students.

Independent Practice

For additional practice, have students take turns saying and then writing a number between 10 and 99. Have students use the Base 10 Blocks Tool to construct each number using the appropriate number of tens rods and unit cubes. Have students state the value of the tens and ones digit for each number.

Supported Practice

For additional support, use math-link cubes to show the connection between the digits in a two-digit number and the values they represent. Point to a number, such as 35, on a number chart. Have students count out 35 math-link cubes and then make groups of ten.

▶ **How many groups of ten do you have? How many ones do you have?** 3 tens, 5 ones

▶ **Which digit in the number 35 do the groups of ten represent? Which digit in the number 35 do the ones represent?** 3 and 5

Guide students as they make the connection between the digits they see in a number and the values the digits represent. Discuss the placement of the ones and tens digits within the number.

▶ **What number do you hear first when I say the number *thirty-five*?** 30

Point out that the first number students hear tells the value of the first digit.

REFLECT

Think Critically

Review students' answers to the Reflect prompt at the bottom of **Student Workbook** page 19, and then review the Engage activity.

▶ **How did the different models look different?** Each was made of different blocks.

▶ **How were they alike?** All represented the same number.

▶ **How do you know that each model represents the same number?** Answers may vary. Possible answer: I counted the blocks in each model.

ASSESS

Informal Assessment

Use the online or print Student Record, **Assessment,** page 128, to record informal observations.

Block Trading
Did the student
- ☐ make important observations?
- ☐ provide insightful answers?
- ☐ extend or generalize learning?
- ☐ pose insightful questions?

Additional Practice

For additional practice, have students complete **Practice,** page 32.

Practice, p. 32

Week 2 • Constructing Whole Numbers to 999

Lesson 1

Key Idea
10 ones = 1 ten

Try This

Circle groups of 10 unit blocks. Write how many tens and ones there are.

1. __1__ tens __7__ ones
2. __2__ tens __3__ ones

Draw a model of the number on the number construction mat.

3. 46

Hundreds	Tens	Ones
	Students should draw 4 base-ten rods in the tens column and 6 base-ten unit blocks in the ones column.	

Practice

Tell how many tens and ones are in each number.

4. 90 __9__ tens __0__ ones
5. 7 __0__ tens __7__ ones
6. 86 __8__ tens __6__ ones
7. 36 __3__ tens __6__ ones

18 Level D Unit 2 **Number Sense to 1,000**

Make trades. Write each number.

8. __36__

9. __81__

Reflect

Draw 32 using only unit blocks.

Students should draw 32 unit blocks.

Draw 32 using the fewest blocks possible.

Students should draw 3 rods and 2 unit blocks.

Week 2 Constructing Whole Numbers to 999 • Lesson 1 19

Student Workbook, pp. 18–19

WEEK 2
Constructing Whole Numbers to 999

Lesson 2

Objective
Students can identify the number of hundreds, tens, and ones in a three-digit number.

Standard
2.NBT.3 Read and write numbers to 1000 using base-ten numerals, number names, and expanded form.

Vocabulary
- digit
- hundreds
- ones
- tens

Creating Context
Understanding place value is necessary for success in mathematics. This lesson uses mats designed to clearly separate the amounts in each place value and help students to easily identify what trades are needed. This is an excellent strategy for English Learners, especially those at early English proficiency levels who might not be able to verbally explain their understanding. On the Number Construction Mat, students can manipulate the objects and point to the answer.

Materials
Program Materials
- Number Construction Mat, 1 per student
- Number 1–6 Cubes, 1 per group

Additional Materials
- base-ten blocks
- Vocabulary Card 10, *digit*
- Vocabulary Card 23, *hundreds*
- Vocabulary Card 32, *ones*
- Vocabulary Card 45, *tens*

1 WARM UP

Prepare
Gather students around a Number Construction Mat. Have students slowly skip count by tens. As they say each number, have a student volunteer construct the number using base-ten blocks. When the group gets to 100, ask: "How many tens are on our mat?" Explain that because 10 tens are the same as 1 hundred, we can make a trade. Construct the numbers 130, 180, and 200 this way.

Just the Facts
Have students use mental math to add. Tell them to put their hands up when they think they have the right answer. Use questions such as the following:

▶ What does 7 and 4 more ones make? 11
▶ What does 19 and 3 more ones make? 22
▶ What does 51 and 1 more ten make? 61

2 ENGAGE

Develop: The 500 Game

"Today we are going to identify the number of hundreds, tens, and ones in a three-digit number." Follow the instructions on the Activity Card **The 500 Game**. As students complete the activity, be sure to use the Questions to Ask.

Activity Card 2E

Alternative Groupings

Whole Class: Divide the class into three or four teams and have them play together.

Individual: Play the game with the student.

Progress Monitoring

If... a student is struggling to make trades at the appropriate times,

Then... encourage the student to record the result of each roll so you can review the final score with him or her.

Practice
Have students complete **Student Workbook,** pages 20–21. Guide students through the Key Idea example and the Try This exercises.

Interactive Differentiation
Consult the **Teacher Dashboard** for grouping suggestions. You can also use performance on the Engage activity to guide students.

Independent Practice
For additional practice, have students play Book Stacks to gain a visual understanding of how new groups of ten are made from adding on additional ones.

Supported Practice
For additional support, have students use Base-Ten Blocks or the Base 10 Blocks Tool to help them connect the digits to the values the digits represent.

- Have students roll a number cube two times. The first roll represents the tens place and the second roll the ones place.
- Tell students to write down each digit they roll.
- Help students make models of the digits they rolled with Base-Ten Blocks and the Number Construction Mat. Students may also use the Base 10 Blocks Tool.
- Have students name each number they modeled. Then ask them how to show the value of the tens digit using the Base-Ten Blocks. Use tens rods.
- Make sure students can identify the value of each digit in the numbers they built.

REFLECT

Think Critically

Review students' answers to the Reflect prompt at the bottom of **Student Workbook** page 21, and then review the Engage activity.

▶ **How can you be sure that you are representing a number with the fewest number of base-ten materials possible? Explain your reasoning.** By making sure that appropriate trades are always made; ten unit blocks for one tens rod, ten tens rods for one hundred flat.

Real-World Application

Have students design a sports stadium scoreboard that will allow for two teams' scores to each reach 999.

Students should draw a picture of the scoreboard showing each team name and a different score for each team.

ASSESS

Informal Assessment

Use the online or print Student Record, **Assessment,** page 128, to record informal observations.

The 500 Game

Did the student
- ☐ respond accurately?
- ☐ respond quickly?
- ☐ respond with confidence?
- ☐ self-correct?

Additional Practice

For additional practice, have students complete **Practice,** page 33.

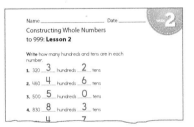

Practice, p. 33

Week 2 • Constructing Whole Numbers to 999

Lesson 2

Key Idea

10 tens = 1 hundred

Try This

Write the number of hundreds and tens shown.

1. __1__ hundreds __4__ tens

Draw a model of the number on each number construction mat.

2. 170

Hundreds	Tens	Ones
Students should draw 1 flat in the hundreds column and 7 rods in the tens column.		

Practice

Write how many hundreds, tens, and ones there are in each number.

3. 280 __2__ hundreds __8__ tens and __0__ ones
4. 1000 __10__ hundreds __0__ tens and __0__ ones
5. 410 __4__ hundreds __1__ tens and __0__ ones

Make trades. Write each number in standard form.

6. 120

Hundreds	Tens	Ones

7. 230

Hundreds	Tens	Ones

Reflect

Start by constructing 50 with base-ten blocks. Add 73 more to your mat. Write how many hundreds, tens, and ones there are altogether. Explain any trading you need to do.

1 hundred, 2 tens, 3 ones; 10 tens must be traded for 1 hundred.

Student Workbook, pp. 20–21

WEEK 2
Constructing Whole Numbers to 999

Lesson 3

Objective
Students can make trades between ones, tens, and hundreds.

Standard
2.NBT.3 Read and write numbers to 1000 using base-ten numerals, number names, and expanded form.

Vocabulary
- digit
- hundreds
- ones
- tens

Creating Context
The number *1* means a single item, yet *ones* refers to single-digit numbers fewer than ten. Ask English Learners to draw a labeled illustration to show the difference between *1* and *ones*.

Materials
Program Materials
- Number Construction Mat, 1 per student
- Number 1–6 Cubes, 1 per group

Additional Materials
- base-ten blocks
- Vocabulary Card 10, *digit*
- Vocabulary Card 23, *hundreds*
- Vocabulary Card 32, *ones*
- Vocabulary Card 45, *tens*

1 WARM UP

Prepare
Gather students around a Number Construction Mat, or draw one on a large sheet of paper with columns for ones, tens, and hundreds. Construct a variety of three-digit numbers using base-ten blocks. Ask students to name each number. Use numbers such as 400, 263, 350, 194, 710, and 502.

Reverse the process. Name a variety of three-digit numbers, and have students construct the numbers. Use numbers such as 500, 623, 280, 999, 720, and 107.

Just the Facts
Tell students to put their hands up if they could trade 10 ones for 1 ten and their heads down if they would not have enough ones to make a trade. Use prompts such as the following:

▶ **You have 1 one and 7 ones.** *heads down*
▶ **You have 5 ones and 8 ones.** *hands up*
▶ **You have 3 ones and 7 ones.** *hands up*

2 ENGAGE

Develop: The 100 Game
"Today we are going to make trades between ones, tens, and hundreds." Follow the instructions on the Activity Card **The 100 Game**. As students complete the activity, be sure to use the Questions to Ask.

Activity Card 2F

Alternative Groupings
Small Group: Each group member makes sure the player whose turn it is makes correct trades.

Individual: During his or her turn, the player will say the new number represented on the mat aloud, as well as any trades made.

Progress Monitoring

If... a student is struggling to understand the build-on concept,

Then... give him or her a set of base-ten blocks and a Number Construction Mat so he or she can practice with an adult outside of class.

Practice
Have students complete **Student Workbook,** pages 22–23. Guide students through the Key Idea example and the Try This exercises.

Interactive Differentiation
Consult the **Teacher Dashboard** for grouping suggestions. You can also use performance on the Engage activity to guide students.

Independent Practice
For additional practice, have students work in pairs. Explain that they will use the Base Ten Tool to count on to make the next hundred. One partner will choose a number less than 1,000, and the other will build that number using the Base Ten Tool. Together, students will decide how many more tens they need in order to make another hundred. Have students add the appropriate number of tens rods and make a trade for 1 hundred flat.

Supported Practice
For additional support, work with students in a small group as they make trades and write two- and three-digit numbers in standard form.

- Model 64 using 6 tens rods and 4 ones. Ask students whether you need to make a trade. No. Then have students write the number in standard form.
- Model 25 using 1 ten rod and 15 ones. Ask students whether you need to make a trade. Yes; trade 10 ones for 1 tens rod. Then have students write the number in standard form.
- Model 112 using 11 tens rods and 2 ones. Ask students whether you need to make a trade. Yes; trade 10 tens rods for 1 hundred flat. Then have students write the number in standard form.
- Model 209 using 1 hundred flat, 9 tens rods, and 19 ones. Ask students whether you need to make a trade. Yes; trade 10 ones for 1 tens rod and 10 tens rods for 1 hundred flat. Then have students write the number in standard form.

REFLECT

Think Critically

Review students' answers to the Reflect prompt at the bottom of *Student Workbook* page 23, and then review the Engage activity.

- **How were the trades similar?** Trades resulted in one flat.
- **How were the trades different?** The first and second trades used only rods. The third trade used unit blocks and rods.

Real-World Application

The fourth-grade classes are going on a field trip. Each train carriage has 100 seats and must be full before the next carriage can begin loading. The classes board the train carriages in the following order.

Class A—28 students, Class B—24 students, Class C—31 students, Class D—29 students, and Class E—28 students.

Students may use base-ten blocks to solve this problem.

- **Which class will be split between Carriage 1 and Carriage 2?** Class D will be split.
- **How many students will be on Carriage 2?** 40

ASSESS

Informal Assessment

Use the online or print Student Record, *Assessment,* page 128, to record informal observations.

The 100 Game

Did the student

☐ respond accurately?   ☐ respond with confidence?

☐ respond quickly?   ☐ self-correct?

Additional Practice

For additional practice, have students complete *Practice,* page 34.

Practice, p. 34

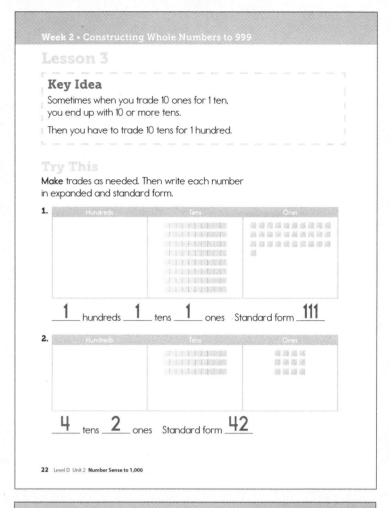

Student Workbook, pp. 22–23

WEEK 2
Constructing Whole Numbers to 999

Lesson 4

Objective
Students can make connections between the base-ten place-value system and our monetary system.

Standard
2.NBT.3 Read and write numbers to 1000 using base-ten numerals, number names, and expanded form.

Vocabulary
- digit
- hundreds
- ones
- tens

Creating Context
When a sentence describes more than one object, the noun is plural. In some languages there is no inflected ending. In English, an *s* is usually added to the end of the word, but some words also require spelling changes. Help English Learners make a chart of the types of U.S. currency and their plurals.

Materials
Program Materials
- Money Construction Mat, 1 per student
- Number 1–6 Cubes, 1 per group

Additional Materials
- play money
- math-link cubes
- Vocabulary Card 10, *digit*
- Vocabulary Card 23, *hundreds*
- Vocabulary Card 32, *ones*
- Vocabulary Card 45, *tens*

1 WARM UP

Prepare
Gather students around a Money Construction Mat, or draw one on a large sheet of paper with columns for pennies, dimes, one-dollar bills, and ten-dollar bills. Use the coin and bill models to make an amount of money. Ask students to name the amount.

Then reverse the process. Name an amount of money. Have a student construct the amount on the mat.

Just the Facts
Have students say the value, in cents, of each coin or bill as you show it to them. Use questions such as the following:

▶ **How much is a penny worth?** 1 cent

▶ **How much is a dime worth?** 10 cents

▶ **How much is a dollar worth?** 100 cents

2 ENGAGE

Develop: The One-Dollar Game

"Today we are going to make connections between the base-ten place-value system and our monetary system." Follow the instructions on the Activity Card **The One-Dollar Game.** As students complete the activity, be sure to use the Questions to Ask.

Activity Card 2G

Alternative Groupings
Whole Class: Display a Money Construction Mat, and have one student demonstrate each example for the class.

Individual: Take turns modeling problems with the student.

Progress Monitoring

If... students do not know the relationships between the coins and dollar bills,

Then... spend time with those students reviewing the relationships using play money.

Practice
Have students complete **Student Workbook,** pages 24–25. Guide students through the Key Idea example and the Try This exercises.

Interactive Differentiation
Consult the **Teacher Dashboard** for grouping suggestions. You can also use performance on the Engage activity to guide students.

Independent Practice

For additional practice, have students use the Coins and Money Tool to practice making trades between pennies and dimes and dimes and dollars. Have students place a set of coins, such as 20 pennies or 13 dimes, on the Number Construction Mat. Tell them to select the appropriate number of coins and combine them to make the trade.

Supported Practice

For additional support, use math-link cubes and play money to help students make connections between the base ten place-value system and our monetary system. Have children count out 16 pennies and 16 math-link cubes. Tell them to place one penny on top of each cube. Point out that the value of each is 1.

▶ **How many math-link cubes would you join together to make a group of 10?** 10

Tell students to link together 10 cubes.

▶ **How many pennies is 1 dime worth?** 10

Have students trade the 10 pennies on top of the linked cubes for 1 dime. Continue practicing making trades with pennies and dimes as well as dimes and dollars as time permits.

126 Level D Unit 2 **Number Sense to 1,000**

REFLECT

Think Critically

Review students' answers to the Reflect prompt at the bottom of **Student Workbook** page 13, and then review the Engage activity.

▶ Can you justify all the trades you make?

▶ How does money use the base-ten money system?

Real-World Application

Select two students to act out the exchange of 100 pennies and 10 dimes. One student is the banker, and the other student is the customer.

For example, 100 pennies for 1 one-dollar bill, 100 dimes for 1 ten-dollar bill, or 10 one-dollar bills for 1 ten-dollar bill.

▶ How would you explain your reasoning to someone who does not understand this concept?

As time permits, try different trading combinations.

ASSESS

Informal Assessment

Use the online or print Student Record, **Assessment,** page 128, to record informal observations.

The One-Dollar Game

Did the student

☐ make important observations?   ☐ provide insightful answers?

☐ extend or generalize learning?   ☐ pose insightful questions?

Additional Practice

For additional practice, have students complete **Practice,** page 35.

Practice, p. 35

WEEK 2
Constructing Whole Numbers to 999

Lesson 5 Review

Objective
Students review skills learned this week and complete the weekly assessment.

Standard
2.NBT.3 Read and write numbers to 1000 using base-ten numerals, number names, and expanded form.

Vocabulary
Review vocabulary introduced during the week.

Creating Context
In English, the letter *c* has several pronunciations. At the beginning of words, it can have a hard /k/ or soft /s/ sound. When the letter *c* comes before *a, o,* or *u*, it makes the hard sound, as in *coin, cat,* and *cut*. When the letter *c* comes before *e* or *i*, it makes the soft sound, as in *cent, cipher,* and *cycle*. Have students make a two-column chart labeled *Hard c* and *Soft c*. Write examples you find during the week.

1 WARM UP

Prepare

▶ **How many tens are in 160?** 16

Name other multiples of 10, and ask students to tell how many tens are in the number.

2 ENGAGE

Practice
Have students complete **Student Workbook,** pages 26–27.

Week 2 • Constructing Whole Numbers to 999

Lesson 5 Review

This week, you explored different ways to represent numbers. You regrouped smaller units into larger units.

Lesson 1 Make trades. Write each number.
1. 72

Hundreds	Tens	Ones

Lesson 2 Make trades. Write each number.
2. 94

Hundreds	Tens	Ones

Lesson 3 Use a number construction mat and base-ten blocks to build each number. Then write how many hundreds, tens, and ones there are altogether. Then write each number.

3. 37 and 28 __0__ hundreds __6__ tens __5__ ones __65__
4. 309 and 88 __3__ hundreds __9__ tens __7__ ones __397__

26 Level D Unit 2 **Number Sense to 1,000**

Lesson 4 Make trades. Write how many one-dollar bills, dimes, and pennies there are altogether.

5. 2 dollars, 30 dimes, 12 pennies
   __5__ one-dollar bills
   __1__ dimes
   __2__ pennies

6. 1 dollar, 50 dimes, 30 pennies
   __6__ one-dollar bills
   __3__ dimes
   __0__ pennies

7. 3 dollars, 24 dimes, 26 pennies
   __5__ one-dollar bills
   __6__ dimes
   __6__ pennies

8. 4 dollars, 40 dimes, 44 pennies
   __8__ one-dollar bills
   __4__ dimes
   __4__ pennies

Reflect
Would you rather have one ten-dollar bill or 1,000 pennies? Use trades to explain your answer.

Answers may vary. Possible answers: Because 1,000 pennies = 10 dollars, it doesn't matter. A ten-dollar bill, because it is easier to carry a ten-dollar bill than 1,000 pennies.

Week 2 Constructing Whole Numbers to 999 • Lesson 5 27

Student Workbook, pp. 26–27

3 REFLECT

Think Critically

Review students' answers to the Reflect prompt at the bottom of *Student Workbook* page 27.

Discuss the answer with the group to reinforce Week 2 concepts

4 ASSESS

Formal Assessment

Students may take the weekly assessment online.

As an alternative, students may complete the weekly test on *Assessment,* pages 27–28. Record progress using the Student Assessment Record, *Assessment,* page 128.

Going Forward

Use the *Teacher Dashboard* to view results of the online assessments, to input the results of print student assessments, and to review progress before making decisions about next steps. Use the weekly test results and observations to determine the next steps for each student.

Retention	
Student displays good grasp of this week's concepts and skills.	Students should use the Coins and Money Tool to practice making trades between pennies and dimes and dimes and dollars for amounts to $9.99.
Remediation	
Student is still struggling with the week's concepts and skills.	Work with students in a small group as they make trades and write two- and three-digit numbers in standard form.
	Make a model of 227 using 22 tens rods and 7 ones. Ask students to make a trade. Then have students write the number in standard form.
	Make a model of 319 using 2 hundred flats, 9 tens rods, and 19 ones. Ask students to make a trade. Then have students write the number in standard form.

Suggestions for Re-Evaluation: If a student has struggled without success for several weeks, use observations and test results to place the student at a level in which he or she can find success and build confidence to move forward.

Name _____ Date _____ **WEEK 2**
Constructing Whole Numbers to 999

Write the tens and ones for each number.

1. 28   __2__ tens and __8__ ones

2. 53   __5__ tens and __3__ ones

Write the hundreds and tens for each number.

3. 680   __6__ hundreds and __8__ tens

4. 720   __7__ hundreds and __2__ tens

Write the hundreds, tens, and ones for the sum.

5. 28 + 43
   __0__ hundreds, __7__ tens, __1__ ones

WEEK 2   Name _____ Date _____
Constructing Whole Numbers to 999

Write the hundreds, tens, and ones for each sum.

6. 59 + 62
   __1__ hundreds, __2__ tens, __1__ ones

7. 87 + 75
   __1__ hundreds, __6__ tens, __2__ ones

Show each amount using the least number of coins and bills.

8. 45 pennies and 3 dimes
   __0__ dollars, __7__ dimes, __5__ pennies

9. 62 pennies and 14 dimes
   __2__ dollars, __0__ dimes, __2__ pennies

10. 3 dollar bills, 27 pennies, 19 dimes
    __5__ dollars, __1__ dimes, __7__ pennies

Assessment, pp. 27–28

Project Preview

This week, students learned the value of each digit within two- and three-digit numbers, how to make trades between ones and tens and tens and hundreds, and how understanding place value can help them understand money. The project for this unit requires students to use what they know about place value to count and trade groups of coins. This week, students will use this knowledge to trade pennies for dimes and dimes for dollars.

Project-Based Learning

Standards-driven Project-Based Learning is effective in building deep content understanding. Project-Based Learning increases long-term retention of concepts and has been shown to be more effective than traditional instruction. By completing a project to answer an essential question, students are challenged to apply and demonstrate mastery of concepts and skills by expressing understanding through discussion, research, and presentation.

Essential Question

WHY do I need to understand place value to use money?

Project Evaluation Criteria

Exceeds Expectations
☐ Project result is explained and can be extended.
☐ Project result is explained in context and can be applied to other situations.
☐ Project result is explained using advanced mathematical vocabulary.
☐ Project result is described, and mathematics are used correctly and can be extended.
☐ Project result is explained and extended, and shows advanced knowledge of mathematical concepts and skills.

Meets Expectations
☐ Project result is explained.
☐ Project result is explained in context.
☐ Project result is explained using mathematical vocabulary.
☐ Project result is described, and mathematics are used correctly.
☐ Project result is explained, and shows satisfactory knowledge of mathematical concepts and skills.

Does Not Meet Expectations
☐ Project result is not explained.
☐ Project result is explained, but out of context.
☐ Project result is explained, but mathematical vocabulary is oversimplified.
☐ Project result is described, but mathematics are not used correctly.
☐ Project result is not explained and/or extended, or shows less than satisfactory knowledge of mathematical concepts and skills.

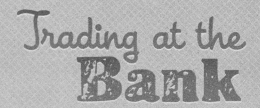

Trading at the Bank

Objective
Students can use place value to make trades between pennies and dimes and dimes and dollars.

Standard
2.NBT.3 Read and write numbers to 1000 using base-ten numerals, number names, and expanded form.

Materials
Additional Materials
- large pieces of construction paper or poster board
- markers or crayons
- play money (sets of pennies, dimes, and dollars)

Best Practices
- Check for student understanding frequently.
- Make efficient use of cooperative learning groups.
- Attend to the varying cognitive styles of individual students.

Introduce

- People who work at a bank use place value just like we do in the classroom.
- How many ones units have the same value as 1 ten rod? 10
- How many pennies have the same value as 1 dime? 10
- How many tens rods have the same value as 1 hundred flat? 10
- How many dimes have the same value as 1 dollar? 10

Explore

- Today you will practice working at a bank by making trades between pennies, dimes, and dollars.
- How do you know when you need to make a trade? I need to make a trade when there are ten or more pennies or dimes.
- Have students work with the same partner they worked with in Week 1.
- Choose one person to be the banker and one person to be the customer. Trade places as you answer the questions.
- Encourage students to use play money or the Coins and Money Tool to model the coins the customer brings to the bank.
- Guide students as they practice trading 10 pennies for 1 dime and 10 dimes for 1 dollar.
- Allow students sufficient time to model and count the given set of coins in each problem.
- Point out that it may be necessary to make trades for both dimes and dollars within the same practice problem.
- Complete *Student Workbook,* page 28, to practice using place value to make trades using coins and dollars.

Wrap Up

- Make sure students are carefully counting sets of 10 before they make a trade.
- If students struggle to make trades, encourage them to break trades into smaller parts by counting sets of 10 one at a time.
- Discuss students' answers to the Reflect prompt at the bottom of *Student Workbook,* page 28.
- Have students work on the sign for their bank. Students might use an electronic design program, or distribute a large piece of construction paper or poster board to each student pair.
- If time permits, have students make a key or chart showing trade equivalents for pennies, dimes, and dollars. For example, they might make a simple key that showed that 10 pennies = 1 dime and that 10 dimes = 1 dollar. Or, partners might make a more detailed chart showing that 10 pennies = 1 dime, 20 pennies = 2 dimes, and so on. Add students' keys or charts to their project folders.

Week 2 • Constructing Whole Numbers to 999

Project
Trading at the Bank

Work with your partner to make trades. Take turns being the banker and the customer.

1. A customer brings 82 pennies to the bank. She wants to trade her pennies for dimes and pennies. How many of each should the banker give her?

   8 dimes and 2 pennies

2. A customer brings in 14 dimes and 20 pennies. He wants to trade for dollars and dimes. How many of each should the banker give him?

   1 dollar and 6 dimes

3. A customer brings in 25 dimes and 60 pennies. She wants to trade for dollars and dimes. How many of each should the banker give her?

   3 dollars and 1 dime

Reflect

How is trading with ones, tens, and hundreds like trading with pennies, dimes, and dollars?

Answers may vary. Possible answer: We trade 10 ones for 1 ten just as we trade 10 pennies for 1 dime. We trade 10 tens for 1 hundred just as we trade 10 dimes for 1 dollar.

28 Level D Unit 2 Number Sense to 1,000

Student Workbook, p. 28

Teacher Reflect

☐ Did I explain what students had to find, make, or do before they began the project?

☐ Did students use their time wisely and effectively?

☐ Was I able to answer questions when students did not understand?

WEEK 3
Representing Number Systems

Week at a Glance

This week, students continue with **Number Worlds,** Level D, Number Sense to 1,000, by examining the place value of two- and three-digit numbers as well as extending place value knowledge to the monetary system.

Skills Focus

- Identify the value of a digit based on its position.
- Apply knowledge of the base-ten system to money.
- Make trades with pennies, dimes, and dollars.

How Students Learn

The grouping concepts that students have been developing can be extended to a variety of everyday applications. Money equivalents are based on groupings of 5 and 10: 5 pennies = 1 nickel, 10 pennies = 1 dime, 10 dimes = 1 dollar. Help students make connections between the place-value concepts they are studying and everyday applications of the concepts.

English Learners ELL

For language support, use the **English Learner Support Guide,** pages 44–45, to preview lesson concepts and teach academic vocabulary.

Math at Home

Give one copy of the Letter to Home, page 9, to each student. Encourage students to share and complete the activity with their caregivers.

Weekly Planner

Lesson	Learning Objectives
1 pages 134–135	Students can determine the value of each digit in a two-digit number.
2 pages 136–137	Students can determine the value of each digit in a three-digit number.
3 pages 138–139	Students can relate base-ten relationships to money by modeling amounts with one-dollar bills, dimes, and pennies.
4 pages 140–141	Students can connect the basics of the base-ten number system to money.
5 pages 142–143	**Review and Assess** Students review skills learned this week and complete the weekly assessment.
Project pages 144–145	Students can use place value to model a given value of money using the fewest number of coins possible.

132 Level D Unit 2 **Number Sense to 1,000**

Key Standard for the Week

Domain: Number and Operations in Base Ten
Cluster: Understand place value.
2.NBT.1 Understand that the three digits of a three-digit number represent amounts of hundreds, tens, and ones; e.g., 706 equals 7 hundreds, 0 tens, and 6 ones.

Materials		Technology
Program Materials • *Student Workbook*, pp. 29–31 • *Practice*, p. 36 • Activity Card 2H, **Make the Most of It** • Number Cards (0–9) • Number Construction Mat	**Additional Materials** base-ten blocks*	*Teacher Dashboard* Building Blocks Number Compare 4 Base 10 Blocks Tool
Program Materials • *Student Workbook*, pp. 32–33 • *Practice*, p. 37 • Activity Card 2I, **Hold My Place** • Number Cards (0–9) • Number Construction Mat	**Additional Materials** base-ten blocks*	*Teacher Dashboard* Base 10 Blocks Tool
Program Materials • *Student Workbook*, pp. 34–35 • *Practice*, p. 38 • Activity Card 2J, **Piggy Bank** • Place-Value Mat • Number Cube (1–6)	**Additional Materials** • base-ten blocks* • play money*	*Teacher Dashboard* Coins and Money Tool
Program Materials • *Student Workbook*, pp. 36–37 • *Practice*, p. 39 • Activity Card 2K, **Trading Cents**	**Additional Materials** • base-ten blocks* • classroom objects • play money* • price tags	*Teacher Dashboard* Coins and Money Tool
Program Materials • *Student Workbook*, pp. 38–39 • Weekly Test, *Assessment*, pp. 29–30		Review previous activities.
Program Materials • *Student Workbook*, p. 40 • Number Construction Mat • Withdrawal Slips, 2 per pair of students	**Additional Materials** play money*	

*Available from McGraw-Hill Education

WEEK 3
Representing Number Systems

Find the Math

In this week, students will learn the value of each digit in two- and three-digit numbers and extend their place-value knowledge to the monetary system.

Use the following to begin a guided discussion:

▶ **How could you trade rolls of pennies for dimes or dollar bills?**
Answers may vary. Possible answer: Trade two rolls of pennies for one dollar bill or one roll of pennies for five dimes.

Have students complete **Student Workbook**, page 29.

Student Workbook, p. 29

Lesson 1

Objective
Students can determine the value of each digit in a two-digit number.

Standard
2.NBT.1 Understand that the three digits of a three-digit number represent amounts of hundreds, tens, and ones; e.g., 706 equals 7 hundreds, 0 tens, and 6 ones.

Vocabulary
- expanded form
- standard form

Creating Context
In English, there are many words that have more than one meaning, such as *greatest*. This can confuse some English Learners when using math terms. An excellent strategy to use with English Learners is to preview the lesson to identify any potentially confusing words or phrases. Then you can introduce those words or explain them while teaching the lesson.

Materials
Program Materials
- Number Cards (0–9), 1 set per student
- Number Construction Mat, 1 per student

Additional Materials
- base-ten blocks

1 WARM UP

Prepare

▶ **Who can show me how to write the number *nineteen*?**

Ask a volunteer to write it on the board. If the standard form of this number (19) is written, tell students that this is the *standard form*.

▶ **Can anyone write this number in a different way that shows how many tens and how many ones are in this number?** 10 + 9

Call on volunteers to write the expanded form on the board and demonstrate it if no one volunteers. Tell students that this is the *expanded form* of writing the number because we are expanding it—making it bigger—to describe how many tens and ones are in the number.

▶ **What is the standard form of 20 + 9?** 29

▶ **What is the expanded form of 74?** 70 + 4

2 ENGAGE

Develop: Make the Most of It

"Today we are going to learn different ways to write numbers." Follow the instructions on the Activity Card **Make the Most of It.** As students complete the activity, be sure to use the Questions to Ask.

Activity Card 2H

Alternative Groupings

Individual: Partner with the student and complete the activity as written.

Progress Monitoring

If... students are not making the greater number,

Then... use base-ten blocks to display the two options.

Practice

Have students complete **Student Workbook,** pages 30–31. Guide students through the Key Idea example and the Try This exercises.

Interactive Differentiation

Consult the **Teacher Dashboard** for grouping suggestions. You can also use performance on the Engage activity to guide students.

Independent Practice

For additional practice, have students use Number Compare 4. In this activity, students compare two cards and choose the one with the greater value.

Supported Practice

For additional support, use the Number Cards and Number Construction Mat from the **Make the Most of It** activity. Have students choose two number cards, such as 4 and 8, and place them on the Number Construction Mat in either order.

▶ **What number did you make?** Answers may vary. Possible answer: 48

Have students construct the number using the counting mat in the Base 10 Blocks Tool. Remind them to combine groups of ten blocks. Then ask questions such as the following:

▶ **The digit 4 is in the tens place. What does that mean? What is the value of the 4?** There are 4 tens in the number 48. Four tens are equal to 40.

▶ **The digit 8 is in the ones place. What does that mean? What is the value of the 8?** There are 8 ones in the number 48. Eight ones are equal to 8.

Guide students as they write the expanded form of each number they create; for example, 40 + 8.

REFLECT

Think Critically

Review students' answers to the Reflect prompt at the bottom of **Student Workbook,** page 31, and then review the Engage activity.

▶ **If I want to make a big number, would I want to use more rods or more blocks?** Answers may vary. Possible answer: more rods

▶ **If I ask you to make the smallest number using the 8 Number Card and the 2 Number Card, which would be in the tens place?** Answers may vary. Possible answer: The 2 Number Card would be in the tens place because that would make 28. Placing the 8 in the tens place would make the number 82.

ASSESS

Informal Assessment

Use the online or print Student Record, **Assessment,** page 128, to record informal observations.

Make the Most of It
Did the student
☐ make important observations?   ☐ provide insightful answers?
☐ extend or generalize learning?   ☐ pose insightful questions?

Additional Practice
For additional practice, have students complete **Practice,** page 36.

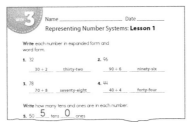
Practice, p. 36

Week 3 • Representing Number Systems

Lesson 1

Key Idea
A place-value chart can help you determine the value of a number.

The number 58 in standard form is 58.

The number 58 in expanded form is 50 + 8.

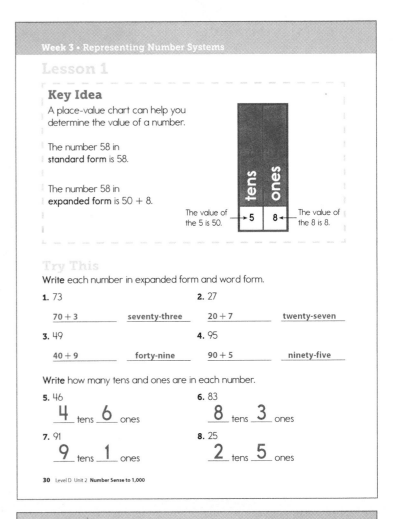

The value of the 5 is 50.   The value of the 8 is 8.

Try This
Write each number in expanded form and word form.

1. 73   2. 27
   70 + 3   seventy-three   20 + 7   twenty-seven

3. 49   4. 95
   40 + 9   forty-nine   90 + 5   ninety-five

Write how many tens and ones are in each number.

5. 46   6. 83
   4 tens **6** ones   **8** tens **3** ones

7. 91   8. 25
   9 tens **1** ones   **2** tens **5** ones

30 Level D Unit 2 Number Sense to 1,000

Practice
Write the value of each digit that makes up the number. Put the largest value on the top line and the smallest value on the bottom line.

9. 76   10. 18   11. 94   12. 82
   70   10   90   80
   6   8   4   2

Write the place value of each underlined digit. Then write the value of that digit.

13. 6<u>7</u>   14. <u>8</u>6   15. 9<u>4</u>   16. <u>4</u>3
    tens   ones   ones   tens
    60   6   4   40

Fill in each blank with the correct value.

17. 30 + **4** = 34   18. **70** + 5 = 75

Reflect
In 2012, Ryan Braun hit 41 home runs. Write the number of home runs Braun hit in expanded and word form.

40 + 1
forty-one

Week 3 Representing Number Systems • Lesson 1 31

Student Workbook, pp. 30–31

WEEK 3
Representing Number Systems

Lesson 2

Objective
Students can determine the value of each digit in a three-digit number.

Standard
2.NBT.1 Understand that the three digits of a three-digit number represent amounts of hundreds, tens, and ones; e.g., 706 equals 7 hundreds, 0 tens, and 6 ones.

Vocabulary
- expanded form
- standard form

Creating Context
English Learners need plenty of opportunity to practice speaking the new language. An excellent way to provide additional practice is through small cooperative groups. When English Learners are at beginning proficiency levels, it may be helpful to have another student who speaks the same primary language as a partner in order to encourage discussion of concepts in a low-stress environment.

Materials
Program Materials
- Number Cards (0–9), 1 set per student
- Number Construction Mat, 1 per student

Additional Materials
base-ten blocks

Prepare Ahead
Prepare the Number Construction Mat for classroom display.

1 WARM UP

Prepare
Display a Number Construction Mat for students to see.

▶ **We can use charts to help us find values of each digit in a number quickly. Columns labeled *hundreds*, *tens*, and *ones* can help you quickly see the value of each place.**

Using the number 235, have students tell you how to fill out the chart.

Just the Facts
Have students use mental math to answer questions such as the following:

▶ What is 50 + 6? 56
▶ What is 72 in expanded form? 70 + 2
▶ What is 51 in expanded form? 50 + 1

2 ENGAGE

Develop: Hold My Place
"Today we are going to look at how important 0 can be." Follow the instructions on the Activity Card **Hold My Place.** As students complete the activity, be sure to use the Questions to Ask.

Activity Card 21

Alternative Groupings
Whole Class: Complete the activity as written.

Progress Monitoring

If… students do not understand that 0 is a placeholder,

Then… use three students standing side by side to demonstrate a three-digit number. Have the "hundreds" student hold up a 4, the "tens" student a 10, and the "ones" student a 7. Trade the tens for a hundred. Now the "hundreds" student holds a 5 and the "tens" student holds a 0.

Practice
Have students complete **Student Workbook,** pages 32–33. Guide students through the Key Idea example and the Try This exercises.

Interactive Differentiation
Consult the **Teacher Dashboard** for grouping suggestions. You can also use performance on the Engage activity to guide students.

Independent Practice
For additional practice, have students use the Base 10 Blocks Tool.

- Have students draw three Number Cards and construct the greatest and least possible numbers. They must use all three cards.
- Then have students show each number on the counting mat in the Base 10 Blocks Tool.
- Students should write each number in standard and expanded form.
- Encourage students to pay attention to the digit 0 as a placeholder and the value it represents.

Supported Practice
For additional support, have students use the Base 10 Blocks Tool to model the number 248.

▶ **There are 8 ones. Let's add two more. How many are there now?** 10
▶ **Since we have 10 ones, what can we do?** Trade 10 ones for 1 tens rod.

Guide students as they make a trade.

▶ **How many ones units are there now?** 0

Have students write the new number, 250, in both standard and expanded form. Point out that the ones place value cannot be left blank because 25 and 250 do not have the same value. Show students how to use a 0 to indicate that there are no ones units in the number 250. If time permits, have students add tens rods until they can trade 10 tens rods for 1 hundred flat. Have them write the number 300 in standard and expanded form. Discuss the function of each zero in 300.

136 Level D Unit 2 **Number Sense to 1,000**

REFLECT

Think Critically

Review students' answers to the Reflect prompt at the bottom of *Student Workbook,* page 33, and then review the Engage activity.

Display a Number Construction Mat for students to see.

▶ We can use charts to help us find values of each digit in a number quickly. Columns labeled *hundreds, tens,* and *ones* can help you quickly see the value of each place.

Using the number 235, have students tell you where to place each digit on the mat.

Real-World Application

▶ Sometimes it is helpful to use a Number Construction Mat to think of an amount in expanded form. Envelopes sometimes come in boxes of 100. If you need 142 envelopes, you could think of 142 as 1 hundred, 40 tens, and 2 ones. Then you know that you need to buy 1 pack of 100 and 1 more pack for the 42 additional envelopes.

ASSESS

Informal Assessment

Use the online or print Student Record, *Assessment,* page 128, to record informal observations.

Hold My Place

Did the student

☐ make important observations? ☐ provide insightful answers?

☐ extend or generalize learning? ☐ pose insightful questions?

Additional Practice

For additional practice, have students complete *Practice,* page 37.

Practice, p. 37

Week 3 • Representing Number Systems

Lesson 2

Key Idea

A place-value chart can include the hundreds place.

The number 143 in standard form is 143.

The number 143 in expanded form is $100 + 40 + 3$.

The number 143 in word form is one hundred forty-three.

The value of the 1 is 100. → 1 | 4 | 3 ← The value of the 3 is 3.
The value of the 4 is 40.

Try This

Write each number in expanded form and word form.

1. 347
   $300 + 40 + 7$
   three hundred forty-seven

2. 934
   $900 + 30 + 4$
   nine hundred thirty-four

3. 289
   $200 + 80 + 9$
   two hundred eighty-nine

4. 563
   $500 + 60 + 3$
   five hundred sixty-three

32 Level D Unit 2 Number Sense to 1,000

Write the hundreds, tens, and ones in each number.

5. 854
   __8__ hundreds __5__ tens __4__ ones

6. 576
   __5__ hundreds __7__ tens __6__ ones

Practice

Write the place value of each underlined digit. Then write the value of each underlined digit.

7. 5 4̲ 6   8. 3̲ 2 5   9. 4 7̲ 5   10. 7̲ 9 7

   ones   hundreds   tens   hundreds
   __6__   __300__   __70__   __700__

Fill in each blank with the correct value.

11. $700 + 80 + \underline{1} = 781$
12. $\underline{400} + 50 + 2 = 452$

Reflect

The revolution of Mars around the sun lasts 687 days. Write the value of each digit in the number 687.

__600__ __80__ __7__

Which digit is in the tens place in the number 687?
__8__

Week 3 Representing Number Systems • Lesson 2 33

Student Workbook, pp. 32–33

WEEK 3
Representing Number Systems

Lesson 3

Objective
Students can relate base-ten relationships to money by modeling amounts with one-dollar bills, dimes, and pennies.

Standard
2.NBT.1 Understand that the three digits of a three-digit number represent amounts of hundreds, tens, and ones; e.g., 706 equals 7 hundreds, 0 tens, and 6 ones.

Creating Context
In the United States, we place the decimal point between the dollars and cents when we write a number. Some families who have lived outside the United States may write amounts of money using a comma instead of a decimal point. Make sure to share with families how dollar amounts are written in the United States.

Materials
Program Materials
- Place-Value Mat
- Number Cube (1–6)

Additional Materials
- base-ten blocks
- play money, 10 dollars, 40 dimes, and 262 pennies per group

1 WARM UP

Prepare
Show the class a penny, a dime, and a dollar bill.

▶ **These units of money represent the base-ten system.**
▶ **How much is a penny worth?** one cent
▶ **How much is a dime worth?** ten cents
▶ **How much is a dollar worth?** one hundred cents
▶ **How many pennies are in a dime?** 10 pennies
▶ **How many dimes are in a dollar?** 10 dimes

Just the Facts
Play a game of "Simon Says" using questions such as the following:

▶ **Simon says if 43 = 40 + 3, clap your hands.**
Students should clap.

▶ **Simon says if 312 = 300 + 10 + 2, hold your nose.**
Students should hold their noses.

▶ **Simon says if 506 = 500 + 60, clap your hands.**
Students should be still.

2 ENGAGE

Develop: Piggy Bank
"Today we are going to solve problems that involve money." Follow the instructions on the Activity Card **Piggy Bank**. As students complete the activity, be sure to use the Questions to Ask.

Activity Card 2J

Alternative Groupings
Individual: Assist the student with the problems as needed.

Progress Monitoring
If... students cannot associate money with the base-ten system,

Then... use the Place-Value Mat to trade pennies for a dime and dimes for one dollar.

Practice
Have students complete **Student Workbook,** pages 34–35. Guide students through the Key Idea example and the Try This exercises.

Interactive Differentiation
Consult the **Teacher Dashboard** for grouping suggestions. You can also use performance on the Engage activity to guide students.

Independent Practice
For additional practice, provide pairs of students with play money. Give two dollars, 14 dimes, and 42 pennies to each pair. One student will name an amount between one and two dollars, such as $1.53. The other student will use play money to model it. Together, pairs will then model the same amount using a different group of coins or bills. Then have partners switch roles.

Supported Practice
For additional support, have students use base-ten blocks and the Coins and Money Tool. Have students roll a number cube three times. The first roll will represent the hundreds, the second the tens, and the third the ones. Instruct students to first build the number using base-ten blocks.

▶ **How many hundred flats do you have? What is the value of the flats?**
Answers may vary. Possible answer: Three hundred flats have a value of 300.

▶ **Each hundred flat has a value of 100. Does a penny, a dime, or a dollar have a value of 100?** A dollar is worth 100 cents.

Help students use the Coins and Money Tool to model the amount using dollar bills, dimes, and pennies.

3 REFLECT

Think Critically

Review students' answers to the Reflect prompt at the bottom of **Student Workbook,** page 35, and then review the Engage activity.

Discuss the way monetary amounts are written. The decimal is placed between the dollars and cents.

▶ **How is counting money similar to what we have done with numbers base-ten?** Answers will vary. Possible answer: We found hundreds, tens, and ones.

▶ **What would one dollar be equal to in the base-ten system?** Answers will vary. Possible answer: It would be equal to one flat.

Real-World Application

▶ **It is important to know the value of each unit of money and how many of each unit to give to make a certain amount. How could you pay exact change for a $3.82 gallon of milk using one-dollar bills, dimes, and pennies?** 3 one-dollar bills, 8 dimes, 2 pennies

4 ASSESS

Informal Assessment

Use the online or print Student Record, **Assessment,** page 128, to record informal observations.

Piggy Bank

Did the student

☐ respond accurately?   ☐ respond with confidence?

☐ respond quickly?   ☐ self-correct?

Additional Practice

For additional practice, have students complete **Practice,** page 38.

Practice, p. 38

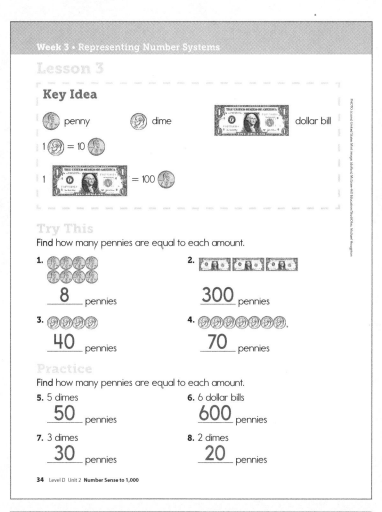

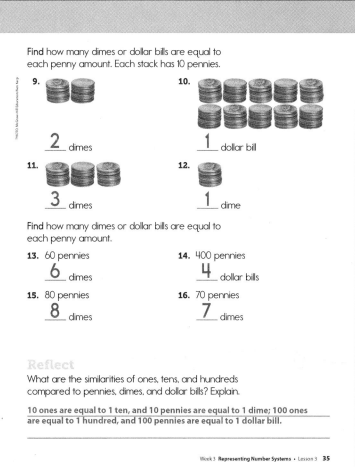

Student Workbook, pp. 34–35

WEEK 3
Representing Number Systems

Lesson 4

Objective
Students can connect the basics of the base-ten number system to money.

Standard
2.NBT.1 Understand that the three digits of a three-digit number represent amounts of hundreds, tens, and ones; e.g., 706 equals 7 hundreds, 0 tens, and 6 ones.

Creating Context
Work with English Learners to identify the dollar sign and the cent sign. Find out what sign is used in another country when money values are written.

Materials
Additional Materials
- base-ten blocks
- classroom objects, 1 per student
- play money, 3 dollars, 18 dimes, and 30 pennies per pair
- price tags, 1 per student

Prepare Ahead
Collect several copies of a weekly grocery store circular. Be sure to remove any inappropriate advertisements.

1 WARM UP

Prepare
Use base-ten blocks to review the fundamentals of regrouping and the concept of trading ones for tens in the base-ten system.

Just the Facts
Have students practice counting up and down by ten cents. Tell them to finish the pattern by saying the next number. Use prompts such as the following:

▶ 10¢, 20¢, 30¢, 40¢, _____ 50¢
▶ 60¢, 50¢, 40¢, 30¢, _____ 20¢
▶ 25¢, 35¢, 45¢, 55¢, _____ 65¢

2 ENGAGE

Develop: Trading Cents
"Today we are going to solve problems using coins and dollar bills." Follow the instructions on the Activity Card **Trading Cents**. As students complete the activity, be sure to use the Questions to Ask.

Activity Card 2K

Alternative Groupings
Individual: Partner with the student and complete the activity as written.

Progress Monitoring
If... students understand trading pennies, dimes, and dollar bills,

Then... play a game of "Guess the Question." The answer is 15 pennies. Guess the Question. "What is the value of 1 dime and 5 pennies?"

Practice
Have students complete **Student Workbook,** pages 36–37. Guide students through the Key Idea example and the Try This exercises.

Interactive Differentiation
Consult the **Teacher Dashboard** for grouping suggestions. You can also use performance on the Engage activity to guide students.

Independent Practice
Provide each student with a weekly circular from a local grocery store. Instruct students to circle the sales price of several items in the circular. Then have students use the Coins and Money Tool to show the price. Encourage them to model the price using the fewest possible one-dollar bills, dimes, and pennies.

Supported Practice
For additional support, have students use the Coins and Money Tool. Give each student a weekly circular from a local grocery store.

- Have each student circle the sales price of one item in the circular.
- Guide students as they use the Coins and Money Tool to show the price. At this point, do not ask students whether they have used the fewest number of coins possible.
- ▶ **Can you make a group of ten with your pennies? How many groups can you make?** Answers may vary. Possible answer: Yes; 1 group of ten pennies
- ▶ **Which coin has the same value as ten pennies?** a dime
- Have students who can make a group of ten pennies use the arrow to select the pennies. Then they should use the combine button to trade the pennies for a dime.
- ▶ Continue the activity by guiding students as they trade dimes for dollars. Have students check to make sure that they are using the fewest number of coins possible.

140 Level D Unit 2 **Number Sense to 1,000**

REFLECT

Think Critically

Review students' answers to the Reflect prompt at the bottom of **Student Workbook,** page 37, and then review the Engage activity.

▶ **Which is larger, 2 one-dollar bills or 12 dimes?** 2 one-dollar bills
Possible answer: 12 dimes is the same as 1 one-dollar bill and 2 dimes

▶ **What does trading dimes for dollars remind you of in the base-ten number system?** Possible answer: trading tens for hundreds

Real-World Application

▶ **Imagine that an arcade game you want to play takes only dimes. Explain the trade you need to make with the cashier to have dimes to play the game if you have 50 pennies.** Trade 50 pennies for 5 dimes.

ASSESS

Informal Assessment

Use the online or print Student Record, **Assessment,** page 128, to record informal observations.

Trading Cents

Did the student

☐ make important observations?   ☐ provide insightful answers?

☐ extend or generalize learning?   ☐ pose insightful questions?

Additional Practice

For additional practice, have students complete **Practice,** page 39.

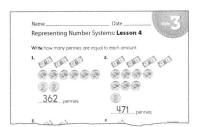

Practice, p. 39

Week 3 • Representing Number Systems

Lesson 4

Key Idea

= 347 pennies

Try This
Find how many pennies are equal to each amount.

1. __253__ pennies
2. __524__ pennies
3. __186__ pennies
4. __476__ pennies

Practice
Find how many pennies are equal to each amount.

5. 4 dollar bills, 3 dimes, and 8 pennies
__438__ pennies

6. 9 dimes and 2 pennies
__92__ pennies

36 Level D Unit 2 **Number Sense to 1,000**

Draw dollar bills, dimes, and pennies to equal each amount.

7. 162 pennies

> Students should draw 1 dollar bill, 6 dimes, and 2 pennies.

8. 304 pennies

> Students should draw 3 dollar bills and 4 pennies.

Fill in each blank with the correct value.

9. 3 dollar bills + __9__ dimes + 2 pennies = 392 pennies

10. __6__ dollar bills + 3 dimes + 4 pennies = 634 pennies

Reflect
Write the number of dollar bills, dimes, and pennies shown.

__2__ dollar bills, __6__ dimes, __3__ pennies

How many pennies does this amount equal?
__263__ pennies

Week 3 Representing Number Systems • Lesson 4   37

Student Workbook, pp. 36–37

Week 3 **Representing Number Systems** • Lesson 4   141

WEEK 3
Representing Number Systems

Lesson 5 Review

Objective
Students review skills learned this week and complete the weekly assessment.

Standard
2.NBT.1 Understand that the three digits of a three-digit number represent amounts of hundreds, tens, and ones; e.g., 706 equals 7 hundreds, 0 tens, and 6 ones.

Vocabulary
Review vocabulary introduced during the week.

Creating Context
Review with English Learners the name of each United States coin and its value. Show students how to do a rubbing of the coin by placing the coin under paper and then rubbing over the face of the coin. Make a sheet for each student with the various values listed.

1 WARM UP

Prepare
Select students to model three-digit numbers.

- Show me the number 415 using base-ten blocks.
- Show me 415 cents using dollars, dimes, and pennies.
- Show me the number 169 using base-ten blocks.
- Show me 169 cents using dollars, dimes, and pennies.
- Show me the number 827 using base-ten blocks.
- Show me 827 cents using dollars, dimes, and pennies.

2 ENGAGE

Practice
Have students complete **Student Workbook,** pages 38–39.

Week 3 • Representing Number Systems

Lesson 5 Review

This week you explored the value of each digit in a number. You learned how skip counting can help you name the value of a digit. You also learned that you can make a model of numbers using money.

Lesson 1 Write how many tens and ones are in each number.

1. 64
   __6__ tens __4__ ones
2. 27
   __2__ tens __7__ ones
3. 82
   __8__ tens __2__ ones
4. 36
   __3__ tens __6__ ones

Lesson 2 Write the place of each underlined digit. Then write the value of each underlined digit.

5. 12<u>3</u>
   __ones__
   __3__
6. <u>4</u>63
   __hundreds__
   __400__

Lesson 3 Write how many pennies are equal to each amount.

7. 6 dimes __60__ pennies
8. 7 dollar bills __700__ pennies
9. 9 dimes __90__ pennies
10. 4 dollar bills __400__ pennies

Write how many dimes or dollar bills are equal to each amount.

11. 10 pennies __1__ dime
12. 800 pennies __8__ dollar bills

Lesson 4 Fill in each blank with the correct value.

13. 9 dollar bills + __3__ dimes + 8 pennies = 938 pennies
14. __4__ dollar bills + 1 dime + 6 pennies = 416 pennies

Draw dollar bills, dimes, and pennies to equal each amount.

15. 395 pennies

   Students should draw 3 dollar bills, 9 dimes, and 5 pennies.

16. 462 pennies

   Students should draw 4 dollar bills, 6 dimes, and 2 pennies.

Reflect
Write the amount for 8 pennies, 3 dimes, and 4 dollars in pennies.
__438__ pennies

Student Workbook, pp. 38–39

3 REFLECT

Think Critically

Review students' answers to the Reflect prompt at the bottom of **Student Workbook,** page 39.

Discuss the answer with the group to reinforce Week 3 concepts.

4 ASSESS

Formal Assessment

Students may take the weekly assessment online.

As an alternative, students may complete the weekly test on **Assessment,** pages 29–30. Record progress using the Student Assessment Record, **Assessment,** page 128.

Going Forward

Use the **Teacher Dashboard** to view results of the online assessments, to input the results of print student assessments, and to review progress before making decisions about next steps. Use the weekly test results and observations to determine the next steps for each student.

Retention	
Student displays good grasp of this week's concepts and skills.	Have each student imagine they have a $350 online store gift card. Instruct students to select three different gifts they could afford. Then have students model the price of each gift using the fewest dollar bills, dimes, and pennies possible.

Remediation	
Student is still struggling with the week's concepts and skills.	Have students write the number 350, in both standard and expanded form. Remind students that the ones place value cannot be left blank because 35 and 350 do not have the same value. Show students how to use a 0 to indicate that there are no ones units in the number 350. Have students add base-ten rods until they can trade 10 base-ten tens rods for a 1 base-ten hundred flat. Have them write the number 400 in standard and expanded form. Discuss the function of each zero in 400.

Suggestions for Re-Evaluation: If a student has struggled without success for several weeks, use observations and test results to place the student at a level in which he or she can find success and build confidence to move forward.

Name _____ Date _____
Representing Number Systems

1. Write the number forty-six. __46__

2. Write the number eighty-seven. __87__

3. Write the name of the number 28. __twenty-eight__

4. Write the name of the number 92. __ninety-two__

5. Write the number that is the same as 30 + 5. __35__

6. Write the expanded form of 54. __50 + 4__

7. What is the value of the digit 7 in 79? __70__

8. What is the value of the digit 6 in 46? __6__

9. Write the number six hundred fourteen. __614__

10. Write the number two hundred fifty-eight. __258__

Level D Unit 2 Week 3 29

Name _____ Date _____
Representing Number Systems

11. Write the name of the number 631.
    __six hundred thirty-one__

12. What is the value of the digit 4 in 543? __40__

13. What is the value of the digit 5 in 543? __500__

14. What is the value of the digit 3 in 543? __3__

15. Write the value of each digit in the number 382.
    __300__ __80__ __2__

16. How many pennies are in 7 dimes? __70__

17. How many dollar bills are equal to 600 pennies? __6__

18. How many dollar bills are equal to 90 dimes? __9__

19. What is the value of 532 pennies in dollar bills, dimes, and pennies?
    __5__ dollar bills, __3__ dimes, __2__ pennies

20. How many pennies are equal to 6 dollar bills, 4 dimes, and 7 pennies? __647__

30 Level D Unit 2 Week 3

Assessment, pp. 29–30

Project Preview

This week, students learned how to determine the values of digits in two- and three-digit numbers and relate the values to money. In the project for this unit, students must use what they know about place value to count and make trades with pennies, dimes, and dollars. Students will count out a given amount of money using the fewest number of coins possible.

Project-Based Learning

Standards-driven Project-Based Learning is effective in building deep content understanding. Project-Based Learning increases long-term retention of concepts and has been shown to be more effective than traditional instruction. By completing a project to answer an essential question, students are challenged to apply and demonstrate mastery of concepts and skills by expressing understanding through discussion, research, and presentation.

Essential Question

WHY do I need to understand place value to use money?

Project Evaluation Criteria

Review project evaluation criteria with students prior to beginning the project.

Exceeds Expectations
- ☐ Project result is explained and can be extended.
- ☐ Project result is explained in context and can be applied to other situations.
- ☐ Project result is explained using advanced mathematical vocabulary.
- ☐ Project result is described, and mathematics are used correctly and can be extended.
- ☐ Project result is explained and extended, and shows advanced knowledge of mathematical concepts and skills.

Meets Expectations
- ☐ Project result is explained.
- ☐ Project result is explained in context.
- ☐ Project result is explained using mathematical vocabulary.
- ☐ Project result is described, and mathematics are used correctly.
- ☐ Project result is explained, and shows satisfactory knowledge of mathematical concepts and skills.

Does Not Meet Expectations
- ☐ Project result is not explained.
- ☐ Project result is explained, but out of context.
- ☐ Project result is explained, but mathematical vocabulary is oversimplified.
- ☐ Project result is described, but mathematics are not used correctly.
- ☐ Project result is not explained and/or extended, or shows less than satisfactory knowledge of mathematical concepts and skills.

Counting at the Bank

Objective
Students can use place value to model a given value of money using the fewest number of coins possible.

Standard
2.NBT.1 Understand that the three digits of a three-digit number represent amounts of hundreds, tens, and ones.

Materials

Program Materials
- Number Construction Mat
- Withdrawal Slip

Additional Materials
- play money

Prepare Ahead
Cut apart the Withdrawal Slips so that you can hand six slips to each student pair.

Best Practices
- Provide students the opportunity to set goals, share information, and self-evaluate.
- Check for student understanding frequently.
- Evaluate through observational records and performance assessment.

Introduce

▶ **You have been learning about what it is like to work at a bank.**

▶ **What do customers keep at a bank?** money

▶ **What do you think customers can do if they want to use some of their money in the bank to buy something they want?** Answers may vary. Possible answer: They can take money out of the bank.

▶ **When customers take money out of an account at the bank, they withdraw money from the bank.** *Withdraw* **means that the bank gives them back some of their money. Customers write down the amount of money they want to take out of their account on a withdrawal slip.**

• Show students a blank withdrawal slip and explain how to complete it.

Explore

▶ **Today you will practice giving customers bills and coins to match the value on their withdrawal slip.**

• Hand six blank withdrawal slips to each pair of students (two sheets cut into individual slips). Have students write down any three-digit number in the blank box next to "cents" on each slip.

▶ **Take turns with your partner being the banker and the customer. When you are the customer, hand your filled-out withdrawal slip to the banker. When you are the banker, give your customer the amount of money he or she would like to withdraw using the fewest coins possible.**

• Instruct students who are the customers to check to make sure their banker gave them the correct amount of money using the fewest number of coins possible.

• Allow students adequate time to practice being both the banker and the customer.

▶ **Complete** *Student Workbook,* **page 40, to practice using the fewest number of coins possible.**

Wrap Up

• Make sure students are checking their work as they go along.

• If students have difficulty using the fewest number of coins possible, have them place the coins on a Number Construction Mat. Guide students to make groups of ten and then make the appropriate trades.

▶ **Bankers and other people who work with money need to understand place value in order to count and trade coins and bills.**

• Discuss students' answers to the Reflect prompt at the bottom of *Student Workbook,* page 40.

• Students may use the Coins and Money Tool to show the correct number of bills and coins as they complete the workbook page.

• If time permits, have pairs of students exchange their withdrawal slips with another student pair. They can practice counting out the coins for the withdrawals on the other slips using the fewest number of coins possible.

Week 3 • Representing Number Systems

Project
Counting at the Bank

Practice being a banker by giving customers the amount of money they want to withdraw using the fewest number of bills and coins possible. Use dollars, dimes, and pennies.

1. A customer wants to withdraw 219 cents. How can you make 219 cents using the fewest number of bills and coins possible?

   2 dollars, 1 dime, and 9 pennies

2. A customer wants to withdraw 305 cents. How can you make 305 cents using the fewest number of bills and coins possible?

   3 dollars and 5 pennies

3. A customer wants to withdraw 558 cents. Your banking partner gives the customer 5 dollars and 58 pennies. Did your partner give the customer the fewest number of bills and coins possible? How do you know?

   No; 58 pennies can be traded for 5 dimes and 8 pennies.

Reflect
How can you check to make sure you have the fewest number of coins possible?

Answers may vary. Possible answer: If there are no groups of ten, I have the fewest number of coins possible.

40 Level D Unit 2 Number Sense to 1,000

Student Workbook, p. 40

Teacher Reflect

☐ Did I define vocabulary words clearly and correctly?

☐ Did I explain the directions before students began their projects?

☐ Did students focus on the major concept of the activity?

WEEK 4: Place Value to 1,000

Week at a Glance

This week, students continue with **Number Worlds,** Level D, Number Sense to 1,000 by investigating the meaning of place value to the thousands place. Students will name and make models of numbers shown with base-ten blocks and continue to gain understanding of the role that regrouping plays in place value.

Skills Focus

- Use base-ten blocks to model place value.
- Trade and regroup numbers through the hundreds place.
- Make a model of a number in two or more ways using base-ten blocks.

How Students Learn

Students' knowledge of number and quantity becomes more integrated as their experiences in mathematics continue. As students continue to use manipulatives and write number representations, they will begin to link numbers to quantities and realize that questions about numbers can be answered with or without the use of concrete objects.

English Learners ELL

For language support, use the **English Learner Support Guide,** pages 46–47, to preview lesson concepts and teach academic vocabulary. **Number Worlds** Vocabulary Cards are listed as additional materials in many lessons and can be used to preteach and reinforce academic vocabulary.

Math at Home

Give one copy of the Letter to Home, page 10, to each student. Encourage students to share and complete the activity with their caregivers.

Weekly Planner

Lesson	Learning Objectives
1 pages 148–149	Students can name and make models of three-digit numbers with base-ten blocks.
2 pages 150–151	Students can trade ones for tens and tens for hundreds.
3 pages 152–153	Students can compare numbers with the same number of digits and can identify the greater or lesser value.
4 pages 154–155	Students can compare numbers with different numbers of digits and can identify the greater or lesser value.
5 pages 156–157	**Review and Assess** Students review skills learned this week and complete the weekly assessment.
Project pages 158–159	Students can use place value to count and compare two sets of coins to find the greater value.

Key Standard for the Week

Domain: Number and Operations in Base Ten
Cluster: Understand place value.
2.NBT.3 Read and write numbers to 1000 using base-ten numerals, number names, and expanded form.

Materials		Technology
Program Materials • **Student Workbook**, pp. 41–43 • **Practice**, p. 40 • Activity Card 2L, **Number Construction** • Number Construction Mat • Number Cards (0–9)	*Additional Materials* • base-ten blocks* • math-link cubes*	*Teacher Dashboard* Base 10 Blocks Tool
Program Materials • **Student Workbook**, pp. 44–45 • **Practice**, p. 41 • Activity Card 2M, **Trading Post** • Number Cards (0–40)	*Additional Materials* base-ten blocks, 1 flat, 9 rods, and 26 units per pair*	*Teacher Dashboard* Base 10 Blocks Tool
Program Materials • **Student Workbook**, pp. 46–47 • **Practice**, p. 42 • Activity Card 2N, **Comparison Modeling** • Number Construction Mat	*Additional Materials* • base-ten blocks, 4 flats, 10 rods, and 30 units per pair* • Vocabulary Card 21, *greater than* • Vocabulary Card 24, *less than*	*Teacher Dashboard* Base 10 Blocks Tool
Program Materials • **Student Workbook**, pp. 48–49 • **Practice**, p. 43 • Activity Card 2O, **Comparing Numbers** • Number Construction Mat	*Additional Materials* • unlined paper • Vocabulary Card 21, *greater than* • Vocabulary Card 24, *less than*	*Teacher Dashboard* Base 10 Blocks Tool
Program Materials • **Student Workbook**, pp. 50–51 • Weekly Test, **Assessment**, pp. 31–32		Review previous activities.
Program Materials • **Student Workbook**, p. 52 • Deposit Slips • Place Value Mat	*Additional Materials* play money*	

*Available from McGraw-Hill Education

WEEK 4
Place Value to 1,000

Find the Math

In this week, students use place value to compare and order three-digit numbers.

Use the following to begin a guided discussion:

▶ **How can you use place value to find whether there is more money on the table or in the jar?** Answers may vary. Possible answer: Count the money on the table and the money in the jar. Look at the hundreds place first because hundreds have the greatest value. The value with more hundreds is greater.

Have students complete **Student Workbook,** page 41.

Student Workbook, p. 41

Lesson 1

Objective
Students can name and make base-ten block models of three-digit numbers.

Standard
2.NBT.3 Read and write numbers to 1000 using base-ten numerals, number names, and expanded form.

Creating Context
Some English Learners and others who have traveled outside the United States may enjoy sharing coins from other countries. Invite students to ask their caregivers for permission to bring in these coins. Ask them to identify the names of the currencies and the denominations. Have students discuss how they are similar to and different from United States coins.

Materials
Program Materials
- Number Construction Mat, 1 per student
- Number Cards (0–9)

Additional Materials
- base-ten blocks, 2 flats, 5 rods, and 30 units per student
- math-link cubes, 20 per student

1 WARM UP

Prepare
Show students a flat from the base-ten blocks.

▶ **This is called a flat. It shows one hundred. How many rods does it take to make a flat?** ten

Have students check their answers by placing rods on top of the flat. Point out that ten rods equal one flat.

Show students a cube from the base-ten blocks.

▶ **This is called a cube. It shows 1,000. How many flats does it take to make a cube?** ten

Have students check their answers by placing flats on top of one another and comparing them with the cube. Point out that ten flats equal one cube.

2 ENGAGE

Develop: Number Construction

"Today we are going to model three-digit numbers with base-ten blocks." Follow the instructions on the Activity Card **Number Construction.** As students complete the activity, be sure to use the Questions to Ask.

Alternative Groupings

Individual: Partner with the student and complete the activity as written.

Activity Card 2L

Progress Monitoring

If... students need the game to be more challenging,

Then... restrict the type of questions that can be asked to "yes or no" questions.

Practice

Have students complete **Student Workbook,** pages 42–43. Guide students through the Key Idea example and the Try This exercises.

Interactive Differentiation

Consult the **Teacher Dashboard** for grouping suggestions. You can also use performance on the Engage activity to guide students.

Independent Practice

For additional practice, have students use the Base 10 Blocks Tool to model three-digit numbers. Give each student a set of Number Cards (0–9). Have students draw three Number Cards from the pile and place them in any order. Students will model the number in the Base 10 Blocks Tool on the counting mat in the format area. Tell students to check their models by looking at the bottom of the screen. The Base 10 Blocks Tool shows the modeled number in expanded and standard form.

148 Level D Unit 2 **Number Sense to 1,000**

Supported Practice

For additional support, give pairs of students a set of Number Cards (0–9), a Number Construction Mat, and base-ten blocks.

- Show students how to model 217 using base-ten blocks. Write 217 on the board in standard and expanded form: 217 = 200 + 10 + 7.
- Explain that one partner will draw three Number Cards and place them in any order. That partner will then build the number with base-ten blocks on the Number Construction Mat.
- The other will write down the number that the first partner constructed.
- Have partners work together to determine the value of each digit. Then they should write the number in expanded form.
- Compare students' models to the number they wrote in expanded and standard form.

REFLECT

Think Critically

Review students' answers to the Reflect prompt at the bottom of **Student Workbook,** page 43, and then review the Engage activity.

Count with students the blocks they have drawn by hundreds, then tens, then ones.

▶ **How did you know how many flats to draw?** Answers may vary. Possible answer: Flats stand for the hundreds place, and there are 3 hundreds, so I drew 3 flats.

▶ **How many rods do you draw when there is a 0 in the tens place, such as in the number 309?** Answers may vary. Possible answer: None, a 0 in the tens place means you have no tens.

ASSESS

Informal Assessment

Use the online or print Student Record, **Assessment,** page 128, to record informal observations.

Number Construction

Did the student

☐ respond accurately? ☐ respond with confidence?

☐ respond quickly? ☐ self-correct?

Additional Practice

For additional practice, have students complete **Practice,** page 40.

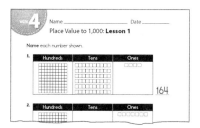

Practice, p. 40

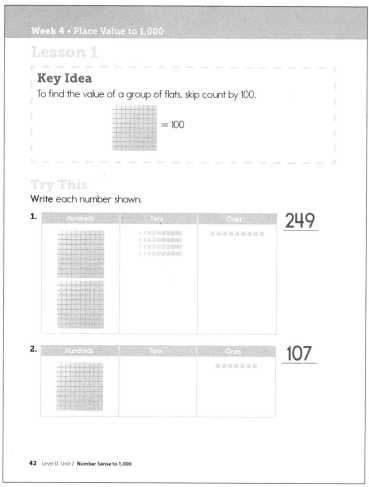

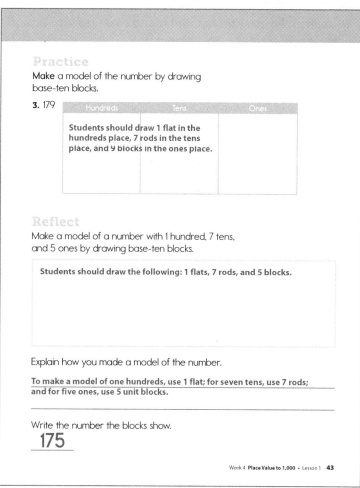

Student Workbook, pp. 42–43

Week 4 **Place Value to 1,000** • Lesson 1 **149**

WEEK 4
Place Value to 1,000

Lesson 2

Objective
Students can trade ones for tens and tens for hundreds.

Standard
2.NBT.3 Read and write numbers to 1000 using base-ten numerals, number names, and expanded form.

Creating Context
Some word parts can help us figure out new words that we encounter. The metric system has many measures with the affix *-meter*. Ask English Learners to work together to list as many words they can think of that have *-meter* in them to see if they are related to the metric system. Start with *millimeter* and *centimeter*.

Materials
Program Materials
Number Cards (0–40)

Additional Materials
base-ten blocks, 1 flat, 19 rods, and 26 units per pair

1 WARM UP

Prepare
- You can make quantities simpler and easier to visualize if you group smaller units into a bigger unit.
- Suppose a game cost 999 pennies. Which is a more common way to say that amount of money: 9 dollars and 99 cents or 999 pennies? 9 dollars and 99 cents
- How many pennies did we replace to make each dollar? 100 pennies

Just the Facts
Have students hold up the correct number of fingers to model trades. Use questions such as the following:

- **I can trade 30 ones for how many tens?** Students hold up 3 fingers.
- **I can trade 20 tens for how many hundreds?** Students hold up 2 fingers.
- **I can trade 80 ones for how many tens?** Students hold up 8 fingers.

2 ENGAGE

Develop: Trading Post
"Today we are going to try to find the fewest number of base-ten blocks we can use to model a number." Follow the instructions on the Activity Card **Trading Post**. As students complete the activity, be sure to use the Questions to Ask.

Activity Card 2M

Alternative Groupings
Individual: Partner with the student and complete the activity as written.

Progress Monitoring
If... students trade tens for hundreds before they trade ones for tens, **Then...** point out that they should regroup the smallest units first so they won't have to regroup larger units twice.

Practice
Have students complete **Student Workbook,** pages 44–45. Guide students through the Key Idea example and the Try This exercises.

Interactive Differentiation
Consult the **Teacher Dashboard** for grouping suggestions. You can also use performance on the Engage activity to guide students.

Independent Practice

For additional practice provide students with Number Cards (0–40). Tell students to draw three cards from the pile and put them in a row, side by side. Students should put a one-digit number on the left, in the hundreds place. The order and placement of the other two cards does not matter. Have students model the number using the Base 10 Blocks Tool. Tell them to use the Combine button to make trades until they have the fewest number of base-ten blocks possible.

Supported Practice

For additional support, have students use the Base 10 Blocks Tool. Tell them to place 5 hundreds flats, 8 tens rods, and 24 ones blocks on the counting mat.

- **Would it make sense to write the number like this?** Write 5824 on the board. **Why or why not?** Answers may vary. Possible answer: No; the number *eight* represents 8 tens rods, but it is in the hundreds place. The number *five* represents five hundreds flats, but it is in the thousands place.
- **What can we do to make sure each digit in the number has the correct place value?** Combine groups of ones, tens, or hundreds if there are more than nine in any group and trade.
- **What do you need to trade?** I need to trade 20 ones for 2 tens rods. That gives me 10 tens rods. Then I can trade the tens rods for a hundreds flat.
- **How do you know when you are finished trading?** There are 9 or fewer blocks in each place value group.
- **How should we write this number in standard form?** 604

REFLECT

Think Critically

Review students' answers to the Reflect prompt at the bottom of ***Student Workbook,*** page 45, and then review the Engage activity.

Ask which model of 235 the students prefer and why. Invite students to share their preferences and reasons with the class.

▸ **What do you do when you have 10 or more rods in the tens place? Why?** Answers may vary. Possible answer: Trade the 10 rods for 1 flat because 10 rods is the same as 1 flat.

Real-World Application

The metric system uses the base-ten system. Ten of one unit can be traded for one of the next larger-sized unit. These trades are like the trades made with base-ten blocks. There are 10 millimeters in 1 centimeter.

▸ **How many centimeters can you make with 20 millimeters?** 2 centimeters

▸ **Explain the trade.**

ASSESS

Informal Assessment

Use the online or print Student Record, ***Assessment,*** page 128, to record informal observations.

Trading Post
Did the student
☐ respond accurately?   ☐ respond with confidence?
☐ respond quickly?   ☐ self-correct?

Additional Practice

For additional practice, have students complete ***Practice,*** page 41.

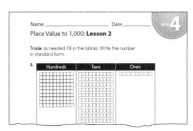

Practice, p. 41

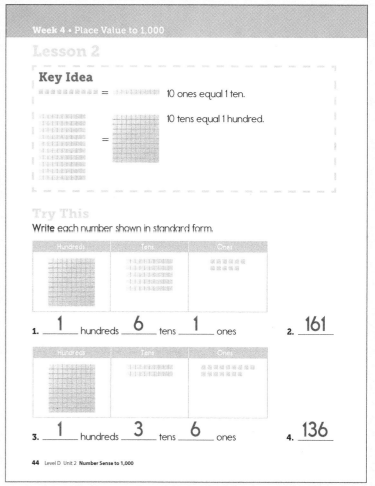

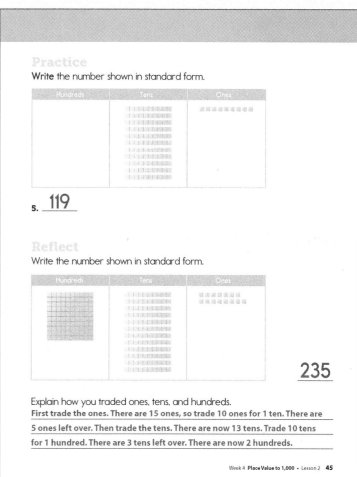

Student Workbook, pp. 44–45

Week 4 **Place Value to 1,000** • Lesson 2   **151**

WEEK 4
Place Value to 1,000

Lesson 3

Objective
Students can compare numbers with the same number of digits and can identify the greater or lesser value.

Standard
2.NBT.3 Read and write numbers to 1000 using base-ten numerals, number names, and expanded form.

Vocabulary
- greater than
- less than
- place value

Creating Context
In this lesson, the concept of greater than or less than is built by having students ask questions to find clues to the missing number. Giving English Learners a template for asking questions as we do in this lesson is very helpful for those at the early proficiency levels who may not yet have question production skills in oral English.

Materials
Program Materials
- Number Construction Mat, 3 per student

Additional Materials
- base-ten blocks, 4 flats, 10 rods, and 30 units per pair
- Vocabulary Card 21, *greater than*
- Vocabulary Card 24, *less than*

1 WARM UP

Prepare
Select a volunteer from the class. Write any two-digit number on a sheet of paper and tape it onto the student's back. Have the student turn around so other students can see the number.

The student should ask questions to find out the number. Questions must be in the following form: Is it greater than _____? Is it less than _____?

The student should ask these questions until he or she can identify the number.

Just the Facts
Have students hold up an appropriate number of fingers to answer questions such as the following:

- **Eight is greater than what number?** Answers may vary. Possible answer: 5
- **Ten is greater than what number?** Answers may vary. Possible answer: 3
- **Four is greater than what number?** Answers may vary. Possible answer: 2

2 ENGAGE

Develop: Comparison Modeling
"Today we are going to compare numbers using base-ten blocks." Follow the instructions on the Activity Card **Comparison Modeling**. As students complete the activity, be sure to use the Questions to Ask.

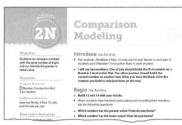
Activity Card 2N

Alternative Groupings
Small Group: Partner with a student if there is an odd number of students, and complete the activity as written.

Individual: Have the student build and record both numbers.

Progress Monitoring
If... students can compare the modeled values easily,

Then... substitute other numbers for them to compare.

Practice
Have students complete **Student Workbook,** pages 46–47. Guide students through the Key Idea example and the Try This exercises.

Interactive Differentiation
Consult the **Teacher Dashboard** for grouping suggestions. You can also use performance on the Engage activity to guide students.

Independent Practice

For additional practice, have students build 2 two-digit numbers with the Base 10 Blocks Tool. You can provide the numbers or have students think of their own. Have students use the addition mat, which displays the numbers in separate boxes and allows students to compare the numbers. Then students should write comparisons, such as, "57 is more than 23 because 5 tens is more than 2 tens." Have students repeat the activity with three-digit numbers.

Supported Practice

For additional support, use the Base 10 Blocks Tool to have students compare two- and three-digit numbers. Organize students into pairs. Have students use the addition mat, which displays the amount on each mat at the bottom of the screen. They can take turns using the same computer.

- ▶ **One of you should build 15 and the other should build 27.**
- ▶ **We can use place value to decide who has the greater number. Because both numbers have a tens and a ones place, look at the tens. Which number has more tens?** 27 has 2 tens, and 15 only has 1.
- ▶ **Because 2 tens is greater than 1 ten, we know that 27 is greater than 15. Now, one of you should build 48 and the other should build 43.**
- ▶ **Which number has more tens?** They both have 4 tens.
- ▶ **Because both numbers have the same number of tens, where do we look to find out which number is greater?** We look at the ones place. **Which number is greater and why?** 48 is greater than 43 because 8 is greater than 3.

Continue the activity by having students build three-digit numbers with and without the same number of hundreds and tens.

152 Level D Unit 2 **Number Sense to 1,000**

REFLECT

Think Critically

Review students' answers to the Reflect prompt at the bottom of *Student Workbook,* page 47, and then review the Engage activity.

Challenge students to evaluate the score and say, in their opinion, if the score of the game was close or not.

▶ What did you find interesting?

▶ Can you think of ways you can use the terms *greater than* or *less than* outside class?

Real-World Application

Have students create this table:

	Tens	Ones
Route A		
Route B		

▶ Justin needs to drive to downtown for a meeting. There are two ways he can get there.

Route A is thirty-nine miles, and Route B is forty-four miles. Fill in the table, and decide which route is shorter.

	Tens	Ones
Route A	3	9
Route B	4	4

Route A is the shorter route.

ASSESS

Informal Assessment

Use the online or print Student Assessment Record, *Assessment,* page 128, to record informal observations.

Comparison Modeling

Did the student

☐ respond accurately?   ☐ respond with confidence?

☐ respond quickly?   ☐ self-correct?

Additional Practice

For additional practice, have students complete *Practice,* page 42.

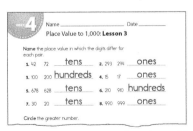

Practice, p. 42

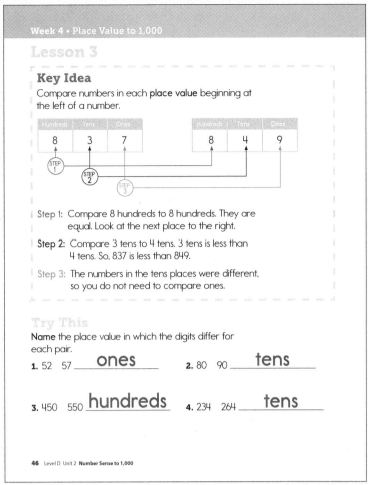

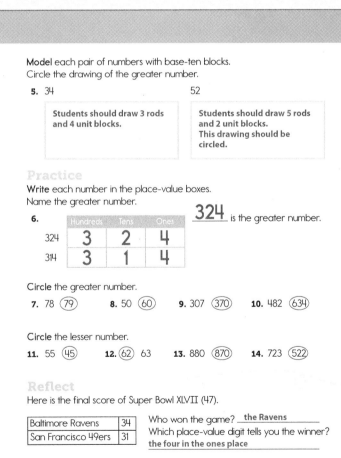

Student Workbook, pp. 46–47

WEEK 4
Place Value to 1,000

Lesson 4

Objective
Students can compare numbers with different numbers of digits and can identify the greater or lesser value.

Standard
2.NBT.3 Read and write numbers to 1000 using base-ten numerals, number names, and expanded form.

Vocabulary
- greater than
- less than

Creating Context
Mathematics has many specialized terms that are necessary to understand the subject. English Learners may know the math concepts in their primary language but need to accelerate their acquisition of English academic language to advance in mathematics. Help English Learners use the glossary to look up math terms and phrases that are unfamiliar.

Materials
Program Materials
- Number Construction Mat, 5 per student

Additional Materials
- unlined paper
- Vocabulary Card 21, *greater than*
- Vocabulary Card 24, *less than*

1 WARM UP

Prepare
- **Think about base-ten blocks. Does a flat have greater value than a rod?** yes
- **Draw a model of a flat and two rods. Then draw a model of three rods.**
- **Which model has a greater value? How do you know?** the flat and two rods; one flat is worth the same as 10 rods
- **Draw one flat. Then draw nine rods. Which model has a greater value? How do you know?** the flat; a flat is worth ten rods
- **Write the numbers that the models you drew represent.** 100, 90
- **What numbers should we look at first when we want to compare numbers to see which is the biggest?** the numbers in the hundreds place

Just the Facts
Instruct students to raise their hands when they hear a correct statement and keep them down when they hear an incorrect statement. Use statements such as the following:

- **Fifteen ones is greater than 1 ten and 2 ones.** hands up
- **Three tens and 5 ones is greater than 4 tens and 2 ones.** hands down
- **Two tens and 13 ones is greater than 3 tens.** hands up

2 ENGAGE

Develop: Comparing Numbers
"Today we are going to use place value to compare numbers." Follow the instructions on the Activity Card **Comparing Numbers**. As students complete the activity, be sure to use the Questions to Ask.

Activity Card 20

Alternative Groupings
Individual: Have the student write numbers and compare them with your numbers. Help him or her as necessary.

Progress Monitoring
If… students are not capable of standing or sitting,

Then… change the directions to have students sit upright or put their heads down on their desks.

Practice
Have students complete **Student Workbook,** pages 48–49. Guide students through the Key Idea example and the Try This exercises.

Interactive Differentiation
Consult the **Teacher Dashboard** for grouping suggestions. You can also use performance on the Engage activity to guide students.

Independent Practice
For additional practice, have students collect two books, such as a math book and a reading book. Have students compare the total number of pages in each book using place value. Tell them to write each number in place-value charts, such as those on **Student Workbook** page 49. Then have them circle the greater number. Encourage students to continue the activity with several books as time permits.

Supported Practice
For additional support, have students build a number less than 1,000 using the Base 10 Blocks Tool.

- **We can use place value to find out who built the greatest number. Which place value should we look at first, the ones, tens, or hundreds?** hundreds

Take a moment to point out that a two-digit number has 0 hundreds. So, if a student built a two-digit number, comparing the first digit of that number to the first digit in a three-digit number would be pointless. Explain that it is important to understand the value of each digit before you compare.

- **Show the number of hundreds in your number on your fingers. Look around the room. Who has the greatest number of hundreds?**
- **What should we do if two or more numbers have the same number of hundreds?** Compare the numbers in the tens place.
- **What do we do if two numbers have the same number of hundreds *and* tens?** Compare the numbers in the ones place.

Determine which student built the greatest number and have that student take a bow.

154 Level D Unit 2 **Number Sense to 1,000**

REFLECT

Think Critically

Review students' answers to the Reflect prompt at the bottom of **Student Workbook,** page 49, and then review the Engage activity.

Discuss the similarity and the difference between the numbers 250 and 25 using base-ten blocks.

▶ **If I asked you to compare a three-digit number to a two-digit number, what could you put in the hundreds place to make it easier to compare?**
Possible answer: a zero

▶ **Tell me something you know about two-digit numbers and three-digit numbers.** Possible answer: A three-digit number is larger than a two-digit number.

ASSESS

Informal Assessment

Use the online or print Student Record, **Assessment,** page 128, to record informal observations.

Comparing Numbers
Did the student
☐ respond accurately?   ☐ respond with confidence?
☐ respond quickly?   ☐ self-correct?

Additional Practice

For additional practice, have students complete **Practice,** page 43.

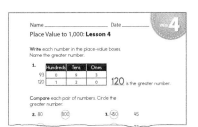
Practice, p. 43

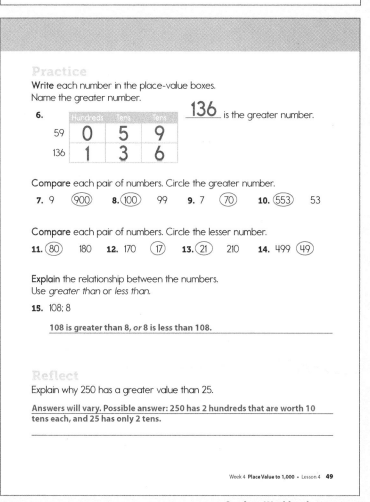

Student Workbook, pp. 48–49

WEEK 4
Place Value to 1,000

Lesson 5 Review

Objective
Students review skills learned this week and complete the weekly assessment.

Standard
2.NBT.3 Read and write numbers to 1000 using base-ten numerals, number names, and expanded form.

Vocabulary
Review vocabulary introduced during the week.

Creating Context
Teachers often report that English Learners do well in mathematics. Many mathematical symbols, including numbers, are often the same in different languages. English Learners may only need to learn new names for these symbols.

1 WARM UP

Prepare

▶ Draw pictures of models for 34. Show at least three different models for this number.

2 ENGAGE

Practice
Have students complete **Student Workbook,** pages 50–51.

Week 4 • Place Value to 1,000

Lesson 5 Review

This week you explored place value. You learned how made models of numbers using base-ten blocks. You also learned how to trade ones, tens, and hundreds.

Lesson 1 Draw a model of the number using groups of blocks.

1. 149

Hundreds	Tens	Ones
Answers may vary. Sample answer: Students could draw 1 flat in the hundreds place, 4 rods in the tens place, and 9 blocks in the ones place.		

Lesson 2 Trade ones and tens or tens and hundreds as needed.

2.

Hundreds	Tens	Ones

__1__ hundreds __1__ tens __3__ ones

Lesson 3 Compare each pair of numbers. Circle the greater number.

3. (42) 41
4. 77 (770)

Circle the lesser number. Name the place you used to make your decision.

5. (914) 945 __tens__
6. 872 (871) __ones__

Lesson 4 Compare each pair of numbers. Circle the lesser number.

7. (2) 6
8. 1000 (999)

Circle the greater number. Name the place you looked at to make your decision.

9. 46 (48) __ones__
10. 23 (123) __hundreds__

Reflect

Write a number that is less than 145 but has equal values in the hundreds and tens place.

Students should write a number from 140 to 144.

Write a number that is greater than 383 but has equal values in the hundreds and tens place.

Students should write a number from 384 to 389.

Student Workbook, pp. 50–51

3 REFLECT

Think Critically

Review students' answers to the Reflect prompt at the bottom of **Student Workbook,** page 51.

Discuss the answer with the group to reinforce Week 4 concepts.

4 ASSESS

Formal Assessment

Students may take the weekly assessment online.

As an alternative, students may complete the weekly test on **Assessment,** pages 31–32. Record progress using the Student Assessment Record, **Assessment,** page 128.

Going Forward

Use the **Teacher Dashboard** to view results of the online assessments, to input the results of print student assessments, and to review progress before making decisions about next steps. Use the weekly test results and observations to determine the next steps for each student.

Retention	
Student displays good grasp of this week's concepts and skills.	Have students build three-digit numbers with the Base 10 Blocks Tool as time permits.

Remediation	
Student is still struggling with the week's concepts and skills.	Have students place 7 hundred flats, 9 tens rods, and 19 ones blocks on the counting mat in the Base 10 Blocks Tool. Guide students to make trades to build 809 using the fewest blocks. ▶ **What can we do to make sure each digit in the number has the correct place value?** Combine groups of ones, tens, or hundreds if there are more than nine in any group and trade. ▶ **What do you need to trade?** I need to trade 10 ones for 1 tens rod. That gives me 10 tens rods. Then I can trade the tens rods for a hundred flat. ▶ **How do you know when you are finished trading?** 9 or fewer blocks are in each place. ▶ **How do you write this number in standard form?** 809

Suggestions for Re-Evaluation: If a student has struggled without success for several weeks, use observations and test results to place the student at a level in which he or she can find success and build confidence to move forward.

Name _____ Date _____ WEEK 4

Place Value to 1,000

Write the standard form of each number.

1. 5 hundreds, 7 tens, 3 ones **573**

2. 3 hundreds, 12 tens, 9 ones **429**

3. 4 hundreds, 31 tens, 18 ones **728**

4. Circle the number with a 7 in the hundreds place.
   397   478   **(726)**

5. Circle the number with a 6 in the ones place.
   (916)   604   263

Level D Unit 2 Week 4 **31**

WEEK 4 Name _____ Date _____

Place Value to 1,000

6. Circle the number with a 4 in the tens place.
   491   **(548)**   734

7. Circle the greater number.
   (934)   493

8. Circle the lesser number.
   857   **(758)**

9. Circle the number with the higher tens value.
   514   **(451)**

10. Circle the number with the higher hundreds value.
    (836)   683

11. Circle the statement that is true.
    50 is greater than 500   **(50 is less than 500)**

32 Level D Unit 2 Week 4

Assessment, pp. 31–32

Week 4 **Place Value to 1,000** • Lesson 5 **157**

Project Preview

This week, students learned how to make trades between ones, tens, and hundreds as well as how to use place value to compare two- and three-digit numbers. The project for this unit requires students to extend their knowledge of place value in order to understand our monetary system. Students will use place value to compare two sets of coins to find the greater value.

Project-Based Learning

Standards-driven Project-Based Learning is effective in building deep content understanding. Project-Based Learning increases long-term retention of concepts and has been shown to be more effective than traditional instruction. By completing a project to answer an essential question, students are challenged to apply and demonstrate mastery of concepts and skills by expressing understanding through discussion, research, and presentation.

Essential Question

WHY do I need to understand place value to use money?

Project Evaluation Criteria

Review project evaluation criteria with students prior to beginning the project.

Exceeds Expectations
☐ Project result is explained and can be extended.
☐ Project result is explained in context and can be applied to other situations.
☐ Project result is explained using advanced mathematical vocabulary.
☐ Project result is described, and mathematics are used correctly and can be extended.
☐ Project result is explained and extended, and shows advanced knowledge of mathematical concepts and skills.

Meets Expectations
☐ Project result is explained.
☐ Project result is explained in context.
☐ Project result is explained using mathematical vocabulary.
☐ Project result is described, and mathematics are used correctly.
☐ Project result is explained, and shows satisfactory knowledge of mathematical concepts and skills.

Does Not Meet Expectations
☐ Project result is not explained.
☐ Project result is explained, but out of context.
☐ Project result is explained, but mathematical vocabulary is oversimplified.
☐ Project result is described, but mathematics are not used correctly.
☐ Project result is not explained and/or extended, or shows less than satisfactory knowledge of mathematical concepts and skills.

Who Has More?

Objective
Students can use place value to count and compare two sets of coins to find the greater value.

Standard
2.NBT.3 Read and write numbers to 1000 using base-ten numerals, number names, and expanded form.

Materials

Program Materials
- Deposit Slip, 2 per student pair
- Place Value Mat

Additional Materials
play money

Prepare Ahead
Cut apart the Deposit Slips so that you can hand six slips to each student pair.

Best Practices
- Clearly enunciate instructions.
- Pair oral directions with accessible pictures, icons, or written words for student needs.
- Organize the classroom as an activity-based space.

Introduce

▶ When you do chores or other jobs for your family or neighbors, you might earn money for the work you do.

▶ **What are some jobs you could do around your house or neighborhood to earn money?** Answers may vary. Possible answers: walk dogs, rake leaves, clean house.

▶ **Where can you keep the money you earn?** Answers may vary. Possible answers: in a piggy bank, at a bank

▶ When you earn money, you might want to take your money to a bank to keep it safe. When you want to give the bank your money, you fill out a *deposit slip*. A deposit slip tells the bank how much money you will put in your account.

Explore

▶ Today you will use place value to compare the amount of money you earned with the amount of money your partner earned.

▶ One of you will be Partner A, and the other will be Partner B.

- Hand six blank deposit slips to each pair of students (two sheets cut into individual slips).
- Guide students as they fill out the deposit slips with the appropriate dollar amounts found on *Student Workbook* page 52.
- Encourage students to model the money they earned by placing the appropriate number of coins on the Place Value Mat.

▶ **Use place value to explain who earned the greater amount of money. Which place value should you look at first?** the hundreds place

▶ Complete *Student Workbook,* page 52, to practice using place value to compare sets of coins.

Wrap Up

- Encourage students to use play money to model the amount each partner earned.
- Make sure each student uses place value to explain how to determine who earned more money.
- If students have difficulty using place value to compare amounts of money, have them first compare dollars, then dimes, then pennies.
- Discuss students' answers to the Reflect prompt at the bottom of *Student Workbook,* page 52.
- Students may also use the Coins and Money Tool to model the appropriate money amounts.
- If time permits, allow students to decorate withdrawal and deposit slips with their bank logos.

Week 4 • Place Value to 1,000

Project
Who Has More?

Write down the amount of money each partner earned on a deposit slip. Use place value to explain who earned more money. For example, 32 cents are greater than 25 cents because 3 tens are greater than 2 tens.

1. Partner A earned 451 cents walking the neighbor's dog, and Partner B earned 468 cents. Use place value to explain who earned more money walking the dog.

   Answers may vary. Possible answer: Partner B earned more because 6 tens is greater than 5 tens.

2. Partner A earned 64 cents washing dishes, and partner B earned 121 cents. Use place value to explain who earned more money washing dishes.

   Answers may vary. Possible answer: Partner B earned more because 1 hundred is greater than 0 hundreds.

3. Partner A earned 325 cents raking leaves, and Partner B earned 322 cents. Use place value to explain who earned more money raking leaves.

   Answers may vary. Possible answer: Partner A earned more because 5 ones is greater than 2 ones.

Reflect

How do you use place value to compare two amounts of money?

Answers may vary. Possible answer: First look at the dollars. If they are the same, look at the dimes. If they are the same, look at the pennies.

52 Level D Unit 2 **Number Sense to 1,000**

Student Workbook, p. 52

Teacher Reflect

☐ Did my students use the appropriate available materials?

☐ Did I clearly explain how to organize the activity?

☐ Did students tell or show the steps when they explained how to do something?

5 Skip Counting within 1,000

Week at a Glance

This week, students continue **Number Worlds,** Level D, Number Sense to 1,000 by skip counting within 1,000. Students will use linear and non-linear models to complete number sequences.

Skills Focus

- Skip count by 5, 10s, and 100s within 1,000.
- Find numbers within 1,000 by skip counting by 5, 10s, and 100s.
- Complete sequences by counting by twos, fives, tens, and hundreds.

How Students Learn

Students' knowledge of numbers and quantity becomes more integrated as their experiences in mathematics continue. Students begin to link numbers to quantities and realize that questions about numbers can be answered with or without the use of concrete objects.

English Learners ELL

For language support, use the **English Learner Support Guide,** pages 48–49, to preview lesson concepts and teach academic vocabulary.

Math at Home

Give one copy of the Letter to Home, page 11, to each student. Encourage students to share and complete the activity with their caregivers.

Weekly Planner

Lesson	Learning Objectives
1 pages 162–163	Students can skip count by twos and fives within twenty.
2 pages 164–165	Students can skip count by fives and tens within 100.
3 pages 166–167	Students can skip count by tens within 1,000.
4 pages 168–169	Students can complete sequences by counting by twos, fives, tens, and hundreds.
5 pages 170–171	**Review and Assess** Students review skills learned this week and complete the weekly assessment.
Project pages 172–173	Students can use skip counting by fives and tens to count sets of nickels and dimes.

160 Level D Unit 2 **Number Sense to 1,000**

Key Standard for the Week

Domain: Number and Operations in Base Ten
Cluster: Understand place value.
2.NBT.2 Count within 1000; skip-count by 5s, 10s, and 100s.

Materials		Technology
Program Materials • **Student Workbook,** pp. 53–55 • **Practice,** p. 44 • Activity Card 2P, **Skip Down the Line** • Counters	**Additional Materials** • dark-colored marker • wide masking tape	**Teacher Dashboard** Building Blocks Tire Recycling Number Line Tool
Program Materials • **Student Workbook,** pp. 56–57 • **Practice,** p. 45 • Activity Card 2Q, **Race to 100** • 1–100 Chart • Number Line to 100 Game Board • Number Cube (1–6) • Pawns		**Teacher Dashboard** Building Blocks School Supply Shop
Program Materials • **Student Workbook,** pp. 58–59 • **Practice,** p. 46 • Activity Card 2R, **Mystery Number**	**Additional Materials** index cards	**Teacher Dashboard** 100 Table Tool
Program Materials • **Student Workbook,** pp. 60–61 • **Practice,** p. 47 • Activity Card 2R, **Mystery Number Variation: Skipping to the Mystery**	**Additional Materials** beach ball or other soft object that can be thrown	**Teacher Dashboard** Number Line Tool
Program Materials • **Student Workbook,** pp. 62–63 • Weekly Test, **Assessment,** pp. 33–34		Review previous activities.
Program Materials **Student Workbook,** p. 64 • Deposit Slips • Withdrawal Slips	**Additional Materials** play money*	

*Available from McGraw-Hill Education

WEEK 5
Skip Counting within 1,000

Find the Math

In this week, students will use skip counting to find the value of a set of coins that includes nickels and dimes.

Use the following to begin a guided discussion:

▶ A roll of nickels is worth $2. A roll of dimes is worth $5. A roll of quarters is worth $10. How could you skip count rolls of nickels? Dimes? Quarters?

Skip count rolls of nickels by twos, dimes by fives, and rolls of quarters by tens.

Have students complete **Student Workbook,** page 53.

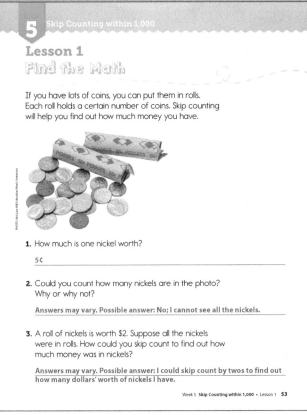

Student Workbook, p. 53

Lesson 1

Objective
Students can skip count by twos and fives within twenty.

Standard
2.NBT.2 Count within 1000; skip-count by 5s, 10s, and 100s.

Creating Context
Help English Learners understand the concept of skip counting by having them jump or skip over one or more small objects. Explain that when we skip count, we jump over a number or numbers just like we jump over the objects.

Materials
Program Materials
Counters

Additional Materials
- dark-colored marker
- wide masking tape

Prepare Ahead
Lay down long pieces of masking tape on the floor of your classroom, hallway, or gym. Using a dark marker, write the numbers 0–20 on the tape. The numbers should be far enough apart that students can stand on each number but close enough that students can easily jump across five numbers. Make one number line for each small group of students.

1 WARM UP

Prepare
- Have students name classroom objects that come in sets of two and sets of five, such as desks arranged in pairs or the fingers on their hands.
- Guide students to count the objects they have named using skip counting.

2 ENGAGE

Develop: Skip Down the Line

"Today we are going to practice skip counting by twos and fives on a giant number line." Follow the instructions on the Activity Card **Skip Down the Line.** As students complete the activity, be sure to use the Questions to Ask.

Activity Card 2P

Alternative Groupings
Whole Class: Instead of jumping from one number to the next, students should stand on the appropriate numbers as the whole class counts aloud. For example, students can stand on 2, 4, 6, and so on when skip counting by twos and on numbers 5, 10, 15, and so on when skip counting by fives.

Progress Monitoring

If... students have difficulty determining the next number in the skip-counting sequence,

Then... place a self-sticking note on 0. As students count on by ones, place self-sticking notes on each number in the skip-counting sequence.

Practice
Have students complete **Student Workbook,** pages 54–55. Guide students through the Key Idea example and the Try This exercises.

Interactive Differentiation

Consult the **Teacher Dashboard** for grouping suggestions. You can also use performance on the Engage activity to guide students.

Independent Practice

For additional practice skip counting by twos and fives, have students play Tire Recycling. Students will skip count by twos and fives to load a truck with the appropriate number of tires.

Supported Practice

For additional support, students can use the Number Line Tool to practice skip counting by twos and fives.

- Instruct students to begin the number line on 0. Make sure they have selected a positive number and set the number line to count forward.
- Have students practice skip counting by twos as they look at the jumps on the number line.
- ▶ **Look at the number line. What pattern do you notice?** Answers may vary. Possible answer: There is a jump to every other number.
- Repeat the activity, having students select 5 on the "count by" menu.
- Have students practice skip counting by fives and identify any patterns they notice.

REFLECT

Think Critically

Review students' answers to the Reflect prompt at the bottom of **Student Workbook,** page 55, and then review the Engage activity.

▶ **When you are skip counting, how do you know which number comes next?** Answers may vary. Possible answer: I think about the pattern. When I am skip counting by twos, the next number is two more than the number before. When I am skip counting by fives, the next number is five more than the number before.

ASSESS

Informal Assessment

Use the online or print Student Record, **Assessment,** page 128, to record informal observations.

Skip Down the Line
Did the student
☐ make important observations?   ☐ provide insightful answers?
☐ extend or generalize learning?   ☐ pose insightful questions?

Additional Practice

For additional practice, have students complete **Practice,** page 44.

Practice, p. 44

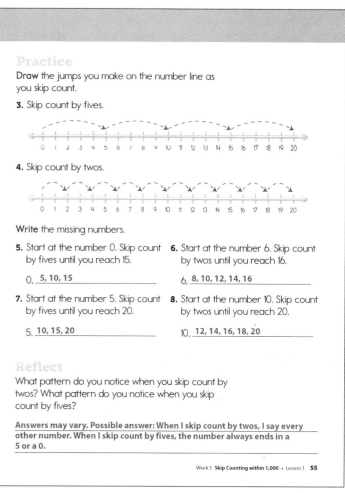

Student Workbook, pp. 54–55

Week 5 **Skip Counting within 1,000** • Lesson 1   **163**

WEEK 5
Skip Counting within 1,000

Lesson 2

Objective
Students can skip count by fives and tens within 100.

Standard
2.NBT.2 Count within 1000; skip-count by 5s, 10s, and 100s.

Creating Context
On the 1–100 Chart, point out the number 10 to English Learners. Draw attention to the 0 in the ones place. Have students count 10 more on the chart. Again, point out the 0 in the ones place. Repeat with 30 and 40. Then repeat the entire process beginning at 14 and counting by tens up to 44. Explain that when you are counting by tens, the digit in the ones place will always be the same number.

Materials
Program Materials
- 1–100 Chart
- Number Line to 100 Game Board
- Number Cube (1–6)
- Pawns

 WARM UP

Prepare
- Have students look around the classroom and identify objects that come in sets of five and sets of ten, such as school days on the calendar or toes on a foot.
- Guide students to count the objects they have named using skip counting.

Just the Facts
Tell students that you are going to skip count. Instruct them to stand up when they hear an incorrect number in each sequence. Use sequences such as the following:

▶ **25, 30, 35, 40, 50, 55, 60** Students should stand when they hear the number 50.

▶ **40, 50, 60, 65, 70, 80** Students should stand when they hear the number 65.

▶ **55, 60, 65, 70, 75, 80, 85, 95, 100** Students should stand when they hear the number 95.

2 ENGAGE

Develop: Race to 100
"Today we are going to play a game to practice skip counting by fives and tens." Follow the instructions on the Activity Card **Race to 100**. As students complete the activity, be sure to use the Questions to Ask.

Activity Card 2Q

Alternative Groupings
Small Group: Instead of the Number Line to 100 Game Board, provide each group with a 1–100 Chart. Then complete the activity as written.

Progress Monitoring

If... a student loses his or her place on the Game Board,

Then... have a classmate place a finger on the Pawn's starting position.

Practice
Have students complete **Student Workbook,** pages 56–57. Guide students through the Key Idea example and the Try This exercises.

Interactive Differentiation
Consult the **Teacher Dashboard** for grouping suggestions. You can also use performance on the Engage activity to guide students.

Independent Practice

For additional practice counting by tens, students can use School Supply Shop. Students count by tens to load carts with the appropriate number of school supplies.

Supported Practice

For additional support, provide each student with two 1–100 Charts.

- Have students start at 0, count on five more, and color the number 5.
- Have students continue counting on five more and coloring each number they land on.
- ▶ **Do you notice a pattern?** Answers may vary. Possible answers: All the colored numbers end in either a 5 or 0; all the numbers in the fifth and the last columns are colored.
- Have students count by tens and use the second 1–100 Chart. Point out that the digit in the ones place stays the same.

164 Level D Unit 2 **Number Sense to 1,000**

REFLECT

Think Critically

Review students' answers to the Reflect prompt at the bottom of *Student Workbook,* page 57, and then review the Engage activity.

▶ **When you are skip counting, how do you know which number comes next?** Answers may vary. Possible answers: Count on by ones to the next number. Look at a number line. Look at a 100 table.

Real-World Application

Your soccer team is selling raffle tickets for 10¢ each.

▶ **How can you find out how much money you will collect if you sell 8 raffle tickets?** Count by tens 8 times: 10, 20, 30, 40, 50, 60, 70, 80. You will collect 80¢.

ASSESS

Informal Assessment

Use the online or print Student Record, *Assessment,* page 128, to record informal observations.

Race to 100

Did the student

☐ make important observations? ☐ provide insightful answers?

☐ extend or generalize learning? ☐ pose insightful questions?

Additional Practice

For additional practice, have students complete *Practice,* page 45.

Practice, p. 45

Week 5 • Skip Counting within 1,000

Lesson 2

Key Idea
When you skip count by fives and tens on a 1–100 Chart, you make a pattern with numbers.

1	2	3	4	5	6	7	8	9	10
11	12	13	14	15	16	17	18	19	20
21	22	23	24	25	26	27	28	29	30
31	32	33	34	35	36	37	38	39	40
41	42	43	44	45	46	47	48	49	50
51	52	53	54	55	56	57	58	59	60
61	62	63	64	65	66	67	68	69	70
71	72	73	74	75	76	77	78	79	80
81	82	83	84	85	86	87	88	89	90
91	92	93	94	95	96	97	98	99	100

Try This
Fill in the missing numbers. Look at the 1–100 Chart.

1. 5, 10, __15__, 20, 25, __30__
2. 10, __20__, 30, 40, __50__, 60
3. 45, 55, __65__, 75, __85__
4. 30, 40, __50__, __60__, 70

56 Level D Unit 2 **Number Sense to 1,000**

Practice
Draw the jumps you make as you skip count. Fill in the missing numbers along the number line.

5. Skip count by fives.

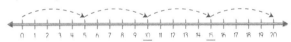

6. Skip count by tens.

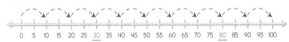

Fill in the missing numbers.

7. Start at the number 40. Skip count by tens until you reach 100.

   40, __50, 60, 70, 80, 90, 100__

8. Start at the number 55. Skip count by fives until you reach 80.

   55, __60, 65, 70, 75, 80__

Reflect
How is the pattern for skip counting by fives different from the pattern for skip counting by tens?

Answers may vary. Possible answer: When I skip count by tens, the ones digit always stays the same. When I skip count by fives, the ones digit does not stay the same.

Week 5 **Skip Counting within 1,000** • Lesson 2 57

Student Workbook, pp. 56–57

WEEK 5
Skip Counting within 1,000

Lesson 3

Objective
Students can skip count by tens within 1,000.

Standard
2.NBT.2 Count within 1000; skip-count by 5s, 10s, and 100s.

Creating Context
Some English Learners may want to say 200 as "two hundreds" rather than "two hundred." In many languages, including Spanish, one hundred is singular (*cien*), but multiple hundreds are plural (*doscientos*). Explain that in English, the word *hundred* is never plural when it is used to express a number.

Materials
Additional Materials
index cards

Prepare Ahead
Write Mystery Number clues on individual index cards; for example, I am 50 more than 325. What number am I? Write the answer to each clue on a separate card. There should be enough clues for one half of the class and enough answers for the other half. Match the difficulty level of the Mystery Number clues to the abilities of the students.

1 WARM UP

Prepare

▶ **How many fingers do you have? How many toes do you have?**
10 and 10

- Have students count by tens to find the total number of fingers and toes in the classroom.

Just the Facts

Have students practice skip counting by tens as they finish each sequence. Use sequences such as the following:

▶ 50, 60, 70, ___, ___, ___ 80, 90, 100
▶ 210, 220, 230, ___, ___, ___ 240, 250, 260
▶ 746, 756, 766, ___, ___, ___ 776, 786, 796

2 ENGAGE

Develop: Mystery Number

"Today we are going to skip count by tens to find mystery numbers." Follow the instructions on the Activity Card **Mystery Number**. As students complete the activity, be sure to use the Questions to Ask.

Activity Card 2R

Alternative Groupings

Individual: Prepare ten Mystery Number clues and ten answers. Have students mix up the cards and lay them facedown in a 4 × 5 grid. Explain that students will play a memory game, matching each clue with its answer.

Pair: Prepare ten Mystery Number clues and ten answers. Have students mix up the cards and lay them facedown in a 4 × 5 grid. Explain that they will play a memory game, matching each clue with its answer. Students will take turns. If one player makes a match, he or she can take another turn. The player with the most matches wins.

Progress Monitoring

If... students have difficulty finding the answers to match the clues,

Then... help students determine the number of tens in the first number of the clue.

Practice

Have students complete **Student Workbook,** pages 58–59. Guide students through the Key Idea example and the Try This exercises.

Interactive Differentiation

Consult the **Teacher Dashboard** for grouping suggestions. You can also use performance on the Engage activity to guide students.

Independent Practice

For additional practice, have students use the 100 Table Tool. At the bottom of the screen, have them click the *Add Table* icon and set the *Grid Type* to "0–100." At the left, have students set up *Skip Count A* as follows: *Start* = "0," *End* = "100," and *Count By* = "10." Have students click *Start* and see the result. Then have students explore other combinations.

Supported Practice

- For additional support, show students a skip-counting number pattern when they skip count by tens.

- First, point to the number 10 on a 1–100 Chart. Have students skip count aloud to 100 as you point to each number.

▶ **Look at the tens place. What do you notice as we skip count by tens?**
The tens place increases by 1 each time.

▶ **What do you think will happen to the hundreds place if we continue skip counting by tens to 1,000?** The hundreds place will stay the same until we reach the next hundred. Then it will go up by 1.

- In the 100 Table Tool, have students select 10 on the yellow "Count by" menu and 0 on the "Start on" menu. Then have them select 10 on the blue "Count by" menu and any number 1–9 on the "Start on" menu.

▶ **What pattern does the tool create?** The last column is all yellow, and one of the other columns is all blue.

3 REFLECT

Think Critically

Review students' answers to the Reflect prompt at the bottom of **Student Workbook,** page 59, and then review the Engage activity.

▶ **What patterns do you see when you skip count on a number line?** Answers may vary. Possible answer: The jumps are always the same distance apart on the number line.

Real-World Application

A dime is worth 10¢, so you can skip count by tens to find the value of a group of dimes.

▶ **Pretend you have 23 dimes in your piggy bank. How would you count the dimes? How much money do you have?** Answers may vary. Possible answer: I can count the dimes by tens. If there are 23 dimes, I have 230¢ or $2.30.

4 ASSESS

Informal Assessment

Use the online or print Student Record, **Assessment,** page 128, to record informal observations.

Mystery Number

Did the student
- ☐ make important observations?
- ☐ extend or generalize learning?
- ☐ provide insightful answers?
- ☐ pose insightful questions?

Additional Practice

For additional practice, have students complete **Practice,** page 46.

Practice, p. 46

Week 5 • Skip Counting within 1,000

Lesson 3

Key Idea
When you skip count by tens, the ones digit always stays the same. The tens digit increases by 1 each time you skip count to the next number.

1	2	3	4	5	6	7	8	9	10
11	12	13	14	15	16	17	18	19	20
21	22	23	24	25	26	27	28	29	30
31	32	33	34	35	36	37	38	39	40
41	42	43	44	45	46	47	48	49	50
51	52	53	54	55	56	57	58	59	60
61	62	63	64	65	66	67	68	69	70
71	72	73	74	75	76	77	78	79	80
81	82	83	84	85	86	87	88	89	90
91	92	93	94	95	96	97	98	99	100

Try This

Skip count by tens to find the number. Fill in the missing numbers along the number line.

1. Start at 450. Make 3 skip-counting jumps. Write each number you land on.

   380 390 400 410 420 430 440 450 <u>460</u> <u>470</u> <u>480</u>

2. Start at 375. Make 2 skip-counting jumps. Write each number you land on.

   295 305 315 325 335 345 355 365 375 <u>385</u> <u>395</u>

Practice

Skip count by tens to continue each pattern.

3. 30, 40, 50, <u>60</u> <u>70</u> <u>80</u>
4. 26, 36, 46, <u>56</u> <u>66</u> <u>76</u>
5. 125, 135, 145, <u>155</u> <u>165</u> <u>175</u>
6. 531, 541, 551, <u>561</u> <u>571</u> <u>581</u>
7. 204, 214, <u>224</u> <u>234</u> <u>244</u> <u>254</u>
8. 632, 642, <u>652</u> <u>662</u> <u>672</u> <u>682</u>

Find each number. Write the number on the line.

9. I am 10 more than 870. What number am I? <u>880</u>
10. I am 30 more than 610. What number am I? <u>640</u>
11. I am 50 more than 723. What number am I? <u>773</u>
12. I am 20 more than 399. What number am I? <u>419</u>

Reflect
What pattern do you notice when you skip count by tens?

Answers may vary. Possible answers: Each number is always 10 more; the numbers in the 1–100 Chart are all in one column; the tens place keeps increasing by 1.

Student Workbook, pp. 58–59

WEEK 5
Skip Counting within 1,000

Lesson 4

Objective
Students can complete sequences by counting by twos, fives, tens, and hundreds.

Standard
2.NBT.2 Count within 1000; skip-count by 5s, 10s, and 100s.

Creating Context
Help English Learners understand the terms *before*, *after*, and *between* by having several students stand in a line. Have students physically move to a space that is before a friend or after a friend as well as between two friends.

Materials
Additional Materials
beach ball or other soft object that can be thrown

1 WARM UP

Prepare
- Use a beach ball or other soft object that you can throw to play a game of "What's Next."
- Say the first three numbers in a skip-counting sequence, such as 35, 40, 45, and toss the ball to a student.
- The student holding the ball will say the next number in the sequence and then toss the ball to another student. The next student continues the sequence.
- After students have supplied three or four numbers, take the ball back. Start again with a new skip-counting sequence.

Just the Facts
Have students hold their hands up as you slowly say a skip-counting sequence. Instruct students to put their hands down when they hear a number that does not follow the sequence. Use sequences such as the following:

▶ **8, 10, 12, 14, 18, 20...** Students should put their hands down when they hear 18.

▶ **426, 436, 446, 456, 457, 458...** Students should put their hands down when they hear 457.

▶ **385, 390, 395, 300, 305, 310...** Students should put their hands down when they hear 300.

2 ENGAGE

Develop: Mystery Number Variation: Skipping to the Mystery Number

"Today we are going to use what we know about skip counting to complete sequences." Follow the instructions on the Activity Card **Mystery Number Variation: Skipping to the Mystery Number.** As students complete the activity, be sure to use the Questions to Ask.

Activity Card 2R

Alternative Groupings

Individual: Prepare nine Mystery Number clues and nine answers, three pairs each for skip counting by twos, fives, and hundreds. Include one Mystery Number clue and answer from the first Mystery Number activity. Have students mix up the cards and lay them facedown in a 4 × 5 grid. Explain that they will play a game of Memory, matching each clue with its answer.

Pair: Prepare nine Mystery Number clues and nine answers, three pairs each for skip counting by twos, fives, and hundreds. Include one Mystery Number clue and answer from the first Mystery Number activity. Have students mix up the cards and lay them facedown in a 4 × 5 grid. Explain that they will play a game of Memory, matching each clue with its answer. If one player makes a match, he or she can take another turn. The player with the most matches wins.

Progress Monitoring

If... students have difficulty finding a match,

Then... organize students in small groups to figure out each mystery number. Once students have found the numbers, have them try again to find the matching answer cards.

Practice
Have students complete **Student Workbook,** pages 60–61. Guide students through the Key Idea example and the Try This exercises.

Interactive Differentiation
Consult the **Teacher Dashboard** for grouping suggestions. You can also use performance on the Engage activity to guide students.

Independent Practice
For additional practice, students can use the Number Line Tool to count by twos, fives, tens, and hundreds. Have students skip count aloud, very quietly, up each sequence and then back down.

Supported Practice
For additional support, use the Number Line Tool.

- Draw a number line from 0 to 100 on the board, marking intervals of 5. Circle three numbers in a skip-counting sequence, such as 45, 50, and 55, on the number line.

▶ **Count by ones from 45 to 50. How many numbers did you count?** 5

▶ **Count by ones from 50 to 55. How many numbers did you count?** 5

▶ **If you count 45, 50, and 55, what are you skip counting by?** fives

- Have students open the Number Line Tool. Tell them to select 0 on the "Number Line Begin" menu and 5 on the "Skip Count by" menu. Ask students to tell which numbers come before and after the sequence.

168 Level D Unit 2 **Number Sense to 1,000**

REFLECT

Think Critically

Review students' answers to the Reflect prompt at the bottom of **Student Workbook,** page 61, and then review the Engage activity.

▶ **How do you use the number pattern to find the numbers that come before and after a set of numbers?** Answers may vary. Possible answer: If the pattern is skip counting by twos, then count forward 2 to find the number that comes after and count backward 2 to find the number that comes before.

Real-World Application

Skip counting can help you count more quickly. Coins are great to skip count.

▶ **How would you skip count to find the amount of money in a pile of nickels?** I would skip count by fives.

ASSESS

Informal Assessment

Use the online or print Student Assessment Record, **Assessment,** page 128, to record informal observations.

Skipping to the Mystery Number

Did the student
- ☐ make important observations?
- ☐ extend or generalize learning?
- ☐ provide insightful answers?
- ☐ pose insightful questions?

Additional Practice

For additional practice, have students complete **Practice,** page 47.

Practice, p. 47

Week 5 • Skip Counting within 1,000

Lesson 4

Key Idea

What number is missing? When you know what the skip-counting pattern is, you can figure it out.

The missing number is 2 more than 4 and 2 less than 8.

The missing number is 6!

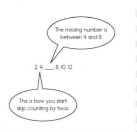

2, 4, ___, 8, 10, 12

The missing number is between 4 and 8

This is how you start skip counting by twos.

Try This

Skip count by fives. Complete the sequence.

1. 45, 50, __55__, __60__
2. 20, 25, __30__, __35__

Skip count by twos. Complete the sequence.

3. __6__, __8__, 10, 12
4. __11__, __13__, 15, 17

Skip count by tens. Complete the sequence.

5. 40, __50__, 60
6. 131, __141__, 151

60 Level D Unit 2 **Number Sense to 1,000**

Practice

Circle the number that does not belong in each sequence.

7. 4, 6, 8, 10, (11), 14, 16, 18, 20
8. 35, 40, (54), 50, 55, 60, 65, 70
9. 14, 24, 34, 44, (48), 54, 64, 74
10. 309, 409, 509, (909), 609, 709, 809

Write the number that comes before and the number that comes next. Name the pattern.

11. __52__, 54, 56, 58, __60__ Skip count by __twos__
12. __60__, 65, 70, 75, __80__ Skip count by __fives__
13. __427__, 437, 447, 457, __467__ Skip count by __tens__
14. __292__, 302, 312, 322, __332__ Skip count by __tens__
15. __13__, 15, 17, 19, __21__ Skip count by __twos__
16. __15__, 20, 25, 30, __35__ Skip count by __fives__

Reflect

How can you find the pattern for a set of numbers?

Answers may vary. Possible answer: Look at a number line and count how many are between each number.

Week 5 **Skip Counting within 1,000** • Lesson 4 61

Student Workbook, pp. 60–61

WEEK 5
Skip Counting within 1,000

Lesson 5 Review

Objective
Students review skills learned this week and complete the weekly assessment and project.

Standard
2.NBT.2 Count within 1000; skip-count by 5s, 10s, and 100s.

Vocabulary
Review vocabulary introduced during the week.

Creating Context
Help English Learners review skip counting. Have students start at 0 on the Number Line to 100 Game Board. Using their fingers or game pieces, students should skip count by twos, fives, or tens. Point out that when we skip count, we are skipping over numbers on the number line.

1 WARM UP

Prepare

▶ **What are some things that come in sets of twos, fives, or tens?** Answers may vary. Possible answers: pairs of shoes, days in a school week, markers in a box

▶ **Pretend you have 10 pairs of shoes. How can you find how many shoes you have in all?** Count by twos 10 times: 2, 4, 6, 8, 10, 12, 14, 16, 18, 20. There are 20 shoes.

▶ **Five friends have 10 markers each. How can we find how many markers there are in all?** Count by tens 5 times: 10, 20, 30, 40, 50. There are 50 markers in all.

2 ENGAGE

Practice
Have students complete **Student Workbook,** pages 62–63.

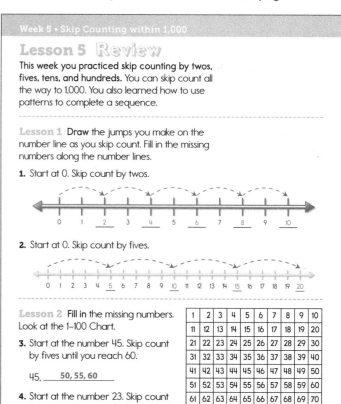

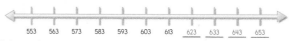

Student Workbook, pp. 62–63

3 REFLECT

Think Critically

Review students' answers to the Reflect prompt at the bottom of *Student Workbook,* page 63.

Discuss the answer with the group to reinforce Week 5 concepts.

4 ASSESS

Formal Assessment

Students may take the weekly assessment online.

As an alternative, students may complete the weekly test on *Assessment,* pages 33–34. Record progress using the Student Assessment Record, *Assessment,* page 128.

Going Forward

Use the **Teacher Dashboard** to view results of the online assessments, to input the results of print student assessments, and to review progress before making decisions about next steps. Use the weekly test results and observations to determine the next steps for each student.

Retention	
Student displays good grasp of this week's concepts and skills.	Have students practice skip counting by twos, fives, and tens with a partner. Encourage students to start at a number other than 0.
Remediation	
Student is still struggling with the week's concepts and skills.	Use the Number Line to 100 Game Board. Have students begin at 0 and count on two. Then have them place a translucent counter on the number 2. Repeat the process, students counting on by two and placing a translucent counter, until they reach 20. Have students practice skip counting by saying the numbers under each counter. Repeat the activity to practice skip counting by fives and tens.

Suggestions for Re-Evaluation: If a student has struggled without success for several weeks, use observations and test results to place the student at a level in which he or she can find success and build confidence to move forward.

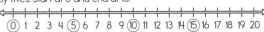

Assessment, pp. 33–34

Week 5 **Skip Counting within 1,000** • Lesson 5   **171**

Project Preview

This week, students learned how to identify and use patterns when skip counting by twos, fives, and tens. Students will need to be able to use what they learned this week to count on from a given number to find the total value of a set of coins.

Project-Based Learning

Standards-driven Project-Based Learning is effective in building deep content understanding. Project-Based Learning increases long-term retention of concepts and has been shown to be more effective than traditional instruction. By completing a project to answer an essential question, students are challenged to apply and demonstrate mastery of concepts and skills by expressing understanding through discussion, research, and presentation.

Essential Question

WHY do I need to understand place value to use money?

Project Evaluation Criteria

Review project evaluation criteria with students prior to beginning the project.

Exceeds Expectations
☐ Project result is explained and can be extended.
☐ Project result is explained in context and can be applied to other situations.
☐ Project result is explained using advanced mathematical vocabulary.
☐ Project result is described, and mathematics are used correctly and can be extended.
☐ Project result is explained and extended, and shows advanced knowledge of mathematical concepts and skills.

Meets Expectations
☐ Project result is explained.
☐ Project result is explained in context.
☐ Project result is explained using mathematical vocabulary.
☐ Project result is described, and mathematics are used correctly.
☐ Project result is explained, and shows satisfactory knowledge of mathematical concepts and skills.

Does Not Meet Expectations
☐ Project result is not explained.
☐ Project result is explained, but out of context.
☐ Project result is explained, but mathematical vocabulary is oversimplified.
☐ Project result is described, but mathematics are not used correctly.
☐ Project result is not explained and/or extended, or shows less than satisfactory knowledge of mathematical concepts and skills.

Saving Money

Objective
Students can use skip counting by fives and tens to count sets of nickels and dimes.

Standard
2.NBT.2 Count within 1000; skip count by 5s, 10s, and 100s.

Materials

Program Materials
- Deposit Slip
- Withdrawal Slip
- Number Line to 100 Game Board

Additional Materials
play money

Best Practices
- Pair oral directions with accessible pictures, icons, or written words for student needs.
- Coach, demonstrate, and model.
- Provide meaning and organization to the lessons and concepts.

Introduce

▶ You want to keep your money safe by keeping it in a piggy bank or at a bank. When you earn more money from doing chores or if someone gives you money, you might want to know the total amount of money you have. You can use skip counting to find the total value of all of your coins.

▶ How would you use skip counting to count a set of dimes? **skip count by tens**

▶ How would you use skip counting to count a set of nickels? **skip count by fives**

Explore

▶ Today you will use skip counting to count on from a starting number to find the total value of a set of coins.

▶ Pretend that you have 20¢ in your piggy bank. You have 4 more dimes you want to put into your piggy bank. How would you use skip counting to find the total value of the coins you have? **I would skip count by tens 4 times from 20: 30, 40, 50, 60. I would have 60¢.**

- Write the amount 20¢ on the board and draw 4 circles, representing dimes, after the number. Write 10¢ inside each circle. Under the circles write 30¢, 40¢, 50¢, and 60¢, respectively, in order to model how to count on from 20¢ using skip counting.

▶ Pretend that you have 75¢. You earn 3 more nickels for walking your dog. You put the 3 nickels into your piggy bank. How would you use skip counting to find the total value of the coins in your piggy bank? **I would skip count by fives 3 times starting at 75: 80, 85, 90. I would have 90¢.**

- Write the number 75¢ on the board and draw 3 circles, representing nickels, after the number. Write 5¢ inside each circle. Under the circles write 80¢, 85¢, and 90¢, respectively, in order to model how to count on from 75¢ using skip counting.

▶ Complete *Student Workbook,* page 64, to practice using skip counting to count on from a given number.

Wrap Up

- Encourage students to model each problem with play money.
- Make sure students are using skip counting to count on from the starting number.
- If students struggle to count on from the starting number, have them find the starting number on the Number Line to 100 Game Board. Then have students count on by ones until they find the next number in the skip counting sequence.
- Discuss students' answers to the Reflect prompt at the bottom of *Student Workbook,* page 65.
- Have students prepare several deposit and withdrawal slips with their bank logo to use during the final week of the project.
- **If time permits,** allow each student to practice his or her skip counting skills by counting additional sets of coins.

Week 5 • Skip Counting Within 1,000

Project
Saving Money

Draw each of the coins you earn. Then skip count to find the total value of all your coins. Write each number you would say when you skip count. Part of the first one has been done for you.

1. You earn 1 dime, 2 nickels, and 6 pennies helping your neighbor plant her garden. How much money do you have?

10¢   15¢   20¢   21¢   22¢   23¢   24¢   25¢

2. You find 2 dimes and 3 nickels under your bed. You want to put that with the money you earned from helping your neighbor plant her garden. How much money do you have now?

25¢   35¢   45¢   50¢   55¢   60¢

Reflect
When do bank tellers use number patterns? Describe the number patterns they use.

Answers may vary. Possible answer: Bank tellers count by tens when they count dimes, and they count by fives when they count nickels.

Student Workbook, p. 64

Teacher Reflect

☐ Did students tell or show the steps when they explained how to do something?

☐ Did students correctly use art, objects, graphics, or posters to explain their solution?

☐ Did students use their time wisely and effectively?

WEEK 6 Comparing Whole Numbers

Week at a Glance

This week students conclude **Number Worlds,** Level D, Number Sense to 1,000, by examining relationships among numbers and comparing the values of whole numbers through the hundreds place.

Skills Focus

- Compare and order whole numbers to 999.
- Order numbers from least to greatest or greatest to least.
- Use the signs <, >, and = to express *less than, greater than,* and *equal to.*

How Students Learn

As students' understanding of place value deepens, so too does their understanding of the relative magnitude of numbers. Students begin to move away from using a mental number list or mental number line for comparing and ordering numbers, and rely more on their knowledge of place value to compare and order quantities.

English Learners ELL

For language support, use the **English Learner Support Guide,** pages 50–51, to preview lesson concepts and teach academic vocabulary. **Number Worlds** Vocabulary Cards are listed as additional materials in many lessons and can be used to preteach and reinforce academic vocabulary.

Math at Home

Give one copy of the Letter to Home, page 12, to each student. Encourage students to share and complete the activity with their caregivers.

Weekly Planner

Lesson	Learning Objectives
1 pages 176–177	Students can compare numbers with the same number of digits and identify the greater or lesser value.
2 pages 178–179	Students can compare numbers with different numbers of digits and identify the greater or lesser value.
3 pages 180–181	Students can order a series of three or four numbers from least to greatest and greatest to least.
4 pages 182–183	Students can correctly use the symbols <, >, and = to relate two numbers.
5 pages 184–185	**Review and Assess** Students review skills learned this week and complete the weekly assessment.
Project pages 186–187	Students can use the Number Sense skills they have gained to act out what it is like to work at a bank.

174 Level D Unit 2 **Number Sense to 1,000**

Key Standard for the Week

Domain: Number and Operations in Base Ten
Cluster: Understand place value.
2.NBT.4 Compare two three-digit numbers based on meanings of the hundreds, tens, and ones digits, using >, =, and < symbols to record the results of comparisons.

Materials		Technology
Program Materials • *Student Workbook,* pp. 65–67 • *Practice,* p. 48 • Activity Card 2N, **Comparison Modeling** • Number Construction Mat • Number Cube (1–6)	Additional Materials • base-ten blocks* • Vocabulary Card 13, *equal* • Vocabulary Card 21, *greater than* • Vocabulary Card 24, *less than*	*Teacher Dashboard*
Program Materials • *Student Workbook,* pp. 68–69 • *Practice,* p. 49 • Activity Card 2S, **Number Line-Up** • Number Construction Mat • Number Cube (1–6)	Additional Materials • base-ten blocks* • Vocabulary Card 13, *equal* • Vocabulary Card 21, *greater than* • Vocabulary Card 24, *less than*	*Teacher Dashboard* Base 10 Blocks Tool
Program Materials • *Student Workbook,* pp. 70–71 • *Practice,* p. 50 • Activity Card 2T, **Numeral Shuffle** • Number Cards (0–9)	Additional Materials • index cards • Vocabulary Card 13, *equal* • Vocabulary Card 21, *greater than* • Vocabulary Card 24, *less than*	*Teacher Dashboard* Base 10 Blocks Tool
Program Materials • *Student Workbook,* pp. 72–73 • *Practice,* p. 51 • Activity Card 2U, **More Hungry Alligators** • Number Cards (0–9)	Additional Materials • construction paper • marker • Vocabulary Card 13, *equal* • Vocabulary Card 21, *greater than* • Vocabulary Card 24, *less than*	*Teacher Dashboard*
Program Materials • *Student Workbook,* pp. 74–75 • Weekly Test, *Assessment,* pp. 35–36		Review previous activities.
Program Materials • *Student Workbook,* pp. 76 • Deposit Slip • Withdrawal Slip	Additional Materials • *greater than*, *less than*, and *equal* symbol cards (construction paper and marker) • play money*	

*Available from McGraw-Hill Education

WEEK 6
Comparing Whole Numbers

Find the Math

In this week, model for students how to use place value to determine the greatest value of money.

Use the following to begin a guided discussion:

▶ **If you wanted to save money for new music, what are some ways you could earn money?** Answers may vary. Possible answers: walk the neighbor's dog, clean out the garage, help clean the house, wash the dishes

Have students complete **Student Workbook,** page 65.

Student Workbook, p. 65

Lesson 1

Objective
Students can compare numbers with the same number of digits and identify the greater or lesser value.

Standard
2.NBT.4 Compare two three-digit numbers based on meanings of the hundreds, tens, and ones digits, using >, =, and < symbols to record the results of comparisons.

Vocabulary
- equal
- greater than
- less than

Creating Context
As English Learners work compare place values, review with them the English pattern of comparatives and superlatives, such as *small, smaller,* and *smallest*.

Materials
Program Materials
- Number Construction Mat
- Number Cube (1–6), 1 per student

Additional Materials
- base-ten blocks
- Vocabulary Card 13, *equal*
- Vocabulary Card 21, *greater than*
- Vocabulary Card 24, *less than*

1 WARM UP

Prepare
Select a volunteer from the class. Write any two-digit number on a sheet of paper and tape it onto the student's back. Have the student turn around so other students can see the number. The student asks questions to find out the number. Questions must be in the following form: *Is it greater than ___? Is it less than ___?*

The student should continue to ask these questions until he or she can identify the number.

2 ENGAGE

Develop: Comparison Modeling

"Today we will use base-ten blocks to compare numbers." Follow the instructions on the Activity Card **Comparison Modeling**. As students complete the activity, be sure to use the Questions to Ask.

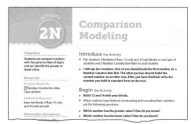

Activity Card 2N

Alternative Groupings

Small Group: Partner with a student if there is an odd number of students and complete the activity as written.

Individual: Have the student build and record both numbers.

Progress Monitoring

| If… students can compare the modeled values easily, | Then… substitute numbers, either written or represented on Number Cards, for them to compare. |

Practice

Have students complete **Student Workbook,** pages 66–67. Guide students through the Key Idea example and the Try This exercises.

Interactive Differentiation

Consult the **Teacher Dashboard** for grouping suggestions. You can also use performance on the Engage activity to guide students.

Independent Practice

For additional practice, organize students in pairs and give each pair a Number Cube. Tell them to take turns rolling the Number Cube three times. Have students use the three numbers they rolled to construct a three-digit number. Then student pairs use place value to compare their number to their partner's number. Instruct students to repeat the activity several times, using a tally chart to keep track of who has the greater number in each round. The partner with the greater number of tallies is the winner.

Supported Practice

For additional support, use a Number Construction Mat to help students understand how to use place value when comparing numbers.

- Have students roll a Number Cube and record the number in the hundreds place on a Number Construction Mat. Instruct students to roll the Number Cube two more times and record the numbers in the tens place and the ones place respectively.
- Allow students to look at one another's mats to find the number with the greatest number of hundreds.
- Draw students' attention to numbers with equal hundreds.
- ▶ **Do we know which is the greater number? Why or why not? What do we need to do?** Answers may vary. Possible answer: No. The number of hundreds is the same, so we need to look at the tens place.
- ▶ **What should we do if the digits in the tens place are the same?** Compare the digits in the ones place.
- Have students repeat the activity. Ask them to explain their reasoning as they identify the greatest number.

REFLECT

Think Critically

Review students' answers to the Reflect prompt at the bottom of **Student Workbook,** page 67, and then review the Engage activity.

Challenge students to evaluate the score and say in their opinion if the score of the game was close or not.

Ask the following questions:

- ▶ **What did you find interesting?**
- ▶ **Can you think of ways you can use the terms *greater than* or *less than* outside of class?** Answers may vary. Possible answer: to buy the cheapest item when shopping

ASSESS

Informal Assessment

Use the online or print Student Record, **Assessment,** page 128, to record informal observations.

Comparison Modeling

Did the student

☐ respond accurately? ☐ respond with confidence?

☐ respond quickly? ☐ self-correct?

Additional Practice

For additional practice, have students complete **Practice,** page 48.

Practice, p. 48

Week 6 • Comparing Whole Numbers

Lesson 1

Key Idea
Compare numbers in each place value beginning at the left of a number.

Step 1: Compare 9 hundreds to 9 hundreds. They are equal. Look at the next place to the right.

Step 2: Compare 5 tens to 2 tens. 5 tens is greater than 2 tens. So, 952 is greater than 927. Write this using the > symbol.

952 > 927

Step 3: The numbers in the tens places were different, so you do not need to compare ones.

Try This

Name the place value in which the digits differ for each pair.

1. 231   237   __ones__   2. 540   530   __tens__

3. 623   523   __hundreds__   4. 827   837   __tens__

Practice
Circle the greater number.

5. 748   (758)
6. (257)   207
7. 524   (534)
8. (482)   399

Circle the lesser number.

9. 455   (445)
10. (730)   831
11. 362   (263)
12. (322)   325

Reflect
In the 2013 NCAA Basketball Championship Game, Michigan scored 76 points. Louisville scored 82 points.

Who won the game?

__Louisville__

Which place-value digit tells you the winner?

__the tens__

Student Workbook, pp. 66–67

Week 6 **Comparing Whole Numbers** • Lesson 1   **177**

WEEK 6
Comparing Whole Numbers

Lesson 2

Objective
Students can compare numbers with different numbers of digits and identify the greater or lesser value.

Standard
2.NBT.4 Compare two three-digit numbers based on meanings of the hundreds, tens, and ones digits, using >, =, and < symbols to record the results of comparisons.

Vocabulary
- equal
- greater than
- less than

Creating Context
A place-value chart is an excellent visual cue for English Learners to support their understanding. They might already have a conceptual understanding of place value but may not have the English fluency to say number names quickly.

Materials
Program Materials
- Number Construction Mat, multiple copies per student
- Number Cube (1–6), 1 for each student pair

Additional Materials
- base-ten blocks
- Vocabulary Card 13, *equal*
- Vocabulary Card 21, *greater than*
- Vocabulary Card 24, *less than*

1 WARM UP

Prepare
▶ **Think about Base-Ten Blocks. Does a flat have greater value than a rod?** yes
▶ **Make a model of a flat and two rods. Then make a model of three rods.**
▶ **Which model has a greater value?** the flat and two rods
▶ **One flat, or hundred, is worth more than no flats or no hundreds.**

Just the Facts
Tell students that you are going to say two numbers. If the first number is the greater number, students should hold up one finger. If the second number is the greater number, students should hold up two fingers. Use sets of numbers such as the following:

▶ **47 or 42** *Students hold up one finger.*
▶ **312 or 337** *Students hold up two fingers.*
▶ **451 or 415** *Students hold up one finger.*

2 ENGAGE

Develop: Number Line-Up
"Today we are going to compare numbers with different numbers of digits." Follow the instructions on the Activity Card **Number Line-Up.** As students complete the activity, be sure to use the Questions to Ask.

Alternative Groupings
Small Group: Complete activity using 2 student-created numbers at a time.

Activity Card 2S

Progress Monitoring
If... students are having difficulty comparing place values,

Then... make sure they understand that a place value may be equal to zero.

Practice
Have students complete **Student Workbook,** pages 68–69. Guide students through the Key Idea example and the Try This exercises.

Interactive Differentiation
Consult the **Teacher Dashboard** for grouping suggestions. You can also use performance on the Engage activity to guide students.

Independent Practice
For additional practice, organize students into pairs. Instruct students to take turns snapping (or clapping) as many times as they can in one minute while their partner helps count the number of snaps. Use a timer so that each student gets the same amount of time. Then have students use place value to determine who was able to snap the greater number of times. Have students repeat the activity to see if they can snap more times than they did before.

Supported Practice
For additional support, instruct students to clap as many times as they can for 60 seconds.

- Have students write down their number of claps.
- Instruct students to clap for another 60 seconds and record their numbers.
- Have students model both numbers using the Base 10 Blocks Tool.
- ▶ **How can you tell which number is greater?** Answers may vary. Possible answer: 105 is greater than 82 because 105 has 1 hundred flat and 82 does not have any.
- If time permits, have students compare their numbers with each other and use place value to determine the greatest number of claps.
- Make sure students understand that a two-digit number has an invisible 0 in the hundreds place.

178 Level D Unit 2 **Number Sense to 1,000**

REFLECT

Think Critically

Review students' answers to the Reflect prompt at the bottom of **Student Workbook,** page 69, and then review the Engage activity.

Discuss the similarities and the differences between the numbers 250 and 25 using base-ten blocks. Ask the following questions:

▶ **If I asked you to compare a three-digit number to a two-digit number, what could you do to make it easier to compare?** I could put a zero in the hundreds place of the two-digit number.

▶ **Tell me something you know about two-digit numbers and three-digit numbers.** Answers may vary. Possible answer: A three-digit number is larger than a two-digit number.

Real-World Application

To determine if a date is before or after another date, you must compare the numbers that tell the day of the month or the number that is assigned to the month. Which date comes first, April 19 or April 9? April 9; the months are the same, so you compare the number of days.

ASSESS

Informal Assessment

Use the online or print Student Record, **Assessment,** page 128, to record informal observations.

Number Line-Up

Did the student

☐ respond accurately?   ☐ respond with confidence?

☐ respond quickly?   ☐ self-correct?

Additional Practice

For additional practice, have students complete **Practice,** page 49.

Practice, p. 49

Week 6 • Comparing Whole Numbers

Lesson 2

Key Idea
When comparing numbers that have different numbers of digits, use zeros in the tens and hundreds places as placeholders so each number has the same number of digits.

Hundreds	Tens	Ones
0	5	7
1	2	4

Compare the hundreds.
1 hundred is greater than 0 hundreds.
So, 124 is greater than 57.

Try This
Compare each pair of numbers. Write the greater number.

1. 224   25   **224**
2. 367   42   **367**
3. 929   842 **929**
4. 899   900 **900**

Make a model of each number using Base-Ten Blocks. Draw the models in the boxes, and then circle the drawing of the greater number.

5. 281   98

Students should draw 2 flats, 8 rods, and 1 block circled.	Students should draw 9 rods and 8 blocks.

68 Level D · Unit 2 · Number Sense to 1,000

Practice
Write each number in the place value boxes. Name the greater number and explain how you know.

6. 439
   524

Hundreds	Tens	Ones
4	3	9
5	2	4

524 is the greater value.

5 hundreds are greater than 4 hundreds.

Compare each pair of numbers. Circle the greater number.

7. 89   (890)
8. (325) 253
9. 367   (376)
10. 582 (692)

Compare each pair of numbers. Circle the lesser number.

11. (295) 925
12. 875   (822)
13. 999   (900)
14. 541   (539)

Reflect
Explain why 999 has a greater value than 99.

Answers will vary. Sample answers: 999 has 9 hundreds that are worth 10 tens each, and 99 has only 9 tens; or hundreds is the greater place value, 999 has 9 hundreds, and 99 has 0 hundreds.

Week 6 Comparing Whole Numbers • Lesson 2   69

Student Workbook, pp. 68–69

WEEK 6
Comparing Whole Numbers

Lesson 3

Objective
Students can order a series of three or four numbers from least to greatest and greatest to least.

Standard
2.NBT.4 Compare two three-digit numbers based on meanings of the hundreds, tens, and ones digits, using >, =, and < symbols to record the results of comparisons.

Vocabulary
- equal
- greater than
- less than

Creating Context
In the Real-World Application in this lesson, we talk about a school election. Not all English Learners at this grade level may be familiar with this school tradition. Explain that *junior* means students in their third year of high school. Review the meaning of *voting*, *election*, and *candidate*.

Materials
Program Materials
- Number Cards (0–9)

Additional Materials
- index cards
- Vocabulary Card 13, *equal*
- Vocabulary Card 21, *greater than*
- Vocabulary Card 24, *less than*

1 WARM UP

Prepare
Give two students each a card with a different two-digit number written on it. Ask them to arrange the cards from least to greatest. Give a third student a card, and have the students figure out where that card should be placed. Does the card go to the right of the first two cards, to the left, or between the cards? Repeat the exercise with three three-digit numbers.

Just the Facts
Write or say groups of three or four numbers. Have students say the place value they should use to compare the numbers. Use sets of numbers such as the following:

▶ **128, 54, 219** hundreds
▶ **245, 237, 222, 208** tens
▶ **623, 621, 629** ones

2 ENGAGE

Develop: Numeral Shuffle
"Today we are going to put three or four numbers in order from least to greatest or greatest to least." Follow the instructions on the Activity Card **Numeral Shuffle** As students complete the activity, be sure to use the Questions to Ask.

Activity Card 2T

Alternative Groupings
Whole Class: Complete the activity as written.

Progress Monitoring

| If... students feel that the combined list has too many numbers to deal with at one time, | Then... have students make an ordered list of one-digit numbers, a list of two-digit numbers, and a list of three-digit numbers. They can then order the lists to make one complete list. |

Practice
Have students complete **Student Workbook,** pages 70–71. Guide students through the Key Idea example and the Try This exercises.

Interactive Differentiation
Consult the **Teacher Dashboard** for grouping suggestions. You can also use performance on the Engage activity to guide students.

Independent Practice
For additional practice, have students count the total number of coloring utensils in their desk (crayons, markers, colored pencils, etc.). Have students write the number on an index card. If the class is large, organize students into small groups. Then have group members compare the numbers on their cards. Instruct students to use place value to order the index cards from least to greatest. If time permits, have students rearrange the cards in order from greatest to least.

Supported Practice
For additional support, write 425, 452, 557, and 21 on the board.

- Have students use the Base 10 Blocks Tool to model each of the numbers.
- ▶ **We want to put these numbers in order from least to greatest. Which place value should we look at first? Why?** Look at the hundreds place first because it is the greatest place value.
- ▶ **Which is the least number? Explain how you know.** The least number is 21 because it does not have any hundred flats.
- Instruct students to look at the remaining numbers. Have them identify the number of hundreds in each number.
- ▶ **425 and 452 each have 4 hundreds flats. We are ordering from least to greatest, so how do we know which number is less?** We need to look at the tens rods. 425 is less than 452 because 2 tens is less than 5 tens.
- Guide students to continue using place value as they finish ordering the numbers from least to greatest.

180 Level D Unit 2 **Number Sense to 1,000**

3 REFLECT

Think Critically

Review students' answers to the Reflect prompt at the bottom of **Student Workbook,** page 71, and then review the Engage activity.

Point out that the lowest number is actually the highest number with the digits written backward. In the highest number, the numerals go down in value (7-5-2), and in the lowest number, the numerals go up in value (2-5-7).

Ask the following questions:

▶ **Tell me something you know about ordering numbers.** Answers may vary. Possible answer: A three-digit number will come before a two-digit number if you are ordering from largest to smallest.

▶ **Can you think of a way we could use ordering in class?** Answers may vary. Possible answer: to line up from oldest to youngest

Real-World Application

▶ **To judge which candidate has won an election, the votes must be compared and ordered. In an election for the junior class president at East High School, there were four candidates. Brian received 98 votes, Carly 132 votes, Michelle 125 votes, and Jack 119 votes. Who won the election?** Carly

4 ASSESS

Informal Assessment

Use the online or print Student Record, **Assessment,** page 128, to record informal observations.

Numeral Shuffle

Did the student

☐ respond accurately?   ☐ respond with confidence?

☐ respond quickly?   ☐ self-correct?

Additional Practice

For additional practice, have students complete **Practice,** page 50.

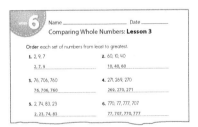

Practice, p. 50

Week 6 • Comparing Whole Numbers

Lesson 3

Key Idea
Ordering numbers means listing them from least to greatest or greatest to least.
36, 45, 80 are ordered least to greatest.
700, 500, 300 are ordered greatest to least.

Try This
Order each set of numbers from least to greatest.

1. 6, 3, 5   __3, 5, 6__
2. 700, 300, 900, 100   __100, 300, 700, 900__
3. 349, 348, 350   __348, 349, 350__
4. 230, 130, 30   __30, 130, 230__

Practice
Order the numbers from greatest to least.

5. 5, 8, 7   __8, 7, 5__
6. 501, 511, 491   __511, 501, 491__
7. 600, 6, 60   __600, 60, 6__
8. 43, 177, 431, 290   __431, 290, 177, 43__
9. 36, 98, 48, 70   __98, 70, 48, 36__
10. 123, 99, 141, 107   __141, 123, 107, 99__

70 Level D Unit 2 **Number Sense to 1,000**

Order these distances from greatest to least. Write the distance that received first, second, third, and fourth place in the track and field meet.

11. Shot Put   30 feet, 34 feet, 28 feet, 12 feet

    First Place   __34 feet__

    Second Place   __30 feet__

    Third Place   __28 feet__

    Fourth Place   __12 feet__

12. Discus   115 feet, 108 feet, 160 feet, 157 feet

    First Place   __160 feet__

    Second Place   __157 feet__

    Third Place   __115 feet__

    Fourth Place   __108 feet__

Reflect
Use the digits 2, 7, and 5 to make every three-digit number that can be written using the digits. What is the largest number you can make? What is the smallest number you can make?

__257, 275, 527, 572, 725, 752; largest is 752; smallest is 257__

Week 6 Comparing Whole Numbers • Lesson 3 71

Student Workbook, pp. 70–71

WEEK 6
Comparing Whole Numbers

Lesson 4

Objective
Students can correctly use the symbols <, >, and = to relate two numbers.

Standard
2.NBT.4 Compare two three-digit numbers based on meanings of the hundreds, tens, and ones digits, using >, =, and < symbols to record the results of comparisons.

Vocabulary
- equal
- greater than
- less than

Creating Context
In this lesson, an analogy to an alligator mouth is used to help students better understand the meaning of the directionality of the relationship symbols. Some English Learners may be unfamiliar with the alligator but may know other animals with large beaks or snouts, such as a toucan or pelican.

Materials
Program Materials
Number Cards (0–9)

Additional Materials
- construction paper
- marker
- Vocabulary Card 13, *equal*
- Vocabulary Card 21, *greater than*
- Vocabulary Card 24, *less than*

Preparing Ahead
Use construction paper and a marker to create three sets of symbol <, >, and = cards. Write the text shown under each symbol.

1 WARM UP

Prepare
Brainstorm as a class signs or symbols you see in math or in the world. Write them on the board. Explain that these symbols usually stand for a word or words. Some examples include **$ (dollar)**, **+ (plus)**, **& (and)**, and **= (equal value)**.

▶ **Do you remember what the symbol** (display >) **means?**
greater than

▶ **Do you remember what the symbol** (display <) **means?**
less than

Just the Facts
Write two numbers with similar place values. Have students point to the greater number. Use numbers such as the following:

▶ **56 and 52** *Students should point to 56.*
▶ **102 and 120** *Students should point to 120.*
▶ **471 and 461** *Students should point to 471.*

2 ENGAGE

Develop: More Hungry Alligators

"Today we are going to use comparison symbols to show which numbers are greater than and less than." Follow the instructions on the Activity Card **More Hungry Alligators**. As students complete the activity, be sure to use the Questions to Ask.

Activity Card 2U

Alternative Groupings
Small Group: Use index cards instead of construction paper. Complete the activity as written placing the numbers and symbol on a desk or table that allows all members of the group to participate.

Pair: Use two or three sets of the 0–9 Number Cards to generate the two numbers. Each member of the pair creates a number and writes it on a piece of paper with room between the numbers for a symbol. Together they decide the correct symbol and complete the number sentence.

Progress Monitoring
If… students have trouble remembering that < means "less than" and > means "greater than,"

Then… provide students with a number line so that they can see which number is farthest to the right for *greater than* and farthest to the left for *less than*.

Practice
Have students complete **Student Workbook**, pages 72–73. Guide students through the Key Idea example and the Try This exercises.

Interactive Differentiation
Consult the **Teacher Dashboard** for grouping suggestions. You can also use performance on the Engage activity to guide students.

Independent Practice
For additional practice, provide students with construction paper and coloring utensils. Have students draw the comparison symbol (< or >) and decorate it to look like a hungry alligator. Instruct students to write two numbers and then correctly position the hungry alligator to compare the two numbers. Provide students with the following pairs of numbers: 15 and 41, 88 and 206, 318 and 875, 23 and 427, and 930 and 582.

Supported Practice
For additional support, provide students with construction paper and coloring utensils. Have students draw the comparison symbol (< or >) and decorate it to look like a hungry alligator.

- Have students identify the number of pages in their math book as well as another large book such as a science textbook or teacher's manual. Instruct students to write the two numbers.

▶ **Do hungry alligators like to eat the greater or the lesser number?**
the greater number

▶ **How can we use place value to find the number the alligator wants to eat?** Answers may vary. Possible answer: Look at the greatest place value and compare the values of the digits.

- Have students position their hungry alligators to correctly compare the two numbers.

182 Level D Unit 2 **Number Sense to 1,000**

3 REFLECT

Think Critically

Review students' answers to the Reflect prompt at the bottom of *Student Workbook*, page 73, and then review the Engage activity.

Discuss with the class what they have learned in previous weeks to help them find the answer to the third model. Trading and modeling numbers in different ways should be addressed. Ask the following question:

▶ **Can you ever put the equal sign between two different numbers?**
No, because *equal* means "the same."

Real-World Application

▶ **Parents use "greater than" and "less than" when they look through ads to see which store has the best deals. For instance one store may sell milk for $4.20 and another may sell it for $3.90. Which store would your parents likely go to? Why?** Answers may vary. Possible answer: the second one; milk is cheaper there

4 ASSESS

Informal Assessment

Use the online or print Student Record, *Assessment,* page 128, to record informal observations.

More Hungry Alligators
Did the student
☐ respond accurately?   ☐ respond with confidence?
☐ respond quickly?   ☐ self-correct?

Additional Practice

For additional practice, have students complete *Practice,* page 51.

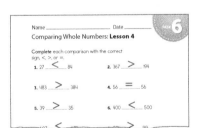
Practice, p. 51

Week 6 • Comparing Whole Numbers

Lesson 4

Key Idea
< means "less than"   Example: 13 < 49
> means "greater than" Example: 5 > 2
= means "equal to"   Example: 450 = 450

Try This
Use Base-Ten Blocks to make a model of each number. Complete the comparison using the correct sign, <, >, or =.

1. 365 < 833   2. 500 < 501
3. 645 > 644   4. 16 > 11
5. 200 < 300   6. 98 > 89
7. 7 = 7   8. 924 > 915

Practice
Use Base-Ten Blocks to make a model of each number. Mark these comparisons *true* or *false*.

9. 72 > 62 true   10. 100 < 50 false
11. 8 < 13 true   12. 424 > 390 true
13. 123 = 321 false   14. 76 < 77 true

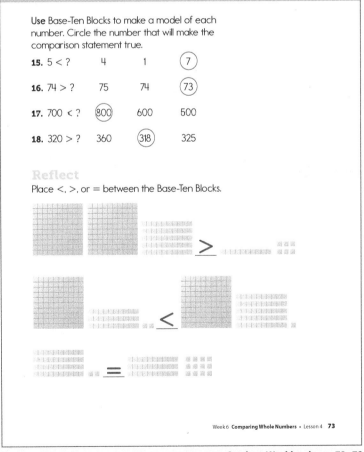

Use Base-Ten Blocks to make a model of each number. Circle the number that will make the comparison statement true.

15. 5 < ?   4   1   ⑦
16. 74 > ?   75   74   ㉓
17. 700 < ? ⑧⓪⓪ 600 500
18. 320 > ? 360 ㉛⑧ 325

Reflect
Place <, >, or = between the Base-Ten Blocks.

Student Workbook, pp. 72–73

WEEK 6
Comparing Whole Numbers

Lesson 5 Review

Objective
Students review skills learned this week and complete the weekly assessment.

Standard CCSS
2.NBT.4 Compare two three-digit numbers based on meanings of the hundreds, tens, and ones digits, using >, =, and < symbols to record the results of comparisons.

Vocabulary
Review vocabulary introduced during the week.

Creating Context
English Learners can accelerate their acquisition of English by knowing some of the ways that their language relates to English. Languages that share Greek, Latin, and Arabic roots with English may have many root words that are the same. For speakers of Spanish, for example, words that end with *-tion* in English, almost always end with *-ción* or *-sión* in Spanish. In this lesson, *comparison* is an example.

1 WARM UP

Prepare

▶ Write three number sentences that use the < or > symbol. At least one of the sentences must include a three-digit number.

▶ Trade with a neighbor and check each other's sentences to be sure they are true.

2 ENGAGE

Practice
Have students complete **Student Workbook,** pages 74–75.

Week 6 • Comparing Whole Numbers
Lesson 5 Review

This week you explored the value of each digit in a number. You learned how to compare numbers. You also learned the meaning of *greater than* and *less than*.

Lesson 1 Compare each pair of numbers. Circle the greater number.
1. 411 (421)
2. 129 (921)

Circle the lesser number. Name which place you used to decide.
3. 217 (127) __hundreds__
4. (418) 481 __tens__

Lesson 2 Compare each pair of numbers. Circle the lesser number.
5. 525 (255)
6. (728) 982

Circle the greater number. Name which place you used to decide.
7. 207 (702) __hundreds__
8. 413 (431) __tens__

Reflect
Write a number that is less than 439 but has equal values in the hundreds and tens places.

Answers will vary but must be any number from 430 to 438.

Lesson 3 Order these numbers from greatest to least.
9. 20, 8, 17 __20, 17, 8__
10. 475, 435, 450 __475, 450, 435__

Order these numbers from least to greatest.
11. 52, 50, 54 __50, 52, 54__
12. 25, 125, 5, 105 __5, 25, 105, 125__
13. 679, 355, 420 __355, 420, 679__
14. 219, 265, 202, 243 __202, 219, 243, 265__

Lesson 4 Complete the comparison using <, >, or =.
15. 299 __<__ 301
16. 48 __>__ 39
17. 7 __=__ 7
18. 170 __>__ 70
19. 3 __<__ 4
20. 470 __<__ 472
21. 97 __=__ 97
22. 100 __>__ 99

Reflect
Complete each statement so that it states a true comparison.

524 > __any number less than 524__

__any number less than 62__ < 62

Student Workbook, pp. 74–75

3 REFLECT

Think Critically

Review students' answers to the Reflect prompt at the bottom of **Student Workbook,** page 75.

Discuss the answer with the group to reinforce Week 6 concepts.

4 ASSESS

Formal Assessment

Students may take the weekly assessment online.

As an alternative, students may complete the weekly test on **Assessment,** pages 35–36. Record progress using the Student Assessment Record, **Assessment,** page 128.

Going Forward

Use the **Teacher Dashboard** to view results of the online assessments, to input the results of print student assessments, and to review progress before making decisions about next steps. Use the weekly test results and observations to determine the next steps for each student.

Retention	
Student displays good grasp of this week's concepts and skills.	Have students time one another performing various repetitive tasks. Use a timer so that each student gets the same amount of time. Then have students use place value to determine who was able to perform the task the greater number of times. Have students repeat the activity to see if they can perform the task more times.

Remediation	
Student is still struggling with the week's concepts and skills.	Instruct students to write two numbers on opposite ends of a piece of paper and then add the comparison symbol that correctly compares the two numbers. Provide students with the following pairs of numbers: 28 and 56, 306 and 360, 825 and 852, 910 and 901, and 582 and 527.

Suggestions for Re-Evaluation: If a student has struggled without success for several weeks, use observations and test results to place the student at a level in which he or she can find success and build confidence to move forward.

Name_____ Date_____
Comparing Whole Numbers — WEEK 6

1. Fill in the place values boxes for each number.

	Hundreds	Tens	Ones
73	0	7	3
98	0	9	8
256	2	5	6
410	4	1	0
681	6	8	1

Circle the greater number in each group.

2. 61 (62)

3. (612) 512

4. 374 (384)

Level D Unit 2 Week 6   35

WEEK 6 Name_____ Date_____
Comparing Whole Numbers

Circle the lesser number in each group.

5. 47 (37)
6. (382) 392
7. 926 (916)

Order each set of numbers from smallest to largest.

8. 405, 45, 540, 54   **45, 54, 405, 540**

9. 683, 38, 83, 368   **38, 83, 368, 683**

Order each set of numbers from largest to smallest.

10. 712, 21, 217, 12   **712, 217, 21, 12**

11. 19, 913, 31, 319   **913, 319, 31, 19**

Write the correct sign for each comparison.

12. 48 $>$ 37   13. 93 $<$ 112

14. 80 $=$ 80   15. 747 $>$ 574

16. 609 $<$ 699

36 Level D Unit 2 Week 6

Assessment, pp. 35–36

Week 6 **Comparing Whole Numbers** • Lesson 5   **185**

Project Preview

This week, students learned how to compare and order whole numbers based on place value. They will use what they have learned during this unit to model what it is like to work at and visit a bank. They will use place value to model and compare sets of dollars and coins.

Project-Based Learning

Standards-driven Project-Based Learning is effective in building deep content understanding. Project-Based Learning increases long-term retention of concepts and has been shown to be more effective than traditional instruction. By completing a project to answer an essential question, students are challenged to apply and demonstrate mastery of concepts and skills by expressing understanding through discussion, research, and presentation.

Essential Question

WHY do I need to understand place value to use money?

Project Evaluation Criteria

Review project evaluation criteria with students prior to beginning the project.

Exceeds Expectations
☐ Project result is explained and can be extended.
☐ Project result is explained in context and can be applied to other situations.
☐ Project result is explained using advanced mathematical vocabulary.
☐ Project result is described, and mathematics are used correctly and can be extended.
☐ Project result is explained and extended, and shows advanced knowledge of mathematical concepts and skills.

Meets Expectations
☐ Project result is explained.
☐ Project result is explained in context.
☐ Project result is explained using mathematical vocabulary.
☐ Project result is described, and mathematics are used correctly.
☐ Project result is explained, and shows satisfactory knowledge of mathematical concepts and skills.

Does Not Meet Expectations
☐ Project result is not explained.
☐ Project result is explained, but out of context.
☐ Project result is explained, but mathematical vocabulary is oversimplified.
☐ Project result is described, but mathematics are not used correctly.
☐ Project result is not explained and/or extended, or shows less than satisfactory knowledge of mathematical concepts and skills.

Be the Banker

Objective
Students can use the Number Sense skills they have gained to act out what it is like to work at a bank.

Standard
2.NBT.4 Compare two three-digit numbers based on meanings of the hundreds, tens, and ones digits, using >, =, and < symbols to record the results of the comparisons.

Materials

Program Materials
- Deposit Slip
- Withdrawal Slip

Additional Materials
- *greater than, less than, equal* symbols on paper
- play money

Prepare Ahead
For the final activity, students will need their completed bank logo poster, blank deposit and withdrawal slips with their bank's logo, and the greater than and less than symbols drawn on two pieces of paper.

Best Practices
- Set clear expectations, rules, and procedures.
- Focus students on their work to maintain engagement.
- Create an energetic environment.

Introduce

Today students will practice using the skills they learned in previous weeks as they pretend to be bank workers and bank customers.

▶ **What are some things you need to be able to do to work at a bank?**
Answers may vary. Possible answer: I need to be able to count by fives and tens to count nickels and dimes and trade coins for other coins with the same value.

▶ **What are some things you might do if you are a bank customer?**
Answers may vary. Possible answer: I might deposit money in my bank account or withdraw money from my bank account so I can spend it at a store.

Explore

▶ **Today you and your partner will set up your bank. Sometimes you will be the banker at your bank and other times you will be the customer at another bank.**

- Allow students time to set up their banks. Help them display their bank logo posters somewhere close nearby. Remind them to set out deposit and withdrawal slips for customers to use.

- Provide each bank and each customer with play money. Include pennies, nickels, dimes, and dollars.

▶ **Partners, decide who will be the banker first and who will be the customer.**

- Allow customers time to visit other banks. Instruct them to use a deposit slip and a set of coins to deposit money into a bank account. Bankers should count the coins to make sure their value matches the number on the deposit slip. Students may also withdraw money from their accounts by filling out a withdrawal slip. Bankers should provide customers with the fewest number of coins possible. Customers may also trade several coins for fewer coins with the same value.

▶ **Complete Student Workbook, page 76, to record what you did as a banker and what you did as a customer.**

Wrap Up

- Once each student has had a turn to be both banker and customer, tell students to return to their own banks.

▶ **Find any two deposit slips that your customers filled out. Set them on opposite ends of your desk. Use greater than symbol to show the greater number.**

- Have students continue to use the greater than and less than symbols to compare several completed deposit and withdrawal slips.

- Engage students in a discussion about how they used place value to perform their roles as bankers and customers.

- Discuss students' answers to the Reflect prompts at the bottom of **Student Workbook**, page 76.

- Before students break down their banks, take a photo of each pair of students with their banks. Add the photos to students' project folders.

- If time permits, allow each student to work with his or her partner to write a brief journal entry about their bank. They could write about how they chose the name of their bank and what inspired their bank logo design, or they could describe what they learned about banking over the course of the unit.

Week 6 • Comparing Whole Numbers

Project
Be the Banker

Describe two jobs you did as a banker and two jobs you did as a customer. Tell about each job you did. Write about the money you counted or used to do your job. For example:
"When I was the banker, Ethan wanted to trade 25 pennies. I gave him 2 dimes and 1 nickel."

1. When I was the banker,
   Answers may vary. Possible answer: I traded 230 pennies for 23 dimes.

2. When I was the banker,
   Answers may vary. Possible answer: I skip counted to find 465 cents.

3. When I was the customer,
   Answers may vary. Possible answer: I deposited 693 pennies.

4. When I was the customer,
   Answers may vary. Possible answer: I withdrew 490 cents in dimes.

Reflect
When did you need to use place value to do your job as a banker or as a customer?

Answers may vary. Possible answer: I used place value when I counted a group of dimes by tens. The ones digit stayed the same, but the tens digit went up by 1 for each dime I counted.

76 Level D Unit 2 Number Sense to 1,000

Student Workbook, p. 76

Teacher Reflect

☐ Did I clearly explain how to organize the activity?

☐ Did students use their time wisely and effectively?

☐ Did students show knowledge of how their project related to the major concept?

UNIT 3 Addition

Unit at a Glance
This **Number Worlds** unit builds on prior knowledge of addition. Students will apply this knowledge to extend their understanding of single-digit addition to develop strategies for adding multi-digit numbers with regrouping and to solve addition equations using the standard algorithm.

Skills Trace

Before Level D	Level D	After Level D
Level C Students can add single-digit numbers and can also solve multiple addend problems.	By the end of this unit, students should be able to solve one- and two-digit addition equations within 1,000, including word problems. They can also add within 1,000 using concrete models and drawings.	**Moving on to Level E** Students will be able to add two- and three-digit numbers with regrouping and will master mental addition strategies.

Learning Technology
The following activities are available online to support the learning goals in this unit.

Building Blocks
- Barkley's Bones 1–10
- Barkley's Bones 1–20
- Bright Idea: Counting On Game
- Dinosaur Shop 3
- Dinosaur Shop 4
- Easy as Pie
- Eggcellent: Addition Choice
- Figure the Fact
- Lots O' Socks
- Number Compare 4
- Number Snapshots 5
- Number Snapshots 6
- Number Snapshots 7
- Number Snapshots 8
- Numeral Train
- Off the Tree
- Tidal Tally
- Word Problems with Tools 1
- Word Problems with Tools 4
- Word Problems with Tools 7

Digital Tools
- The Addition Table
- Base 10 Blocks Tool
- Number Line Tool
- Sets Former Tool

Unit Overview

Week	Focus
1	**Addition Fundamentals** • *Teacher Edition,* pp. 190–203 • *Activity Cards,* 3A, 3B, 3C, 3D • *Student Workbook,* pp. 5–16 • *English Learner Support Guide,* pp. 58–59 • *Assessment,* pp. 37–38
2	**Mastering the Basic Facts** • *Teacher Edition,* pp. 204–217 • *Activity Cards,* 3E, 3F, 3G, 3H • *Student Workbook,* pp. 17–28 • *English Learner Support Guide,* pp. 60–61 • *Assessment,* pp. 39–40
3	**Solving Addition Problems** • *Teacher Edition,* pp. 218–231 • *Activity Cards,* 3I, 3J, 3K, 3L • *Student Workbook,* pp. 29–40 • *English Learner Support Guide,* pp. 62–63 • *Assessment,* pp. 41–42
4	**Addition Tools and Strategies** • *Teacher Edition,* pp. 232–245 • *Activity Cards,* 3M, 3N, 3O, 3P • *Student Workbook,* pp. 41–52 • *English Learner Support Guide,* pp. 64–65 • *Assessment,* pp. 43–44
5	**Addition Word Problems within 100** • *Teacher Edition,* pp. 246–259 • *Activity Cards,* 3Q, 3R, 3S, 3T • *Student Workbook,* pp. 53–64 • *English Learner Support Guide,* pp. 66–67 • *Assessment,* pp. 45–46
6	**Solving Addition Word Problems within 1,000** • *Teacher Edition,* pp. 260–273 • *Activity Cards,* 3U, 3V, 3W, 3X • *Student Workbook,* pp. 65–76 • *English Learner Support Guide,* pp. 68–69 • *Assessment,* pp. 47–48

Essential Question

HOW can I use addition to solve real-world problems?

In this unit, students will explore how addition can be used to solve real-world problems by creating a poster and presentation about a collection.

Learning Goals	CCSS Key Standards
Students can add a series of single-digit numbers, understand the meaning of the plus sign and the equal sign, and recognize and create addition equations. **Project:** Students can use addition fundamentals to describe their collections or parts of their collections.	**Domain:** Operations and Algebraic Thinking **Cluster:** Add and subtract within 20. **2.OA.2:** Fluently add and subtract within 20 using mental strategies. By end of Grade 2, know from memory all sums of two one-digit numbers.
Students can use an addition table to solve problems, recognize and solve doubles and near-doubles facts up to 20, and choose a strategy to solve addition facts with sums up to 20. **Project:** Students can identify how many items they have in their collections by counting, counting on, or adding.	**Domain:** Operations and Algebraic Thinking **Cluster:** Add and subtract within 20. **2.OA.2:** Fluently add and subtract within 20 using mental strategies. By end of Grade 2, know from memory all sums of two one-digit numbers.
Students can write and solve addition problems with more than two addends, reorder and group addends to make addition easier, and solve addition word problems. **Project:** Students can write addition word problems and the equations to solve them within 100.	**Domain:** Number and Operations in Base Ten **Cluster:** Use place value understanding and properties of operations to add and subtract. **2.NBT.5:** Fluently add and subtract within 100 using strategies based on place value, properties of operations, and/or the relationship between addition and subtraction.
Students can add successive single-digit quantities, write equations, and find missing addends. **Project:** Students can identify tools and strategies that can be used to solve addition problems.	**Domain:** Number and Operations in Base Ten **Cluster:** Use place value understanding and properties of operations to add and subtract. **2.NBT.5:** Fluently add and subtract within 100 using strategies based on place value, properties of operations, and/or the relationship between addition and subtraction.
Students can write and solve one- and two-step equations with unknown sums and addends. **Project:** Students can create and solve word problems with unknowns.	**Domain:** Operations and Algebraic Thinking **Cluster:** Represent and solve problems involving addition and subtraction. **2.OA.1:** Use addition and subtraction within 100 to solve one- and two-step word problems involving situations of adding to, taking from, putting together, taking apart, and comparing, with unknowns in all positions, e.g., by using drawings and equations with a symbol for the unknown number to represent the problem.
Students can add one- and two-digit addition word problems within 1,000. **Project:** Students can add multi-digit numbers with regrouping.	**Domain:** Number and Operations in Base Ten **Cluster:** Use place value understanding and properties of operations to add and subtract. **2.NBT.7:** Add and subtract within 1000, using concrete models or drawings and strategies based on place value, properties of operations, and/or the relationship between addition and subtraction; relate the strategy to a written method. Understand that in adding or subtracting three-digit numbers, one adds or subtracts hundreds and hundreds, tens and tens, ones and ones; and sometimes it is necessary to compose or decompose tens or hundreds.

CCSS Daily lesson activities emphasize using communication, logic, reasoning, modeling, tools, precision, structure, and patterns to solve problems. All student activities, reflections, and assessments require application of the **Common Core Standards for Mathematical Practice.**

WEEK 1: Addition Fundamentals

Week at a Glance

This week, students begin **Number Worlds,** Level D, Addition. Students explore different ways to visualize and represent addition problems using manipulatives and write addition equations.

Skills Focus

- Add a series of single-digit numbers.
- Understand the meaning of the plus sign and the equal sign.
- Recognize and create addition equations.

How Students Learn

Begin to introduce the strategy of doubles facts with easier doubles facts, such as $2 + 2$. Throughout Unit 3, examine the near-doubles facts related to each doubles fact.

English Learners ELL

For language support, use the **English Learner Support Guide,** pages 58–59, to preview lesson concepts and teach academic vocabulary. **Number Worlds** Vocabulary Cards are listed as additional materials in many lessons and can be used to preteach and reinforce academic vocabulary.

Math at Home

Give one copy of the Letter to Home, page 13, to each student. Encourage students to share and complete the activity with their caregivers.

Weekly Planner

Lesson	Learning Objectives
1 pages 192–193	Students can understand the relative magnitude of double-digit numbers.
2 pages 194–195	Students can understand the connection between counting and addition.
3 pages 196–197	Students can use the plus sign and the equal sign to create addition equations.
4 pages 198–199	Students can understand that the order in which you add addends does not affect the sum.
5 pages 200–201	**Review and Assess** Students review skills learned this week and complete the weekly assessment and project.
Project pages 202–203	Students can use addition fundamentals to describe their collections or parts of their collections.

Key Standard for the Week

Domain: Operations and Algebraic Thinking
Cluster: Add and subtract within 20.

2.OA.2 Fluently add and subtract within 20 using mental strategies. By end of Grade 2, know from memory all sums of two one-digit numbers.

Materials		Technology
Program Materials • **Student Workbook,** pp. 5–7 • **Practice,** p. 52 • Activity Card 3A, **Count and Compare** • Count and Compare	**Additional Materials** • glue • scissors	**Teacher Dashboard** Building Blocks Number Compare 4 Sets Former Tool
Program Materials • **Student Workbook,** pp. 8–9 • **Practice,** p. 53 • Activity Card 3B, **Meet the Dragon** • Dragon Quest Game Boards • Dragon Quest Cards • Counters • Pawns • Spinners	**Additional Materials** • Vocabulary Card 1, *add* • Vocabulary Card 13, *equal* • Vocabulary Card 43, *sum*	**Teacher Dashboard** Building Blocks Lots O' Socks
Program Materials • **Student Workbook,** pp. 10–11 • **Practice,** p. 54 • Activity Card 3C, **Making and Writing Equations** • Counters	**Additional Materials** • index cards • Vocabulary Card 1, *add* • Vocabulary Card 13, *equal* • Vocabulary Card 43, *sum*	**Teacher Dashboard** Building Blocks Off The Tree; Numeral Train
Program Materials • **Student Workbook,** pp. 12–13 • **Practice,** p. 55 • Activity Card 3D, **Drop and Add** • Counters	**Additional Materials** • coffee cans or other containers • Vocabulary Card 1, *add* • Vocabulary Card 13, *equal* • Vocabulary Card 43, *sum*	**Teacher Dashboard** Building Blocks Dinosaur Shop 3 Sets Former Tool
Program Materials • **Student Workbook,** pp. 14–15 • Weekly Test, **Assessment,** pp. 37–38		Review previous activities.
Program Materials **Student Workbook,** p. 16	**Additional Materials** • large poster paper • personal collections (rocks, books, stamps, stickers, coins) • pictures of different types of collections	

WEEK 1
Addition Fundamentals

Find the Math

Introduce students to counting objects in groups and finding the sum of the groups.

Use the following to begin a guided discussion:

▶ **Do you have a collection? What do you collect? Do you know how many items are in your collection? How do you organize your collection?** Answers may vary. Possible answer: I keep seashells in boxes.

Have students complete **Student Workbook**, page 5.

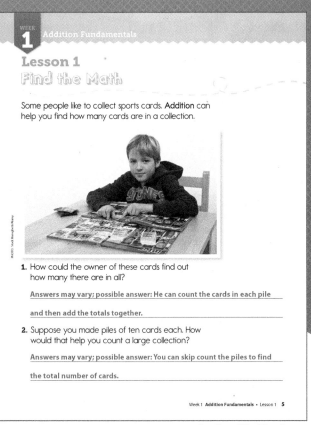

Student Workbook, p. 5

Lesson 1

Objective
Students can understand the relative magnitude of two-digit numbers.

Standard CCSS
2.OA.2 Fluently add and subtract within 20 using mental strategies. By end of Grade 2, know from memory all sums of two one-digit numbers.

Creating Context
What Number Am I? is a game in which students pretend they are a mystery number and others must guess which number they are. When you explain the rules of the game, spend a few minutes explaining role-playing.

Materials
Program Materials
 Count and Compare, 1 per student

Additional Materials
- glue
- scissors

1 WARM UP

Prepare
Follow the directions on Activity Card 1R to present students with **What Number Am I?** riddles. Allow volunteers to guess the answers, and ask them to explain their reasoning.

▶ I am greater than 22 and less than 24. What number am I? 23
▶ I am less than 49 and greater than 47. What number am I? 48
▶ I am less than 17 and greater than 15. What number am I? 16
▶ I am less than 84 and greater than 82. What number am I? 83
▶ I am greater than 54 and less than 56. What number am I? 55
▶ I am less than 72 and greater than 70. What number am I? 71
▶ I am greater than 98 and less than 100. What number am I? 99

2 ENGAGE

Develop: Count and Compare

"Today we are going to use numbers to tell how many objects are in a set." Follow the instructions on the Activity Card **Count and Compare**. As students complete the activity, be sure to use the Questions to Ask.

Activity Card 3A

Alternative Grouping
Individual: Have each student complete the activity as written.

Progress Monitoring	
If... students need extra practice matching numerals to set size and using numerals to make relative quantity predictions,	▶ **Then...** have them cut out each set of objects with the corresponding number and sequence the pictures.

Practice
Have students complete **Student Workbook**, pages 6–7. Guide students through the Key Idea example and the Try This exercises.

192 Level D Unit 3 **Addition**

Interactive Differentiation

Consult the Teacher Dashboard for grouping suggestions. You can also use performance on the Engage activity to guide students.

Independent Practice

For additional practice with comparing numbers, have children use Number Compare 4: Numerals to 100. Limit the values to a maximum of 20.

Supported Practice

For additional support, use the Sets Former Tool with one student or a small group of students.

- At the bottom of the screen, add the tool, and adjust the workspace low enough so the totals are not visible.
- Put a number of marbles on the mat and have students count the total number of marbles. Then move the workspace up so students can check the totals.
- Next, move the workspace down again, and have students put a specific number of marbles on the mat.
- Once again, move the workspace up, so students can check the totals.

3 REFLECT

Think Critically

Review students' answers to the Reflect prompt at the bottom of *Student Workbook,* page 7, and then review the Engage activity.

Discuss how comparing parts of a set is similar to and different from comparing multiple sets.

▶ **How was comparing the number of colored balloons like comparing the number of objects in Count and Compare?** Answers may vary. Possible answer: You have to count all the objects when you make both comparisons.

▶ **How was it different?** Answers may vary. Possible answer: In **Count and Compare,** you can see which lines are longer because the objects you are comparing are on different lines. All the balloons are on the same line.

▶ **Which one was easier? Why?** Answers may vary. Possible answer: **Count and Compare** was easier because I could quickly see which row was longer.

4 ASSESS

Informal Assessment

Use the online or print Student Record, *Assessment,* page 128, to record informal observations.

Count and Compare
Did the student
- ☐ make important observations?
- ☐ provide insightful answers?
- ☐ extend or generalize learning?
- ☐ pose insightful questions?

Additional Practice

For additional practice, have students complete *Practice,* page 52.

Practice, p. 52

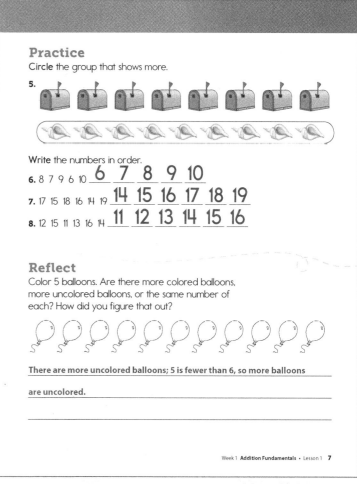

Student Workbook, pp. 6–7

WEEK 1
Addition Fundamentals

Lesson 2

Objective
Students can understand the connection between counting and addition.

Standard
2.OA.2 Fluently add and subtract within 20 using mental strategies. By end of Grade 2, know from memory all sums of two one-digit numbers.

Vocabulary
- add
- addition
- sum
- addends
- equal

Creating Context
English Learners may benefit from a review of the language patterns used to compare. Review comparative and superlative adjectives, such as *large, larger,* and *largest*.

Materials
Program Materials
- Dragon Quest Game Board, 1 per group of 3 or 4 students
- Dragon Quest Cards (+1 through +4 only), 1 set per group
- Counters (optional), 10 per student
- Pawns, 1 per student
- Spinners, 1 per group

Additional Materials
- Vocabulary Card 1, *add*
- Vocabulary Card 13, *equal*
- Vocabulary Card 43, *sum*

1 WARM UP

Prepare
Distribute two to four Counters to four students.

▶ **How can we figure out how many Counters were given out altogether?**

Have students put their Counters on a table, making sure each group is distinct and count all the Counters. Write an addition equation on the board to represent the problem. Show students how to solve the problem mentally by counting up from the first addend for each additional addend. Demonstrate how to group addends as doubles facts to make addition easier. Write another four-addend problem on the board. Have students solve the problem mentally, and invite volunteers to solve the problem aloud.

Just the Facts
Have students nod their heads "yes" for correct statements and shake their heads "no" for incorrect statements. Use prompts such as the following:

▶ **Twelve is more than thirteen.** shake "no"
▶ **Nineteen is more than eighteen.** nod "yes"
▶ **Eleven is more than twelve.** shake "no"

2 ENGAGE

Develop: Meet the Dragon
"Today we are going to add by counting on." Follow the instructions on the Activity Card **Meet the Dragon.** As students complete the activity, be sure to use the Questions to Ask.

Alternative Grouping
Pair: Pair students and have them play the game as written.

Activity Card 3B

Progress Monitoring

| **If…** students have trouble tracking how many buckets of water they have, | ▶ **Then…** encourage them to write the corresponding numeral after each turn. |

Practice
Have students complete **Student Workbook,** pages 8–9. Guide students through the Key Idea example and the Try This exercises.

Interactive Differentiation
Consult the Teacher Dashboard for grouping suggestions. You can also use performance on the Engage activity to guide students.

Independent Practice

For additional practice with addition, have students play Lots O' Socks: Adding Game. Students use the number line to help them add.

Supported Practice

For additional support, show students how to use Counters to solve a problem with more than two addends.

- Write $3 + 2 + 3$ on the board.
- From a pile of Counters, group Counters into piles representing each addend.
- Model how to count up from the first addend to find the sum. For example, say, "Here are three Counters. I am going to count up two: four, five. Then I am going to count up three more: six, seven, eight." Tap each Counter as you count.
- Show students that you can group the two piles of three Counters together to make a doubles fact, $3 + 3$. Then demonstrate how to count up two in order to add the last addend.

Have students use Counters or their fingers to find the sums of similar problems with more than two addends.

3 REFLECT

Think Critically

Review students' answers to the Reflect prompt at the bottom of **Student Workbook,** page 9, and then review the Engage activity.

▶ **When adding, which numbers did you add first? Why?** Answers may vary. Possible answer: I added the biggest numbers because it is easier to add smaller numbers last.

▶ **How did you find the answer?** Answers may vary. Possible answers: I counted on my fingers; I added 4 + 3 which makes 7 and added 2 to 7 to get 9.

Real-World Application

▶ **Has anyone ever asked you to see how many people want to drink water with dinner? How do you keep track of how many people want water?** Possible answer: I keep count in my head of the number of people who want water.

4 ASSESS

Informal Assessment

Use the online or print Student Record, **Assessment,** page 128, to record informal observations.

Meet the Dragon
Did the student
☐ respond accurately?   ☐ respond with confidence?
☐ respond quickly?   ☐ self-correct?

Additional Practice

For additional practice, have students complete **Practice,** page 53.

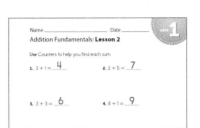

Practice, p. 53

Week 1 • Addition Fundamentals

Lesson 2

Key Idea
You can find a larger number by putting smaller amounts together and counting. The smaller amounts are **addends**. The larger number is the **sum**.

$$3 + 2 = 5$$
addend + addend = sum

Try This
Write the addends and sum.

1.

$\underline{1} + \underline{3} = \underline{4}$

2.

$\underline{4} + \underline{2} = \underline{6}$

3.

$\underline{3} + \underline{3} = \underline{6}$

Practice
Use Counters to find the sum of each problem.

4. $1 + 4 = \underline{5}$ 5. $3 + 5 = \underline{8}$

6. $2 + 2 = \underline{4}$ 7. $6 + 1 = \underline{7}$

8. $5 + 4 = \underline{9}$ 9. $1 + 3 = \underline{4}$

10. $6 + 2 = \underline{8}$ 11. $3 + 7 = \underline{10}$

Reflect
How many buckets of water do you have if you have 3 + 4 + 2? How did you figure that out?

9; Possible answers: I counted on my fingers; I added 4 + 3 which makes 7 and added 2 to 7 to get 9.

Student Workbook, pp. 8–9

WEEK 1
Addition Fundamentals

Lesson 3

Objective
Students can use the plus sign and the equal sign to create addition equations.

Standard
2.OA.2 Fluently add and subtract within 20 using mental strategies. By end of Grade 2, know from memory all sums of two one-digit numbers.

Vocabulary
- add
- addends
- addition
- equal
- equation
- sum

Creating Context
Teachers often report that English Learners do well in mathematics. Many of the symbols used, including numbers, are used in many languages. English Learners may need to learn only the English names for these symbols.

Materials
Program Materials
Counters, 10 per pair

Additional Materials
- index cards, 2
- Vocabulary Card 1, *add*
- Vocabulary Card 13, *equal*
- Vocabulary Card 43, *sum*

Prepare Ahead
Make symbol cards by writing a plus sign on one index card and an equal sign on the other index card.

1 WARM UP

Prepare
Place the equal sign card on the table. Place six objects on each side of the equal sign.

▶ **Does anyone know what this sign means?**

▶ **Who can describe the equation I've created?**

Separate one group into two groups of three.

▶ **Does the equation still make sense? Why?** Possible answer: yes, because 3 and 3 make 6

▶ **What do we need to add to make an equation that shows that 3 and 3 make 6?** a plus sign between the two groups of 3

▶ **Describe the equation we've created.** $3 + 3 = 6$

Repeat this process with larger quantities up to 10 and by occasionally creating unequal groups on one side of the equal sign.

Just the Facts
Replace *plus* or *equal* with a nonsense word. Tell students to correct you after you say each equation. Use equations such as the following:

▶ **Three splat two makes five.** Students say *plus*.

▶ **Four and four splat eight.** Students say *equal*.

▶ **Two splat five makes seven.** Students say *plus*.

2 ENGAGE

Develop: Making and Writing Equations

"Today we are going to learn to use symbols to create equations." Follow the instructions on the Activity Card **Making and Writing Equations**. As students complete the activity, be sure to use the Questions to Ask.

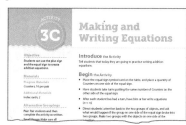
Activity Card 3C

Alternative Groupings

Pair: Pair students and complete the activity as written.

Small Group: Make sure all students have several opportunities to create equations.

Progress Monitoring

If… students have trouble remembering the plus sign and equal sign,

Then… post the signs in the classroom and model making equations with them. Invite volunteers to stand in front of the class, and have the rest of the class organize the volunteers into groups to make simple equations.

Practice
Have students complete **Student Workbook,** pages 10–11. Guide students through the Key Idea example and the Try This exercises.

Interactive Differentiation
Consult the Teacher Dashboard for grouping suggestions. You can also use performance on the Engage activity to guide students.

Independent Practice

For additional practice with adding, have students play Off the Tree: Add Apples. Students will use the number line to help them add.

Supported Practice

For additional support, use the Numeral Train Game with one student or a small group of students. As students work, write an expression on the board that matches their turn. For example, write $2 + 5$ if a student is on the second space and gets a 5.

▶ **What does $2 + 5$ equal?** 7

▶ **Count from the beginning of the track to the space where you are now. What number do you count up to?** 7

3 REFLECT

Think Critically

Review students' answers to the Reflect prompt at the bottom of **Student Workbook,** page 11, and then review the Engage activity.

Discuss that in an addition equation, you must have two things: an equal sign and a plus sign.

▶ **What is needed to make an equation out of 6 + 1?** = 7

▶ **What is needed to make an equation out of 5 □ 2 = 7?** +

▶ **How did you figure that out?** Answers may vary. Possible answer: I know that every equation must have a plus sign and equal sign.

Real-World Application

▶ **When adding, we need to be sure that we use symbols so someone who is reading our work can understand what we mean. When else do we use symbols to write something so people can understand what we are talking about?** Possible answers: when we refer to money and temperature

4 ASSESS

Informal Assessment

Use the online or print Student Record, **Assessment,** page 128, to record informal observations.

Making and Writing Equations

Did the student

☐ respond accurately? ☐ respond with confidence?

☐ respond quickly? ☐ self-correct?

Additional Practice

For additional practice, have students complete **Practice,** page 54.

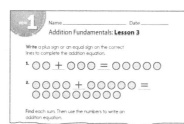

Practice, p. 54

Week 1 • Addition Fundamentals

Lesson 3

Key Idea
- \+ is a "plus sign." It tells you to **add**.
- = is an "equal sign." It tells you that the numbers on each side of it **equal** the same amount.
- All **equations** have an equal sign.

1+1=2

Try This

Decide whether Set 1 equals Set 2.
If yes, circle the =.
If no, put an X over the =.

Set 1   Set 2

1. ●●●   =   ●●
                ●●

2. ●●●●●   X   ●●●
                ●●●

Write *yes* or *no* to tell whether each is an equation.

3. 6 + 1 ___no___   4. 3 + 2 = 5 ___yes___

10 Level D Unit 3 Addition

Practice
Place a plus sign or an equal sign on the correct lines to complete each addition equation.

5. ● __+__ ●● __=__ ●●●

6. ●● __+__ ●●● __=__ ●●●●●

7. ●●●● __+__ ●● __=__ ●●●●●●

Use the numbers below to write an equation that includes the sum.

8. 3 and 4 __3 + 4 = 7__   9. 5 and 1 __5 + 1 = 6__
10. 4 and 4 __4 + 4 = 8__   11. 6 and 2 __6 + 2 = 8__

Reflect
For Problems 3 and 4, rewrite any problems that aren't equations, including the sum.

Problem 3: 6 + 1 = 7

Week 1 Addition Fundamentals • Lesson 3 11

Student Workbook, pp. 10–11

WEEK 1
Addition Fundamentals

Lesson 4

Objective
Students can understand that the order in which you add addends does not affect the sum.

Standard
2.OA.2 Fluently add and subtract within 20 using mental strategies. By end of Grade 2, know from memory all sums of two one-digit numbers.

Vocabulary
- add
- addends
- addition
- equal
- equation
- sum

Creating Context
In English, many words used in mathematics have more than one meaning. In this lesson, the terms *table*, *counters*, *point*, and *round* are used. Have English Learners draw a picture for each of the meanings of these words.

Materials
Program Materials
Counters, 10 per student

Additional Materials
- coffee can or other container, 1 per student pair
- Vocabulary Card 1, *add*
- Vocabulary Card 13, *equal*
- Vocabulary Card 43, *sum*

1 WARM UP

Prepare
Show students a group of three Counters and a group of five Counters.

▶ **How many Counters are there altogether?** 8

Count aloud with students to find 8.

▶ **What operation are we performing?** addition

▶ **When we are adding, does the order we use matter?** no

▶ **If we start with 5, will we get the same sum we would get if we start with 3?** yes

Ask students to add 5 + 3 and 3 + 5 to verify that the order in which you add the addends doesn't affect the sum. Pose several addition problems, and have students model the problems using the Counters. Then have them write an equation.

Just the Facts
Remind students that it does not matter in which order addends are added. Tell them to show the number of fingers that match the missing addends. Ask questions such as the following:

▶ **Five plus two is the same as two plus _____?**
Students show five fingers.

▶ **Three plus four is the same as four plus _____?**
Students show three fingers.

▶ **Six plus one is the same as one plus _____?**
Students show six fingers.

2 ENGAGE

Develop: Drop and Add
"Today we are going to add the Counters we drop into a can." Follow the instructions on the Activity Card **Drop and Add**. As students complete the activity, be sure to use the Questions to Ask.

Alternative Grouping
Individual: Have each student complete the activity as written.

Activity Card 3D

Progress Monitoring
If... one team loses numerous rounds,
▶ **Then...** make sure students understand that the more Counters they drop into the can, the higher their sums will be.

Practice
Have students complete **Student Workbook,** pages 12–13. Guide students through the Key Idea example and the Try This exercises.

Interactive Differentiation
Consult the **Teacher Dashboard** for grouping suggestions. You can also use performance on the Engage activity to guide students.

Independent Practice
For additional practice with adding, have students play Dinosaur Shop 3: Add Dinosaurs.

Supported Practice
For additional support, use the Sets Former Tool with one student or a small group of students.

- Change the format of the mat to the Addition Mat (third button).
- Move the workspace down, so that students cannot see the totals.
- Drag five marbles onto the top mat and four marbles onto the bottom mat.
- Have students add the marbles. Then move the mat upward, so students can check the totals.
- Again, move the workspace down, so that students cannot see the totals.
- Next, put four marbles on the top mat and five on the bottom mat.
- Again, have students add the marbles, and then move the workspace up to check the totals.

Ask students if it matters in what order these marbles were added. Guide them to realize that the sum was nine marbles no matter which set of marbles was the first addend. Continue experimenting with the mat so that students can see that the order of the addends does not affect the sum.

3 REFLECT

Think Critically

Review students' answers to the Reflect prompt at the bottom of **Student Workbook,** page 13, and then review the Engage activity.

Discuss how you get the same answer when adding two numbers no matter which number comes first.

▶ **Does it matter which number comes first when you add?** no

▶ **Do you think this is true every time you add? Why or why not?**
Answers may vary. Possible answer: Yes, you have the same amount.

Real World Application

▶ **When playing a game, we often want the highest score. What is the highest score possible when playing Drop and Add?** $10 + 10 = 20$

▶ **What is the lowest score possible when playing Drop and Add?**
$0 + 0 = 0$

4 ASSESS

Informal Assessment

Use the online or print Student Record, **Assessment,** page 128, to record informal observations.

Drop and Add

Did the student

☐ pay attention to the contributions of others?

☐ contribute information and ideas?

☐ improve on a strategy?

☐ reflect on and check accuracy of work?

Additional Practice

For additional practice, have students complete **Practice,** page 55.

Practice, p. 55

Week 1 • Addition Fundamentals

Lesson 4

Key Idea
You can add the same addends in a different order and get the same sum.
$1 + 4 = 5 \qquad 4 + 1 = 5$

Try This
Write two equations for the groups shown. Put the addends in a different order each time.

1. $2 + 1 = 3$
   $1 + 2 = 3$

2. $7 + 3 = 10$
   $3 + 7 = 10$

3. $4 + 9 = 13$
   $9 + 4 = 13$

4. $2 + 6 = 8$
   $6 + 2 = 8$

Practice
If you were playing Drop and Add, what would the score be after each turn? Use Counters to make a model of each problem, and write the answer.

	Drop 1	Drop 2	Score
5.	9	5	14
6.	6	10	16
7.	8	2	10
8.	3	7	10

Use Counters to find each sum.

9. $4 + 3 = 7$
10. $6 + 4 = 10$
11. $5 + 4 = 9$
12. $2 + 9 = 11$
13. $4 + 7 = 11$
14. $9 + 9 = 18$

Reflect
What is the sum of $3 + 4$? What is the sum of $4 + 3$? Are the answers the same?

7; 7; yes

Student Workbook, pp. 12–13

WEEK 1
Addition Fundamentals

 Review

Objective
Students review skills learned this week and complete the weekly assessment and project.

Standard
2.OA.2 Fluently add and subtract within 20 using mental strategies. By end of Grade 2, know from memory all sums of two one-digit numbers.

Vocabulary
Review vocabulary introduced during the week.

Creating Context
In written directions, English Learners may encounter the passive voice. For example, some instructions say, "Counters may be used." Another way that this instruction could be written is "You may use Counters." Make sure that English Learners understand the directions.

1 WARM UP

Prepare
Present students with **What Number Am I?** mystery problems. Allow volunteers to guess the answers, and ask them to explain their reasoning. Base your problems on the following questions:

▶ I am 50 plus 2. What number am I? 52

▶ I am 19 plus 2. What number am I? 21

▶ When you add 2 to my number, you have 10. What number am I? 8

After students are familiar with the activity and have a good understanding of the term *plus,* have volunteers give clues to you or to one another.

2 ENGAGE

Practice
Have students complete **Student Workbook,** pages 14–15.

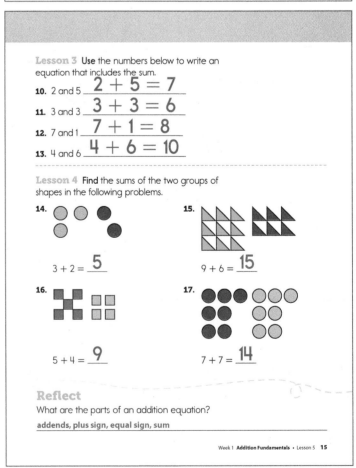

Student Workbook, pp. 14–15

200 Level D Unit 3 **Addition**

3 REFLECT

Think Critically

Review students' answers to the Reflect prompt at the bottom of **Student Workbook,** page 15.

Discuss the answer with the group to reinforce Week 1 concepts.

4 ASSESS

Formal Assessment

Students may take the weekly assessment online.

As an alternative, students may complete the weekly test on **Assessment,** pages 37–38. Record progress using the Student Assessment Record, **Assessment,** page 128.

Going Forward

Use the **Teacher Dashboard** to view results of the online assessments, to input the results of print student assessments, and to review progress before making decisions about next steps. Use the weekly test results and observations to determine the next steps for each student.

Retention	
Student displays good grasp of this week's concepts and skills.	For additional practice, students should use the Building Blocks game Dinosaur Shop 3: Add Dinosaurs.
Remediation	
Student is still struggling with the week's concepts and skills.	Have students continue to work with multiple addends. • Write 3 + 4 + 2 on the board. • From a pile of Counters, group Counters into piles representing each addend. • Model how to count on from the first addend to find the sum. For example, say, "Here are three Counters. I am going to count on four: four, five, six, seven. Then I am going to count on two more: eight, nine." • Tap each Counter as you count.

Have students use Counters or their fingers to solve similar problems with more than two addends.

Suggestions for Re-Evaluation: If a student has struggled without success for several weeks, use observations and test results to place the student at a level in which he or she can find success and build confidence to move forward.

Name _____ Date _____

WEEK 1

Addition Fundamentals

1. How many letters are in this word?

    house
    __5__

2. Circle the word that has more letters.

    back   (because)

Write these numbers in the correct order.

3. 6 8 5 9 7
   __5, 6, 7, 8, 9__

4. 17 14 12 16 15 13
   __12, 13, 14, 15, 16, 17__

5. How many apples are there in all?

    🍎🍎🍎 + 🍎🍎🍎🍎 = __7__

Level D Unit 3 Week 1 **37**

WEEK 1

Name _____ Date _____

Addition Fundamentals

Find the following sums.

6. 2 and 3 is __5__

7. 4 + 3 = __7__

8. 8 + 2 = __10__

9. 9 + 6 = __15__

38 Level D Unit 3 Week 1

Assessment, pp. 37–38

Week 1 **Addition Fundamentals** • Lesson 5   **201**

Project Preview

This week, students learned the fundamentals of addition. They learned about two-digit numbers, made connections between counting and addition, used the plus and equal sign to make equations, and learned that the order of addends does not affect the sum. The project for this unit requires students to apply each week's objectives to create a presentation about a collection. This week, students will begin to make a poster that they will present at the end of the unit.

Project-Based Learning

Standards-driven Project-Based Learning is effective in building deep content understanding. Project-Based Learning increases long-term retention of concepts and has been shown to be more effective than traditional instruction. Completing a project to answer an essential question challenges students to apply and demonstrate mastery of concepts and skills by expressing understanding through discussion, research, and presentation.

Essential Question

HOW can I use addition to solve real-world problems?

Project Evaluation Criteria

Review project evaluation criteria with students prior to beginning the project.

Exceeds Expectations
☐ Project result is explained and can be extended.
☐ Project result is explained in context and can be applied to other situations.
☐ Project result is explained using advanced mathematical vocabulary.
☐ Project result is described, and mathematics are used correctly and can be extended.
☐ Project result is explained and extended, and shows advanced knowledge of mathematical concepts and skills.

Meets Expectations
☐ Project result is explained.
☐ Project result is explained in context.
☐ Project result is explained using mathematical vocabulary.
☐ Project result is described, and mathematics are used correctly.
☐ Project result is explained, and shows satisfactory knowledge of mathematical concepts and skills.

Does Not Meet Expectations
☐ Project result is not explained.
☐ Project result is explained, but out of context.
☐ Project result is explained, but mathematical vocabulary is oversimplified.
☐ Project result is described, but mathematics are not used correctly.
☐ Project result is not explained and/or extended, or shows less than satisfactory knowledge of mathematical concepts and skills.

Pick a Collection

Objective
Students can use addition fundamentals to describe their collections or parts of their collections.

Standard
2.OA.2 Fluently add and subtract within 20 using mental strategies. By the end of Grade 2, know from memory all sums of two one-digit numbers.

Materials
Additional Materials
- large poster paper, one for each pair of students
- personal collections (rocks, books, stamps, stickers, coins)
- pictures of different types of collections

Best Practices
- Make efficient use of cooperative learning groups.
- Allow active learning with noise and movement.
- Create an energetic environment.

Introduce

Many people have collections.

▶ **Do you have any collections at home? If so, what do you collect?** Answers may vary. Possible answer: Yes, I collect cards.

▶ **Why did you start your collection?** Answers may vary. Possible answer: People gave me some cards, and then I decided I wanted the whole set.

▶ **How do you organize and keep track of your collection?** Answers may vary. Possible answer: I put them in order from first to last in a notebook.

Explore

- Show students pictures of different types of collections. You might wish to bring in your personal collection to show to students. Have students discuss their own collections or the collections of friends or family.

▶ **Today you will begin making a poster describing a collection. You will work with a partner for this project.**

- Place students in groups of two.

▶ **You and your partner need to choose a collection that you will present to the class at the end of this unit. It may be something that you both enjoy collecting, it may be just one of your collections, or you may even decide to start a collection for this project.**

- Give students 5 minutes to decide what they will collect. Allow student pairs to share their decisions with the class.
- As students make their announcements, write the subject of each collection on the board. If more than one pair has chosen the same subject, use tally marks to show how many pairs have chosen the same type of item.
- Use the data you collected to create two or three simple word problems such as the following:

▶ **There are 2 groups collecting leaves and 3 groups collecting rocks. How many rock and leaf collections are there in all?** 5 rock and leaf collections

- Encourage students to count up to solve these addition word problems.

▶ **Complete Student Workbook, page 16, to practice adding sums up to 20.**

Wrap Up

- Discuss how students solved the word problems you presented with the collection data.
- Discuss students' answers to the Reflect prompt at the bottom of **Student Workbook,** page 16.

Distribute a piece of poster paper to each group. Have students write their names on the back. On the front, in the center, have students write in large, neat letters the name of the collection they will focus on for the project. Set aside students' poster papers and **Student Workbook** sheets for next week.

If time permits, allow each student to share more about his or her own collection.

Week 1 • Addition Fundamentals

Project
Pick a Collection

Answer the following questions.

1. What is the title of your collection?

   Answers may vary; possible answer: My Card Collection

2. Describe how you will organize your collection.

   Answers may vary; possible answer: in order, from first to last

Solve these word problems.

3. There are 3 trading cards in one store display, 2 trading cards in another display, and 1 trading card in the last display. How many trading cards are in all of the store displays together?

   6 trading cards

4. Andy has 5 trading cards. He adds 4 more. How many trading cards does he have now? Write the equation that shows the answer.

   Andy now has 9 trading cards; $5 + 4 = 9$.

Reflect

How can people use addition with their collections?

Answers may vary; possible answer: When people add more items to a collection, they can count up to find the new total.

Student Workbook, p. 16

Teacher Reflect

☐ Did students use their time wisely and effectively?

☐ Was I able to answer questions when students did not understand?

☐ Did students focus on the major concept of the activity?

WEEK 2
Mastering the Basic Facts

Week at a Glance
This week, students continue **Number Worlds,** Level D, Addition. Students will explore different strategies to solve addition problems with sums to 20.

Skills Focus
- Use an addition table to solve problems.
- Recognize and solve doubles facts up to 20.
- Use doubles facts to help solve near-doubles facts.
- Choose a strategy to solve addition facts with sums up to 20.

How Students Learn
Drill and rote memorization do not help students understand the relationships among addition facts. When students focus on using doubles facts, there are fewer addition facts for them to remember. Students should develop this strategy to quickly come up with an answer if they forget an addition fact. As they become fluent with facts to 20, they will extend doubles strategies to larger addends.

English Learners ELL
For language support, use the *English Learner Support Guide,* pages 60–61, to preview lesson concepts and teach academic vocabulary. **Number Worlds** Vocabulary Cards are listed as additional materials in many lessons and can be used to preteach and reinforce academic vocabulary.

Math at Home
Give one copy of the Letter to Home, page 14, to each student. Encourage students to share and complete the activity with their caregivers.

Weekly Planner

Lesson	Learning Objectives
1 pages 206–207	Students can identify patterns in an addition table and find addition facts for addends of 0, 1, 2, 3, or 10.
2 pages 208–209	Students can find sums of doubles facts for addends up to 10.
3 pages 210–211	Students can find sums of near-doubles facts for addends up to 10.
4 pages 212–213	Students can solve addition facts with sums up to 20.
5 pages 214–215	**Review and Assess** Students review skills learned this week and complete the weekly assessment and project.
Project pages 216–217	Students can identify how many items they have in their collections by counting, counting on, or adding.

204 Level D Unit 3 **Addition**

Key Standard for the Week

Domain: Operations and Algebraic Thinking
Cluster: Add and subtract within 20.

2.OA.2 Fluently add and subtract within 20 using mental strategies. By end of Grade 2, know from memory all sums of two one-digit numbers.

Materials		Technology
Program Materials • **Student Workbook,** pp. 17–19 • **Practice,** p. 56 • Activity Card 3E, **Using the Addition Table** • Addition Table		*Teacher Dashboard* Building Blocks Number Snapshots 5 The Addition Table
Program Materials • **Student Workbook,** pp. 20–21 • **Practice,** p. 57 • Activity Card 3F, **Even or Odd** • Addition Table • Counters	**Additional Materials** • unlined paper • Vocabulary Card 14, *even number* • Vocabulary Card 30, *odd number*	*Teacher Dashboard* Building Blocks Dinosaur Shop 3
Program Materials • **Student Workbook,** pp. 22–23 • **Practice,** p. 58 • Activity Card 3G, **One More, One Less** • Counters • Number 1–6 Cubes		*Teacher Dashboard* Building Blocks Number Snapshots 6, 7, or 8
Program Materials • **Student Workbook,** pp. 24–25 • **Practice,** p. 59 • Activity Card 3H, **Make It Easier** • Addition Table • Number 1–6 Cubes		*Teacher Dashboard* Building Blocks Barkley's Bones The Addition Table
Program Materials • **Student Workbook,** pp. 26–27 • Weekly Test, **Assessment,** pp. 39–40		Review previous activities.
Program Materials **Student Workbook,** p. 28	**Additional Materials** posters from Week 1	

WEEK 2
Mastering the Basic Facts

Find the Math

Develop student strategies to master addition facts with sums up to 20.

Use the following to begin a guided discussion:

▶ **How can organizing a collection into small groups be helpful?** Answers may vary. Possible answers: It can help you know how many items you have in the collection; you can group the same types of items together; you can compare the numbers of different types of items in your collection.

Have students complete **Student Workbook,** page 17.

Student Workbook, p. 17

Lesson 1

Objective
Students can identify patterns in an addition table and find addition facts for addends of 0, 1, 2, 3, or 10.

Standard
2.OA.2 Fluently add and subtract within 20 using mental strategies. By end of Grade 2, know from memory all sums of two one-digit numbers.

Creating Context
Explain to students that *table* can mean "a kind of chart to display data," and not just a piece of furniture. Help students see that the context can often tell them the meaning of a word.

Materials
Program Materials
- Addition Table, 1 per student

Prepare Ahead
Prepare the Addition Table for classroom display.

1 WARM UP

Prepare

▶ **What happens when we add zero to a number?**

Allow several students to answer to assess their understanding. Write each of the problems on the board, and model the first few problems with any readily available classroom objects.

▶ **What is 1 + 0?** If I have 1 object and I add zero, or nothing, to it, how many will I have altogether? 1

▶ **What is 2 + 0?** If I have 2 objects and I add zero to them, how many will I have altogether? 2

▶ **What is 4 + 0? 5 + 0? 10 + 0?** 4; 5; 10

Ask students to help you write a rule on the board to describe what happens when we add 0 to any number. Then write the rule in more formal terms—the sum of any number plus 0 is equal to that number.

2 ENGAGE

Develop: Using the Addition Table

"Today we are going to explore a tool that can help us remember addition facts." Follow the instructions on the Activity Card **Using the Addition Table.** As students complete the activity, be sure to use the Questions to Ask.

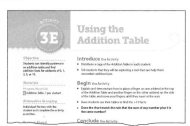

Activity Card 3E

Alternative Grouping

Individual: Partner with the student and complete the activity as written.

Progress Monitoring	
If… students get incorrect sums using the Addition Table,	▶ **Then…** make sure they are moving their fingers correctly. Have them use index cards as placeholders if necessary.

Practice

Have students complete **Student Workbook,** pages 18–19. Guide students through the Key Idea example and the Try This exercises.

Interactive Differentiation

Consult the **Teacher Dashboard** for grouping suggestions. You can also use performance on the Engage activity to guide students.

Independent Practice

For additional practice adding numbers with sums to 5, have students use Number Snapshots 5: Dot Sums to Numerals up to 5. Students will view an image to identify the addends. Then they will determine the sum from among four multiple-choice answers.

Supported Practice

For additional support, use the Addition Table with a small group of students to practice finding sums and checking answers.

- Have a student choose two numbers from 0 to 10 (for example, 4 and 1).
- Tell students to use the numbers as addends in an addition sentence and find the sum.
  $4 + 1 = 5$
- Invite a student to the Addition Table. Tell the student to click the two addends.
- Demonstrate how two addends intersect to show the answer.
- Point out the number sentence in the upper right corner of the board for students to check their work.

3 REFLECT

Think Critically

Review students' answers to the Reflect prompt at the bottom of **Student Workbook,** page 19, and then review the Engage activity.

Discuss why the +10 facts are easy to remember.

▶ **What are some other facts that are easy to remember?**

▶ **What strategies can you use to remember those facts?**

Real-World Application

▶ **Do you know any games or sports that add points to a score every time a player completes a task or answers a question correctly?**
Answers may vary. Possible answers: trivia games, basketball

▶ **Does the player always earn the same number of points?** Answers may vary. Possible answer: No; in basketball, players can earn from 1–3 points for making a basket.

4 ASSESS

Informal Assessment

Use the online or print Student Record, **Assessment,** page 128, to record informal observations.

Using the Addition Table
Did the student
☐ make important observations?   ☐ provide insightful answers?
☐ extend or generalize learning?   ☐ pose insightful questions?

Additional Practice

For additional practice, have students complete **Practice,** page 56.

Practice, p. 56

Week 2 • Mastering the Basic Facts

Lesson 1

Key Idea
Looking for patterns in the Addition Table can help you memorize the basic facts. If you forget, counting on is a good strategy for addends 1, 2, and 3.

Try This
Find the sums.

1. $1 + 0 = \underline{1}$
2. $2 + 0 = \underline{2}$
3. $2 + 1 = \underline{3}$
4. $2 + 3 = \underline{5}$
5. $3 + 0 = \underline{3}$
6. $3 + 2 = \underline{5}$
7. $4 + 2 = \underline{6}$
8. $5 + 1 = \underline{6}$
9. $5 + 2 = \underline{7}$
10. $6 + 1 = \underline{7}$
11. $5 + 3 = \underline{8}$
12. $6 + 3 = \underline{9}$
13. $1 + 0 = \underline{1}$
14. $1 + 10 = \underline{11}$

Practice
Find the sums.

15. $0 + 2 = \underline{2}$
16. $10 + 2 = \underline{12}$
17. $0 + 3 = \underline{3}$
18. $10 + 3 = \underline{13}$
19. $4 + 0 = \underline{4}$
20. $4 + 10 = \underline{14}$
21. $5 + 0 = \underline{5}$
22. $5 + 10 = \underline{15}$
23. $0 + 6 = \underline{6}$
24. $10 + 6 = \underline{16}$
25. $0 + 7 = \underline{7}$
26. $7 + 10 = \underline{17}$
27. $8 + 0 = \underline{8}$
28. $8 + 10 = \underline{18}$
29. $9 + 0 = \underline{9}$
30. $10 + 9 = \underline{19}$

Reflect
What strategy can you use to remember +10 facts? Explain how it works.

Answers may vary; possible answer: The sum for a +10 fact is the same single-digit number with a 1 in front of it.

Student Workbook, pp. 18–19

WEEK 2
Mastering the Basic Facts

Lesson 2

Objective
Students can find sums of doubles facts for addends up to 10.

Standard
2.OA.2 Fluently add and subtract within 20 using mental strategies. By end of Grade 2, know from memory all sums of two one-digit numbers.

Vocabulary
- even number
- odd number

Creating Context
Explain to English Learners that outside of mathematics, the word *odd* can also mean "strange" and the word *even* has other meanings as well. Review the strategy of using context to make sure that English Learners know when *odd* and *even* are referring to numbers and when they are being used in some other way.

Materials

Program Materials
- Addition Table, 1 per student
- Counters

Additional Materials
- unlined paper, 2 sheets per student
- Vocabulary Card 14, *even number*
- Vocabulary Card 30, *odd number*

1 WARM UP

Prepare
Have students find and shade the doubles facts on their Addition Tables.

▶ **What is 2 + 2?** 4

▶ **What is 6 + 6?** 12

▶ **What is 5 + 5?** 10

▶ **Who can tell me what a doubles fact is?**

Write an example on the board, such as 5 + 5. Also note on the board that doubles facts use the same addend twice.

Just the Facts
Remind students that they are working with doubles facts today. Explain that you are going to give them an answer to a doubles fact. Using their fingers, students will show the addend that was doubled. Use questions such as the following:

▶ **What number is doubled to get a sum of 6?** Students show three fingers.

▶ **What number is doubled to get a sum of 10?** Students show five fingers.

▶ **What number is doubled to get a sum of 12?** Students show six fingers.

2 ENGAGE

Develop: Even or Odd
"Today we are going to explore addition patterns." Follow the instructions on the Activity Card **Even or Odd.** As students complete the activity, be sure to use the Questions to Ask.

Alternative Grouping
Individual: Have the student verbally identify whether the number is even or odd.

Activity Card 3F

Progress Monitoring

If... a student is struggling with the difference between even and odd,	▶ Then... have the student make a number line at the top of the paper and circle the even numbers in one color and the odd numbers in a different color.

Practice
Have students complete **Student Workbook,** pages 20–21. Guide students through the Key Idea example and the Try This exercises.

Interactive Differentiation
Consult the Teacher Dashboard for grouping suggestions. You can also use performance on the Engage activity to guide students.

Independent Practice

For additional practice with adding numbers, have students use Dinosaur Shop 3: Add Dinosaurs. Students will combine items from two different boxes into a third box. They must find the sum of the combined items.

Supported Practice

For additional support, use Counters with students to solve problems with doubles facts.

- Write 4 + 4 on the board.
- Have students place two groups of four Counters in front of them.
- Model how to count on from the first addend to find the sum.
- ▶ **Here are four Counters. I am going to count on four more: five, six, seven, eight.**
- Tap each Counter as you count.
- Continue practicing with other doubles facts.

3 REFLECT

Think Critically

Review students' answers to the Reflect prompt at the bottom of **Student Workbook**, page 21, and then review the Engage activity.

Discuss that the sums of doubles facts form a diagonal line and each doubles sum is 2 greater than the one before it.

▶ **What are some other patterns you can see in the Addition Table?**

Real-World Application

▶ **Suppose you were playing a game of soccer with your friend and he was winning 5 to 3. If he said, "I'll let you take one more shot—double or nothing," what does he mean?** If you make it, you will double 3 points to 6 points; if you don't make it, you lose them all.

4 ASSESS

Informal Assessment

Use the online or print Student Record, **Assessment,** page 128, to record informal observations.

Even or Odd

Did the student

☐ make important observations?   ☐ provide insightful answers?

☐ extend or generalize learning?   ☐ pose insightful questions?

Additional Practice

For additional practice, have students complete **Practice,** page 57.

Practice, p. 57

Week 2 • Mastering the Basic Facts

Lesson 2

Key Idea
Doubles facts are easy to remember because when both addends are the same, the sum of a doubles fact is always an **even number**.
$5 + 5 = 10$

Try This
Shade the doubles facts sums on the Addition Table.

+	0	1	2	3	4	5	6	7	8	9	10
0	0	1	2	3	4	5	6	7	8	9	10
1	1	2	3	4	5	6	7	8	9	10	11
2	2	3	4	5	6	7	8	9	10	11	12
3	3	4	5	6	7	8	9	10	11	12	13
4	4	5	6	7	8	9	10	11	12	13	14
5	5	6	7	8	9	10	11	12	13	14	15
6	6	7	8	9	10	11	12	13	14	15	16
7	7	8	9	10	11	12	13	14	15	16	17
8	8	9	10	11	12	13	14	15	16	17	18
9	9	10	11	12	13	14	15	16	17	18	19
10	10	11	12	13	14	15	16	17	18	19	20

20 Level D Unit 3 Addition

Practice
Find the sum of each doubles fact.

1. $2 + 2 = \underline{4}$ 2. $6 + 6 = \underline{12}$
3. $4 + 4 = \underline{8}$ 4. $7 + 7 = \underline{14}$
5. $5 + 5 = \underline{10}$ 6. $10 + 10 = \underline{20}$
7. $8 + 8 = \underline{16}$ 8. $3 + 3 = \underline{6}$
9. $1 + 1 = \underline{2}$ 10. $9 + 9 = \underline{18}$

Complete the doubles fact. Draw a picture if you need help.

11.

$3 + \underline{3} = \underline{6}$

12.

$2 + \underline{2} = \underline{4}$

13.

$5 + \underline{5} = \underline{10}$

14.

$1 + \underline{1} = \underline{2}$

Reflect
What patterns do you see in the doubles facts on the addition table?

They form a diagonal line; each doubles sum is 2 greater than the one before it.

Week 2 Mastering the Basic Facts • Lesson 2 21

Student Workbook, pp. 20–21

WEEK 2
Mastering the Basic Facts

Lesson 3

Objective
Students can find sums of near-doubles facts for addends up to 10.

Standard
2.OA.2 Fluently add and subtract within 20 using mental strategies. By end of Grade 2, know from memory all sums of two one-digit numbers.

Creating Context
A phrase that is sometimes used in English to describe an approximate answer is the expression *more or less*. Give a few examples for English Learners, such as "There are around 500 students in this school. We serve 350 school lunches, more or less."

Materials

Program Materials
- Counters
- Number 1–6 Cubes, 2 per pair

Additional Materials
- blank sheets of paper

1 WARM UP

Prepare

▶ **What are doubles facts?** two of the same number used as addends

▶ **Can anyone figure out what a near-doubles fact is?** Allow discussion, and prompt students if needed.

▶ **Here's a doubles fact: $4 + 4 = 8$. Can anyone suggest what a near-doubles fact would be for this problem?** $4 + 3 = 7$; $4 + 5 = 9$

▶ **Here's a rule for near-doubles facts: One of the numbers in the problem is one number bigger or one number smaller than the other number.**

Ask students to suggest near-doubles addition problems for the following doubles problems:

- $3 + 3$
- $6 + 6$
- $9 + 9$

Just the Facts

Have students put their hands up for correct statements and keep their hands down for incorrect statements. Use prompts such as the following:

▶ **Five plus six equals ten.** *hands down*
▶ **Eight plus seven equals fifteen.** *hands up*
▶ **Nine plus ten equals eighteen.** *hands down*

2 ENGAGE

Develop: One More, One Less

"Today we are going to see how many doubles problems and near-doubles problems we can think of." Follow the instructions on the Activity Card **One More, One Less**. As students complete the activity, be sure to use the Questions to Ask.

Activity Card 3G

Alternative Grouping

Individual: Act as the opposing "team" and complete the activity.

Progress Monitoring

| If... students are having difficulty deciding whether an equation is correct, | Then... make sure you are available to solve any disagreements between teams. |

Practice

Have students complete **Student Workbook,** pages 22–23. Guide students through the Key Idea example and the Try This exercises.

Interactive Differentiation

Consult the **Teacher Dashboard** for grouping suggestions. You can also use performance on the Engage activity to guide students.

Independent Practice

For additional practice with adding numbers, have students use Number Snapshots 6, 7, or 8. Students will view an image to identify the addends. Then they will determine the sum from among four multiple choice items.

Supported Practice

For additional support, use Counters with students to solve problems with near-doubles facts.

- Write $4 + 4$ on the board.
- Have students place two groups of four Counters in front of them.
- Tell students to solve the problem.
- Write $4 + 4 = 8$ on the board.
- Under the original problem write $4 + 5$.
- ▶ **How can the first problem help solve the second problem?**
- ▶ **I already know that 4 plus 4 equals 8.**
- ▶ **If I'm adding 4 plus 5, the only thing that is different is there is one more in the second addend.**
- ▶ **Because the problem has only one more, the answer will have just one more.**
- ▶ $4 + 5 = 9$
- Continue practicing with other near-doubles facts.

3 REFLECT

Think Critically

Review students' answers to the Reflect prompt at the bottom of **Student Workbook,** page 23, and then review the Engage activity.

Discuss the pattern of the sums of near-doubles facts.

▶ **Is the sum of 3 + 4 an even number or an odd number?** Seven is an odd number.

▶ **Is the sum of 4 + 5 an even number or an odd number?** Nine is an odd number.

▶ **Is the sum of a near-doubles fact always an even number or an odd number?** The sum of a near-doubles fact is always an odd number.

Real-World Application

Doubles facts are often used with traveling.

▶ **If it took me 10 minutes to walk to my cousin's house, how long would it take me to walk home?** 10 minutes

▶ **How much time do I need to travel to my cousin's house and back home?** 10 + 10 = 20 minutes

▶ **If I travel 5 miles from home to get to soccer practice, how many miles do I travel altogether going there and back again?** 10 miles

4 ASSESS

Informal Assessment

Use the online or print Student Record, **Assessment,** page 128, to record informal observations.

One More, One Less

Did the student
- ☐ respond accurately?
- ☐ respond quickly?
- ☐ respond with confidence?
- ☐ self-correct?

Additional Practice

For additional practice, have students complete **Practice,** page 58.

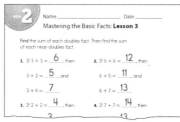

Practice, p. 58

Week 2 • Mastering the Basic Facts

Lesson 3

Key Idea

Doubles facts have the same addend.

5 + 5 = 10

Near-doubles facts have addends and sums that are 1 more or 1 less.

+1 Near-Doubles Fact	−1 Near-Doubles Fact
5 + 6 = 11	5 + 4 = 9

Try This

Find the sum of each doubles fact. Use that to find the sum of each near-doubles fact.

1. 1 + 1 = __2__   1 + 2 = __3__
2. 3 + 3 = __6__   3 + 2 = __5__
3. 4 + 4 = __8__   4 + 5 = __9__
4. 2 + 2 = __4__   2 + 1 = __3__
5. 6 + 6 = __12__   6 + 5 = __11__
6. 7 + 7 = __14__   7 + 8 = __15__
7. 7 + 7 = __14__   7 + 6 = __13__

22 Level D Unit 3 Addition

Practice

Find the sum of each doubles fact to help you find the sum of each near-doubles fact.

8. If 5 + 5 = __10__, then
   5 + 4 = __9__, and
   5 + 6 = __11__

9. If 8 + 8 = __16__, then
   8 + 7 = __15__, and
   8 + 9 = __17__

10. If 4 + 4 = __8__, then
    4 + 3 = __7__, and
    4 + 5 = __9__

11. If 9 + 9 = __18__, then
    9 + 8 = __17__, and
    9 + 10 = __19__

Find each sum.

12. 4 + 5 = __9__   13. 7 + 8 = __15__
14. 1 + 2 = __3__   15. 6 + 7 = __13__

Reflect

What pattern do you see with the sums of doubles facts?

They are all even.

What pattern do you see with the sums of near-doubles facts?

They are all odd.

Week 2 Mastering the Basic Facts • Lesson 3 23

Student Workbook, pp. 22–23

WEEK 2
Mastering the Basic Facts

Lesson 4

Objective
Students can solve addition facts with sums up to 20.

Standard
2.OA.2 Fluently add and subtract within 20 using mental strategies. By end of Grade 2, know from memory all sums of two one-digit numbers.

Creating Context
In the Real-World Application, prior knowledge is needed to solve problems. This excellent strategy is especially important for English Learners who may have different experiences than those being discussed in class. Before beginning a lesson, determine whether there are experts on a topic or whether some students may have no previous knowledge of or experience with the topic.

Materials
Program Materials
- Addition Table, 1 per group
- Number 1–6 Cubes, 4 per group

1 WARM UP

Prepare
Explain that addition facts can help us remember other facts.

▶ **How can we use 10 + 7 to help find the sum of 9 + 7?** Because 9 is 1 less than 10, the sum must be 1 less than 17, which is 16.

▶ **Look at the Addition Table. What pattern do you notice about the +9 facts and the +10 facts?** The +9 facts are always 1 less than the +10 facts.

Just the Facts
Have students put their hands on their heads for a correct answer and their hands on the table for an incorrect answer. Use prompts such as the following:

▶ **Four plus ten equals fourteen.** *hands on heads*
▶ **Seven plus ten equals seventeen.** *hands on heads*
▶ **Eight plus ten equals twenty.** *hands on table*

2 ENGAGE

Develop: Make It Easier
"Today we are going to practice our addition facts." Follow the instructions on the Activity Card **Make It Easier.** As students complete the activity, be sure to use the Questions to Ask.

Alternative Groupings
Pair: Have the students roll the Number Cubes and solve the problems while you check their answers and suggest alternate strategies.

Individual: Have the student roll the Number Cubes and solve the problems. Check his or her answers and suggest alternate strategies.

Activity Card 3H

Progress Monitoring

| **If...** students have difficulty finding a related addition fact, | ▶ **Then...** remind them to think of the doubles and near-doubles facts. If these do not work for the quantities rolled, they should try another strategy, such as counting on. |

Practice
Have students complete **Student Workbook,** pages 24–25. Guide students through the Key Idea example and the Try This exercises.

Interactive Differentiation
Consult the **Teacher Dashboard** for grouping suggestions. You can also use performance on the Engage activity to guide students.

Independent Practice

For additional practice with adding numbers, have students use Barkley's Bones. Students count on from an addend to find the given sum. Concrete objects are available to find the missing addend.

Supported Practice

For additional support, use the Addition Table with a small group of students to practice finding sums and answers.

- Have a student create an addition problem for group members to solve (for example, 6 + 7).
- Tell students to solve the problem in their notebooks.
- Invite a student to the Addition Table to click the addends of the problem to find the sum.
- Continue to create addition problems and check answers using the Addition Table.

212 Level D Unit 3 **Addition**

3 REFLECT

Think Critically
Review students' answers to the Reflect prompt at the bottom of **Student Workbook,** page 25, and then review the Engage activity.

Discuss with students that they may use different strategies for remembering the addition facts. Many strategies work, and each student should use the ones he or she likes best.

Real-World Application
We often use what we know to help us with something more difficult. Discuss with students when they have used prior knowledge to help them do something they did not know how to do—for example, figuring out a puzzle.

4 ASSESS

Informal Assessment
Use the online or print Student Record, **Assessment,** page 128, to record informal observations.

Make It Easier
Did the student
- ☐ respond accurately?
- ☐ respond quickly?
- ☐ respond with confidence?
- ☐ self-correct?

Practice
For additional practice, have students complete **Practice,** page 59.

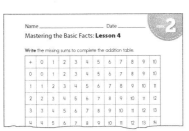
Practice, p. 59

Week 2 • Mastering the Basic Facts

Lesson 4

Key Idea
Use facts you know to find sums for facts you don't know.

Try This
Find the sums.

1. 10 + 4 = **14**
2. 9 + 4 = **13**
3. 7 + 10 = **17**
4. 9 + 7 = **16**
5. 5 + 10 = **15**
6. 5 + 9 = **14**
7. 10 + 6 = **16**
8. 6 + 9 = **15**
9. 8 + 9 = **17**
10. 10 + 8 = **18**

Find each sum. Circle the best strategy to find it.

11. 5 + 6 = **11**
   (Near Doubles)   Counting On   +10 and +9   None of these

12. 9 + 7 = **16**
   Near Doubles   Counting On   (+10 and +9)   None of these

13. 8 + 9 = **17**
   (Near Doubles)   Counting On   +10 and +9   None of these

24 Level D Unit 3 **Addition**

Practice
Fill in the missing sums on the Addition Table.

14.

+	0	1	2	3	4	5	6	7	8	9	10
0	0	1	2	3	4	5	6	7	8	9	10
1	1	2	3	4	5	6	7	8	9	10	11
2	2	3	4	5	6	7	8	9	10	11	12
3	3	4	5	6	7	8	9	10	11	12	13
4	4	5	6	7	8	9	10	11	12	13	14
5	5	6	7	8	9	10	11	12	13	14	15
6	6	7	8	9	10	11	12	13	14	15	16
7	7	8	9	10	11	12	13	14	15	16	17
8	8	9	10	11	12	13	14	15	16	17	18
9	9	10	11	12	13	14	15	16	17	18	19
10	10	11	12	13	14	15	16	17	18	19	20

Reflect
What strategies could you use to help you remember other facts?

Answers may vary; possible answer: The +8 facts are 1 less than the +9 facts.

Week 2 Mastering the Basic Facts • Lesson 4 **25**

Student Workbook, pp. 24–25

WEEK 2
Mastering the Basic Facts

Lesson 5 Review

Objective
Students review skills learned this week and complete the weekly assessment and project.

Standard
2.OA.2 Fluently add and subtract within 20 using mental strategies. By end of Grade 2, know from memory all sums of two one-digit numbers.

Vocabulary
Review vocabulary introduced during the week.

Creating Context
Student Workbook Problems 15–18 use "If _____, then _____." This may be a new grammatical construction for some English Learners. Practice some conditional statements using examples from your class.

1 WARM UP

Prepare

▶ **What is a doubles fact?** a number plus itself

▶ **What is a near-doubles fact?** a number plus the number 1 larger or 1 smaller than itself

▶ **What other strategies can you use to solve addition problems?**
Answers may vary. Possible answer: counting on

2 ENGAGE

Practice
Have students complete *Student Workbook,* pages 26–27.

Week 2 • Mastering the Basic Facts

Lesson 5 Review

This week you continued to solve addition problems. You used strategies and the Addition Table to master addition facts to 20.

Lesson 1 Find the sums.

1. 1 + 0 = __1__
2. 6 + 1 = __7__
3. 4 + 2 = __6__
4. 2 + 3 = __5__
5. 6 + 0 = __6__
6. 10 + 6 = __16__

Lesson 2 Use doubles facts to find each sum.

7. 4 + 4 = __8__
8. 2 + 2 = __4__
9. 3 + 3 = __6__
10. 7 + 7 = __14__
11. 5 + 5 = __10__
12. 6 + 6 = __12__
13. 8 + 8 = __16__
14. 0 + 0 = __0__

26 Level D Unit 3 Addition

Lesson 3 Find the sum of each doubles fact. Then find the sum of each near-doubles fact.

15. If 3 + 3 = __6__, then
    3 + 2 = __5__, and
    3 + 4 = __7__

16. If 7 + 7 = __14__, then
    7 + 6 = __13__, and
    7 + 8 = __15__

17. If 2 + 2 = __4__, then
    2 + 1 = __3__, and
    2 + 3 = __5__

18. If 9 + 9 = __18__, then
    9 + 8 = __17__, and
    9 + 10 = __19__

Lesson 4 Find each sum.

19. 6 + 9 = __15__
20. 10 + 2 = __12__
21. 3 + 9 = __12__
22. 9 + 10 = __19__
23. 10 + 5 = __15__
24. 9 + 5 = __14__

Reflect
What strategy can you use to remember the +9 facts?
Answers may vary; possible answer: +9 facts are 1 less than +10 facts.

Week 2 **Mastering the Basic Facts** • Lesson 5 27

Student Workbook, pp. 26–27

3 REFLECT

Think Critically

Review students' answers to the Reflect prompt at the bottom of *Student Workbook,* page 27.

Discuss the answer with the group to reinforce Week 2 concepts.

4 ASSESS

Formal Assessment

Students may take the weekly assessment online.

As an alternative, students may complete the weekly test on *Assessment,* pages 39–40. Record progress using the Student Assessment Record, *Assessment,* page 128.

Going Forward

Use the *Teacher Dashboard* to view results of the online assessments, to input the results of print student assessments, and to review progress before making decisions about next steps. Use the weekly test results and observations to determine the next steps for each student.

Retention	
Student displays good grasp of this week's concepts and skills.	Have students continue to use the Barkley's Bones Building Blocks activity with increasingly higher numbers. Students count on from an addend to find the given sum. Concrete objects are available to find the missing addend.

Remediation	
Student is still struggling with the week's concepts and skills.	Use Counters with students to solve problems with near-doubles facts. • Write 3 + 3 on the board. • Have students place two groups of three Counters in front of them. • Tell students to solve the problem. • Write 3 + 3 = 6 on the board. • Under the original problem write 3 + 4. ▶ **How can the first problem help solve the second problem?** ▶ I already know that 3 plus 3 equals 6. ▶ If I'm adding 3 plus 4, the only thing that is different is that there is one more in the second addend. ▶ Because the problem has only one more, the sum will, too. 3 + 4 = 7 • Practice with additional near-doubles facts.

Suggestions for Re-Evaluation: If a student has struggled without success for several weeks, use observations and test results to place the student at a level in which he or she can find success and build confidence to move forward.

Name _____ Date _____ WEEK 2
Mastering the Basic Facts

Find the following sums.

1. 5 + 0 = __5__

2. 4 + 3 = __7__

3. 10 + 7 = __17__

4. 2 + 2 = __4__

5. 9 + 9 = __18__

6. 4 + 5 = __9__

7. 5 + 6 = __11__

WEEK 2 Name _____ Date _____
Mastering the Basic Facts

8. Circle the two problems that have the same answer.

   (7 + 6)   8 + 6   (8 + 5)

9. Write a doubles fact using the number 8.
   __8 + 8__

10. Write a near-doubles fact using the number 9.
    __8 + 9 or 9 + 10__

Assessment, pp. 39–40

Week 2 **Mastering the Basic Facts** • Lesson 5 **215**

Project Preview

This week, students learned basic addition facts. They used addition tables, worked with doubles and near-doubles facts, and solved addition facts with sums up to 20. This week, students will apply this new knowledge to their collections by counting on or adding together the numbers of items in each category.

Project-Based Learning

Standards-driven Project-Based Learning is effective in building deep content understanding. Project-Based Learning increases long-term retention of concepts and has been shown to be more effective than traditional instruction. Completing a project to answer an essential question challenges students to apply and demonstrate mastery of concepts and skills by expressing understanding through discussion, research, and presentation.

Essential Question

HOW can I use addition to solve real-world problems?

Project Evaluation Criteria

Review project evaluation criteria with students prior to beginning the project.

Exceeds Expectations
☐ Project result is explained and can be extended.
☐ Project result is explained in context and can be applied to other situations.
☐ Project result is explained using advanced mathematical vocabulary.
☐ Project result is described, and mathematics are used correctly and can be extended.
☐ Project result is explained and extended, and shows advanced knowledge of mathematical concepts and skills.

Meets Expectations
☐ Project result is explained.
☐ Project result is explained in context.
☐ Project result is explained using mathematical vocabulary.
☐ Project result is described, and mathematics are used correctly.
☐ Project result is explained, and shows satisfactory knowledge of mathematical concepts and skills.

Does Not Meet Expectations
☐ Project result is not explained.
☐ Project result is explained, but out of context.
☐ Project result is explained, but mathematical vocabulary is oversimplified.
☐ Project result is described, but mathematics are not used correctly.
☐ Project result is not explained and/or extended, or shows less than satisfactory knowledge of mathematical concepts and skills.

Organize a Collection

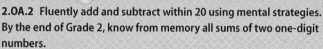

Objective
Students can identify how many items they have in their collections by counting, counting on, or adding.

Standard CCSS
2.OA.2 Fluently add and subtract within 20 using mental strategies. By the end of Grade 2, know from memory all sums of two one-digit numbers.

Materials
Additional Materials
posters from Week 1

Prepare Ahead
- Prepare a collection, or share one of your own with students.
- If possible, have students bring in their personal collections. If that is not possible, tell students to bring pictures of their collections to class to share.

Best Practices
- Check for student understanding frequently.
- Coach, demonstrate, and model.
- Organize the classroom as an activity-based space.

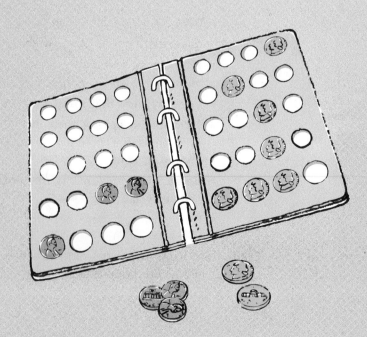

Introduce

People who have collections often want to know how many items they have collected. They may count how many items they have. If they add to their collections, they use addition to find the new total.

▶ **What have you learned about addition this week?** Answers may vary. Possible answer: I learned about patterns in addition tables, doubles facts, and near-doubles facts.

Explore

▶ **Today you will find out how many items you have in your collection. You will show the total on your poster.**

- Arrange students in the same pairs as last week. Give students their posters and *Student Workbooks* with answers from last week.

▶ **If you collect coins, cards, or stickers, you might keep them in a special book. If you collect rocks, you might have a box to keep them in. This is how you organize your collection.**

- Show students your own collection. Describe how you organize it. On a piece of poster paper, model how you could present this information. For example, if you collect books, write *Books* in the center of the poster. Then show students how you could represent your organizational scheme. First, write the heading *How I Organize My Collection* on the poster. If you organize the books according to genre, write the genre names in a list under the heading. Or, if you organize the books in alphabetical order, write *ABC Order* under the heading.

▶ **I also want to know how many (item name) are in my collection.**

- Count the number of items in your collection. Find a place on the poster board and write *My Collection Total*. Write the total under the heading.

▶ **Complete *Student Workbook,* page 28, to help you add this information to your own posters. You will also practice adding sums to 20.**

Wrap Up

- Allow students sufficient time to count the items in their collections.
- Discuss students' answers to the Reflect prompts at the bottom of *Student Workbook,* page 28.

Have students record on their posters how they have organized their collections and the number of items in the collections.

If time permits, allow student pairs to share the totals for their collections and how they arrived at that amount.

Student Workbook, p. 28

Teacher Reflect

☐ Did I clearly explain how to organize the activity?

☐ Did students organize their ideas?

☐ Did students finish all of the steps required by the project?

WEEK 3
Solving Addition Problems

Week at a Glance

This week, students continue **Number Worlds,** Level D, Addition. Students will explore creating equations using formal notation to solve addition problems. They will solve addition equations using two or more addends.

Skills Focus

- Solve addition problems with more than two addends.
- Write equations for addition problems with more than two addends.
- Reorder and group addends to make addition easier.
- Solve addition word problems.

How Students Learn

Encourage students to explore the principles of commutativity and associativity. Give them plenty of opportunities to experiment with adding numbers in different orders and then judge the relative ease or difficulty of the problem.

English Learners ELL

For language support, use the **English Learner Support Guide,** pages 62–63, to preview lesson concepts and teach academic vocabulary. **Number Worlds** Vocabulary Cards are listed as additional materials in many lessons and can be used to preteach and reinforce academic vocabulary.

Math at Home

Give one copy of the Letter to Home, page 15, to each student. Encourage students to share and complete the activity with their caregivers.

Weekly Planner

Lesson	Learning Objectives
1 pages 220–221	Students can solve addition problems with more than two addends and can reorder addends to make addition easier.
2 pages 222–223	Students can solve addition problems with three addends and write addition equations.
3 pages 224–225	Students can add and subtract quantities to perform successive operations.
4 pages 226–227	Students can solve addition word problems and write equations to describe them.
5 pages 228–229	**Review and Assess** Students review skills learned this week and complete the weekly assessment and project.
Project pages 230–231	Students can write addition word problems and the equations to solve them within 100.

218 Level D Unit 3 **Addition**

Key Standard for the Week

Domain: Number and Operations in Base Ten
Cluster: Use place value understanding and properties of operations to add and subtract.
2.NBT.5 Fluently add and subtract within 100 using strategies based on place value, properties of operations, and/or the relationship between addition and subtraction.

Materials		Technology
Program Materials • **Student Workbook,** pp. 29–31 • **Practice,** p. 60 • Activity Card 3I, **Grab and Add** • Grab and Add • Counters	**Additional Materials** paper lunch bags	*Teacher Dashboard* Building Blocks Easy as Pie Sets Former Tool
Program Materials • **Student Workbook,** pp. 32–33 • **Practice,** p. 61 • Activity Card 3J, **Three-Area Counter Drop** • Counters	**Additional Materials** • cardboard shirt box lid • construction paper	*Teacher Dashboard* Building Blocks Dinosaur Shop 4
Program Materials • **Student Workbook,** pp. 34–35 • **Practice,** p. 62 • Activity Card 3K, **The Fiercest Dragon** • Dragon Quest Record Form • Dragon Quest Game Boards	• Dragon Quest Cards • Pawns • Spinners	*Teacher Dashboard* Building Blocks Eggcellent Addition
Program Materials • **Student Workbook,** pp. 36–37 • **Practice,** p. 63 • Activity Card 3L, **Addition Action** • Counters		*Teacher Dashboard* Building Blocks Word Problems with Tools 1
Program Materials • **Student Workbook,** pp. 38–39 • Weekly Test, **Assessment,** pp. 41–42		Review previous activities.
Program Materials **Student Workbook,** p. 40	**Additional Materials** • business-sized envelopes • index cards • poster from Week 2 • tape or glue sticks	

WEEK 3
Solving Addition Problems

Find the Math
In this week, students solve addition problems with two or more addends. Students also use addition to solve word problems.

Use the following to begin a guided discussion:

▶ **Has someone ever given you things to add to a collection? Have you ever shared your collection with others?** Answers may vary; possible answer: My cousin knew I collected marbles. Every time I saw my cousin, she gave me ten marbles to add to my collection.

Have students complete **Student Workbook,** page 29.

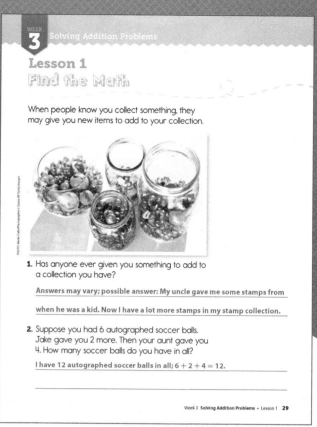

Student Workbook, p. 29

Lesson 1

Objective
Students can solve addition problems with more than two addends and can reorder addends to make addition easier.

Standard
2.NBT.5 Fluently add and subtract within 100 using strategies based on place value, properties of operations, and/or the relationship between addition and subtraction.

Vocabulary
addends

Creating Context
Review the ordinal numbers *first, second,* and *third* to remind English Learners that these words describe the order of objects or numbers when we are referring to a sequence.

Materials
Program Materials
- Grab and Add, 1 per student
- Counters, 10 per group of 4 students

Additional Materials
paper lunch bag, 1 per group of 4 students

1 WARM UP

Prepare
Write "2 + 4 + 5 + 3 = ?" on the board. Ask students to solve the problem mentally, and invite volunteers to solve the problem aloud. Remind students that numbers can be added in any order, and it is sometimes easier to add certain numbers first.

▶ **How can we reorder these numbers to make addition easier?**
5 + 3 + 2 + 4

Allow discussion, and then suggest two strategies: Put the largest number first, and group numbers that make 5 or that make 10 together. Use these strategies to rewrite the problem.

Ask volunteers to share their answers aloud.

▶ **Did reordering make the addition easier?**

2 ENGAGE

Develop: Grab and Add
"Today we are going to solve addition problems with more than two addends." Follow the instructions on the Activity Card **Grab and Add.** As students complete the activity, be sure to use the Questions to Ask.

Alternative Groupings
Pair: Have one student grab and one student record. Alternate roles.

Individual: Act as the grabber throughout the activity while the student records, reorders, and adds the quantities.

Activity Card 31

Progress Monitoring

If… students only add the numbers in the order given when working with more than two addends,	▶ Then… explain that grouping addends can make the problem easier: with 2 + 3 + 8, it is easier to first add the 2 and the 8 to make 10.

Practice
Have students complete **Student Workbook,** pages 30–31. Guide students through the Key Idea example and the Try This exercises.

Interactive Differentiation

Consult the **Teacher Dashboard** for grouping suggestions. You can also use performance on the Engage activity to guide students.

Independent Practice

For additional practice, have students play Easy as Pie. Students will practice counting on from the first addend.

Supported Practice

For additional support, use the Sets Former Tool with a small group of students to practice adding more than two addends.

- Add the tool, and drag the workspace down so the totals are not visible.
- Write the problem $4 + 6 + 3$ on the board.
- Have students place 4 blue, 6 red, and 3 yellow marbles on the mat.
▶ **We want to add all of these numbers together.**
▶ **What numbers would you add together first?**
▶ **I would add 4 and 6 first because they equal 10.**
▶ **Then I would add 3 to 10 to make 13.**
- Drag the workspace up, so that students can see the totals.

3 REFLECT

Think Critically

Review students' answers to the Reflect prompt at the bottom of **Student Workbook,** page 31, and then review the Engage activity.

How do students decide which addends to add first?

Discuss their strategies.

▶ **How do you add the third number if 2 and 3 are added first?** $5 + 8 = 13$
▶ **How do you add the third number if 3 and 8 are added first?** $11 + 2 = 13$
▶ **How do you add the third number if 2 and 8 are added first?** $10 + 3 = 13$
▶ **Which is easiest? Why?** Answers may vary. Possible answer: Add $2 + 8$ first; it is easiest to add 3 to 10.

4 ASSESS

Informal Assessment

Use the online or print Student Record, **Assessment,** page 128, to record informal observations.

Grab and Add

Did the student

☐ respond accurately? ☐ respond with confidence?

☐ respond quickly? ☐ self-correct?

Additional Practice

For additional practice, have students complete **Practice,** page 60.

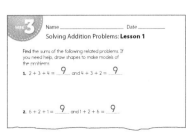

Practice, p. 60

Week 3 • Solving Addition Problems

Lesson 1

Key Idea
You can add more than two groups to find a sum.

$4 + 2 + 3 = 9$

Try This
Write an addition equation to describe how to add the groups shown. Include the sum.

1. $4 + 1 + 3 = 8$
2. $1 + 5 + 4 = 10$
3. $3 + 2 + 4 = 9$
4. $2 + 1 + 2 = 5$
5. $5 + 3 + 5 = 13$
6. $4 + 4 + 4 = 12$

30 Level D Unit 3 Addition

Practice
Find the sums of the related problems. Draw shapes to make models of the problems if you need help.

7. $1 + 2 + 3 = \underline{6}$ and $2 + 3 + 1 = \underline{6}$

8. $5 + 3 + 1 = \underline{9}$ and $1 + 3 + 5 = \underline{9}$

9. $2 + 4 + 1 = \underline{7}$ and $4 + 2 + 1 = \underline{7}$

10. $2 + 3 + 5 = \underline{10}$ and $5 + 2 + 3 = \underline{10}$

Reflect
When adding $2 + 3 + 8$, in which order would it be easiest to add the numbers?

$8 + 2 = 10$, then $10 + 3 = 13$

Week 3 Solving Addition Problems • Lesson 1   31

Student Workbook, pp. 30–31

WEEK 3
Solving Addition Problems

Lesson 2

Objective
Students can solve addition problems with three addends and write addition equations.

Standard
2.NBT.5 Fluently add and subtract within 100 using strategies based on place value, properties of operations, and/or the relationship between addition and subtraction.

Creating Context
Using manipulatives helps English Learners because students can see the concept as they hear the words associated with it. *Number Worlds* uses hands-on learning to support concept development for all students.

Materials
Program Materials
Counters, 10 per group of 2 or 3 students

Additional Materials
- cardboard shirt box lid
- construction paper, various colors

Prepare Ahead
For each group of two or three students, create a three-color mat by cutting three different-colored sheets of construction paper into thirds and taping one-third of each color together. Place the mat in a shirt box lid so the counters stay on the mat when dropped.

1 WARM UP

Prepare
▸ Which numbers would you add first in the problem 2 + 9 + 8? **Why?** 2 and 8; 2 + 8 = 10
▸ Numbers that add up to 10 are called *nice numbers* because it is easy to add them first and then add other numbers to them. What other nice numbers can you think of?

Create a list of addends that are nice numbers.

Just the Facts
Remind students that numbers that add to 10 are nice numbers because they are easy to add in your head. Have students stand up if they agree with your statement and sit or stay seated if they do not agree. Use statements such as the following:

▸ **The numbers 3 and 7 are nice numbers.** *stand up*
▸ **The numbers 6 and 8 are nice numbers.** *stay seated*
▸ **The numbers 2 and 8 are nice numbers.** *stand up*

2 ENGAGE

Develop: Three-Area Counter Drop
"Today we are going to write and solve addition equations." Follow the instructions on the Activity Card **Three-Area Counter Drop**. As students complete the activity, be sure to use the Questions to Ask.

Activity Card 3J

Alternative Grouping
Pair: Act as the student's partner, and complete the activity as written.

Progress Monitoring
If... students have had plenty of experience with multiple addends to 10,

▸ **Then...** have them drop fifteen to twenty Counters.

Practice
Have students complete **Student Workbook,** pages 32–33. Guide students through the Key Idea example and the Try This exercises.

Interactive Differentiation
Consult the Teacher Dashboard for grouping suggestions. You can also use performance on the Engage activity to guide students.

Independent Practice

For additional practice, have students work on Dinosaur Shop 4: Make It Right. Students will use concrete objects to find missing addends.

Supported Practice

For additional support, use Counters with students to write a problem with more than two addends.

- Tell the following story.
  ▸ There are 3 students playing at the playground. Soon 7 more students come over to play. Then 4 more students join them. How many students are at the playground?
- Repeat the story, but this time use Counters to represent students. Demonstrate that at first there were 3 students by using 3 Counters. Next, show 7 more Counters in a new pile. Do not combine the groups of Counters yet. Then, show a group of 4 Counters.
- Write the problem 3 + 7 + 4 on the board.
- Demonstrate why you add 3 and 7 first—because they are addends that equal 10.
- Tell students a new story. Have them use Counters to model and solve the story.

222 Level D Unit 3 **Addition**

3 REFLECT

Think Critically

Review students' answers to the Reflect prompt at the bottom of *Student Workbook,* page 33, and then review the Engage activity.

Discuss strategies students have used. Emphasize that students should first add the addends they find easiest to work with.

▶ **Which numbers would you add first in the problem 2 + 9 + 8? Why?**
2 and 8; their sum is 10.

▶ **Which numbers would you add first in the problem 6 + 3 + 3? Why?**
3 and 3; they are doubles.

Real-World Application

▶ **The order in which the Counters were added together in Three-Area Counter Drop does not matter. The sum will always be the same. Can you think of foods with ingredients that can be combined in any order?**
Possible answers: stir fry, pizza toppings

4 ASSESS

Informal Assessment

Use the online or print Student Record, *Assessment,* page 128, to record informal observations.

Three-Area Counter Drop

Did the student
- ☐ pay attention to the contributions of others?
- ☐ contribute information and ideas?
- ☐ improve on a strategy?
- ☐ reflect on and check accuracy of work?

Additional Practice

For additional practice, have students complete *Practice,* page 61.

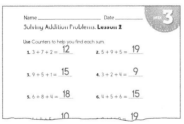

Practice, p. 61

Week 3 • Solving Addition Problems

Lesson 2

Key Idea
When you have more than two addends, add two numbers together at a time. Choose numbers that make adding easier.
Then add the third number.

Try This
Find the sum of the two connected addends. Then add the third addend to find the total sum.

1. 6 + 3 + 2 =
   9 + 2 = 11

2. 8 + 5 + 6 =
   8 + 11 = 19

3. 7 + 7 + 4 =
   14 + 4 = 18

4. 9 + 6 + 1 =
   10 + 6 = 16

5. 12 + 3 + 5 =
   15 + 5 = 20

6. 6 + 6 + 8 =
   12 + 8 = 20

7. 2 + 5 + 1 =
   2 + 6 = 8

8. 3 + 3 + 4 =
   6 + 4 = 10

Practice
Make a model of each problem with Counters to find the sum.

9. 6 + 4 + 2 = 12
10. 5 + 6 + 4 = 15
11. 7 + 4 + 4 = 15
12. 3 + 9 + 6 = 18
13. 8 + 1 + 2 = 11
14. 8 + 4 + 6 = 18
15. 2 + 9 + 4 = 15
16. 7 + 6 + 3 = 16
17. 3 + 7 + 3 = 13
18. 8 + 6 + 3 = 17
19. 4 + 4 + 6 = 14
20. 9 + 7 + 1 = 17
21. 6 + 6 + 8 = 20
22. 3 + 3 + 5 = 11

Reflect
How do you decide which numbers to add first when you have more than two addends?

Answers will vary; possible answers: I make ten, or I add doubles.

Student Workbook, pp. 32–33

WEEK 3
Solving Addition Problems

Lesson 3

Objective
Students can add and subtract quantities to perform successive operations.

Standard
2.NBT.5 Fluently add and subtract within 100 using strategies based on place value, properties of operations, and/or the relationship between addition and subtraction.

Creating Context
Help English Learners make a chart listing collocations (word sequences) that use the word *count* and suggest that they illustrate each one to help them remember.

Materials
Program Materials
- Dragon Quest Record Form, 1 per student
- Dragon Quest Game Board, 1 per group of 3 or 4
- Dragon Quest Cards, 1 set per group
- Pawns, 1 per student
- Spinner, 1 per group

1 WARM UP

Prepare
Copy or display the Dragon Quest Record Form onto the board.

Draw two cards from a shuffled set of Dragon Quest Cards. Write the quantities shown on the cards in the top two boxes on the record form. Ask a volunteer whether you need to add buckets or take buckets away, and put the corresponding operation symbol in the blank between the numbers.

▶ **How many buckets do I have? Now that I know the answer, I can follow the arrow to put the answer on the next line. This way I know what to start from on each turn.**

Draw another card and repeat the demonstration until you believe students understand how to use the form.

Just the Facts
Have students show the answer by holding up the appropriate number of fingers. Use questions such as the following:

▶ **Five plus four minus three equals what number?** Students show six fingers.

▶ **Ten minus four plus one equals what number?** Students show seven fingers.

▶ **Two plus eight minus one equals what number?** Students show nine fingers.

2 ENGAGE

Develop: The Fiercest Dragon
"Today we are going to play another Dragon Quest game." Follow the instructions on the Activity Card **The Fiercest Dragon**. As students complete the activity, be sure to use the Questions to Ask.

Activity Card 3K

Alternative Grouping
Individual: Partner with the student to play the game.

Progress Monitoring

| If… students have trouble doing the subtraction problems, | Then… allow them to use Counters. |

Practice
Have students complete **Student Workbook,** pages 34–35. Guide students through the Key Idea example and the Try This exercises.

Interactive Differentiation
Consult the Teacher Dashboard for grouping suggestions. You can also use performance on the Engage activity to guide students.

Independent Practice

For additional practice, have students work on Eggcellent Addition. Students choose numbers whose sum gets them the farthest on the game board. Students must choose carefully, as some spaces send players backward on the game board.

Supported Practice

For additional support, use Counters and number cards in a small group to practice adding and subtracting in successive operations.

- Write on the board $4 + 3 - 2$.
- Model the problem using Counters.
- Next, have a student choose three cards. Have the student create a problem with those three cards using addition and/or subtraction. For example, the student may choose 5, 8, and 2. Then he or she may write $5 + 8 - 2$.
- Have students use Counters to find the answer.
- Watch as students create problems so that they don't create a problem with a negative answer.

REFLECT

Think Critically

Review students' answers to the Reflect prompt at the bottom of *Student Workbook,* page 35, and then review the Engage activity.

Review the meaning of the addends and the sum.

▶ **How did the record form help you keep track of the number of buckets?**

▶ **Would you rather reorder your cards before using the form? Why or why not?**

Real-World Application

▶ **If you emptied your piggy bank to count your coins, what might be the first thing you would do?** I might organize them by amounts.

▶ **Why would you organize them this way?** It would be easier to count.

▶ **Organizing coins into groups makes it easier to count, just like organizing addends into groups makes it easier to add.**

4 ASSESS

Informal Assessment

Use the online or print Student Record, **Assessment,** page 128, to record informal observations.

The Fiercest Dragon
Did the student
☐ make important observations?   ☐ provide insightful answers?
☐ extend or generalize learning?   ☐ pose insightful questions?

Additional Practice

For additional practice, have students complete **Practice,** page 62.

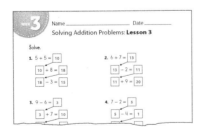
Practice, p. 62

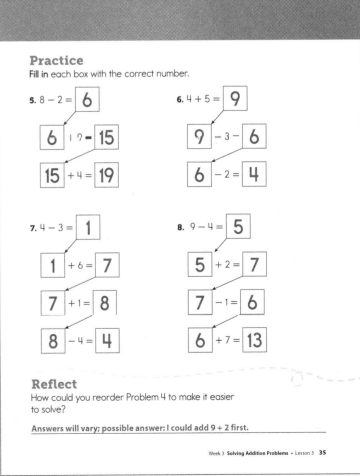

Student Workbook, pp. 34–35

Week 3 **Solving Addition Problems** • Lesson 3   **225**

WEEK 3
Solving Addition Problems

Lesson 4

Objective
Students can solve addition word problems and write equations to describe them.

Standard
2.NBT.5 Fluently add and subtract within 100 using strategies based on place value, properties of operations, and/or the relationship between addition and subtraction.

Creating Context
To prepare students for this activity, organize them into small groups and have them talk about things they do at home that require addition.

Materials
Program Materials
Counters, 20 per pair

1 WARM UP

Prepare
- **Jackie went to the candy shop 2 days in a row. The first day she bought 2 pieces of candy. The second day she bought 5 pieces of candy to share with her friends. How many pieces of candy did Jackie buy over the 2 days? What is being asked?** 7; How many pieces of candy did she buy?
- **What are the key words in the question?** pieces of candy

Reread the word problem, and have students raise their hands every time they hear the phrase *pieces of candy*. Read the word problem again, and have students identify the numbers that are associated with pieces of candy. Then form the equation, and solve it. $2 + 5 = 7$

Just the Facts
Have students act out each problem with their fingers. Then solve the problem. Use word problems such as the following:

- **Two rabbits were playing in the garden. Three more rabbits joined them. How many rabbits were in the garden in all?** Students show five fingers.
- **Six ants were eating at a picnic. Three ants joined them. How many ants were at the picnic altogether?** Students show nine fingers.
- **Two dogs were playing catch. One more dog started to play. How many dogs were playing catch?** Students show three fingers.

2 ENGAGE

Develop: Addition Action
"Today we are going to solve problems by acting them out." Follow the instructions on the Activity Card **Addition Action**. As students complete the activity, be sure to use the Questions to Ask.

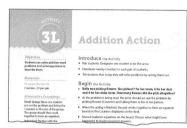
Activity Card 3L

Alternative Groupings
Small Group: Have one student act out the problem and bring the Counters to the rest of the group. The group should then work together to form an equation.

Individual: Partner with the student, and complete the activity as written.

Progress Monitoring
| **If…** students have trouble identifying which numbers should be addends in the word problem, | ▶ **Then…** discuss the meaning of the word *altogether*. Explain that the question "How many altogether?" is a clue that they should add. |

Practice
Have students complete **Student Workbook,** pages 36–37. Guide students through the Key Idea example and the Try This exercises.

Interactive Differentiation
Consult the Teacher Dashboard for grouping suggestions. You can also use performance on the Engage activity to guide students.

Independent Practice
For additional practice, have students work on Word Problems with Tools 1: Find Result or Change. If needed, encourage students to use paper and pencil or Counters to solve problems.

Supported Practice
For additional support, work through the problems in Word Problems with Tools 1: Find Result or Change with a small group of students. Use Counters, drawings, and numbers to guide students as they work.

3 REFLECT

Think Critically

Review students' answers to the Reflect prompt at the bottom of **Student Workbook,** page 37, and then review the Engage activity.

Review with students the question that was being asked in the problem, the key words, and the mistake that Sam made.

▶ **What mistake did Sam make?** Possible answer: He added the $10 with the rides and the games.

▶ **What should he have done instead?** Possible answer: added 7 to 4

Real-World Application

Word problems are usually written about real-world situations. Challenge students to pair up and create word problems relating to something that happens or has happened during a school day. Then have pairs exchange word problems and solve each other's problems.

ASSESS

Informal Assessment

Use the online or print Student Record, **Assessment,** page 128, to record informal observations.

Addition Action	
Did the student	
☐ respond accurately?	☐ respond with confidence?
☐ respond quickly?	☐ self-correct?

Additional Practice

For additional practice, have students complete **Practice,** page 63.

Practice, p. 63

Week 3 • Solving Addition Problems

Lesson 4

Key Idea
When solving a word problem, be sure to answer the question that is being asked. Remember to include labels when writing equations.

Try This
Underline the label. Write an addition equation for each problem.

1. The Waller Family went to a carnival. Each child could bring friends. Connor brought 1 friend, Kameron brought 2 friends, and Heather brought 1 friend. How many <u>friends</u> went to the carnival with the Waller Family?

   1 friend + 2 friends + 1 friend = 4 friends

2. Heather and her friend wanted to ride three rides before dinner. They waited 6 minutes to ride the Scrambler. They waited 4 <u>minutes</u> for the merry-go-round. The Rocket had no wait. How long did they have to wait in total to get on the three rides?

   6 minutes + 4 minutes + 0 minutes = 10 minutes

36   Level D Unit 3 Addition

Practice
Write an addition equation for each problem.

3. Connor and his friend played laser tag. Connor was hit 8 times, and his friend was hit 5 times. How many total <u>times</u> were they hit while playing laser tag?

   8 times + 5 times = 13 times

4. Kameron played the ring toss game to win a basketball jersey. He tossed 5 rings his first turn, 5 rings his second turn, and 4 rings his third turn before he finally won. How many <u>rings</u> did he toss to win?

   5 rings + 5 rings + 4 rings = 14 rings

Reflect
Sam tried to solve the word problem below.

Valerie had $10 to spend at the carnival. She went on 7 rides and played 4 games. How many activities did Valerie do?

10 + 7 + 4 = 21

What did Sam do wrong?

<u>Sam should have added only 7 + 4 because the question is asking for the</u>

<u>number of activities. He also added the amount she spent.</u>

Week 3 Solving Addition Problems • Lesson 4   37

Student Workbook, pp. 36–37

Week 3 **Solving Addition Problems** • Lesson 4   **227**

WEEK 3
Solving Addition Problems

Lesson 5 Review

Objective
Students review skills learned this week and complete the weekly assessment and project.

Standard
2.NBT.5 Fluently add and subtract within 100 using strategies based on place value, properties of operations, and/or the relationship between addition and subtraction.

Vocabulary
Review vocabulary introduced during the week.

Creating Context
Primary language can be a helpful tool for ensuring comprehension of math concepts. Pair English Learners of beginning proficiency with students of the same primary-language background who have greater English proficiency. Have them work through word problems using their primary language to support their understanding.

1 WARM UP

Prepare
- **Do 5 + 3 + 1 and 4 + 1 + 4 have the same sum?** yes
- **Name the addends in each problem.**
- **What does the word *sum* tell you to do?** add

Have a volunteer write the second problem on the board.

- **How did you solve this problem?** Possible answers: I added 4 + 4 to get 8 and added 1 to 8 to get 9; I added 4 + 1 to get 5 and 5 + 4 to get 9.

Write 2 + 5 + 8 on the board.

- **How can we group these numbers to make adding easier?** Possible answer: Group 8 and 2 to make 10, and then add 5 to get 15.

2 ENGAGE

Practice
Have students complete **Student Workbook,** pages 38–39.

Week 3 • Solving Addition Problems

Lesson 5 Review

This week you solved addition problems. You solved and wrote equations with more than two addends. You also solved addition word problems.

Lesson 1 Use Counters to find each sum.

1. 1 + 6 = __7__
2. 7 + 10 = __17__
3. 4 + 5 = __9__
4. 8 + 5 = __13__
5. 3 + 5 + 2 = __10__
6. 4 + 6 + 6 = __16__
7. 5 + 5 + 1 = __11__
8. 1 + 6 + 9 = __16__

Lesson 2 Find the sum of the two connected addends. Then add the third addend to find the total sum.

9. 3 + 10 + 5 =
   __8__ + __10__ = __18__

10. 12 + 4 + 3 =
    __7__ + __12__ = __19__

Lesson 3 Fill in each box with the correct number.

11. 5 + 3 = [8]
    [8] + 5 = [13]
    [13] + 4 = [17]

Lesson 4 Solve the word problem by writing an addition equation.

12. Kurt was selling magazine subscriptions. He sold 2 to his grandmother, 1 to his mother, and 5 to his neighbors. How many magazine subscriptions did Kurt sell?

    2 subscriptions + 1 subscription + 5 subscriptions = 8 subscriptions

Reflect
Write the addends from Problem 12 in a different order and show the sum. Is the sum the same as before? Why?

Possible answer: 5 + 2 + 1 = 8; Yes. The sum is the same because the order of addends does not change the sum.

Student Workbook, pp. 38–39

3 REFLECT

Think Critically

Review students' answers to the Reflect prompt at the bottom of **Student Workbook,** page 39.

Discuss the answer with the group to reinforce Week 3 concepts.

4 ASSESS

Formal Assessment

Students may take the weekly assessment online.

As an alternative, students may complete the weekly test on **Assessment,** pages 41–42. Record progress using the Student Assessment Record, **Assessment,** page 128.

Going Forward

Use the **Teacher Dashboard** to view results of the online assessments, to input the results of print student assessments, and to review progress before making decisions about next steps. Use the weekly test results and observations to determine the next steps for each student.

Retention	
Student displays good grasp of this week's concepts and skills.	For additional practice, have students continue to use the Building Blocks activity Word Problems with Tools 1: Find Result or Change.
Remediation	
Student is still struggling with the week's concepts and skills.	For additional support, have students continue to use **The Fiercest Dragon** activity.

Suggestions for Re-Evaluation: If a student has struggled without success for several weeks, use observations and test results to place the student at a level in which he or she can find success and build confidence to move forward.

Name _____ Date _____

Solving Addition Problems

1. 3 + 2 + 4 = __9__

2. 1 + 5 + 2 = __8__

3. 6 + 7 + 2 =
   __13__ + 2 = __15__

4. 8 + 2 + 9 =
   __10__ + 9 = __19__

5. Circle the problem that has the same answer as 3 + 9 + 4.
   (8 + 2 + 6)   9 + 4 + 5

Name _____ Date _____

Solving Addition Problems

6. Circle the problem that has the same answer as 4 + 7 + 6.
   8 + 6   (10 + 7)

7. 9 + 8 + 4 =
   __17__ + 4 =

8. Circle the two problems that have the same answer.
   (1 + 9 + 8)   9 + 7 + 5   (6 + 7 + 5)

9. For a family cookout, 3 people chose hot dogs, 5 chose hamburgers, and 4 chose chicken. Write an equation to show how many people were at the cookout. Solve the equation.
   3 + 5 + 4 = 12

10. Write a problem with two addends that equals 2 + 7 + 4.
    __any problem that equals 13__

Assessment, pp. 41–42

Project Preview

This week, students learned about word problems. They worked with up to three addends in a problem and wrote equations to describe a word problem. This week, students will write word problems about their own collections.

Project-Based Learning

Standards-driven Project-Based Learning is effective in building deep content understanding. Project-Based Learning increases long-term retention of concepts and has been shown to be more effective than traditional instruction. Completing a project to answer an essential question challenges students to apply and demonstrate mastery of concepts and skills by expressing understanding through discussion, research, and presentation.

Essential Question

HOW can I use addition to solve real-world problems?

Project Evaluation Criteria

Review project evaluation criteria with students prior to beginning the project.

Exceeds Expectations
☐ Project result is explained and can be extended.
☐ Project result is explained in context and can be applied to other situations.
☐ Project result is explained using advanced mathematical vocabulary.
☐ Project result is described, and mathematics are used correctly and can be extended.
☐ Project result is explained and extended, and shows advanced knowledge of mathematical concepts and skills.

Meets Expectations
☐ Project result is explained.
☐ Project result is explained in context.
☐ Project result is explained using mathematical vocabulary.
☐ Project result is described, and mathematics are used correctly.
☐ Project result is explained, and shows satisfactory knowledge of mathematical concepts and skills.

Does Not Meet Expectations
☐ Project result is not explained.
☐ Project result is explained, but out of context.
☐ Project result is explained, but mathematical vocabulary is oversimplified.
☐ Project result is described, but mathematics are not used correctly.
☐ Project result is not explained and/or extended, or shows less than satisfactory knowledge of mathematical concepts and skills.

Solve a Collection Problem

Objective
Students can write addition word problems and the equations to solve them within 100.

Standard
2.NBT.5 Fluently add and subtract within 100 using strategies based on place value, properties of operations, and/or the relationship between addition and subtraction.

Materials
Additional Materials
- business-sized envelopes, 1 per group
- index cards, 3 per group
- poster from Week 2
- tape or glue sticks

Prepare Ahead
Develop two or three word problems about your own collection.

Best Practices
- Provide project directions that are clear and brief.
- Coach, demonstrate, and model.
- Organize the materials before the lesson.

Introduce

Collections provide many real-life examples of when you might need to solve a word problem. You might be solving problems about your collection every day and not even realize you are doing math!

▶ **What numbers can you use to describe your collection? How could these numbers be used in a word problem?** Answers may vary. Possible answer: My friend gave me 12 seashells. I had 29. 12 seashells plus 29 seashells is equal to 41 seashells.

Explore

▶ **Today you will use your collection to write three word problems for other people to solve.**

- Present a word problem about your collection. For example: "Last week I had 3 books about dinosaurs in my collection. I went to the store and bought 2 more books about dinosaurs. How many books about dinosaurs do I have now?"
- Model how to write an equation to solve the word problem.

▶ **You and your partner are going to write three word problems that match your collection. You will write the problems on index cards and put the index cards in an envelope attached to your poster.**

- Referencing the word problem you presented above, show students how to write a word problem on one side of an index card and the answer to the problem on the other side. Demonstrate how to glue a business envelope to the poster. It is simplest to glue the front of the envelope to the poster. Label the envelope flap *Word Problems*. Show students how to put the index card in the envelope.

▶ **Now people will be able to take the index cards from my poster. They can read the word problems, solve them, and check their answers on the back of the card. When they are finished, they return the index cards to the envelope so someone else can try.**

- Distribute three index cards and an envelope to each pair of students.

▶ **Work with your partner to write three word problems that match your collection. Write the problems on the index cards. On the back, write the equation that helps you find the answer.**

▶ **Complete *Student Workbook*, page 40, to practice solving real-world problems.**

Wrap Up

- Give students sufficient time to generate their three word problems.
- You may want students to read their problems to other pairs of students to check to make sure the word problems make sense.
- Discuss students' answers to the Reflect prompts at the bottom of *Student Workbook*, page 40.

Encourage students to add illustrations to their envelopes. Help students tape or glue their envelopes to their posters. Make sure the index cards are tucked securely inside the envelopes before you put the posters aside for next week.

If time permits, allow student pairs to read one of their word problems to the class.

Week 3 • Solving Addition Problems

Project
Solve a Collection Problem

Answer the following questions.

1. How do you add to your collection?

   Answers may vary; possible answers: People give me more items; I find more items; I buy more items.

2. How do you keep track of items in your collection?

   Answers may vary; possible answer: I check each on a checklist.

Solve these word problems.

3. Cory has 25 seashells. He goes to the beach and finds 4 more. How many does he have now?

   29 shells

4. Donna collects buttons. In one jar she has 14 buttons. In another jar she has 6 buttons. How many buttons does she have altogether?

   20 buttons

Reflect

What key words in your word problems tell readers that they need to use addition?

Answers may vary; possible answers: How many are there in all? How many are there altogether? How many do they have now?

Student Workbook, p. 40

Teacher Reflect

☐ Did my use of media explain the project more clearly?

☐ Did students organize their ideas?

☐ Was I able to answer questions when students did not understand?

WEEK 4: Addition Tools and Strategies

Week at a Glance

This week, students continue **Number Worlds,** Level D, Addition. Students will explore multiple-addend addition problems using number lines. They will show equations on number lines and interpret illustrated number lines as equations.

Skills Focus

- Add successive single-digit quantities.
- Illustrate a number line based on an equation.
- Develop an equation based on an illustrated number line.
- Find missing addends.

How Students Learn

The lessons this week include several number-line games. To build a high level of mathematical talk—as opposed to talk about winning, losing, or cheating—focus attention on questions that might prove useful at various points in game play. Such questions include: *Where are you now? Where will you be when you make that move? Who is closer to the goal? How many more do you need? How do you know? How did you figure it out?*

English Learners ELL

For language support, use the **English Learner Support Guide,** pages 64–65, to preview lesson concepts and teach academic vocabulary. **Number Worlds** Vocabulary Cards are listed as additional materials in many lessons and can be used to preteach and reinforce academic vocabulary.

Math at Home

Give one copy of the Letter to Home, page 16, to each student. Encourage students to share and complete the activity with their caregivers.

Weekly Planner

Lesson	Learning Objectives
1 pages 234–235	Students can write equations to describe forward progression on a number line.
2 pages 236–237	Students can add a series of single-digit numbers and write number sentences to record the partial sums.
3 pages 238–239	Students can use counting on and mental math to solve problems with one- and two-digit numbers.
4 pages 240–241	Students can solve problems with one- and two-digit numbers when displayed in a vertical format.
5 pages 242–243	**Review and Assess** Students review skills learned this week and complete the weekly assessment and project.
Project pages 244–245	Students can identify tools and strategies that can be used to solve addition problems.

232 Level D Unit 3 **Addition**

Key Standard for the Week

Domain: Number and Operations in Base Ten
Cluster: Use place value understanding and properties of operations to add and subtract.
2.NBT.5 Fluently add and subtract within 100 using strategies based on place value, properties of operations, and/or the relationship between addition and subtraction.

Materials		Technology
Program Materials • *Student Workbook,* pp. 41–43 • *Practice,* p. 64 • Activity Card 3M, **Secret Number Game** • Neighborhood Number Line • Counters • Number 1–6 Cube	**Additional Materials** envelope	*Teacher Dashboard* Building Blocks Bright Idea Number Line Tool
Program Materials • *Student Workbook,* pp. 44–45 • *Practice,* p. 65 • Activity Card 3N, **Let's Add Some More** • Neighborhood Number Line • Number Cards (1–9) • Plus Cards • Counters	**Additional Materials** small self-sticking notes	*Teacher Dashboard* Building Blocks Lots O' Socks
Program Materials • *Student Workbook,* pp. 46–47 • *Practice,* p. 66 • Activity Card 3O, **Create An Equation** • Neighborhood Number Line • Number 1–6 Cube • Counters		*Teacher Dashboard* Building Blocks Figure the Fact
Program Materials • *Student Workbook,* pp. 48–49 • *Practice,* p. 67 • Activity Card 3P, **Vertical Operations** • Neighborhood Number Line • Double-Digit Number Cards (Addition) • Number 1–6 Cube		*Teacher Dashboard* Building Blocks Easy as Pie
Program Materials • *Student Workbook,* pp. 50–51 • Weekly Test, *Assessment,* pp. 43–44		Review previous activities.
Program Materials *Student Workbook,* p. 52	**Additional Materials** • construction paper • poster from Week 3	

WEEK 4
Addition Tools and Strategies

Find the Math

In this week, students learn to use various tools and strategies to solve more difficult addition problems.

Use the following to begin a guided discussion:

▶ **You add new stickers to your collection. How can you figure out the total number of stickers in your collection?** Answers may vary. Possible answer: First, count how many stickers you already have. Next, count how many you are adding to the collection. Then, add the new amount to the old amount by counting on.

Have students complete *Student Workbook,* page 41.

Student Workbook, p. 41

Lesson 1

Objective
Students can write equations to describe forward progression on a number line.

Standard
2.NBT.5 Fluently add and subtract within 100 using strategies based on place value, properties of operations, and/or the relationship between addition and subtraction.

Creating Context
In English, there are many words that sound the same but are spelled differently and have different meanings. The word *sum* sounds like *some*. It is important that English Learners know that *sum* means "the answer to an addition problem," and that *some* is a general description of "more than one."

Materials
Program Materials
- Neighborhood Number Line (1–30)
- Counters, 25 of 1 color per student
- Number 1–6 Cube, 1 per group of 4 students

Additional Materials
envelope, 1 per group

1 WARM UP

Prepare
Show students the Neighborhood Number Line and Counters they will use in the Engage activity. Model how to write an equation to describe the first move in this game by rolling a Number 1–6 Cube, placing that many Counters on the Neighborhood Number Line, and thinking aloud as you write each step in the equation.

▶ **How many Counters did I have on my number line when I started?**

▶ **What did I do after I rolled the Number Cube?**

▶ **What symbol should I write before I write the sum of the numbers?**

Repeat this process twice, making sure students understand that the total amount for any one turn is the starting number for the next turn.

2 ENGAGE

Develop: Secret Number Game
"Today we are going to use a number line to help us add." Follow the instructions on the Activity Card **Secret Number Game.** As students complete the activity, be sure to use the Questions to Ask.

Activity Card 3M

Alternative Grouping
Individual: Partner with the student to complete the activity as written.

Progress Monitoring

| **If...** students have trouble remembering to include the plus and equal symbols in their equations, | ▶ **Then...** post these symbols on the classroom wall, and practice writing equations throughout the day. You can help the class write equations by describing the number of students in line for recess, the number of books lined up on a bookshelf, and so on. |

Practice
Have students complete *Student Workbook,* pages 42–43. Guide students through the Key Idea example and the Try This exercises.

234 Level D Unit 3 **Addition**

Interactive Differentiation

Consult the **Teacher Dashboard** for grouping suggestions. You can also use performance on the Engage activity to guide students.

Independent Practice

For additional practice, use Bright Idea: Counting On Game. Students find the sum and move forward on the number-line game board.

Supported Practice

For additional support, use the Number Line Tool with a small group of students. Set the Number Line Tool to start with the number 3 and skip count by 4.

- Tell a story using the numbers on the number line.
- ▶ Molly had 3 apples in her refrigerator. At the store, she bought a bag with 4 apples. How many apples will she have altogether?
- Based on the story, have students write and solve the equation using the number line. $3 + 4 = 7$
- On the number line, demonstrate how 4 can continue to be added to the sum if Molly continues to buy bags of apples.

3 REFLECT

Think Critically

Review students' answers to the Reflect prompt at the bottom of **Student Workbook,** page 43, and then review the Engage activity.

Review the meaning of the words *addends* and *sum*. Have students use a number line to explain what each number in the equation stands for.

- ▶ How would you explain what the numbers in this problem stand for using a number line: $2 + 5 = 7$?
- ▶ How do you know which direction to move on the number line when you add?

4 ASSESS

Informal Assessment

Use the online or print Student Record, **Assessment,** page 128, to record informal observations.

Secret Number Game
Did the student
☐ make important observations?   ☐ provide insightful answers?
☐ extend or generalize learning?   ☐ pose insightful questions?

Additional Practice

For additional practice, have students complete **Practice,** page 64.

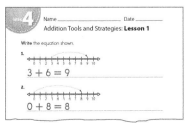

Practice, p. 64

Week 4 • Addition Tools and Strategies

Lesson 1

Key Idea
$3 + 4 = 7$
3 is where the jump starts.
4 is the distance of the jump.
7 is where the jump ends.

Try This
Show the following equations on each number line. Write the sum.

1.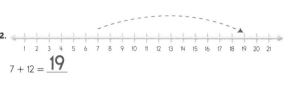

$5 + 5 = \underline{10}$

2.

$7 + 12 = \underline{19}$

42   Level D   Unit 3   Addition

Practice
Write the equation that is shown on each number line.

3.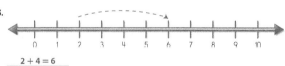

$2 + 4 = 6$

4.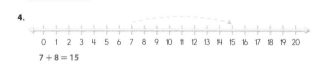

$7 + 8 = 15$

5.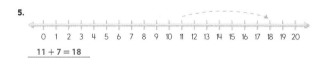

$11 + 7 = 18$

Reflect
Looking at a number line, Carlos wrote the equation $6 + 4 = 10$. Explain what each number stands for when jumping on the number line. You may draw a number line to help you answer the question.

6 is where the jump started. 4 is the distance of the jump. 10 is where the

jump ended.

Week 4 Addition Tools and Strategies • Lesson 1   43

Student Workbook, pp. 42–43

Week 4 Addition Tools and Strategies • Lesson 1   235

WEEK 4
Addition Tools and Strategies

Lesson 2

Objective
Students can add a series of single-digit numbers and write number sentences to record the partial sums.

Standard
2.NBT.5 Fluently add and subtract within 100 using strategies based on place value, properties of operations, and/or the relationship between addition and subtraction.

Creating Context
English Learners may be less worried about making language errors when working with small groups.

Materials
Program Materials
- Neighborhood Number Line (1–20)
- Number Cards (1–9)
- Plus Cards, 1 set per student pair
- Counters, 30 per student pair

Additional Materials
small self-sticking notes, 20

Prepare Ahead
If necessary, create an additional Neighborhood Number Line by covering the numbers of two additional sections with small self-sticking notes. Renumber as 1–20.

1 WARM UP

Prepare
- Use Counters and a student equation from the previous lesson. On another number line, create a display that matches the equation record sheet, and keep this hidden until the end of the activity.
- ▶ **Today we're going to do the opposite of what we did yesterday. We're going to use an equation to recreate the moves on the Number Line that the equation describes.**
- Display the equation, and ask volunteers to use the Counters to show each action the equation describes.
- To check for accuracy, compare the display students created to the one you created before the lesson.

Just the Facts
Have students show the sums for addition problems by holding up the correct number of fingers. Use problems such as the following:
- ▶ 6 + 4   10
- ▶ 5 + 3   8
- ▶ 3 + 6   9

2 ENGAGE

Develop: Let's Add Some More
"Today we are going to write addition equations to match movements on a number line." Follow the instructions on the Activity Card **Let's Add Some More**. As students complete the activity, be sure to use the Questions to Ask.

Activity Card 3N

Alternative Grouping
Individual: Have the student add Counters to the Neighborhood Number Line and write the number and the plus sign while you do the same thing on the other side of the barrier.

Progress Monitoring
| **If...** a team is having difficulty correctly writing equations for each move on the board, | **Then...** remind them that the number they land on during one turn is the number they will start from on the next turn. They should write this number as the first addend in the equation they create for the turn. |

Practice
Have students complete **Student Workbook,** pages 44–45. Guide students through the Key Idea example and the Try This exercises.

Interactive Differentiation
Consult the **Teacher Dashboard** for grouping suggestions. You can also use performance on the Engage activity to guide students.

Independent Practice

For additional practice, use Lots O' Socks. In this game, students solve addition problems and use the number line to move the game piece the appropriate number of spaces on the board.

Supported Practice

For additional support, use this activity with a small group of students. You will need Number Cards (1–9), Counters, and the Neighborhood Number Line.

- Have a student pick two Number Cards. The number on the first card tells the student where to place the Counter on the number line.
- The number on the second card tells the student how many places to move the Counter.
- Have students make up a story problem to go with the numbers.
- ▶ **There were 7 fish in a tank. Then 5 more fish were added to the tank. How many fish are there in all?** 12 fish
- Have the student choose one more card. Tell the student to continue the story problem.
- ▶ **Over the weekend, 2 more fish were added to the tank. Now how many fish are in the tank?** 14 fish

236   Level D Unit 3 **Addition**

3 REFLECT

Think Critically

Review students' answers to the Reflect prompt at the bottom of **Student Workbook,** page 45, and then review the Engage activity.

Discuss the reasons for always going to the right.

▶ **Do you always move the same direction when adding?** yes

▶ **Why?** Numbers increase as you move right, just as a number increases as you add to it.

Real-World Application

▶ **Time only increases; you cannot take away time. A time line is a number line that shows important events during a period of time.**

Have each student make a time line of his or her day.

▶ **What should you include on your time line?** Possible answers: when I get up, what I do before school, what I do after school.

4 ASSESS

Informal Assessment

Use the online or print Student Record, **Assessment,** page 128, to record informal observations.

Let's Add Some More

Did the student
- ☐ respond accurately?
- ☐ respond with confidence?
- ☐ respond quickly?
- ☐ self-correct?

Additional Practice

For additional practice, have students complete **Practice,** page 65.

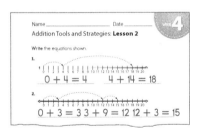

Practice, p. 65

Week 4 • Addition Tools and Strategies

Lesson 2

Key Idea

You can see that this number line shows two equations:

$0 + 8 = 8$ and $8 + 5 = 13$.

OR

You can see that this number line shows one equation:

$0 + 8 + 5 = 13$.

Try This

Draw the jumps on each number line. Use a different-colored pencil for each jump. Find the sum.

1.
$0 + 4 = 4 \qquad 4 + 5 = \underline{9} \qquad 9 + 3 = \underline{12}$

2.
$0 + 2 = 2 \qquad 2 + 10 = \underline{12} \qquad 12 + 6 = \underline{18}$

44 Level D Unit 3 **Addition**

Practice

Write the equations shown.

3.
$0 + 6 = 6$ and $6 + 11 = 17$, or $0 + 6 + 11 = 17$

4.
$0 + 5 = 5, 5 + 4 = 9$, and $9 + 7 = 16$, or $0 + 5 + 4 + 7 = 16$

Show the equation on the number line. Find the sum.

5.
$5 + 4 + 2 = \underline{11}$

Reflect

Look at the problems you have used number lines to solve. When adding numbers on a number line, which direction do you move?

Answers may vary; possible answers: to the right; toward the larger numbers

Week 4 **Addition Tools and Strategies** • Lesson 2 45

Student Workbook, pp. 44–45

WEEK 4
Addition Tools and Strategies

Lesson 3

Objective
Students can use counting on and mental math to solve problems with one- and two-digit numbers.

Standard
2.NBT.5 Fluently add and subtract within 100 using strategies based on place value, properties of operations, and/or the relationship between addition and subtraction.

Creating Context
Teachers often report that English Learners do well in mathematics. Many of the symbols used, including numbers, are used in other languages. English Learners may need to learn only the English names for these symbols.

Materials
Program Materials
- Neighborhood Number Line (1 to 100)
- Number 1–6 Cube, 1 per student pair
- Counters, 5 per student

Prepare
- Demonstrate how to count on. Explain that you will say the first number of a series, and students will add 4 to that number by counting the next four numbers.
- Tell students they can use their fingers to keep track of how many numbers they counted on so they know when to stop counting. Demonstrate by raising a finger for each number counted on.
  - ▶ **2** 3, 4, 5, 6
  - ▶ **18** 19, 20, 21, 22
  - ▶ **37** 38, 39, 40, 41
  - ▶ **55** 56, 57, 58, 59
- Remind students that you always start counting with the next number in the sequence.
- Repeat by asking students to add 3 to numbers you call out by counting the next three numbers.

Just the Facts
Have students use the strategy of counting on to check the answers to addition problems. Have students write the answer to each problem. When everyone has an answer, correct the problem together. Use problems such as the following:
- ▶ $43 + 4 = 47$
- ▶ $52 + 5 = 55$
- ▶ $36 + 3 = 38$

2 ENGAGE

Develop: Create An Equation
"Today we are going to use mental math to add small numbers to larger numbers." Follow the instructions on the Activity Card **Create an Equation.** As students complete the activity, be sure to use the Questions to Ask.

Activity Card 30

Alternative Grouping
Individual: Have the student create and solve equations according to the directions. Then verify that the student's equations are correct.

Progress Monitoring
| **If...** students are struggling with adding on to higher numbers, | **Then...** have the first partner think of a number between 10 and 40. |

Practice
Have students complete **Student Workbook,** pages 46–47. Guide students through the Key Idea example and the Try This exercises.

Interactive Differentiation
Consult the **Teacher Dashboard** for grouping suggestions. You can also use performance on the Engage activity to guide students.

Independent Practice

For additional practice, use Figure the Fact: Two-Digit Adding. Students count on from the last number their game piece landed on.

Supported Practice

For additional support, use Counters to continue working on the concept of counting on.
- Write the problem $34 + 5$ on the board.
- Have each student place five Counters in front of himself or herself.
- Demonstrate how to count on.
- ▶ **First, I say 34. Then, I tap each Counter as I continue counting: 35, 36, 37, 38, 39.**
- Tell students to repeat the problem with you.
- Write the answer on the board.
- Continue with similar problems.

238 Level D Unit 3 **Addition**

3 REFLECT

Think Critically

Review students' answers to the Reflect prompt at the bottom of **Student Workbook,** page 47, and then review the Engage activity.

Discuss to reinforce the strategies students have learned to help them add.

▶ **If you have forgotten the number sequence you use to count on, what tool can you use to help?** a number line

Real-World Application

▶ Suppose you were at a store and decided to buy a drink that cost 49 cents. When you went to pay for it, the grocer said, "The drink is 49 cents, and you have to pay 4 more cents for the tax." How much money do you have to pay the grocer for the drink? How did you figure it out?

4 ASSESS

Informal Assessment

Use the online or print Student Record, **Assessment,** page 128, to record informal observations.

Create an Equation

Did the student

☐ make important observations? ☐ provide insightful answers?

☐ extend or generalize learning? ☐ pose insightful questions?

Additional Practice

For additional practice, have students complete **Practice,** page 66.

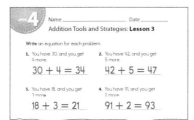

Practice, p. 66

Week 4 • Addition Tools and Strategies

Lesson 3

Key Idea
You can use the count-on strategy to add small numbers to larger numbers.

$48 + 3 = ?$

To add 3 to 48, start at 48 and count on the next three numbers in the sequence: 49, 50, 51. Use your fingers to keep track of how many numbers you count on.

Try This
Solve these problems. Use a number line if you need help.

1. $16 + 3 = \underline{19}$
2. $37 + 4 = \underline{41}$
3. $25 + 5 = \underline{30}$
4. $58 + 4 = \underline{62}$
5. $67 + 2 = \underline{69}$
6. $84 + 3 = \underline{87}$
7. $46 + 1 = \underline{47}$
8. $91 + 5 = \underline{96}$
9. $78 + 3 = \underline{81}$
10. $17 + 4 = \underline{21}$

46 Level D Unit 3 **Addition**

Practice
Write an equation for each problem.

11. You have 54 and you get 3 more. $54 + 3 = 57$
12. You have 19 and you get 2 more. $19 + 2 = 21$
13. You have 11 and you get 6 more. $11 + 6 = 17$
14. You have 44 and you get 4 more. $44 + 4 = 48$
15. You have 72 and you get 3 more. $72 + 3 = 75$
16. You have 51 and you get 2 more. $51 + 2 = 53$
17. You have 87 and you get 5 more. $87 + 5 = 92$
18. You have 36 and you get 1 more. $36 + 1 = 37$

Reflect
Solve the following problem:

$39 + 3 = \underline{42}$

Describe the strategy you used to figure out the answer.

Possible answer: I started at 39 and counted 40, 41, 42. I stopped at 42 because I was only adding three numbers to 39.

Week 4 **Addition Tools and Strategies** • Lesson 3 47

Student Workbook, pp. 46–47

Week 4 **Addition Tools and Strategies** • Lesson 3 **239**

WEEK 4
Addition Tools and Strategies

Lesson 4

Objective
Students can solve problems with one- and two-digit numbers when displayed in a vertical format.

Standard
2.NBT.5 Fluently add and subtract within 100 using strategies based on place value, properties of operations, and/or the relationship between addition and subtraction.

Creating Context
English Learners may benefit from clarification of some common phrases and words that proficient English speakers probably know. Occasionally words have more than one meaning or are used in potentially puzzling idiomatic expressions.

Materials
Program Materials
- Neighborhood Number Line
- Double-Digit Number Cards (Addition), 1 set
- Number 1–6 Cube, 1 per student pair

 WARM UP

Prepare
- Write 53 + 5 = _____ on the board in both a horizontal and a vertical format.
- Point to the horizontal format. Ask students to solve the problem mentally and to describe their thinking aloud.
- Next, point to the vertical format, and tell students that we use a different strategy when the problem is written this way.
- ▶ **First, you add the numbers in the ones column and write that number under the bar in the ones column. Then, you add the numbers in the tens column and write that number under the bar in the tens column. The number under the bar is the answer.**

Just the Facts
Write the following number sentences on the board in vertical format. If the problem is solved correctly, have students clap their hands. If the problem is not solved correctly, have students fold their arms over their chests. Use equations such as the following:

- ▶ **53 + 2 = 54** *fold arms over chest*
- ▶ **22 + 6 = 28** *clap hands*
- ▶ **14 + 9 = 13** *fold arms over chest*

2 ENGAGE

Develop: Vertical Operations
"Today we are going to learn a different way to solve and write addition problems." Follow the instructions on the Activity Card **Vertical Operations.** As students complete the activity, be sure to use the Questions to Ask.

Activity Card 3P

Alternative Groupings
Individual: Partner with the student and complete the activity as written.

Small Group: Complete the activity as written, but make sure all students take a turn recording and creating equations.

Progress Monitoring
If... students have trouble correctly aligning numbers in vertical-format problems, ▶ **Then...** give them graph paper, and demonstrate how to use the paper to keep the places aligned.

Practice
Have students complete **Student Workbook,** pages 48–49. Guide students through the Key Idea example and the Try This exercises.

Interactive Differentiation
Consult the **Teacher Dashboard** for grouping suggestions. You can also use performance on the Engage activity to guide students.

Independent Practice

For additional practice, have students play Easy as Pie. Players add numbers together and then move on from the previous space on the game board.

Supported Practice

For additional support, play Easy as Pie with a small group of students. As the group plays the game, write each problem vertically on the board so students can visualize the number sentence that is being solved. Use Counters to count on from the first addend to the second addend.

240 Level D Unit 3 Addition

3 REFLECT

Think Critically

Review students' answers to the Reflect prompt at the bottom of *Student Workbook,* page 49, and then review the Engage activity.

▶ **What rules do you need to remember when working with number lines?**
Possible answers: When moving left on a number line, the numbers get smaller; when moving right on a number line, the numbers get larger; move forward to find a sum; move backward to find a missing addend.

Real-World Application

▶ **Today we used the information given to determine where we started or finished on the number line. It is like following directions when we do not know where we will end up.**

Give students a set of directions and have them locate something in the room or in another room in the school.

4 ASSESS

Informal Assessment

Use the online or print Student Record, *Assessment,* page 128, to record informal observations.

Vertical Operations

Did the student

☐ respond accurately?   ☐ respond with confidence?

☐ respond quickly?   ☐ self-correct?

Additional Practice

For additional practice, have students complete **Practice,** page 67.

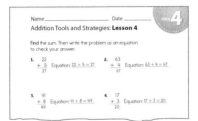

Practice, p. 67

Week 4 • Addition Tools and Strategies

Lesson 4

Key Idea
Addition problems can be written in two ways.

Horizontally: Vertically:

$33 + 4 = 37$ 33
                   $+\ 4$
We count on from 33...   47
34, 35, 36, 37 to find the
answer to equations   We add numbers in the ones and
written horizontally.   tens columns to find the answer to
                        a vertical equation.

Try This
Find each sum.

1. 42   2. 64
   $+\ 2$ $+\ 3$
    44   67

3. 89   4. 61
   $+\ 2$ $+\ 1$
    91   62

5. 18   6. 26
   $+\ 6$ $+\ 5$
    24   31

48 Level D Unit 3 **Addition**

Practice
Find each sum. Then write the problem as an equation and solve to check your answer.

7. 54   8. 26
   $+\ 6$ Equation: $54 + 6 = 60$ $+\ 2$ Equation: $26 + 2 = 28$
    60   28

9. 62   10. 15
   $+\ 3$ Equation: $62 + 3 = 65$ $+\ 4$ Equation: $15 + 4 = 19$
    65   19

11. 31   12. 87
    $+\ 3$ Equation: $31 + 3 = 34$ $+\ 6$ Equation: $87 + 6 = 93$
     34   93

Reflect
In your own words, explain the strategies you used to find the sums for Problems 11 and 12. Which was easier to solve? Why?

Answers will vary. Possible answer: Problem 11 was easier to solve because I had to add fewer than ten ones.

Week 4 Addition Tools and Strategies • Lesson 4 49

Student Workbook, pp. 48–49

WEEK 4
Addition Tools and Strategies

Lesson 5 Review

Objective
Students review skills learned this week and complete the weekly assessment and project.

Standard
2.NBT.5 Fluently add and subtract within 100 using strategies based on place value, properties of operations, and/or the relationship between addition and subtraction.

Vocabulary
Review vocabulary introduced during the week.

Creating Context
The activities this week reinforce the use of terminology associated with number lines. Encourage students to use these words often by having the class work in small groups.

1 WARM UP

Prepare

Choose a number between 8 and 30, and write this number on the board.

▶ **How many different ways can we make this number using addition?**

- Call on volunteers to suggest possible facts, and write these on the board.
- Ask the class to verify the accuracy of each fact that is offered. If the class is unsure, ask the student who offered the fact to explain his or her thinking.
- Encourage students to find several different combinations of numbers that make the target number, and then ask the class to count the number of different combinations they found.

2 ENGAGE

Practice

Have students complete **Student Workbook,** pages 50–51.

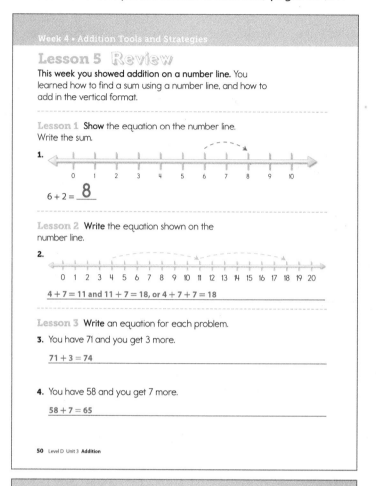

Student Workbook, pp. 50–51

242 Level D Unit 3 **Addition**

3 REFLECT

Think Critically

Review students' answers to the Reflect prompt at the bottom of *Student Workbook,* page 51.

Discuss the answer with the group to reinforce Week 4 concepts.

4 ASSESS

Formal Assessment

Students may take the weekly assessment online.

As an alternative, students may complete the weekly test on *Assessment,* pages 43–44. Record progress using the Student Assessment Record, *Assessment,* page 128.

Going Forward

Use the **Teacher Dashboard** to view results of the online assessments, to input the results of print student assessments, and to review progress before making decisions about next steps. Use the weekly test results and observations to determine the next steps for each student.

Retention	
Student displays good grasp of this week's concepts and skills.	**Building Blocks** Have students use the Building Blocks activity Easy as Pie.

Remediation	
Student is still struggling with the week's concepts and skills.	Use a Number 1–6 Cube, Counters, and the Neighborhood Number Line. • Have the student roll the Number Cube. The number tells the student where to place the Counter on the number line. • Have the student roll the Number Cube again. This number on the Number Cube tells the student how many places to move the Counter. • Have students make up a story problem to go with the numbers. ▶ There were 4 students in the class. Then 5 students were added. How many students are there in all? 9 students • Have the student roll the Number Cube again. Tell the student to continue the story problem. ▶ Over the weekend, 3 more students were added. Now how many students are in the class? 12 students

Suggestions for Re-Evaluation: If a student has struggled without success for several weeks, use observations and test results to place the student at a level in which he or she can find success and build confidence to move forward.

Name _____ Date _____

Addition Tools and Strategies — WEEK 4

1. Write the equation that is shown on this number line.

   $2 + 7 = 9$

2. Write the equation that is shown on this number line.

   $7 + 9 = 16$

3. Write the equation that is shown on this number line.

   $2 + 6 + 10 = 18$

4. Draw a line with an arrow to show this equation on the number line. Solve the equation.

   $4 + 5 + 3 = $ 12; arrow starts at 4, goes to 9, goes to 12

Level D Unit 3 Week 4   43

WEEK 4   Name _____ Date _____

Addition Tools and Strategies

Write an equation for each problem. Solve the equation.

5. You have 35 and get 6 more.

   $35 + 6 = 41$

6. You have 23, you get 4 more, and then you get 7 more.

   $23 + 4 + 7 = 34$

7. 11
   + 8

   $11 + 8 = 19$

8. 9
   13
   + 5

   $9 + 13 + 5 = 27$

44   Level D Unit 3 Week 4

Assessment, pp. 43–44

Week 4 **Addition Tools and Strategies** • Lesson 5   243

Project Preview

This week, students learned about various tools and strategies to solve addition problems. They wrote equations to describe progression on a number line; wrote equations to record partial sums for a series of addends; and used counting on, mental math, and equations set in a vertical format to solve problems with one- and two-digit numbers. Students will add these strategies to their project posters.

Project-Based Learning

Standards-driven Project-Based Learning is effective in building deep content understanding. Project-Based Learning increases long-term retention of concepts and has been shown to be more effective than traditional instruction. Completing a project to answer an essential question challenges students to apply and demonstrate mastery of concepts and skills by expressing understanding through discussion, research, and presentation.

Essential Question

HOW can I use addition to solve real-world problems?

Project Evaluation Criteria

Review project evaluation criteria with students prior to beginning the project.

Exceeds Expectations
☐ Project result is explained and can be extended.
☐ Project result is explained in context and can be applied to other situations.
☐ Project result is explained using advanced mathematical vocabulary.
☐ Project result is described, and mathematics are used correctly and can be extended.
☐ Project result is explained and extended, and shows advanced knowledge of mathematical concepts and skills.

Meets Expectations
☐ Project result is explained.
☐ Project result is explained in context.
☐ Project result is explained using mathematical vocabulary.
☐ Project result is described, and mathematics are used correctly.
☐ Project result is explained, and shows satisfactory knowledge of mathematical concepts and skills.

Does Not Meet Expectations
☐ Project result is not explained.
☐ Project result is explained, but out of context.
☐ Project result is explained, but mathematical vocabulary is oversimplified.
☐ Project result is described, but mathematics are not used correctly.
☐ Project result is not explained and/or extended, or shows less than satisfactory knowledge of mathematical concepts and skills.

Use Addition Tools and Strategies

Objective
Students can identify tools and strategies that can be used to solve addition problems.

Standard
2.NBT.5 Fluently add and subtract within 100 using strategies based on place value, properties of operations, and/or the relationship between addition and subtraction.

Materials
Additional Materials
- construction paper
- poster from Week 3

Best Practices
- Clearly enunciate instructions.
- Coach, demonstrate, and model.
- Allow students to self-monitor.

Introduce

This week you learned about some tools and strategies that help you solve addition problems.

▶ **What tools and strategies did you learn about this week?** Answers may vary. Possible answers: using a number line; writing equations; solving for a partial sum; using mental math; adding vertically

Explore

▶ **Today you will identify tools and strategies you can use to solve addition problems. You will add this information to your poster.**

- Brainstorm with the class the different tools and strategies they used this week to solve addition problems.
- On the board, make a chart with a column for *Tools* and a column for *Strategies*. **Tools:** number lines, equations, addition table; **Strategies:** counting on, mental math, finding partial sums, adding vertically

▶ **I am going to make a chart for my poster showing the tools and strategies I can use to solve addition problems.**

▶ **Write the tools and strategies either on a sheet of paper or, if there is room, on your poster. If you write the tools and strategies on a sheet of paper, attach it to your poster. Remember that you will need to add two more weeks' worth of information to the poster, so leave some extra space.**

▶ **Complete** *Student Workbook,* **page 52, to practice different strategies you can use to solve the same problem.**

Wrap Up

▶ **Work with your partner to add these tools and strategies to your poster. You can use words, pictures, diagrams, or examples.**

- Allow students sufficient time to add the tools and strategies to their posters.
- Allow for creativity, but do not let students spend too much time trying to think of a clever way to add this information to their poster. If a pair of students is struggling, suggest that the students copy the chart on the board onto their poster.
- Discuss students' answers to the Reflect prompt at the bottom of *Student Workbook,* page 52.

Have students add the tools and strategies to their posters. If students attach a sheet of paper to their posters, make sure it is secure before you put the posters aside for next week.

If time permits, allow student pairs to share what they created today.

Week 4 • Addition Tools and Strategies

Project
Use Addition Tools and Strategies

Answer the following questions. Explain different ways to solve the same addition problem.

1. Explain with words or pictures how you can solve 12 + 6.

   Check students' work. Students should explain how to solve 12 + 6.

2. Show another way to solve 12 + 6. For example, if you showed how to solve the problem vertically, show how to solve it with a number line.

   Students should explain how to solve 12 + 6 a different way.

3. Show how to solve 12 + 6 one more way.

   Students should explain how to solve 12 + 6 a third way.

Reflect
Why is it important to have more than one tool or strategy to solve an addition problem?

Answers may vary; possible answer: Sometimes one strategy is easier to use.

Student Workbook, p. 52

Teacher Reflect

☐ Did my use of media explain the project more clearly?

☐ Did students organize their ideas?

☐ Did students focus on the major concept of the activity?

WEEK 5
Addition Word Problems within 100

Week at a Glance
This week students continue **Number Worlds**, Level D, Addition. Students will write and solve addition equations with unknown sums and addends. They will also solve two-step equations with unknown sums and addends.

Skills Focus
- Write addition equations with unknown sums and addends.
- Solve addition equations with unknown sums and addends.
- Solve two-step equations with unknown sums and addends.

How Students Learn
Encourage students to explore algebraic principles. Give them plenty of opportunities to experiment with finding unknown addends and sums, and then step back to judge the relative ease or difficulty of the problem. Students are often more motivated to employ algebraic principles after they become a useful strategy for winning a game.

English Learners ELL
For language support, use the **English Learner Support Guide**, pages 66–67, to preview lesson concepts and teach academic vocabulary. **Number Worlds** Vocabulary Cards are listed as additional materials in many lessons and can be used to preteach and reinforce academic vocabulary.

Math at Home
Give one copy of the Letter to Home, page 17, to each student. Encourage students to share and complete the activity with their caregivers.

Weekly Planner

Lesson	Learning Objectives
1 pages 248–249	Students can solve addition word problems within 100.
2 pages 250–251	Students can solve unknown-addend addition word problems within 100.
3 pages 252–253	Students can solve two-step addition word problems within 100.
4 pages 254–255	Students can solve one- and two-step addition word problems within 100.
5 pages 256–257	**Review and Assess** Students review skills learned this week and complete the weekly assessment and project.
Project pages 258–259	Students create and solve word problems with unknowns.

246 Level D Unit 3 **Addition**

Key Standard for the Week

Domain: Operations and Algebraic Thinking

Cluster: Represent and solve problems involving addition and subtraction.

2.OA.1 Use addition and subtraction within 100 to solve one- and two-step word problems involving situations of adding to, taking from, putting together, taking apart, and comparing, with unknowns in all positions, e.g., by using drawings and equations with a symbol for the unknown number to represent the problem.

Materials		Technology
Program Materials • *Student Workbook,* pp. 53–55 • *Practice,* p. 68 • Activity Card 3Q, **Story Time!** • Story Time! Recording Sheets • Number Cards (0–100)	**Additional Materials** • index cards • Vocabulary Card 43, *sum*	*Teacher Dashboard* Building Blocks Word Problems with Tools 4; Word Problems with Tools 1
Program Materials • *Student Workbook,* pp. 56–57 • *Practice,* p. 69 • Activity Card 3R, **Problems with Unknowns** • Problems with Unknowns Recording Sheets • Number Line to 100 Game Board • Counters	• Number Cards (0–100) • Number 1–6 Cube **Additional Materials** Vocabulary Card 43, *sum*	*Teacher Dashboard* Building Blocks Barkley's Bones 1–10; Dinosaur Shop 4
Program Materials • *Student Workbook,* pp. 58–59 • *Practice,* p. 70 • Activity Card 3S, **Two-Step Word Problems** • Two-Step Word Problems Recording Sheets • Counters	**Additional Materials** Vocabulary Card 43, *sum*	*Teacher Dashboard* Building Blocks Word Problems with Tools 7
Program Materials • *Student Workbook,* pp. 60–61 • *Practice,* p. 71 • Activity Card 3T, **Hotel Mystery** • Hotel Mystery Recording Sheets • Hotel Game Board • Number Cards (0–100)	• Number 1–6 Cube • Pawns **Additional Materials** Vocabulary Card 43, *sum*	*Teacher Dashboard* Building Blocks Barkley's Bones 1–20; Tidal Tally
Program Materials • *Student Workbook,* pp. 62–63 • Weekly Test, *Assessment,* pp. 45–46		Review previous activities.
Program Materials *Student Workbook,* p. 64	**Additional Materials** • construction paper • index cards • poster from Week 4 • tape	

WEEK 5

Addition Word Problems within 100

Find the Math

In this week, students will solve addition word problems.

Use the following to begin a guided discussion:

▶ **A coin collector might use a book such as this one to hold his or her coins. He or she places certain coins in certain spots. How does using a book like this help if you are collecting coins?**
Answers may vary. Possible answer: You know how many coins you have and how many you need to complete your collection.

Have students complete *Student Workbook,* page 53.

Student Workbook, p. 53

Lesson 1

Objective
Students can solve addition word problems within 100.

Standard
2.OA.1 Use addition and subtraction within 100 to solve one- and two-step word problems involving situations of adding to, taking from, putting together, taking apart, and comparing, with unknowns in all positions, e.g., by using drawings and equations with a symbol for the unknown number to represent the problem.

Vocabulary
- addend
- sum

Creating Context
Write the problem $6 + 7 = 13$ on an index card. Draw an arrow pointing to the 13 and label it with the vocabulary word *sum*. Draw an arrow to the 6 and 7 and label each one *addend*.

Materials
Program Materials
- Story Time! Recording Sheets
- Number Cards (0–100)

Additional Materials
- index cards
- Vocabulary Card 43, *sum*

Prepare Ahead
Prepare index cards by labeling them with plus signs, equal signs, and blank boxes for unknown numbers. Each pair of students should have one index card with a plus sign, one index card with an equal sign, and one card with a blank box.

1 WARM UP

Prepare
- Give students this word problem.
- ▶ **There were 23 students in Mrs. Johnson's class. At lunch, 6 more students joined her class. How many students were at lunch?**
- Model how to write an equation to describe the word problem. Write $23 + 6$ on the board.
- ▶ **I don't know the answer to how many students were at lunch. To show this, I am going to write $23 + 6 = \square$. The $\square$ stands for how many students altogether were at lunch.**

- Solve the problem. $23 + 6 = 29$ Model how to show the answer.
- ▶ **There were 29 students at lunch.**
- Continue with additional problems as time permits.

2 ENGAGE

Develop: Story Time!

"Today we are going to write our own addition word problems and use a box to show where an unknown amount is in an addition equation." Follow the instructions on the Activity Card **Story Time!** As students complete the activity, be sure to use the Questions to Ask.

Activity Card 3Q

Alternative Groupings

Pair: Complete the activity the same way as with the whole group, but have students record their word problems and equations on the Recording Sheet. Students may use words or pictures to write their word problems.

Individual: Have each student complete the activity in the same way as with a pair.

Progress Monitoring

| If... students are confused by having a box in an equation, | Then... use a blank line or a question mark instead. |

Practice
Have students complete *Student Workbook,* pages 54–55. Guide students through the Key Idea example and the Try This exercises.

248 Level D Unit 3 Addition

Interactive Differentiation

Consult the **Teacher Dashboard** for grouping suggestions. You can also use performance on the Engage activity to guide students.

Independent Practice

For additional practice, have students complete Word Problems with Tools 4: Multidigit +/− to 100 and Multidigit Solver. Students will use one- and two-digit numbers.

Supported Practice

For additional support, use Word Problems with Tools 1: Find Result or Change, with small groups of students.

- Encourage the use of Counters and drawings as students solve the problems together.
- Model talking through the problems.
- ▶ **This problem tells me there are two different-colored crayons—red and yellow. I want to find out how many crayons there are altogether. That means I need to add the red crayons to the yellow crayons.**
- Demonstrate writing the important information on a piece of paper as you work to solve the problem.

3 REFLECT

Think Critically

Review students' answers to the Reflect prompt at the bottom of **Student Workbook,** page 55, and then review the Engage activity.

▶ **In the addition equation 4 + 7 = ☐, what does the ☐ mean?**
The ☐ shows that there is a number you don't know yet. You have to add 4 + 7 to find it.

4 ASSESS

Informal Assessment

Use the online or print Student Record, **Assessment,** page 128, to record informal observations.

Story Time!
Did the student
☐ respond accurately?   ☐ respond with confidence?
☐ respond quickly?   ☐ self-correct?

Additional Practice

For additional practice, have students complete **Practice,** page 68.

Practice, p. 68

Week 5 • Addition Word Problems within 100

Lesson 1

Key Idea
When you read a word problem, write an equation to go with it so you know how to solve the problem.
Use a ☐ to show the part of the equation that you don't know and that you have to solve for.

Try This
Read the problem. Write an equation to show how you should solve the problem. Solve the problem.

1. Lisa had 25 pennies. She found 6 more pennies. How many pennies does she have altogether?

   25 + 6 = ☐; Lisa has 31 pennies altogether.

2. Max has 15 dimes. Sara has 9 dimes. How many dimes do Max and Sara have in all?

   15 + 9 = ☐; They have 24 dimes in all.

3. Wade has 45 quarters in his collection. His dad gives him 3 more quarters. How many quarters does Wade have in all?

   45 + 3 = ☐; Wade has 48 quarters in all.

Practice
Read the problem. Write an equation to show how you should solve the problem. Solve the problem.

4. Tom had 10 coins. He found 6 more coins. How many coins does Tom have in all?

   10 + 6 = ☐; Tom has 16 coins in all.

5. Ann has 18 pennies. Kim has 11 pennies. How many pennies do the girls have in all?

   18 + 11 = ☐; Ann and Kim have 29 pennies in all.

6. There are 35 nickels on the table. Ryan puts 4 more nickels on the table. How many nickels are on the table now?

   35 + 4 = ☐; There are 39 nickels on the table.

Reflect
How did you know whether to add or subtract in these problems?

Answers may vary; possible answer: In these problems I was adding, or combining one group with another. When you put things together, you add.

Student Workbook, pp. 54–55

WEEK 5
Addition Word Problems within 100

Lesson 2

Objective
Students can solve unknown-addend addition word problems within 100.

Standard
2.OA.1 Use addition and subtraction within 100 to solve one- and two-step word problems involving situations of adding to, taking from, putting together, taking apart, and comparing, with unknowns in all positions, e.g., by using drawings and equations with a symbol for the unknown number to represent the problem.

Vocabulary
- addend
- sum

Creating Context
As you work through addition problems with your students, be sure to continue to use the correct academic vocabulary throughout the lesson.
▶ When I add these two <u>addends</u> together, what is the <u>sum</u>?

Materials

Program Materials
- Problems with Unknowns Recording Sheets
- Number Line to 100 Game Board
- Counters
- Number Cards (0–100)
- Number 1–6 Cube

Additional Materials
Vocabulary Card 43, *sum*

1 WARM UP

Prepare
- Give students this word problem.
▶ Lisa had 13 books on her bookshelf. She brought some books home from the library. Now there are 17 books on the bookshelf. How many books did she bring home from the library?
- Model how to write an equation to describe the word problem. Write $13 + \Box = 17$.
▶ The problem doesn't tell us how many books Lisa brought home from the library. That is the unknown amount. To show that I don't know that addend, I am going to write 13 plus an unknown number is equal to 17. The $\Box$ stands for how many books she brought home from the library.
- Solve the problem. Model how to show the answer. Write $13 + \Box = 17$; $13 + 4 = 17$.
▶ Lisa brought home 4 books from the library.
- Continue with other problems as time permits.

Just the Facts
Have students stand up when they know the answer to these doubles facts. Use questions such as the following:
▶ What does 6 plus 6 equal? 12
▶ What does 5 plus 5 equal? 10

2 ENGAGE

Develop: Problems with Unknowns

"Today we are going to continue to create our own addition word problems. We will use boxes to show an unknown amount in an addition equation. The unknown amount will not always be the sum." Follow the instructions on the Activity Card **Problems with Unknowns.** As students complete the activity, be sure to use the Questions to Ask.

Activity Card 3R

Alternative Grouping

Pair: Complete the activity the same way as with the whole group, but have students record their word problems and equations on the recording sheet. Students may use words or pictures to write and solve their word problems.

Progress Monitoring	
If... students struggle with finding unknown addends using Counters alone,	**Then...** place the Counters on a number line.

Practice

Have students complete **Student Workbook,** pages 56–57. Guide students through the Key Idea example and the Try This exercises.

Interactive Differentiation

Consult the **Teacher Dashboard** for grouping suggestions. You can also use performance on the Engage activity to guide students.

Independent Practice

For additional practice, have students complete Barkley's Bones 1–10. Students must find the unknown number in the equation. There are concrete manipulatives to assist students in finding the missing addend.

Supported Practice

For additional support, use Dinosaur Shop 4: Make It Right with a small group of students. Students are given a total for the order. There are already some dinosaurs in the box. Students must determine how many more dinosaurs need to go in the box to reach the total.

- Model the problem on the board as you work with students. For example, $1 + \Box = 2$
▶ What number makes this equation correct? 1

250 Level D Unit 3 Addition

3 REFLECT

Think Critically

Review students' answers to the Reflect prompt at the bottom of **Student Workbook,** page 57, and then review the Engage activity.

▶ **What addend do you add to 15 to get a sum of 19?** Answers may vary. Possible answer: I count up from 15 until I get to 19, so the addend is 4.

Real-World Application

▶ **You can use almost anything to show an unknown in a problem, but if you use letters, you should try to avoid the letter O and the lowercase letter *l*.**

▶ **Why should these letters be avoided when writing a math problem?** Answers may vary. Possible answer: The letters *O* and *l* look too much like the numerals 0 and 1.

4 ASSESS

Informal Assessment

Use the online or print Student Record, **Assessment,** page 128, to record informal observations.

> **Problems with Unknowns**
> Did the student
> ☐ respond accurately?   ☐ respond with confidence?
> ☐ respond quickly?   ☐ self-correct?

Additional Practice

For additional practice, have students complete **Practice,** page 69.

Practice, p. 69

Week 5 • Addition Word Problems within 100

Lesson 2

> **Key Idea**
> When you read a word problem, write an equation to go with it so you know how to solve the problem.
> Use a ☐ to show the part of the equation that you don't know and that you have to solve for.

Try This

Read the problem. Write an equation to show how you should solve the problem. Solve the problem.

1. Pedro had 38 quarters in his collection. His mom gave him some more quarters. Now he has 42 quarters. How many quarters did his mom give him?

   38 + ☐ = 42; **His mom gave him 4 quarters.**

2. Rachel has 30 buttons in a jar. She buys more buttons at a garage sale. Now she has 50 buttons. How many buttons did she buy at the garage sale?

   30 + ☐ = 50; **Rachel bought 20 buttons at the garage sale.**

3. Jeff has some seashells on his shelf. He adds 7 seashells to the shelf and now he has 14. How many seashells did he have on the shelf to start with?

   ☐ + 7 = 14; **Jeff had 7 seashells on the shelf to start with.**

56 Level D Unit 3 **Addition**

Practice

Read the problem. Write an equation to show how you should solve the problem. Solve the problem.

4. Ms. Larson had 18 books on her desk. She checked out some more books from the library and added them to the pile. She now has 25 books on her desk. How many books did she check out?

   18 + ☐ = 25; **Ms. Larson checked out 7 books.**

5. Tony has 21 toy cars. Friends give him a few more cars for his birthday. Now he has 28 cars. How many cars did he get for his birthday?

   21 + ☐ = 28; **Tony got 7 cars for his birthday.**

6. Mike has a coin collection. He adds 25 more coins to his collection. Now he has 29 coins. How many coins did Mike start with?

   ☐ + 25 = 29; **Mike started with 4 coins.**

Reflect

How did you know where to put the boxes in the equations? Explain how you knew.

Answers may vary; possible answer: The equations each tell you two of the following: what number you start with, what number you add, and what your sum will be. You put the box in where a part of the equation is missing.

Week 5 **Addition Word Problems within 100** • Lesson 2 57

Student Workbook, pp. 56–57

WEEK 5
Addition Word Problems within 100

Lesson 3

Objective
Students can solve two-step addition word problems within 100.

Standard
2.OA.1 Use addition and subtraction within 100 to solve one- and two-step word problems involving situations of adding to, taking from, putting together, taking apart, and comparing, with unknowns in all positions, e.g., by using drawings and equations with a symbol for the unknown number to represent the problem.

Vocabulary
- addend
- sum

Creating Context
Using visual or physical representations of quantities is an excellent way to make new concepts comprehensible. This lesson uses Counters.

Materials

Program Materials
- Two-Step Word Problems Recording Sheets
- Counters

Additional Materials
Vocabulary Card 43, *sum*

1 WARM UP

Prepare

- Give students the following word problem.
- ▶ There were 9 students on the school bus. When the bus stopped, 6 more students got on. The next time the bus stopped, 3 more students got on. How many students are now on the school bus?
- Model how to write an equation to describe the first part of the word problem. Write $9 + 6 = \square$.
- ▶ The problem asks me to find the total number of students on the bus. I will write a blank box to represent the total number of students. First, the problem says there were 9 students on the bus and then 6 more got on; 9 plus 6 is equal to 15. There are now 15 students on the bus.
- Continue talking through the problem and modeling the equation. Write $15 + 3 = \square$.
- ▶ Now there are 15 students on the bus and the bus stops again to pick up 3 more students; 15 plus 3 is equal to 18. There are 18 students on the bus.
- As time permits, continue to solve two-step addition problems, using an empty box to represent the unknown number.

Just the Facts
Have students show you the number of fingers that will correctly complete the problem. Use questions such as the following:

- ▶ What number plus 10 equals 18? *Students show 8 fingers.*
- ▶ What number plus 10 equals 14? *Students show 4 fingers.*
- ▶ What number plus 10 equals 16? *Students show 6 fingers.*

2 ENGAGE

Develop: Two-Step Word Problems

"Today we are going to solve word problems with more than one step." Follow the instructions on the Activity Card **Two-Step Word Problems**. As students complete the activity, be sure to use the Questions to Ask.

Activity Card 3S

Alternative Groupings

Individual: Collect word problems that students create and make a booklet or worksheet of two-step word problems.

Pairs: Have pairs of students make up two-part problems for other students in the class. Use the Two-Step Word Problems Recording Sheet.

Progress Monitoring

| **If...** students are having difficulty with two-step problems, | **Then...** use drawings or pictures to show the progression of the story so they can see how the word problem flows. Show how you need to solve the first part of the problem before you can move on to the second part of the problem. |

Practice

Have students complete **Student Workbook,** pages 58–59. Guide students through the Key Idea example and the Try This exercises.

Interactive Differentiation

Consult the **Teacher Dashboard** for grouping suggestions. You can also use performance on the Engage activity to guide students.

Independent Practice

For additional practice, have students work on Word Problems with Tools 7: Problem Solver +/−. Students will be using one- and two-digit numbers.

Supported Practice

For additional support, use Word Problems with Tools 7: Problem Solver +/− with small groups of students.

- Encourage the use of Counters and drawings as students solve the problems together.
- Discuss each problem together.
- ▶ This problem tells me there are now 8 types of fruit in the bowl. The cook added 4 types of fruit. I need to know how many were in the bowl to start with.
- Demonstrate writing the important information on a piece of paper as you work to solve the problem. For example, write $\square + 4 = 8$.

252 Level D Unit 3 **Addition**

3 REFLECT

Think Critically

Review students' answers to the Reflect prompt at the bottom of *Student Workbook,* page 59, and then review the Engage activity.

▶ **How are these problems different from the ones we did earlier this week?** These problems have more than one step. You have to solve two different problems to answer the questions correctly.

Real-World Application

▶ **Before you try to solve a word problem, you write the problem as an addition equation.**

▶ **How does writing a problem as an addition equation help solve a word problem?** Answers may vary. Possible answer: Writing an addition equation helps organize the problem. You know exactly what you are solving for in the problem.

4 ASSESS

Informal Assessment

Use the online or print Student Record, *Assessment,* page 128, to record informal observations.

Two-Step Word Problems

Did the student

☐ respond accurately?   ☐ respond with confidence?

☐ respond quickly?   ☐ self-correct?

Additional Practice

For additional practice, have students complete *Practice,* page 70.

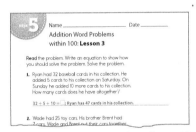
Practice, p. 70

Week 5 • Addition Word Problems within 100

Lesson 3

Key Idea
Some word problems take more than one step to solve. Be sure to read the problem carefully to see how many steps are in the word problem.
Use a ☐ to show the part of the equation that you don't know and that you have to solve for.

Try This
Read the problem. Write an equation to show how you should solve the problem. Solve the problem.

1. Mary has 65 seashells. She adds 10 more to her collection. Later that day, she adds another 5 seashells. How many seashells does Mary have now?

   $65 + 10 + 5 = \square$; Mary has 80 seashells.

2. Dave has 51 coins. He finds a jar with 12 coins and adds them to his collection. Then he finds 3 more coins and adds those to his collection. How many coins does he have now?

   $51 + 12 + 3 = \square$; Dave has 66 coins.

3. Matt has 71 stickers in his book. He buys a package with 8 stickers in it. Then he buys a package with 9 stickers in it. How many stickers does Matt have altogether?

   $71 + 8 + 9 = \square$; Matt has 88 stickers.

58 Level D Unit 3 **Addition**

Practice
Read the problem. Write an equation to show how you should solve the problem. Solve the problem.

4. On Friday Ann checked out 15 books from the library. On Saturday she checked out 5 more books. On Monday she checked out 4 more books. How many books did Ann check out in all?

   $15 + 5 + 4 = \square$; Ann checked out 24 books in all.

5. In one jar there are 50 coins. In another jar there are 20 coins. In a third jar there are 2 coins. How many coins are in the three jars?

   $50 + 20 + 2 = \square$; There are 72 coins in the three jars.

6. Dan had 82 cards. He found 5 more cards and put them with his other cards. Then he found another 7 cards. How many cards does he have altogether?

   $82 + 5 + 7 = \square$; Dan has 94 cards altogether.

Reflect
How does it help to draw a picture of the word problem?

Answers may vary; possible answer: Drawing a picture can help you visualize what is happening in the problem. This gives you a better idea of how to solve the problem.

Week 5 **Addition Word Problems within 100** • Lesson 3   59

Student Workbook, pp. 58–59

WEEK 5
Addition Word Problems within 100

Lesson 4

Objective
Students can solve one- and two-step addition word problems within 100.

Standard
2.OA.1 Use addition and subtraction within 100 to solve one- and two-step word problems involving situations of adding to, taking from, putting together, taking apart, and comparing, with unknowns in all positions, e.g., by using drawings and equations with a symbol for the unknown number to represent the problem.

Vocabulary
- addend
- sum

Creating Context
Continue using the math vocabulary in context throughout the lesson. Encourage students to use the correct terminology.

Materials

Program Materials
- Hotel Mystery Recording Sheets
- Hotel Game Board
- Number Cards (0–100)
- Number 1–6 Cube
- Pawns

Additional Materials
Vocabulary Card 43, *sum*

Prepare Ahead
Pull the 80–100 cards out of the Number Cards set. Set these aside; you will not need them for this activity.

1 WARM UP

Prepare
- Give students the following word problem.
▶ **There were 6 books on the bookshelf. The librarian put 5 more books on the shelf in the morning. In the afternoon, she put a few more books on the shelf. At the end of the day, there were 19 books on the shelf. How many books did the librarian put on the shelf in the afternoon?**
- Model how to write an equation to describe the first part of the word problem; $6 + 5 + \square = 19$.
▶ **The problem tells me there were 6 books to start with and then 5 more books were added.**
- Model solving the problem $11 + \square = 19$.
▶ **I still don't know how many books the librarian added in the afternoon. When I look at this problem, I can figure out that she added 8 books in the afternoon. I can count up: 12, 13, 14, 15, 16, 17, 18, 19. She placed 8 books on the shelf in the afternoon.**
- Discuss additional two-step problems with an unknown number in different positions in the equation.

Just the Facts
Have students put their hands up if the following equation is correct and put their hands down if the equation is incorrect. Use questions such as the following:

▶ $7 + 8 = 15$ *hands up*
▶ $9 + 8 = 16$ *hands down*
▶ $6 + 7 = 13$ *hands up*

2 ENGAGE

Develop: Hotel Mystery
"Today we are going to create two-step word problems with an unknown addend." Follow the instructions on the Activity Card **Hotel Mystery**. As students complete the activity, be sure to use the Questions to Ask.

Activity Card 3T

Alternative Grouping
Individual: Have the student make up two-part problems for you to solve. Use the Hotel Mystery Recording Sheets.

Progress Monitoring

| If... students are having difficulty solving problems with the larger numbers, | ▶ Then... continue with the same content, but use only addends less than ten. |

Practice
Have students complete **Student Workbook,** pages 60–61. Guide students through the Key Idea example and the Try This exercises.

Interactive Differentiation
Consult the **Teacher Dashboard** for grouping suggestions. You can also use performance on the Engage activity to guide students.

Independent Practice

For additional practice, have students work on Barkley's Bones 1–20. Students must find the unknown number in the equation. Manipulatives assist students in finding the missing addend.

Supported Practice

For additional support, use Tidal Tally to work with individual or small groups of students. Students must determine the missing addend in a problem.

- Model the problem on the board as you work with students. For example, $2 + \square = 3$
▶ **What addend will make this a true equation?** 1

254 Level D Unit 3 **Addition**

3 REFLECT

Think Critically
Review students' answers to the Reflect prompt at the bottom of **Student Workbook,** page 61, and then review the Engage activity.

▶ **Why is the unknown number sometimes in the middle of the equation?** Sometimes the unknown number is an addend.

Real-World Application
Remind students that word problems are usually written about real-world situations.

▶ **What happened to you today that could be used in an addition word problem?** Answers may vary. Possible answer: There were students on the school bus. When the bus stopped, a bunch of students got on the bus. I could find out how many students were on the bus after the bus picked up the new students.

4 ASSESS

Informal Assessment
Use the online or print Student Record, **Assessment,** page 128, to record informal observations.

Hotel Mystery
Did the student
☐ respond accurately?   ☐ respond with confidence?
☐ respond quickly?   ☐ self-correct?

Additional Practice
For additional practice, have students complete **Practice,** page 71.

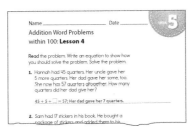
Practice, p. 71

Week 5 • Addition Word Problems within 100

Lesson 4

Key Idea
Some word problems take more than one step to solve. Be sure to read the problem carefully to see how many steps are in the word problem.

Use a ☐ to show the part of the equation that you don't know and that you have to solve for.

Try This
Read the problem. Write an equation to show how you should solve the problem. Solve the problem.

1. Fred saw 15 bicycles go by on the track. He saw 4 more go by. Then he saw some more bicycles in the park. Altogether, 25 bicycles went by. How many bicycles did Fred see in the park?

   $15 + 4 + \square = 25$; Fred saw 6 bicycles in the park.

2. Jenny had 39 quarters. Sydney had 4 quarters. Tara had some more quarters. When they combined their quarters they had 50 quarters. How many quarters did Tara have?

   $39 + 4 + \square = 50$; Tara had 7 quarters.

3. Ben had 62 pencils in a shoe box. He found 10 more pencils at home and some more after school. When he added them to the shoe box he counted 78 pencils. How many pencils did Ben find after school?

   $62 + 10 + \square = 78$; Ben found 6 pencils after school.

60   Level D Unit 3 Addition

Practice
Read the problem. Write an equation to show how you should solve the problem. Solve the problem.

4. Linda has a jar of coins. She puts 7 more coins in the jar. Then her mom adds 8 coins to the jar. Now there are 79 coins in the jar. How many coins did Linda start with?

   $\square + 7 + 8 = 79$; Linda started with 64 coins.

5. Grandma Kelley has 80 buttons in her sewing box. She buys a package of buttons and puts them in the box. Then she adds 10 more buttons. Now she has 99 buttons in the box. How many buttons were in the package?

   $80 + \square + 10 = 99$; The package had 9 buttons in it.

6. Chris has a rock collection with 30 rocks in it. He goes out one day and finds 4 more rocks. On Sunday he adds some more rocks to the collection. Now he has 56 rocks. How many rocks did he add on Sunday?

   $30 + 4 + \square = 56$; On Sunday he added 22 rocks to the collection.

Reflect
Why is it important to read a word problem carefully?

Answers may vary; possible answer: You need to read carefully because you don't know what part of the problem needs to be solved.

Week 5 **Addition Word Problems within 100** • Lesson 4 61

Student Workbook, pp. 60–61

WEEK 5
Addition Word Problems within 100

Lesson 5 Review

Objective
Students review skills learned this week and complete the weekly assessment and project.

Standard
2.OA.1 Use addition and subtraction within 100 to solve one- and two-step word problems involving situations of adding to, taking from, putting together, taking apart, and comparing, with unknowns in all positions, e.g., by using drawings and equations with a symbol for the unknown number to represent the problem.

Vocabulary
Review vocabulary introduced during the week.

Creating Context
The activities this week reinforce addition vocabulary, such as *addend* and *sum*. Use these words often so students are hearing and using these words in real-world contexts.

1 WARM UP

Prepare

▶ **Where do you find the unknown number in an equation?**
The unknown number can be anywhere in the equation; it can be one of the addends or the sum.

▶ **How can we show where the unknown number is in an equation?**
We can use an empty box.

2 ENGAGE

Practice
Have students complete **Student Workbook,** pages 62–63.

Week 5 • Addition Word Problems within 100

Lesson 5 Review

This week you worked on word problems that took one or two steps to solve. You also had unknowns in different parts of the problems that you had to solve for.

Lesson 1 Read the problem. Write an equation to show how you should solve the problem. Solve the problem.

1. Joy had 25 rocks. She found 5 more rocks. How many rocks does she have altogether?

   $25 + 5 = \square$; Joy has 30 rocks.

2. Linda has 80 stickers. Her teacher gives her 6 more stickers. How many stickers does Linda have in all?

   $80 + 6 = \square$; Linda has 86 stickers in all.

Lesson 2 Read the problem. Write an equation to show how you should solve the problem. Solve the problem.

3. Jose had 40 books in his collection. His mom gave him some more books for his birthday. Now he has 50 books. How many books did his mom give him?

   $40 + \square = 50$; His mom gave him 10 books.

4. Gina has 19 pencils. She finds more pencils on the floor after school. Now she has 26 pencils. How many pencils did she find on the floor?

   $19 + \square = 26$; Gina found 7 pencils on the floor after school.

Lesson 3 Read the problem. Write an equation to show how you should solve the problem. Solve the problem.

5. Sean has 65 toy cars. He adds 15 more to his collection. Later that day, he adds another 5. How many cars does Sean have now?

   $65 + 15 + 5 = \square$; Sean has 85 cars.

Lesson 4 Read the problem. Write an equation to show how you should solve the problem. Solve the problem.

6. Sydney has 60 quarters. She collects 8 more quarters. Her mom finds some quarters and adds them to Sydney's collection. Now Sydney has 80 quarters. How many quarters did her mom give her?

   $60 + 8 + \square = 80$; Her mom gave her 12 quarters.

Reflect
Write and solve a two-step word problem.

Answers will vary.

Student Workbook, pp. 62–63

REFLECT

Think Critically

Review students' answers to the Reflect prompt at the bottom of **Student Workbook,** page 63.

Discuss the answer with the group to reinforce Week 5 concepts.

ASSESS

Formal Assessment ✓

Students may take the weekly assessment online.

As an alternative, students may complete the weekly test on **Assessment,** pages 45–46. Record progress using the Student Assessment Record, **Assessment,** page 128.

Going Forward

Use the **Teacher Dashboard** to view results of the online assessments, to input the results of print student assessments, and to review progress before making decisions about next steps. Use the weekly test results and observations to determine the next steps for each student.

Retention	
Student displays good grasp of this week's concepts and skills.	**Building Blocks** Have students independently complete Building Blocks activities, particularly Word Problems with Tools.
Remediation	
Student is still struggling with the week's concepts and skills.	Work step-by-step with students through word problems. Encourage students to use Counters, number lines, manipulatives, or drawings to act out and understand the problems. Have students talk through the unknowns in the problems.

Suggestions for Re-Evaluation: If a student has struggled without success for several weeks, use observations and test results to place the student at a level in which he or she can find success and build confidence to move forward.

Name _____ Date _____ WEEK 5
Addition Word Problems within 100

1. Which number goes in the box to solve the problem?

   $23 + 7 = \square$ ___30___

2. Circle the equation that goes with this problem. Then solve the equation.

   A squirrel collected 6 nuts. Then it collected 8 more. How many nuts did the squirrel collect in all?

   $6 + \square = 8$ $(6 + 8 = \square)$ 14

Write a number sentence that shows how to solve each problem. Solve the problem.

3. Walt has 31 stamps in his collection. He bought more at a yard sale. Now he has 47 stamps. How many stamps did he buy at the yard sale?

   $31 + \square = 47$, 16

4. A well is 84 feet deep. The pump is at 60 feet. How far is the bottom of the well from the pump?

   $60 + \square = 84$, 24

Level D Unit 3 Week 5   45

WEEK 5   Name _____ Date _____
Addition Word Problems within 100

Write a number sentence that shows how to solve each problem. Solve the problem.

5. Juana is getting ready for a race. She ran 3 miles on Monday, 2 miles on Wednesday, and 6 miles on Saturday. How far did she run altogether?

   $3 + 2 + 6 = \square$, 11

6. The amount of snow that fell on a mountain was 28 inches at the end of December. In January, 17 more inches fell. In February, 31 inches fell. How much snow fell on the mountain in those three months?

   $28 + 17 + 31 = \square$, 76

7. In the first basketball game, Ray scored 18 points. He had 9 in the next game, and after three games, he had 40 points. How many points did he score in the third game?

   $18 + 9 + \square = 40$, 13

8. Deb counted geese in a pond for part of her science project. The total number she counted was 87. There were 23 geese on the pond in the morning and 27 at sundown. How many geese were on the pond when she counted them at noon?

   $23 + \square + 27 = 87$, 37

46   Level D Unit 3 Week 5

Assessment, pp. 45–46

Project Preview

This week, students solved one- and two-step addition word problems within 100. Students worked with unknowns in all positions of the equation. For this week's project, students will write a more difficult word problem. The unknown could be the sum or either of the addends.

Project-Based Learning

Standards-driven Project-Based Learning is effective in building deep content understanding. Project-Based Learning increases long-term retention of concepts and has been shown to be more effective than traditional instruction. Completing a project to answer an essential question challenges students to apply and demonstrate mastery of concepts and skills by expressing understanding through discussion, research, and presentation.

Essential Question

HOW can I use addition to solve real-world problems?

Project Evaluation Criteria

Review project evaluation criteria with students prior to beginning the project.

Exceeds Expectations
☐ Project result is explained and can be extended.
☐ Project result is explained in context and can be applied to other situations.
☐ Project result is explained using advanced mathematical vocabulary.
☐ Project result is described, and mathematics are used correctly and can be extended.
☐ Project result is explained and extended, and shows advanced knowledge of mathematical concepts and skills.

Meets Expectations
☐ Project result is explained.
☐ Project result is explained in context.
☐ Project result is explained using mathematical vocabulary.
☐ Project result is described, and mathematics are used correctly.
☐ Project result is explained, and shows satisfactory knowledge of mathematical concepts and skills.

Does Not Meet Expectations
☐ Project result is not explained.
☐ Project result is explained, but out of context.
☐ Project result is explained, but mathematical vocabulary is oversimplified.
☐ Project result is described, but mathematics are not used correctly.
☐ Project result is not explained and/or extended, or shows less than satisfactory knowledge of mathematical concepts and skills.

Solve Collection Problems with Unknowns

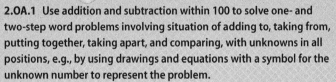

Objective
Students can create and solve word problems with unknowns.

Standard
2.OA.1 Use addition and subtraction within 100 to solve one- and two-step word problems involving situation of adding to, taking from, putting together, taking apart, and comparing, with unknowns in all positions, e.g., by using drawings and equations with a symbol for the unknown number to represent the problem.

Materials
Additional Materials
- construction paper, 1 per student pair
- index cards, 2 per student pair
- poster from Week 4
- tape

Prepare Ahead
Write one of the following equations on each index card. Make one index card for each pair of students.

- ☐ + _____ = _____
- _____ + ☐ = _____
- _____ + _____ = ☐

Best Practices
- Provide project directions that are clear and brief.
- Check for student understanding frequently.
- Provide meaning and organization to the lessons and concepts.

Introduce

This week you solved word problems that took one or two steps to complete. You also worked with having an unknown in part of an equation.

▶ **What is an unknown?** Answers may vary. Possible answers: It is the number that is missing from an equation; it is the number that you need to figure out in an equation.

- Discuss with students the various ways they can solve for an unknown in an addition equation.

Explore

▶ **Today you will write a word problem based on your collection. I will give you and your partner an equation that you will use to write a word problem.**

- Show the class the three different types of equations they may receive.
- Display ☐ + _____ = _____ on the board. Model how to write a word problem using this equation frame.

▶ **In this equation, the unknown is the first number in the problem. I will insert numbers that have to do with my collection in the blanks. The person solving the problem will use these numbers to figure out the value of the unknown and write it in the box.**

- Present a sample problem to students. Use two-digit numbers. For example, a problem about books could be: "I had some books in my collection. I went to the library and checked out 15 more books. Now I have 35 books. How many books did I start with?"
- Write the equation on the board, using numbers that represent your sample collection. For the problem above, the equation is ☐ + 15 = 35. Discuss strategies for solving the problem.
- Distribute an equation card, a sheet of construction paper, and a blank index card to each pair of students.

▶ **Think of a word problem that fits the equation on the card I gave you. Write the word problem at the top of the construction paper. Draw an illustration for the problem in the middle of the paper. At the bottom of the paper, write the answer to the problem.**

- You might wish to have students write their word problems on notebook paper. Then they can write their "final drafts" on the construction paper.

▶ **Complete Student Workbook, page 64, to practice finding the unknown in an addition problem.**

Wrap Up

- Some students find writing a word problem extremely difficult. You might need to let students work in larger groups so that peers can assist one another. You might also need to help students find a connection between their collections and the equation they received.
- Discuss students' answers to the Reflect prompt at the bottom of **Student Workbook,** page 64.

Student Workbook, p. 64

Show students how to tape the blank index card over the answer so that people can lift the card and see the answer to the problem. Provide tape or a glue stick so that students can attach the construction paper to their posters.

If time permits, allow students pairs to share their word problems with the class.

Teacher Reflect

☐ Did I explain the directions before the students began their projects?

☐ Was I able to answer questions when students did not understand?

☐ Did the students' ideas logically relate to one another?

WEEK 6: Solving Addition Word Problems within 1,000

Week at a Glance
This week students conclude **Number Worlds,** Level D, Addition. Students will explore different strategies for adding two-digit numbers and gain facility solving word problems.

Skills Focus
- Model adding one- and two-digit numbers to two-digit numbers with regrouping.
- Add one- and two- digit numbers to two-digit numbers using the standard algorithm.
- Add within 1,000 using the standard algorithm.

How Students Learn
Teaching algorithms solely as a series of procedures deprives children of the opportunity to think flexibly and conceptually about numbers and operations. As students begin to master arithmetical procedures for adding two-digit numbers, continue to have them conceptually approach problems as well. Encourage students to develop a habit of asking, "Can I decompose and tinker with these numbers to make the problem easier to solve?"

English Learners ELL
For language support, use the **English Learner Support Guide,** pages 68–69, to preview lesson concepts and teach academic vocabulary.

Math at Home
Give one copy of the Letter to Home, page 18, to each student. Encourage students to share and complete the activity with their caregivers.

Weekly Planner

Lesson	Learning Objectives
1 pages 262–263	Students can model adding a one-digit number to a two-digit number with regrouping.
2 pages 264–265	Students can model adding a two-digit number to a two-digit number with regrouping.
3 pages 266–267	Students can add a one- or two-digit number to a two-digit number using the standard algorithm.
4 pages 268–269	Students can add within 1,000 using the standard algorithm.
5 pages 270–271	**Review and Assess** Students review skills learned this week and complete the weekly assessment and project.
Project pages 272–273	Students can add multi-digit numbers with regrouping.

Key Standard for the Week

Domain: Number and Operations in Base Ten

Cluster: Use place value understanding and properties of operations to add and subtract.

2.NBT.7 Add and subtract within 1000, using concrete models or drawings and strategies based on place value, properties of operations, and/or the relationship between addition and subtraction; relate the strategy to a written method. Understand that in adding or subtracting three-digit numbers, one adds or subtracts hundreds and hundreds, tens and tens, ones and ones; and sometimes it is necessary to compose or decompose tens or hundreds.

Materials		Technology
Program Materials • *Student Workbook,* pp. 65–67 • *Practice,* p. 72 • Activity Card 3U, **Add with Blocks** • Number Construction Mat	**Additional Materials** base-ten blocks*	*Teacher Dashboard* Base 10 Blocks Tool
Program Materials • *Student Workbook,* pp. 68–69 • *Practice,* p. 73 • Activity Card 3V, **More Adding with Blocks** • Number Construction Mat	**Additional Materials** base-ten blocks*	*Teacher Dashboard* Building Blocks Word Problems with Tools 4
Program Materials • *Student Workbook,* pp. 70–71 • *Practice,* p. 74 • Activity Card 3W, **Add Them Up** • Add Them Up Recording Sheet • Number Cards (0–49)		*Teacher Dashboard* Building Blocks Eggcellent: Addition Choice Base 10 Blocks Tool
Program Materials • *Student Workbook,* pp. 72–73 • *Practice,* p. 75 • Activity Card 3X, **Adding Hundreds** • Adding Hundreds Recording Sheet • Number 1–6 Cube	**Additional Materials** • base-ten blocks* • stickers	*Teacher Dashboard* Base 10 Blocks Tool
Program Materials • *Student Workbook,* pp. 74–75 • Weekly Test, *Assessment,* pp. 47–48		Review previous activities.
Program Materials *Student Workbook,* p. 76	**Additional Materials** • construction paper • poster from Week 5 • tape or glue sticks	

*Available from McGraw-Hill Education

WEEK 6
Solving Addition Word Problems within 1,000

Find the Math

In this week, students will add one-, two-, and three-digit numbers using concrete models and the standard algorithm. Each lesson will incorporate word problems to develop real-world connections to computation skills.

Use the following to begin a guided discussion:

▶ **A book collection may have bins to group books by subject matter. How does sorting books by subject help organize a large collection?** Answers may vary. Possible answers: You can quickly see if you have any books about a certain subject; you can see if you have a greater number of books about one subject than another.

Have students complete **Student Workbook,** page 65.

Student Workbook, p. 65

Lesson 1

Objective
Students can model adding a one-digit number to a two-digit number with regrouping.

Standard
2.NBT.7 Add and subtract within 1000, using concrete models or drawings and strategies based on place value, properties of operations, and/or the relationship between addition and subtraction; relate the strategy to a written method. Understand that in adding or subtracting three-digit numbers, one adds or subtracts hundreds and hundreds, tens and tens, ones and ones; and sometimes it is necessary to compose or decompose tens or hundreds.

Creating Context
Have English Learners create a place-value poster. Tell them to write a three-digit number on a sheet of paper. Then have them draw an arrow pointing to the number in the ones place and write *ones* at the blunt end of the arrow. Have students do the same thing for the tens place and the hundreds place.

Materials
Program Materials
- Number Construction Mat, 1 per student

Additional Materials
base-ten blocks, 1 set per student

Prepare Ahead
- Make a copy of a Number Construction Mat for each student. Keep the copies for use in this lesson as well as other lessons this week.
- Pull out the ones blocks and base-ten rods from the base-ten blocks for today's lesson.

1 WARM UP

Prepare
- Allow students to explore freely with base-ten blocks before the lesson.
- Encourage students to name how many blocks long a base-ten rod is.

2 ENGAGE

Develop: Add with Blocks

"Today we are going to use base-ten blocks to solve addition word problems." Follow the instructions on the Activity Card **Add with Blocks.** As students complete the activity, be sure to use the Questions to Ask.

Activity Card 3U

Alternative Grouping

Pair: Have students work in pairs as you teach the lesson. Encourage students to discuss the problem as they use the blocks to model the word problems.

Progress Monitoring

| **If...** students are having difficulty trading or exchanging 10 ones blocks for 1 base-ten rod during the lesson, | **Then...** have them stop the activity and build models of numbers in the teens using the fewest number of base-ten blocks possible. |

Practice

Have students complete **Student Workbook,** pages 66–67. Guide students through the Key Idea example and the Try This exercises.

Interactive Differentiation

Consult the **Teacher Dashboard** for grouping suggestions. You can also use performance on the Engage activity to guide students.

Independent Practice

For additional practice, have students use the Base 10 Blocks Tool to model and solve the following problems:

- Jon had 14 game cards. His brother gave him 7 new cards for his birthday. How many game cards does Jon have now? 21 game cards
- Tami read 19 books over the summer. Karen read 6 books. How many books did the two girls read over the summer? 25 books
- Lilly has 9 pennies. Sam has 13 pennies. How many pennies do they have in all? 22 pennies

Supported Practice

For additional support, use base-ten blocks and Number Construction Mats to model the following word problems.

▶ **Jon had 14 game cards. His brother gave him 7 new cards for his birthday. How many game cards does Jon have now?**
- Show students how to model 14 and 7 with base-ten blocks on the Number Construction Mat. Regroup the 11 ones blocks into 1 base-ten rod and 1 ones block. Then tell students to find the sum. Remind them to include the regrouped 10 in their sum. 21 game cards

▶ **Tami read 19 books over the summer. Karen read 6 books. How many books did the two girls read over the summer?**
- Show students how to model 19 and 6 with base-ten blocks on the Number Construction Mat. Regroup the 15 ones blocks into 1 base-ten rod and 5 ones blocks. Then have students find the sum. 25 books
- Write the following problem on the board. Have students model and solve the problem on their own.
- Lilly has 9 pennies. Sam has 13 pennies. How many pennies do they have in all? 22 pennies

3 REFLECT

Think Critically

Review students' answers to the Reflect prompt at the bottom of **Student Workbook,** page 67, and then review the Engage activity.

▶ **Using the fewest number of blocks, show me the number 34 with the base-ten blocks.** 3 base-ten rods, 4 ones blocks

4 ASSESS

Informal Assessment

Use the online or print Student Record, **Assessment,** page 128, to record informal observations.

Add with Blocks
Did the student
- ☐ apply learning to a new situation?
- ☐ contribute concepts?
- ☐ contribute answers?
- ☐ connect mathematics to the real world?

Additional Practice

For additional practice, have students complete **Practice,** page 72.

Practice, p. 72

Week 6 • Solving Addition Word Problems within 1,000

Lesson 1

Key Idea
If you have more than 9 ones, regroup 10 of the ones as a ten. Then add 1 to the tens place.

$7 + 24 = $ _____

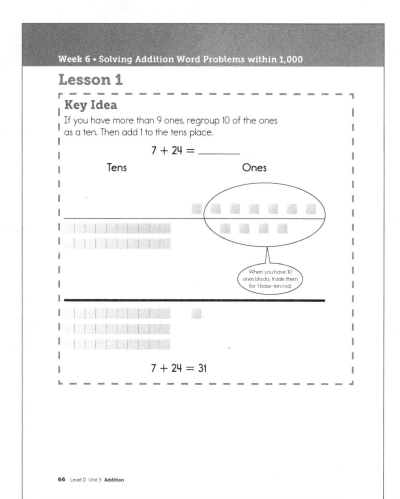

$7 + 24 = 31$

Try This
Solve each problem. Look at the model for help.

1. $25 + 8 = \underline{33}$

2. $6 + 16 = \underline{22}$

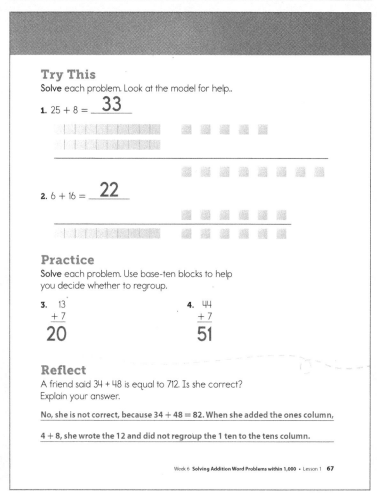

Practice
Solve each problem. Use base-ten blocks to help you decide whether to regroup.

3. $\begin{array}{r} 13 \\ +\ 7 \\ \hline 20 \end{array}$

4. $\begin{array}{r} 44 \\ +\ 7 \\ \hline 51 \end{array}$

Reflect
A friend said $34 + 48$ is equal to 712. Is she correct? Explain your answer.

No, she is not correct, because $34 + 48 = 82$. When she added the ones column, $4 + 8$, she wrote the 12 and did not regroup the 1 ten to the tens column.

Student Workbook, pp. 66–67

Week 6 Solving Addition Word Problems within 1,000 • Lesson 1 **263**

WEEK 6
Solving Addition Word Problems within 1,000

Lesson 2

Objective
Students can model adding a two-digit number to a two-digit number with regrouping.

Standard
2.NBT.7 Add and subtract within 1000, using concrete models or drawings and strategies based on place value, properties of operations, and/or the relationship between addition and subtraction; relate the strategy to a written method. Understand that in adding or subtracting three-digit numbers, one adds or subtracts hundreds and hundreds, tens and tens, ones and ones; and sometimes it is necessary to compose or decompose tens or hundreds.

Creating Context
Pay particular attention to the language you use to describe regrouping. English Learners might become confused if multiple terms or metaphors are used. Remind students of the counting-on strategies they learned in previous lessons.

Materials
Program Materials
Number Construction Mat, 1 per student

Additional Materials
base-ten blocks

 WARM UP

Prepare
- Have students use base-ten blocks to model different numbers.
- ▶ **Show me 58.** 5 base-ten rods, 8 ones blocks
- ▶ **Show me 41.** 4 base-ten rods, 1 ones block
- ▶ **Show me 87.** 8 base-ten rods, 7 ones blocks

Just the Facts
Review addition to 20 with students. Have students complete each addition sentence by saying the sum. Use prompts such as the following:
- ▶ $6 + 7 =$ _____ 13
- ▶ $9 + 8 =$ _____ 17
- ▶ $5 + 7 =$ _____ 12

2 ENGAGE

Develop: More Adding with Blocks
"Today we are going to use base-ten blocks to solve more addition problems." Follow the instructions on the Activity Card **More Adding with Blocks.** As students complete the activity, be sure to use the Questions to Ask.

Activity Card 3V

Alternative Groupings
Pair: After demonstrating the concert-ticket problem, provide word problems for pairs of students to solve together.

Individual: After demonstrating the concert-ticket problem, provide word problems for students to solve on their own.

Progress Monitoring

| If... students forget to include the regrouped ten when adding in the tens column, | Then... point out that the regrouped ten is just as important as the other numbers in the tens column. Tell students to write 1+ above the tens place in the first two-digit number as soon as they regroup. That will help them remember to include the ten in the sum. |

Practice
Have students complete **Student Workbook,** pages 68–69. Guide students through the Key Idea example and the Try This exercises.

Interactive Differentiation
Consult the **Teacher Dashboard** for grouping suggestions. You can also use performance on the Engage activity to guide students.

Independent Practice

For additional practice, have students work on Word Problems with Tools 4: Multidigit +/− to 100 and Multidigit Solver. This Building Block activity includes base-ten materials on mats to help students model these more difficult word problems.

Supported Practice

For additional support, use base-ten blocks and Number Construction Mats to model the following word problems. Have students model the problems with you.

- ▶ **The mother squirrel buried 36 nuts. The father squirrel buried 45 nuts. How many nuts did they bury in all?** 81 nuts
- ▶ **The library has 44 books about plants and 49 books about animals. How many books about plants and animals are there altogether?** 93 books
- ▶ **There are 24 large spools of thread and 17 small spools of thread. How many spools are there in all?** 41 spools

264 Level D Unit 3 **Addition**

REFLECT

Think Critically

Review students' answers to the Reflect prompt at the bottom of **Student Workbook,** page 69, and then review the Engage activity.

▶ **Explain how to solve the problem 38 + 35.** Answers may vary. Possible answer: First, I add the numbers in the ones column, 8 + 5, and that equals 13. I write the 3 in the ones column and regroup the 1 ten to the tens column. Then I add 3 tens plus 3 tens plus 1 ten and get an answer of 7 tens. The solution is 38 + 35 = 73.

Real-World Application

When a carpenter is building a house, she must add measurements together.

▶ **Sue needs one piece of wood that is 23 inches long. She needs another piece that is 47 inches long. How many inches of wood does she need altogether?** 70 inches

ASSESS

Informal Assessment

Use the online or print Student Record, **Assessment,** page 128, to record informal observations.

More Adding with Blocks

Did the student

- ☐ apply learning to a new situation?
- ☐ contribute answers?
- ☐ contribute concepts?
- ☐ connect mathematics to the real world?

Additional Practice

For additional practice, have students complete **Practice,** page 73.

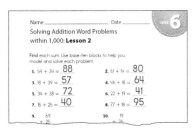

Practice, p. 73

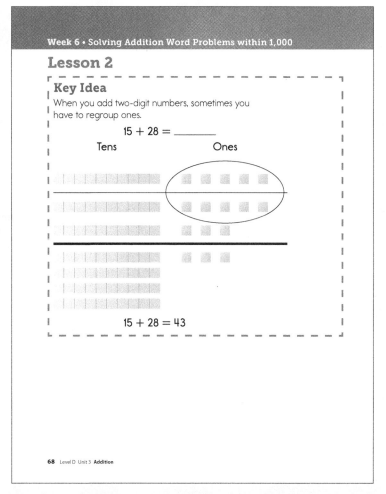

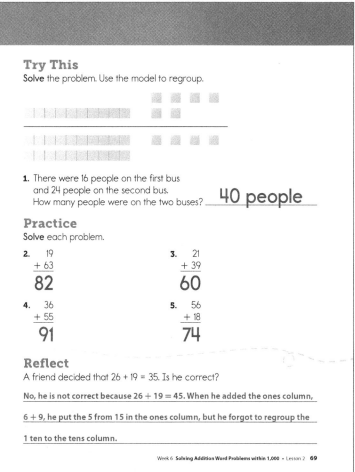

Student Workbook, pp. 68–69

Week 6 **Solving Addition Word Problems within 1,000** • Lesson 2 **265**

WEEK 6
Solving Addition Word Problems within 1,000

Lesson 3

Objective
Students can add a one- or two-digit number to a two-digit number using the standard algorithm.

Standard
2.NBT.7 Add and subtract within 1000, using concrete models or drawings and strategies based on place value, properties of operations, and/or the relationship between addition and subtraction; relate the strategy to a written method. Understand that in adding or subtracting three-digit numbers, one adds or subtracts hundreds and hundreds, tens and tens, ones and ones; and sometimes it is necessary to compose or decompose tens or hundreds.

Creating Context
Have English Learners write the steps for the addition algorithm.
1. Write the problem in vertical form. Line up the numbers in the ones place and tens place.
2. Add the numbers in the ones place.
3. If the sum is greater than 9, regroup. Write 1 above the numbers in the tens place. Write the remaining ones in the ones place below the line.
4. Now add the tens. Write the sum in the tens place below the line.

Materials
Program Materials
- Add Them Up Recording Sheet, two copies per pair
- Number Cards (0–49)

1 WARM UP

Prepare
- Write the problem 56 + 35 on the board in vertical form.
- Have students solve the problem.
- ▶ **Remember to add the ones column first and the tens column second.**
- When students are finished, check the problem together. Make sure that the sum was properly regrouped. *91*

Just the Facts
Write on the board an addition problem with its answer. Have students identify whether the answer is correct. If it is correct, tell them to raise their hands up. If the answer is incorrect, students should keep or put their hands down. Use problems such as the following:
- ▶ 26 + 6 = 22 *hands down*
- ▶ 45 + 7 = 52 *hands up*
- ▶ 69 + 2 = 61 *hands down*

2 ENGAGE

Develop: Add Them Up
"Today we are going to practice adding numbers together. We will add the ones first and then the tens." Follow the instructions on the Activity Card **Add Them Up.** As students complete the activity, be sure to use the Questions to Ask.

Activity Card 3W

Alternative Groupings
Small Group: Have one student on each team select two Number Cards and write the numbers on the recording sheet. Tell teams to work together to solve the problem. Check each team's solution. In a tally chart on the board, keep track of which team has the greatest sum. If a team's solution is incorrect, that sum does not count. After several rounds, the team with the most tally marks wins.

Individual: Complete the activity as written, but have students keep their recording sheets until they have solved several problems. Then have students exchange papers and check each other's work.

Progress Monitoring

If… students are not accurately adding.	▶ **Then…** provide additional practice with concrete manipulatives, such as base-ten blocks.

Practice
Have students complete **Student Workbook,** pages 70–71. Guide students through the Key Idea example and the Try This exercises.

Interactive Differentiation
Consult the **Teacher Dashboard** for grouping suggestions. You can also use performance on the Engage activity to guide students.

Independent Practice
For additional practice, have students work on Eggcellent: Addition Choice. Students use addition to reach the final space on a game board. After students have used the program, tell them to write the equations they used to move on the game board.

Supported Practice
For additional support, use the Base 10 Blocks Tool with students to model the problems they solved during **Add Them Up.**

- Explain how to set up the palette in the tool for addition. Show students how to regroup 10 ones blocks into 1 base-ten rod.
- Show students how they can hide the display of summation lines by dragging the workspace down. First, have students hide the summation lines.
- Have students get out their Add Them Up Recording Sheet from **Add Them Up.** Tell them to use the Base 10 Blocks Tool to model the first problem. Then have students move the workspace up to view the summation lines and check their answers.

3 REFLECT

Think Critically

Review students' answers to the Reflect prompt at the bottom of **Student Workbook,** page 71, and then review the Engage activity.

▶ **Explain the steps to solve 65 + 18.** Answers may vary. Possible answer: First, I add 5 and 8 in the ones column: 5 + 8 = 13. I write the 3 of 13 as the answer in the ones column, then I regroup the 1 ten. Next, I add the numbers in the tens column: 1 + 6 + 1 = 8. The answer is 83.

Real-World Application

Runners keep track of how many miles they run each month.

▶ **Tim ran 25 miles in June and 18 miles in August. How many miles did he run altogether in the two months?** 43 miles

4 ASSESS

Informal Assessment

Use the online or print Student Record, **Assessment,** page 128, to record informal observations.

Add Them Up

Did the student

☐ apply learning to a new situation?   ☐ contribute answers?

☐ contribute concepts?   ☐ connect mathematics to the real world?

Additional Practice

For additional practice, have students complete **Practice,** page 74.

Practice, p. 74

Week 6 • Solving Addition Word Problems within 1,000

Lesson 3

Key Idea
When you add greater numbers, add the ones first and then the tens. Remember to add any regrouped tens to the tens column.

Tens	Ones	Tens	Ones	Tens	Ones	Tens	Ones
				1		1	
3	8	3	8	3	8	3	8
+2	5	+2	5	+2	5	+2	5
			13		3	6	3

Try This
Solve each problem.

1. Tens Ones
   2   4
   +1   9
   4   3

2. Tens Ones
   2   6
   +3   9
   6   5

3. Tens Ones
   7   6
   +1   5
   9   1

4. Tens Ones
   3   8
   +1   7
   5   5

70 Level D Unit 3 **Addition**

Practice
Solve each problem.

5. 17
   + 4
    21

6. 25
   + 7
    32

7. 32
   + 19
    51

8. 53
   + 27
    80

9. 48
   + 17
    65

10. 58
   + 29
    87

11. 36
   + 55
    91

12. 27
   + 27
    54

13. Austin has a collection of 46 pennies. His dad gave him 15 more pennies. How many pennies does Austin have in all?
    61 pennies

Reflect
Explain why you start in the ones column when solving an addition problem.

You start in the ones column first in case you have to regroup.

Week 6 Solving Addition Word Problems within 1,000 • Lesson 3 71

Student Workbook, pp. 70–71

WEEK 6
Solving Addition Word Problems within 1,000

Lesson 4

Objective
Students can add within 1,000 using the standard algorithm.

Standard
2.NBT.7 Add and subtract within 1000, using concrete models or drawings and strategies based on place value, properties of operations, and/or the relationship between addition and subtraction; relate the strategy to a written method. Understand that in adding or subtracting three-digit numbers, one adds or subtracts hundreds and hundreds, tens and tens, ones and ones; and sometimes it is necessary to compose or decompose tens or hundreds.

Creating Context
Throughout the lesson, continue to use the correct terminology as you teach. Encourage students to use the terms as they work.

Materials
Program Materials
- Adding Hundreds Recording Sheet, two copies per pair
- Number 1–6 Cube

Additional Materials
- base-ten blocks
- stickers

Prepare Ahead
Place stickers on the numbers 5 and 6 on the Number Cubes to cover them. For the Warm Up, remove the thousands blocks from the sets of base-ten blocks. Students will only use ones, tens, and hundreds.

1 WARM UP

Prepare
- Have students use base-ten blocks to model different numbers.
▶ **Show me 564.** 5 base-ten flats, 6 base-ten rods, 4 ones blocks
▶ **Show me 167.** 1 base-ten flat, 6 base-ten rods, 7 ones blocks
▶ **Show me 243.** 2 base-ten flats, 4 base-ten rods, 3 ones blocks

Just the Facts
Have students stand when they know the answer. After students have been given a reasonable amount of time to think of the answer, have a volunteer supply the correct sum. Use questions such as the following:
▶ **What is 10 + 5?** 15
▶ **What is 20 + 4?** 24
▶ **What is 50 + 6?** 56

2 ENGAGE

Develop: Adding Hundreds
"Today we are going to add numbers that have digits in the hundreds place." Follow the instructions on the Activity Card **Adding Hundreds.** As students complete the activity, be sure to use the Questions to Ask.

Activity Card 3X

Alternative Groupings
Small Group: Have one student on each team roll the Number Cube and another person record the numbers on the recording sheet. Tell teams to work together to solve the problem. Check each team's solution. In a tally chart on the board, keep track of which team has the greatest sum. If a team's solution is incorrect, that sum does not count. After several rounds, the team with the most tally marks wins.

Individual: Complete the activity as written, but have students keep their recording sheets until they have solved several problems. Then have students exchange papers and check each other's work.

Progress Monitoring
| **If…** students are not using place value to line up the numbers correctly before they add, | ▶ **Then…** provide graph paper so students can write each digit in a single square. |

Practice
Have students complete **Student Workbook,** pages 72–73. Guide students through the Key Idea example and the Try This exercises.

Interactive Differentiation
Consult the **Teacher Dashboard** for grouping suggestions. You can also use performance on the Engage activity to guide students.

Independent Practice
For additional practice, have students use the Base 10 Blocks Tool to model the problems they worked during **Adding Hundreds.** Have students take out their Adding Hundreds Recording Sheet. Tell them to hide the summation lines before they model each problem. After they have modeled the problem, they can toggle the summation lines to check their answer.

Supported Practice
For additional support, use the Base 10 Blocks Tool to model adding a two-digit number and a three-digit number.
▶ **I am going to add 144 and 43.**
- Write 144 + 43 on the board in vertical form. Then launch the Base 10 Blocks Tool so that students can see the screen.
- Hide the summation lines. On the top half of the mat, model 144. On the bottom half, model 43.
▶ **I have one hundred; one, two, three, four, five, six, seven, eight tens; and one, two, three, four, five, six, seven ones. The answer is 187.**
- Reveal the summation lines and check your answer.
- Have students get out their Adding Hundreds Recording Sheet from **Adding Hundreds.** Tell them to use the Base 10 Blocks Tool to model the first problem. Then have students toggle the summation lines to check the answer.

3 REFLECT

Think Critically

Review students' answers to the Reflect prompt at the bottom of *Student Workbook,* page 73, and then review the Engage activity.

▶ **Why is it important to neatly write and correctly align the numbers?** Answers may vary. Possible answer: If you don't line up the digits correctly—the ones under the ones, the tens under the tens—then you will not get the correct answer. You might add 4 tens to 2 ones and get 6, but 4 tens plus 2 ones is actually 42.

Real-World Application

▶ **At a basketball game, one team scored 32 points and the other team scored 6 points. Your friend said that there were 92 points scored altogether. Is he correct?** No, your friend is not correct. He added 32 + 60 and not 32 + 6. There were 38 points scored altogether.

4 ASSESS

Informal Assessment

Use the online or print Student Record, **Assessment,** page 128, to record informal observations.

Adding Hundreds

Did the student

☐ apply learning to a new situation? ☐ contribute answers?

☐ contribute concepts? ☐ connect mathematics to the real world?

Additional Practice

For additional practice, have students complete *Practice,* page 75.

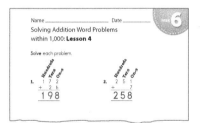

Practice, p. 75

Week 6 • Solving Addition Word Problems within 1,000

Lesson 4

Key Idea
When you add greater numbers, make sure to add the ones to the ones, the tens to the tens, and the hundreds to the hundreds. Start adding in the ones column. Then move to the tens, and then to the hundreds.

Try This
Solve each problem.

1. 106
  + 83
  189

2. 50
  +227
  277

3. 253
  +436
  689

4. 394
  +102
  496

5. 121
  + 45
  166

6. 266
  + 13
  279

72 Level D Unit 3 Addition

Practice
Solve each problem.

7. 509
  + 50
  559

8. 613
  + 35
  648

9. 815
  + 60
  875

10. 156
  + 203
  359

11. 456
  + 123
  579

12. 508
  + 211
  719

13. At the start of the baseball game, there were 253 people. Within twenty minutes, 25 more people had come to the game. How many people were at the game in all?
278 people

Reflect
Explain how you solved one problem above.

Possible answer: For problem 12, I added 8 and 1 ones, 0 and 1 tens, and 5 and 2 hundreds to find the answer: 719.

Week 6 Solving Addition Word Problems within 1,000 • Lesson 4 73

Student Workbook, pp. 72–73

WEEK 6
Solving Addition Word Problems within 1,000

Lesson 5 Review

Objective
Students review skills learned this week and complete the weekly assessment and project.

Standard
2.NBT.7 Add and subtract within 1000, using concrete models or drawings and strategies based on place value, properties of operations, and/or the relationship between addition and subtraction; relate the strategy to a written method. Understand that in adding or subtracting three-digit numbers, one adds or subtracts hundreds and hundreds, tens and tens, ones and ones; and sometimes it is necessary to compose or decompose tens or hundreds.

Creating Context
Have students use these vocabulary words in sentences.
- hundreds
- ones
- sum
- tens

1 WARM UP

Prepare

▶ **When you add numbers with more than one digit, where do you start working on the problem?** You always work from right to left. The ones place is first, then the tens, then the hundreds.

▶ **If you add numbers with more than one digit, what happens if the sum of the digits in the ones place is greater than 9?** You will need to regroup. Move 10 to the tens place. Any remaining ones stay in the ones place.

2 ENGAGE

Practice
Have students complete **Student Workbook,** pages 74–75.

Week 6 • Solving Addition Word Problems within 1,000

Lesson 5 Review

This week you added large numbers together. You added ones, moved to the tens, and then added the numbers in the hundreds place.

Lesson 1 Solve each problem.

1. 45
   + 6
   51

2. 81
   + 9
   90

3. 64
   + 7
   71

4. 53
   + 9
   62

5. 58
   + 6
   64

6. 75
   + 16
   91

Lesson 2 Solve each problem.

7. 36
   + 46
   82

8. 57
   + 18
   75

9. 29
   + 65
   94

10. 18
    + 29
    47

11. 53
    + 29
    82

12. 48
    + 36
    84

74 Level D Unit 3 **Addition**

Lesson 3 Solve each problem.

13. 46
    + 3
    49

14. 66
    + 15
    81

15. 55
    + 39
    94

16. 62
    + 21
    83

17. Marc has 18 baseballs. His cousin has 9 baseballs. How many baseballs do they have altogether? __**27 baseballs**__

Lesson 4 Solve each problem.

18. 194
    + 203
    397

19. 308
    + 161
    469

20. Ms. Sanchez has 212 markers in her classroom. Ms. Meeks has 153 markers in her classroom. How many markers are in both classrooms? __**365 markers**__

Reflect

Solve Problem 5, but first add the tens place instead of the ones place. What happens when you do not start with the ones? Explain your answer.

Answers may vary; possible answer: If you start with the tens place, you will not have the correct answer. You might write 58 + 6 = 514 because you did not regroup the sum of 8 and 6. If you first add the numbers in the ones place, you will regroup 14 and put 1 ten in the tens place and keep 4 in the ones place. The sum of 1 and 5 is 6, so the answer is 64.

Week 6 **Solving Addition Word Problems within 1,000** • Lesson 5 75

Student Workbook, pp. 74–75

270 Level D Unit 3 **Addition**

3 REFLECT

Think Critically

Review students' answers to the Reflect prompt at the bottom of **Student Workbook,** page 75.

Discuss the answer with the group to reinforce Week 6 concepts.

4 ASSESS

Formal Assessment

Students may take the weekly assessment online.

As an alternative, students may complete the weekly test on **Assessment,** pages 47–48. Record progress using the Student Assessment Record, **Assessment,** page 128.

Going Forward

Use the **Teacher Dashboard** to view results of the online assessments, to input the results of print student assessments, and to review progress before making decisions about next steps. Use the weekly test results and observations to determine the next steps for each student.

Retention	
Student displays good grasp of this week's concepts and skills.	**Building Blocks** Have students independently complete the Building Blocks activity Word Problems with Tools 4.
Remediation	
Student is still struggling with the week's concepts and skills.	Work step-by-step with students through multi-digit addition problems. Encourage students to use base-ten blocks.

Suggestions for Re-Evaluation: If a student has struggled without success for several weeks, use observations and test results to place the student at a level in which he or she can find success and build confidence to move forward.

Name _____ Date _____ WEEK 6

Solving Addition Word Problems within 1,000

Solve these problems. Write the answer as tens and ones.

1. 8 + 35 = __4__ tens and __3__ ones

2. 12 + 17 = __2__ tens and __9__ ones

3. 36
  + 18

  __5__ tens and __4__ ones

4. 66
  + 24

  __9__ tens and __0__ ones

Solve these problems. Write the answer as a number.

5. 8 + 33 = __41__

6. 48
  + 28
  __76__

Level D Unit 3 Week 6 47

WEEK 6 Name _____ Date _____

Solving Addition Word Problems within 1,000

Solve these problems. Write the answer as a number.

7. 235
  + 487
  __722__

8. 541
  + 96
  __637__

9. A motorcycle weighs 553 pounds. The rider weighs 162 pounds. How many pounds do they weigh together?
  __715 pounds__

10. The height of a building is 379 feet. The cell phone tower is 103 feet tall. How tall are the building and tower together?
  __482 feet__

48 Level D Unit 3 Week 6

Assessment, pp. 47–48

Project Preview

This week, students used modeling and the standard algorithm to solve addition problems with regrouping. Students also used the standard algorithm to solve word problems within 1,000. This week, students will conclude their projects by showing the steps of the standard algorithm on their posters. They will also present their posters to the class.

Project-Based Learning

Standards-driven Project-Based Learning is effective in building deep content understanding. Project-Based Learning increases long-term retention of concepts and has been shown to be more effective than traditional instruction. Completing a project to answer an essential question challenges students to apply and demonstrate mastery of concepts and skills by expressing understanding through discussion, research, and presentation.

Essential Question

HOW can I use addition to solve real-world problems?

Project Evaluation Criteria

Review project evaluation criteria with students prior to beginning the project.

Exceeds Expectations
☐ Project result is explained and can be extended.
☐ Project result is explained in context and can be applied to other situations.
☐ Project result is explained using advanced mathematical vocabulary.
☐ Project result is described, and mathematics are used correctly and can be extended.
☐ Project result is explained and extended, and shows advanced knowledge of mathematical concepts and skills.

Meets Expectations
☐ Project result is explained.
☐ Project result is explained in context.
☐ Project result is explained using mathematical vocabulary.
☐ Project result is described, and mathematics are used correctly.
☐ Project result is explained, and shows satisfactory knowledge of mathematical concepts and skills.

Does Not Meet Expectations
☐ Project result is not explained.
☐ Project result is explained, but out of context.
☐ Project result is explained, but mathematical vocabulary is oversimplified.
☐ Project result is described, but mathematics are not used correctly.
☐ Project result is not explained and/or extended, or shows less than satisfactory knowledge of mathematical concepts and skills.

Show Your Collection Equation

Objective
Students can add multi-digit numbers with regrouping.

Standard
2.NBT.7 Add and subtract within 1000, using concrete models or drawings and strategies based on place value, properties of operations, and/or the relationship between addition and subtraction; relate the strategy to a written method. Understand that in adding or subtracting three-digit numbers, one adds or subtracts hundreds and hundreds, tens and tens, ones and ones; and sometimes it is necessary to compose or decompose tens or hundreds.

Materials
Additional Materials
- construction paper
- poster from Week 5
- tape or glue sticks

Prepare Ahead
Write a two-digit addition equation on a sheet of construction paper, one for each pair of students. Write the equation in vertical format. The problem should be visible, but allow enough space for students to write out the procedure for solving it.

Best Practices
- Permit the opportunity for students to set goals, share information, and self-evaluate.
- Evaluate through observational records and performance assessment.
- Allow student to think industriously.

Introduce

This week we worked on adding greater numbers. Sometimes you had to regroup numbers to move them from one place value to another.

▶ **Suppose you are adding together two 2-digit numbers. What happens if you add the ones together and the sum is greater than 9?**
  The ones stay in the ones place, but the ten is moved to the tens place.

Explore

▶ **Today you will explain how to solve a two-digit addition problem.**

• Write 45 + 37 on the board in vertical format.

▶ **Can someone come to the board and show us, step by step, how to do this problem?**

• As the volunteer talks through the problem, write the steps needed to solve it on a piece of construction paper. Then reiterate the procedure for the class.

▶ **First, you added the ones in the ones column.**

▶ **Because 5 + 7 = 12, you wrote 2 in the ones column and moved the ten ones to the tens column by writing 1 above the 4 in 45.**

▶ **Then you added all the numbers in the tens column and wrote the answer below the line: 1 + 4 + 3 = 8; 45 + 37 = 82.**

• Distribute the construction paper, one to each pair of students.

• Tell students to write the step-by-step procedure for solving an addition problem about their own collection on the sheet of construction paper. They should also solve the problem and write the answer below the line. When students are finished, they should attach their solutions to their posters with tape or glue.

▶ **Complete *Student Workbook*, page 76, to solve multi-digit addition problems.**

Wrap Up

• To give students more time for their presentations, you can write the steps of the standard algorithm on the board and have students copy them.

• Discuss students' answers to the Reflect prompts at the bottom of ***Student Workbook,*** page 76.

Give student pairs 2 or 3 minutes to present their posters to the class. Encourage them to explain why they chose the item they collected. If possible, display the posters in the classroom or hallway.

If time permits, allow students to share what they liked or did not like about the project.

Project
Show Your Collection Equation

Answer the following questions. The answers will help you solve addition problems.

1. What is the first step in solving a two-digit addition problem?

   Add the numbers in the ones place.

2. What happens if the sum of the numbers in the ones place is less than 10? What happens if the sum is 10 or greater?

   If the sum is less than 10, then it is written below the line in the ones place.

   If the sum is 10 or greater, the value in the ones place of the sum is written

   in the ones place below the line. The 10 is regrouped to the tens place.

3. How can you keep track of a regrouped ten? How do you add the ten to the other numbers?

   Write a 1 above the tens place of the top number that is being added. Then add the 1 to the other numbers in the tens place. Write the sum in the tens place below the line.

Reflect

How does writing out the steps for solving a problem help you add?

Answers may vary; possible answer: Writing out each step helps you to not

make addition mistakes.

Student Workbook, p. 76

Teacher Reflect

☐ Did students talk about the most important thing they learned?

☐ Did students tell or show the steps when they explained how to do something?

☐ Did students show knowledge of how their project related to the major concept?

UNIT 4 Subtraction

Unit at a Glance

This **Number Worlds** unit builds on prior knowledge of addition. Students will apply this knowledge to extend their understanding of basic addition facts to subtraction and writing, solving, and checking the accuracy of equations representing subtraction word problems.

Skills Trace

Before Level D	Level D	After Level D
Level C Students can solve one- and two-digit addition sentences within 100, including problem solving. They can also add and subtract within 1,000 using concrete models and drawings.	By the end of this unit, students will be able to solve subtraction word problems within 1,000 and use addition to check their answers.	**Moving on to Level E** Students subtract numbers up to three digits long with and without regrouping. They often use an equation with a variable. Students use rounding and estimation skills to check their answers.

Learning Technology

The following activities are available online to support the learning goals in this unit.

Building Blocks
- Math-O-Scope
- Pizza Pizzazz 4
- Word Problems with Tools 1
- Word Problems with Tools 2
- Word Problems with Tools 4
- Word Problems with Tools 8

Digital Tools
- Sets Former Tool
- Base 10 Blocks Tool

Unit Overview

Week	Focus
1	**Subtraction Fundamentals** • *Teacher Edition,* pp. 276–289 • *Activity Cards,* 4A, 4B, 4C, 4D • *Student Workbook,* pp. 5–16 • *English Learner Support Guide,* pp. 76–77 • *Assessment,* pp. 49–50
2	**Mastering Basic Subtraction Facts** • *Teacher Edition,* pp. 290–303 • *Activity Cards,* 4E, 4F, 4G, 4H • *Student Workbook,* pp. 17–28 • *English Learner Support Guide,* pp. 78–79 • *Assessment,* pp. 51–52
3	**Solving Subtraction Problems** • *Teacher Edition,* pp. 304–317 • *Activity Cards,* 4I, 4J, 4K, 4L • *Student Workbook,* pp. 29–40 • *English Learner Support Guide,* pp. 80–81 • *Assessment,* pp. 53–54
4	**Subtraction Tools and Strategies** • *Teacher Edition,* pp. 318–331 • *Activity Cards,* 4M, 4N, 3P • *Student Workbook,* pp. 41–52 • *English Learner Support Guide,* pp. 82–83 • *Assessment,* pp. 55–56
5	**Subtraction Word Problems within 100** • *Teacher Edition,* pp. 332–345 • *Activity Cards,* 4O, 4P, 4Q • *Student Workbook,* pp. 53–64 • *English Learner Support Guide,* pp. 84–85 • *Assessment,* pp. 57–58
6	**Solving Subtraction Word Problems within 1,000** • *Teacher Edition,* pp. 346–359 • *Activity Cards,* 4R, 4S, 4T • *Student Workbook,* pp. 65–76 • *English Learner Support Guide,* pp. 86–87 • *Assessment,* pp. 59–60

Essential Question

WHEN would it be useful to write equations outside the classroom?

In this unit, students will explore how multiplication can be used to solve real-world problems by exploring subtraction problems related to running a store, including calculating expected sales, computing change for monetary transactions, subtracting to find final prices, and analyzing inventory.

Learning Goals	CCSS Key Standards
Students can solve basic subtraction facts by counting back and can write subtraction equations. **Project:** Students can use subtraction to find a sale price.	**Domain:** Operations and Algebraic Thinking **Cluster:** Add and subtract within 20. **2.OA.2:** Fluently add and subtract within 20 using mental strategies. By end of Grade 2, know from memory all sums of two one-digit numbers.
Students can use doubles facts and near-doubles facts and can choose a strategy to solve subtraction problems. **Project:** Students can interpret and solve two-step subtraction word problems and write equations to describe each step.	**Domain:** Operations and Algebraic Thinking **Cluster:** Add and subtract within 20. **2.OA.2:** Fluently add and subtract within 20 using mental strategies. By end of Grade 2, know from memory all sums of two one-digit numbers.
Students can perform a series of subtraction operations and write equations to describe them and can interpret and solve word problems. **Project:** Students can use subtraction to solve problems.	**Domain:** Number and Operations in Base Ten **Cluster:** Use place value understanding and properties of operations to add and subtract. **2.NBT.5:** Fluently add and subtract within 100 using strategies based on place value, properties of operations, and/or the relationship between addition and subtraction.
Students can solve subtraction problems with one- and two-digit numbers when displayed in a vertical format. **Project:** Students can use subtraction to determine the correct amount of change a customer should receive.	**Domain:** Number and Operations in Base Ten **Cluster:** Use place value understanding and properties of operations to add and subtract. **2.NBT.5:** Fluently add and subtract within 100 using strategies based on place value, properties of operations, and/or the relationship between addition and subtraction.
Students can solve one- and two-step subtraction word problems within 100 and with unknowns in all positions. **Project:** Students can use what they know about solving subtraction word problems to compose and solve their own subtraction story.	**Domain:** Operations and Algebraic Thinking **Cluster:** Represent and solve problems involving addition and subtraction. **2.OA.1:** Use addition and subtraction within 100 to solve one- and two-step word problems involving situations of adding to, taking from, putting together, taking apart, and comparing, with unknowns in all positions, e.g., by using drawings and equations with a symbol for the unknown number to represent the problem.
Students can subtract within 1,000 using concrete models or drawings and can understand the relationship between addition and subtraction. **Project:** Students can use what they know about solving subtraction problems to explain the parts of a subtraction equation to their peers.	**Domain:** Number and Operations in Base Ten **Cluster:** Use place value understanding and properties of operations to add and subtract. **2.NBT.7:** Add and subtract within 1000, using concrete models or drawings and strategies based on place value, properties of operations, and/or the relationship between addition and subtraction; relate the strategy to a written method. Understand that in adding or subtracting three-digit numbers, one adds or subtracts hundreds and hundreds, tens and tens, ones and ones; and sometimes it is necessary to compose or decompose tens or hundreds.

CCSS Daily lesson activities emphasize using communication, logic, reasoning, modeling, tools, precision, structure, and patterns to solve problems. All student activities, reflections, and assessments require application of the **Common Core Standards for Mathematical Practice**.

WEEK 1: Subtraction Fundamentals

Week at a Glance

This week, students begin **Number Worlds,** Level D, Subtraction. Students will use Counters to perform basic subtraction problems.

Skills Focus

- Identify numbers that are before and after a given number.
- Solve basic subtraction facts by counting back.
- Write subtraction equations.

How Students Learn

As students learn about the operation of subtraction, provide opportunities for them to connect this process to the world around them. Have students act out stories that involve adding and subtracting groups of objects or people. Computer technology can be a useful tool in providing students with "concrete" models. The use of computer activities can also help computational skills by focusing students' attention and increasing motivation.

English Learners ELL

For language support, use the **English Learner Support Guide,** pages 76–77, to preview lesson concepts and teach academic vocabulary. **Number Worlds** Vocabulary Cards are listed as additional materials in many lessons and can be used to preteach and reinforce academic vocabulary.

Math at Home

Give one copy of the Letter to Home, page 19, to each student. Encourage students to share and complete the activity with their caregivers.

Weekly Planner

Lesson	Learning Objectives
1 pages 278–279	Student can identify the position of each number in the 1 to 100 sequence.
2 pages 280–281	Students can understand the connection between counting and subtraction.
3 pages 282–283	Students can use the minus sign and the equal sign to create subtraction equations.
4 pages 284–285	Students can understand that any number minus itself equals zero.
5 pages 286–287	**Review and Assess** Students review skills learned this week and complete the weekly assessment and project.
Project pages 288–289	Students can use subtraction to find a sale price.

276 Level D Unit 4 **Subtraction**

Key Standard for the Week

Domain: Operations and Algebraic Thinking

Cluster: Add and subtract within 20.

2.OA.2 Fluently add and subtract within 20 using mental strategies. By end of Grade 2, know from memory all sums of two one-digit numbers.

Materials		Technology
Program Materials • *Student Workbook*, pp. 5–7 • *Practice*, p. 76 • Activity Card 4A, **Which Is It?** • 1–100 Chart • Neighborhood Number Line • Number Cards (1–20)	**Additional Materials** • Vocabulary Card 9, *difference* • Vocabulary Card 13, *equal* • Vocabulary Card 42, *subtract*	*Teacher Dashboard* Building Blocks Math-O-Scope
Program Materials • *Student Workbook*, pp. 8–9 • *Practice*, p. 77 • Activity Card 4B, **The Dragon's Sister** • Dragon Quest Game Board • Dragon Quest Cards (+1 to +5, and -1 to -2, only) • Counters • Number Cards 1–20	• Pawns • Spinners **Additional Materials** • can or other container • Vocabulary Card 9, *difference* • Vocabulary Card 13, *equal* • Vocabulary Card 42, *subtract*	*Teacher Dashboard* Building Blocks Pizza Pizzazz 4
Program Materials • *Student Workbook*, pp. 10–11 • *Practice*, p. 78 • Activity Card 4C, **They're Gone** • Counters	**Additional Materials** • paper cups • Vocabulary Card 9, *difference* • Vocabulary Card 13, *equal* • Vocabulary Card 42, *subtract*	*Teacher Dashboard* Sets Former Tool
Program Materials • *Student Workbook*, pp. 12–13 • *Practice*, p. 79 • Activity Card 4D, **Subtracting to Zero** • Counters	**Additional Materials** • index cards • Vocabulary Card 9, *difference* • Vocabulary Card 13, *equal* • Vocabulary Card 42, *subtract*	*Teacher Dashboard* Sets Former Tool
Program Materials • *Student Workbook*, pp. 14–15 • Weekly Test, *Assessment*, pp. 49–50		Review previous activities.
Program Materials • *Student Workbook*, p. 16 • Number 1–6 Cube	**Additional Materials** • index cards • scissors	

WEEK 1
Subtraction Fundamentals

Find the Math

In this week, introduce students to subtraction using number lines, counting, and simple problems.

Use the following to begin a guided discussion:

▶ **A baker needs two eggs for a batch of cookies. He will bake several batches of cookies today. How can you find out how many eggs are left in the carton after each batch of cookies?** Answers may vary; possible answer: Count back two after you add each batch of cookies.

Have students complete **Student Workbook,** page 5.

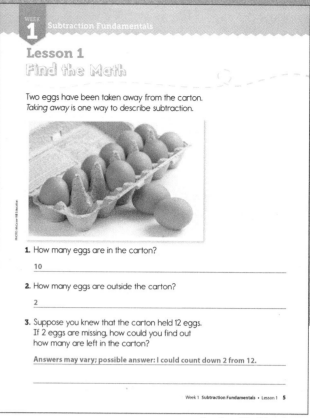

Student Workbook, p. 5

Lesson 1

Objective
Students can identify the position of each number in the 1 to 100 sequence.

Standard
2.OA.2 Fluently add and subtract within 20 using mental strategies. By end of Grade 2, know from memory all sums of two one-digit numbers.

Vocabulary
- difference
- equal
- subtract

Creating Context
Some English Learners may benefit from a bilingual dictionary to help them find English words and spellings. Bilingual dictionaries might serve as important comprehension tools for students or their family members who want to help.

Materials

Program Materials
- 1–100 Chart
- Neighborhood Number Line
- Number Cards (1–20), 1 set per student

Additional Materials
- Vocabulary Card 9, *difference*
- Vocabulary Card 13, *equal*
- Vocabulary Card 42, *subtract*

1 WARM UP

Prepare

Display the Neighborhood Number Line so students can use it as a reference. Ask a series of questions that require students to focus on the position of numbers in the 1 to 100 sequence and the meaning of the terms *before, after,* and *between.* Ask questions such as the following:

▶ **What number comes directly after 77?** 78

▶ **What number comes two numbers after 44?** 46

▶ **What number comes directly before 20?** 19

▶ **What number is between 33 and 35?** 34

▶ **How many numbers are between 5 and 8?** two

2 ENGAGE

Develop: Which Is It?

"Today we are going to listen to clues to find a number." Follow the instructions on the Activity Card **Which Is It?** As students complete the activity, be sure to use the Questions to Ask.

Alternative Grouping

Whole Class: Complete the activity as written.

Activity Card 4A

Progress Monitoring

| If… students are struggling to decide which cards to remove based on the clue, | ▶ Then… have them refer to a number line. |

Practice

Have students complete **Student Workbook,** pages 6–7. Guide students through the Key Idea example and the Try This exercises.

278 Level D Unit 4 **Subtraction**

Interactive Differentiation

Consult the **Teacher Dashboard** for grouping suggestions. You can also use performance on the Engage activity to guide students.

Independent Practice

For additional practice, use Math-O-Scope with students to review number sequence. In this activity, students identify the numbers that surround a given number in the context of the 1–100 Chart.

Supported Practice

For additional support, use Math-O-Scope with students to review number sequence. Before students begin, provide them with the 1–100 Chart or make sure a number line from 1 to 100 is displayed. When students have finished the activity, ask them questions such as the following:

▶ **In Math-O-Scope, which number is immediately after the number in the center of the scope?** the number to the right **Which number is immediately before it?** the number to the left

▶ **When you count on, what happens to the numbers?** Answers may vary. Possible answers: They become greater; they increase. **What happens when you count back?** Answers may vary. Possible answers: They become less; they decrease.

▶ **What three numbers come right after 19 when you count on?** 20, 21, 22

• Ask students other number sequence questions if time permits.

3 REFLECT

Think Critically

Review students' answers to the Reflect prompt at the bottom of **Student Workbook,** page 7, and then review the Engage activity.

▶ **Why would a number line help you answer the questions?** Possible answer: It makes it easier to check whether I am correct.

Explain that even though you may be comfortable with the problems, it is always a good idea to have a visual to double-check your work.

4 ASSESS

Informal Assessment

Use the online or print Student Record, **Assessment,** page 128, to record informal observations.

Which Is It?
Did the student
☐ make important observations?   ☐ provide insightful answers?
☐ extend or generalize learning?   ☐ pose insightful questions?

Additional Practice

For additional practice, have students complete **Practice,** page 76.

Practice, p. 76

Week 1 • Subtraction Fundamentals

Lesson 1

Key Idea
Knowing the number sequence helps you solve subtraction problems.

Try This

Name the three numbers that come *after* the number given.

1. 3 __4, 5, 6__
2. 16 __17, 18, 19__
3. 22 __23, 24, 25__
4. 19 __20, 21, 22__

Name the three numbers that come *before* the number given.

5. 5 __2, 3, 4__
6. 23 __20, 21, 22__
7. 11 __8, 9, 10__
8. 18 __15, 16, 17__

Name the two numbers that the given number is *between*.

9. 10 __9 and 11__
10. 24 __23 and 25__
11. 14 __13 and 15__
12. 8 __7 and 9__

6 Level D Unit 4 Subtraction

Practice
Find the missing numbers in the sequences below.

13. 11, 12, __13__, 14, 15
14. 3, __4__, 5, 6, 7, __8__
15. __46__, 47, 48, 49, __50__, 51
16. 18, 19, __20__, __21__, 22, 23

Write all the numbers that fit the description in Part A, using the numbers 1–15. Write all the numbers from Part A that fit the Part B description. Write the number from Part B that fits the Part C description.

17. A. I come after 7. __8, 9, 10, 11, 12, 13, 14, 15__
    B. I come before 11. __8, 9, 10__
    C. I am between 8 and 10. What number am I? __9__

18. A. I am between 3 and 9. __4, 5, 6, 7, 8__
    B. I come before 6. __4, 5__
    C. I am less than 5. What number am I? __4__

Reflect
How could a number line help you answer the questions above?

Answers will vary. Possible answer: It shows the numbers in order.

Week 1 Subtraction Fundamentals • Lesson 1 7

Student Workbook, pp. 6–7

Week 1 Subtraction Fundamentals • Lesson 1 **279**

WEEK 1
Subtraction Fundamentals

Lesson 2

Objective
Students can understand the connection between counting and subtraction.

Standard
2.OA.2 Fluently add and subtract within 20 using mental strategies. By end of Grade 2, know from memory all sums of two one-digit numbers.

Vocabulary
- difference
- equal
- subtract

Creating Context
Review the basic routines for playing games and math activities. Make sure students understand the rules for taking turns, deciding who goes first, winning the game, and what to do in the event of a tie.

Materials

Program Materials
- Dragon Quest Game Board, 1 per group
- Dragon Quest Cards (+1 to +5, and, -1 to -2, only) 1 set per group
- Counters, 20 per pair
- Number Cards 1–20, 1 set per pair
- Pawns, 4 per group
- Spinner, 1 per group

Additional Materials
- can or other container
- Vocabulary Card 9, *difference*
- Vocabulary Card 13, *equal*
- Vocabulary Card 42, *subtract*

Prepare Ahead
Shuffle the Dragon Quest Cards, making sure that the top six cards in the pile are addition cards so students will not have to subtract from 0.

1 WARM UP

Prepare
Count back as a group from various numbers between 10 and 20, and then call on several students to count back individually.

▶ **Counting back can help you do subtraction.**

Drop 7 Counters, one by one, into a can, counting on as you do so.

▶ **How many Counters are in the can?** 7

Take 3 Counters out of the can, one by one, counting back as you do so.

▶ **The next number back, 4, should tell us how many Counters are left in the can. Reveal the contents of the can. Does it?** yes

- Repeat this process to give students turns solving subtraction problems, such as 6 − 2, 8 − 4, and 9 − 3.

Just the Facts
Have students count back on their fingers to find differences. Use questions such as the following:

▶ **What number is left if you take 5 away from 8?**
Students hold up 3 fingers.

▶ **What number is left if you take 4 away from 10?**
Students hold up 6 fingers.

2 ENGAGE

Develop: The Dragon's Sister
"Today we are going to play another Dragon Quest game." Follow the instructions on the Activity Card **The Dragon's Sister**. As students complete the activity, be sure to use the Questions to Ask.

Activity Card 4B

Alternative Grouping
Pairs: Partner with the student and complete the activity as written.

Progress Monitoring

| If… students do not have totals equaling 15, | ▶ Then… have them work as a group to determine the mistake. |

Practice
Have students complete **Student Workbook**, pages 8–9. Guide students through the Key Idea example and the Try This exercises.

Interactive Differentiation
Consult the **Teacher Dashboard** for grouping suggestions. You can also use performance on the Engage activity to guide students.

Independent Practice — Building Blocks
For additional practice understanding the connection between counting and addition and subtraction, have students use Pizza Pizzazz 4: Count Hidden Pepperoni. A target number of toppings is either added to or taken away from the pizza, and students must determine the resulting number. Tell students to count on or back to reach the answer.

Supported Practice
For additional support, have students work with a partner to solve subtraction problems. Tell them to use the count-down strategy. Give each pair of students 20 Counters and Number Cards 1–20.

- Have students mix up the Number Cards and place them facedown.
- Explain that one partner will draw two cards and make a subtraction problem. For example, a student draws 11 and 7. The subtraction problem is 11 − 7.
- Have the other partner place the number of Counters matching the first number on a sheet of paper. Then both partners should count back from the minuend, sliding one Counter at a time off the paper until they have removed the number of Counters that matches the subtrahend. For example, students would pull 7 Counters from the group while counting back from 11: 11, 10, 9, 8, 7, 6, 5. Remind students that the next number, 4, is the answer. That is also the number of Counters that should remain on the paper.
- Have partners repeat with other pairs of Number Cards.

WEEK 1
Subtraction Fundamentals

Lesson 3

Objective
Students can use the minus sign and the equal sign to create subtraction equations.

Standard
2.OA.2 Fluently add and subtract within 20 using mental strategies. By end of Grade 2, know from memory all sums of two one-digit numbers.

Vocabulary
- difference
- equal
- subtract

Creating Context
In this lesson students solve a problem about pets. Brainstorm with English Learners using a word-web graphic organizer of different types of pets. Help students organize their ideas by broad classifications, such as animals that walk on the ground, swim, or fly.

Materials

Program Materials
Counters, 10 in 4 colors per student (15 for Independent Practice)

Additional Materials
- paper cup, 1 per student
- Vocabulary Card 9, *difference*
- Vocabulary Card 13, *equal*
- Vocabulary Card 42, *subtract*

1 WARM UP

Prepare

▶ **What symbols do we use to write a subtraction number story?**
minus sign, equal sign

▶ **A minus sign means "take away." Does the amount you start with get smaller or larger?**

▶ **What does an equal sign mean?** Allow discussion, and lead students to realize that an equal sign tells us that each side has the same amount.

Write $9 - 3 = 6$ on the board, and use it to demonstrate the meaning of the signs.

▶ **The minus sign on the left side of this equation tells us to take away. If we start with 9 and take away 3, how many do we have left on this side of the equation?** 6

▶ **How many do we have on the other side of the equation?** 6

▶ **Are these sides equal?** yes

Write $9 - 2 = 6$ on the board.

▶ **Is this equation true or false? Explain your thinking.**

Just the Facts

Have students use Counters to determine the answers to subtraction problems. Tell them to hold up the number of Counters that represents the answer. Use prompts such as the following:

▶ **3 minus 2 equals _____.** Students hold up 1 Counter.

▶ **4 minus 2 equals _____.** Students hold up 2 Counters.

▶ **5 minus 4 equals _____.** Students hold up 1 Counter.

2 ENGAGE

Develop: They're Gone

"Today we are going to use counters to talk about subtraction." Follow the instructions on the Activity Card **They're Gone**. As students complete the activity, be sure to use the Questions to Ask.

Activity Card 4C

Alternative Grouping

Whole Class: Complete the activity as written.

Progress Monitoring	
If... students have difficulty creating a subtraction number story,	**Then...** tell them the stories have three parts. First, students should say how many Counters they had at the beginning. Next, they should say what happened to the Counters. Finally, students should say how many Counters remained.

Practice

Have students complete **Student Workbook,** pages 10–11. Guide students through the Key Idea example and the Try This exercises.

Interactive Differentiation

Consult the **Teacher Dashboard** for grouping suggestions. You can also use performance on the Engage activity to guide students.

Independent Practice

For additional practice, provide students with 15 Counters. Tell students you are going to model $10 - 2$. Set out 10 Counters, and then take 2 away. As a class, count the remaining 8 Counters. Write the following equations on the board: $14 - 4 = 10$; $8 - 3 = 5$; $12 - 6 = 6$; $9 - 4 = 5$; $11 - 4 = 7$; $13 - 9 = 4$; $15 - 7 = 8$. Have students model the equations using Counters.

Supported Practice

For additional support, use the Sets Former Tool with students to model subtraction.

- Open the Sets Former Tool. In the Format area of the palette, choose the Subtraction Mat. Place 6 marbles in the top half of the mat. Drag 3 marbles from the top area to the bottom. Tell students that you are subtracting the marbles that you take away. Point out that you have just modeled the problem $6 - 3 = 3$. Write the equation on the board to reinforce equation format.

- Tell students to model $8 - 2$; $8 - 4$; and $8 - 1$. Have them write an equation for each subtraction problem. $8 - 2 = 6$; $8 - 4 = 4$; $8 - 1 = 7$

- Have students make up their own subtraction sentences.

▶ **What do you call the symbol between the first two numbers?**
a minus sign

▶ **What symbol do you add before the answer to show that the two sides of the equation have the same value?** an equal sign

▶ **What do you call the answer to a subtraction equation?**
the difference

282 Level D Unit 4 **Subtraction**

3 REFLECT

Think Critically

Review students' answers to the Reflect prompt at the bottom of *Student Workbook,* page 11, and then review the Engage activity.

▶ Will these answers always be the same? Why?

Real-World Application

▶ Suppose someone in your house makes cupcakes and places 10 of them on a tray in your kitchen. You and a friend walk by and each decide to eat a cupcake.

▶ Would it be necessary to count the cupcakes to know there are fewer of them?

▶ How could someone figure out how many cupcakes were eaten?

4 ASSESS

Informal Assessment

Use the online or print Student Record, *Assessment,* page 128, to record informal observations.

They're Gone

Did the student
- ☐ make important observations?
- ☐ provide insightful answers?
- ☐ extend or generalize learning?
- ☐ pose insightful questions?

Additional Practice

For additional practice, have students complete *Practice,* page 78.

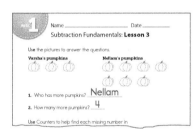

Practice, p. 78

Week 1 • Subtraction Fundamentals

Lesson 3

Key Idea
Draw a picture to help solve the subtraction problem.

● ● ⊗ ⊗ ⊗

$5 - 3 = 2$

The answer to a subtraction problem is called the **difference**. The difference between 5 and 3 is 2.

Try This
Use Counters to solve each problem. Then write the difference.

1. $2 - 1 = \underline{1}$ 2. $4 - 2 = \underline{2}$ 3. $6 - 4 = \underline{2}$

Use the pictures to answer the questions.

4. Pam's pets   Bill's pets

A. Who has more pets? __Pam__

B. How many more pets? __2__

C. Who has fewer pets? __Bill__

D. How many fewer pets? __2__

10 Level D Unit 4 Subtraction

Practice
Use Counters to help find each missing number in the subtraction problems.

5. $\underline{4} - 1 = 3$ 6. $\underline{9} - 1 = 8$

7. $7 - \underline{2} = 5$ 8. $10 - \underline{5} = 5$

9. $14 - 4 = \underline{10}$ 10. $12 - 8 = \underline{4}$

Circle the two problems in each group that have the same difference.

11. ⌈$10 - 4 = \underline{6}$⌉   12. ⌈$11 - 9 = \underline{2}$⌉
    ⌊$8 - 2 = \underline{6}$⌋ $12 - 7 = \underline{5}$
    $6 - 1 = \underline{5}$ ⌊$9 - 4 = \underline{5}$⌋

Reflect
In Problem 4, why are the answers to Part B and Part D the same?

Answers will vary. Possible answer: They describe the same thing, but one person has more and the other fewer.

Week 1 **Subtraction Fundamentals** • Lesson 3 11

Student Workbook, pp. 10–11

WEEK 1
Subtraction Fundamentals

Lesson 4

Objective
Students can understand that any number minus itself equals zero.

Standard
2.OA.2 Fluently add and subtract within 20 using mental strategies. By end of Grade 2, know from memory all sums of two one-digit numbers.

Vocabulary
- difference
- equal
- subtract

Creating Context
Discuss with English Learners some of the verbal clues that may be helpful in understanding word problems. Look at Problem 16 on the student page. Explain that when "Rose quickly opened all 6 gifts . . ." the word *all* tells us there are no more left to open. Find other clues in word problems this week.

Materials
Program Materials
Counters, 20

Additional Materials
- index card, 1 per pair
- Vocabulary Card 9, *difference*
- Vocabulary Card 13, *equal*
- Vocabulary Card 42, *subtract*

Prepare Ahead
Use index cards to create several subtraction problems with an answer of 0 (e.g., 9 − 9 = _____).

1 WARM UP

Prepare
▶ **If 15 students are in the class, all of them would have to leave the room for the classroom to have 0 students. How can I write this in equation form?** 15 − 15 = 0

Just the Facts
Have students applaud if the answer to the question is 0. If the answer is not 0, students should sit still. Use questions such as the following:

▶ **What is 5 minus 3?** Students should sit still.
▶ **What is 4 minus 4?** Students should applaud.
▶ **What is 2 minus 2?** Students should applaud.

2 ENGAGE

Develop: Subtracting to Zero
"Today we are going to work on subtraction problems that equal zero." Follow the instructions on the Activity Card **Subtracting to Zero**. As students complete the activity, be sure to use the Questions to Ask.

Activity Card 4D

Alternative Grouping
Individual: Partner with the student and complete the activity as written.

Progress Monitoring
If... students have trouble translating the formal subtraction problem into a story,

▶ **Then...** remind them to use the Counters or to draw pictures.

Practice
Have students complete **Student Workbook,** pages 12–13. Guide students through the Key Idea example and the Try This exercises.

Interactive Differentiation
Consult the **Teacher Dashboard** for grouping suggestions. You can also use performance on the Engage activity to guide students.

Independent Practice
For additional practice, have students use the Sets Former Tool to model subtracting to 0.

- Have students work in pairs. One student writes a problem that has a difference of 0. The other student models the problem by first placing the marbles on the mat and then dragging them all away. For example, one student writes 3 − 3 = _____. The other student uses the Sets Former Tool to model the problem and the solution, which is 0.
- Tell partners to take turns writing, modeling, and solving problems that subtract to 0.

Supported Practice
For additional support, use the Sets Former Tool to model subtraction to 0.

- Write the following problems on the board: 10 − 10 = _____; 8 − 8 = _____; and 3 − 3 = _____.
- Model each problem with Counters. Talk through the process of counting up and then counting down to zero.
- Have students use the Sets Former Tool to model each problem and solution. Then ask questions such as the following:

▶ **I had 6 grapes. Now I have no grapes. How many grapes did I eat?** 6
How do you know? Answers may vary. Possible answer: Because you had 6 grapes to begin with, you must have eaten 6 grapes to have none left.

▶ **If I subtract a number from itself, will the answer always be 0? How do you know?** Answers may vary. Possible answer: Yes. Because you are taking away the exact amount that is there; there is nothing left.

284 Level D Unit 4 **Subtraction**

3 REFLECT

Think Critically
Review students' answers to the Reflect prompt at the bottom of **Student Workbook,** page 13, and then review the Engage activity.

Discuss the likelihood that all the students would not be picked up at one time.

▶ **Write an equation to record the number of students that remain until all are picked up.** Possible answer: $7 - 3 - 1 - 3 = 0$

▶ **Could more than one equation have been written to show how many students were picked up at a time?** yes

Real-World Application
▶ **Do you have a grocery list in your home?**

▶ **When does an item get written on the grocery list?**

▶ **If a bag of 12 apples was in the refrigerator, how many apples would probably be eaten before writing *apples* on the grocery list?**
Answers may vary. Possible answer: 12

4 ASSESS

Informal Assessment
Use the online or print Student Record, **Assessment,** page 128, to record informal observations.

Subtracting to Zero	
Did the student	
☐ respond accurately?	☐ respond with confidence?
☐ respond quickly?	☐ self-correct?

Additional Practice
For additional practice, have students complete **Practice,** page 79.

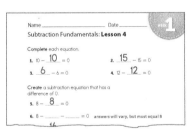
Practice, p. 79

Week 1 • Subtraction Fundamentals

Lesson 4

Key Idea
Any number minus itself always equals 0.
$16 - 16 = 0$

Try This
Find each difference.
1. $6 - 6 = \underline{0}$
2. $10 - 10 = \underline{0}$
3. $3 - 3 = \underline{0}$
4. $15 - 15 = \underline{0}$

Create a subtraction equation that has a difference of 0. **Sample answers are shown.**

5. $10 - \underline{10} = 0$
6. $10 - \underline{3} - \underline{7} = 0$

Practice
Complete each equation.

7. $5 - \underline{5} = 0$
8. $\underline{8} - 8 = 0$
9. $\underline{12} - 12 = 0$
10. $20 - \underline{20} = 0$

12 Level D Unit 4 **Subtraction**

Create a subtraction equation that has a difference of 0. **Sample answers are shown.**

11. $10 - \underline{5} - \underline{5} = 0$
12. $10 - \underline{2} - \underline{8} = 0$
13. $20 - \underline{20} = 0$
14. $20 - \underline{10} - \underline{10} = 0$

Write a subtraction equation for each math story.

15. Rose was celebrating her eighth birthday. She made 7 gift bags. She gave all 7 to her friends. How many bags did she have left over?

    $7 - 7 = 0$

16. Most of Rose's friends arrived with gifts. Rose quickly opened all 6 gifts they brought. How many gifts does she still have to open?

    $6 - 6 = 0$

Reflect
All 7 of Rose's friends left the party by 8:00. Write a subtraction sentence to show how many friends were still at Rose's house after 8:00.

$7 - 7 = 0$

Week 1 **Subtraction Fundamentals** • Lesson 4 13

Student Workbook, pp. 12–13

3 REFLECT

Think Critically

Review students' answers to the Reflect prompt at the bottom of *Student Workbook*, page 15.

Discuss the answer with the group to reinforce Week 1 concepts.

4 ASSESS

Formal Assessment

Students may take the weekly assessment online.

As an alternative, students may complete the weekly test on *Assessment*, pages 49–50. Record progress using the Student Assessment Record, *Assessment*, page 128.

Going Forward

Use the *Teacher Dashboard* to view results of the online assessments, to input the results of print student assessments, and to review progress before making decisions about next steps. Use the weekly test results and observations to determine the next steps for each student.

Retention	
Student displays good grasp of this week's concepts and skills.	Have students use the Sets Former Tool to model subtracting to 0.

Remediation	
Student is still struggling with the week's concepts and skills.	Model 10 − 4. Set out 10 Counters, and then take 4 away. Work with each student to count the remaining 6 Counters. Help students model each subtraction 10 fact with Counters.

Suggestions for Re-Evaluation: If a student has struggled without success for several weeks, use observations and test results to place the student at a level in which he or she can find success and build confidence to move forward.

Name _____ Date _____
Subtraction Fundamentals — WEEK 1

Write the missing numbers.

1. 6, 7, __8__, 9, 10, __11__

2. 26, 27, __28__, 29, __30__, 31

3. What number comes right after 17? __18__

4. What number comes right before 14? __13__

5. 9 − 3 = __6__

6. 14 − 3 = __11__

7. __8__ − 2 = 6

Level D Unit 4 Week 1   49

WEEK 1   Name _____ Date _____
Subtraction Fundamentals

8. 11 − __4__ = 7

9. 8 − 4 = 5 − __1__

10. 12 − 9 = 9 − __6__

11. 15 − __15__ = 0

Write a subtraction equation for this story.

12. Kim saw 7 ducks on a pond. Then the 7 ducks flew away. How many ducks were left on the pond?

    7 − 7 = 0

50   Level D Unit 4 Week 1

Assessment, pp. 49–50

Week 1 **Subtraction Fundamentals** • Lesson 5   **287**

Project Preview

This week, students learned that you can count backward to find the answer to subtraction problems. The project for this unit requires students to extend the knowledge they gained in Find the Math and what they have learned this week. Today students will use subtraction to determine the sale price of an item.

Project-Based Learning

Standards-driven Project-Based Learning is effective in building deep content understanding. Project-Based Learning increases long-term retention of concepts and has been shown to be more effective than traditional instruction. Completing a project to answer an essential question challenges students to apply and demonstrate mastery of concepts and skills by expressing understanding through discussion, research, and presentation.

Essential Question

WHEN would it be useful to write equations outside the classroom?

Project Evaluation Criteria

Review project evaluation criteria with students prior to beginning the project.

Exceeds Expectations
☐ Project result is explained and can be extended.
☐ Project result is explained in context and can be applied to other situations.
☐ Project result is explained using advanced mathematical vocabulary.
☐ Project result is described, and mathematics are used correctly and can be extended.
☐ Project result is explained and extended, and shows advanced knowledge of mathematical concepts and skills.

Meets Expectations
☐ Project result is explained.
☐ Project result is explained in context.
☐ Project result is explained using mathematical vocabulary.
☐ Project result is described, and mathematics are used correctly.
☐ Project result is explained, and shows satisfactory knowledge of mathematical concepts and skills.

Does Not Meet Expectations
☐ Project result is not explained.
☐ Project result is explained, but out of context.
☐ Project result is explained, but mathematical vocabulary is oversimplified.
☐ Project result is described, but mathematics are not used correctly.
☐ Project result is not explained and/or extended, or shows less than satisfactory knowledge of mathematical concepts and skills.

What's the Price?

Objective
Students can use subtraction to find a sale price.

Standard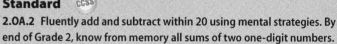
2.OA.2 Fluently add and subtract within 20 using mental strategies. By end of Grade 2, know from memory all sums of two one-digit numbers.

Materials

Program Materials
Number 1–6 Cube

Additional Materials
- index card, 1 per student
- scissors, 1 pair per student

Best Practices
- Provide project directions that are clear and brief.
- Coach, demonstrate, and model.
- Select and provide the appropriate materials.

Introduce

Imagine that you are running a clothing store. What would you like to sell in your store?

▶ **Would you like to sell ties?**

▶ **Would you like to sell shirts?**

- Discuss with students the variety of clothes sold in stores. Help them think of the names of local or national clothing stores.

Explore

▶ **Today you will choose an item to put on sale at your clothing store. Do you know what *on sale* means?** Answers may vary. Possible answers: It means that the clothes are sold for a discount; it means that the price is reduced.

▶ **Have you ever seen a "SALE" sign? Have your parents ever taken you shopping during a big sale?**

- Lead students in a discussion about sale prices and the various ways that stores may discount items: two for the price of one; buy one, get one half off; 25% off; and so on.

▶ **Complete *Student Workbook,* page 16, to help you choose an item and figure out the sale price.**

Wrap Up

- Allow students time to write down which item they want to sell in their store.
- Make sure each student can explain how he or she determined the price of the item when it was put on sale.
- Discuss students' answers to the Reflect prompts at the bottom of *Student Workbook,* page 16.

When students have completed the *Student Workbook* page, distribute an index card and a pair of scissors to each student. Have students make a "sale tag" for the item of clothing that is on sale.

▶ **Write the original price on the tag. Then cross it out and write the sale price.**

If time permits, allow students to draw and color a picture of the clothing item on a sheet of paper. Staple the "sale tag" to the drawing and display the items for sale.

Week 1 • Subtraction Fundamentals

Project
What's the Price?

Choose which item you want to sell.

Item	Regular Price
Shirt	$10
Pants	$15
Skirt	$12
Scarf	$7
Shorts	$8

1. What item did you choose? What is the full price?

   Answers may vary.

2. You have decided to put the item you chose on sale. Roll the Number Cube to see how much money to take off the price. Write the number on the line.

   Answers may vary.

3. Solve an equation that shows the new price.

   Check students' work.

Reflect

What did you need to know before you could write the equation? How did you solve the equation?

Answers may vary; possible answer: I needed to know the regular price of the item and the amount to take off for the sale price. To solve the equation, I counted back the number shown on the Number Cube.

16 Level D Unit 4 **Subtraction**

Student Workbook, p. 16

Teacher Reflect

☐ Was I able to answer questions when students did not understand?

☐ Did students tell or show the steps when they explained how to do something?

☐ Did I explain the directions before the students began their projects?

WEEK 2
Mastering Basic Subtraction Facts

Week at a Glance
This week, students continue with **Number Worlds,** Level D, Subtraction. Students will study different strategies for solving subtraction problems.

Skills Focus
- Recognize and use doubles facts to solve subtraction problems.
- Use doubles facts to help solve near-doubles facts in a subtraction context.
- Choose a strategy to solve subtraction problems.

How Students Learn
Encourage students to use what they know about addition doubles facts to help them construct relationships for subtraction facts. Guide students to use doubles and near-doubles facts to solve subtraction problems by developing related strings of problems. As students expand their repertoire of strategies for subtraction facts beyond doubles and near-doubles, give them many opportunities to play with numbers and relationships between numbers. When relationships are the focus of instruction, students will eventually use these relationships to recall the basic facts.

English Learners ELL
For language support, use the **English Learner Support Guide,** pages 78–79, to preview lesson concepts and teach academic vocabulary.

Math at Home
Give one copy of the Letter to Home, page 20, to each student. Encourage students to share and complete the activity with their caregivers.

Weekly Planner

Lesson	Learning Objectives
1 pages 292–293	Students can use doubles facts to solve subtraction problems.
2 pages 294–295	Students can use near-doubles facts to solve subtraction problems.
3 pages 296–297	Students can use simple known facts to solve problems that involve greater numbers.
4 pages 298–299	Students can write and solve equations to determine the answers to word problems.
5 pages 300–301	**Review and Assess** Students review skills learned this week and complete the weekly assessment and project.
Project pages 302–303	Students can interpret and solve two-step subtraction word problems and write equations to describe each step.

290 Level D Unit 4 **Subtraction**

Key Standard for the Week

Domain: Operations and Algebraic Thinking

Cluster: Add and subtract within 20.

2.OA.2 Fluently add and subtract within 20 using mental strategies. By end of Grade 2, know from memory all sums of two one-digit numbers.

Materials		Technology
Program Materials • **Student Workbook,** pp. 17–19 • **Practice,** p. 80 • Activity Card 4E, **Subtraction Doubles** • Counters	**Additional Materials** math-link cubes*	*Teacher Dashboard*
Program Materials • **Student Workbook,** pp. 20–21 • **Practice,** p. 81 • Activity Card 4F, **Subtraction Near-Doubles**	**Additional Materials** math-link cubes*	*Teacher Dashboard* Sets Former Tool
Program Materials • **Student Workbook,** pp. 22–23 • **Practice,** p. 82 • Activity Card 4G, **Differences around the Room**		*Teacher Dashboard* Sets Former Tool; Base 10 Blocks Tool
Program Materials • **Student Workbook,** pp. 24–25 • **Practice,** p. 83 • Activity Card 4H, **How Many Are Left?** • How Many Are Left? • Counters		*Teacher Dashboard* Building Blocks Word Problems with Tools 1
Program Materials • **Student Workbook,** pp. 26–27 • Weekly Test, **Assessment,** pp. 51–52		Review previous activities.
Program Materials • **Student Workbook,** p. 28 • Number Cards (6–12)		

*Available from McGraw-Hill Education.

WEEK 2
Mastering Basic Subtraction Facts

Find the Math
In this week, students will use doubles facts and near-doubles facts in a subtraction context.

Use the following to begin a guided discussion:

▶ **What other items would you want to share with someone so that you each get half?** Answers may vary. Possible answers: pieces of pizza, toys, markers

Have students complete **Student Workbook,** page 17.

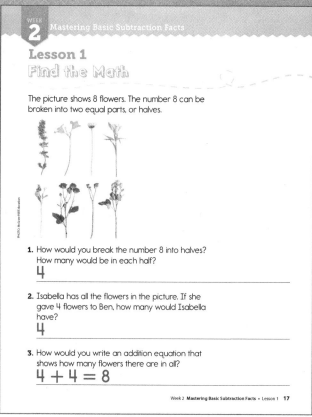

Student Workbook, p. 17

Lesson 1

Objective
Students can use doubles facts to solve subtraction problems.

Standard
2.OA.2 Fluently add and subtract within 20 using mental strategies. By end of Grade 2, know from memory all sums of two one-digit numbers.

Creating Context
In this lesson, the word *double* means "two numbers that are exactly alike." Introduce English Learners to other words that signify a grouping, such as *single, triple,* and *quadruple,* as well as *dozen* and *half dozen*. Have students draw a model of these groupings and label them with the correct word.

Materials
Program Materials
Counters

Additional Materials
math-link cubes, 40 per student

1 WARM UP

Prepare
▶ **Write an addition doubles fact using 3.** $3 + 3 = 6$

▶ **Use this addition fact to write a subtraction fact. This is a subtraction doubles fact. If you take away half of the first number, the other half is the answer.** $6 - 3 = 3$

▶ **Write an addition doubles fact using 8.** $8 + 8 = 16$

▶ **Write a subtraction doubles fact using 8.** $16 - 8 = 8$

▶ **Can you write a different doubles subtraction fact using 8?** Possible answer: $8 - 4 = 4$ (Prompt students by asking them what number is half of 8, if necessary.)

2 ENGAGE

Develop: Subtraction Doubles
"Today we are going to create subtraction doubles facts." Follow the instructions on the Activity Card **Subtraction Doubles**. As students complete the activity, be sure to use the Questions to Ask.

Alternative Grouping
Individual: Partner with the student and complete the activity as written.

Activity Card 4E

Progress Monitoring
If... students are not selecting even numbers when creating halves,	▶ Then... review the sums of doubles facts until they recognize the pattern.

Practice
Have students complete **Student Workbook,** pages 18–19. Guide students through the Key Idea example and the Try This exercises.

Interactive Differentiation
Consult the **Teacher Dashboard** for grouping suggestions. You can also use performance on the Engage activity to guide students.

292 Level D Unit 4 **Subtraction**

Independent Practice

For additional practice making doubles, have students work with a partner.

- Give each pair 20 Counters. Tell students to try to model subtraction doubles facts using 11, 12, 13, 14, 15, 16, 17, 18, 19, and 20 Counters.
- Have students make a list of numbers that they cannot divide into halves.

Supported Practice

For additional support, use Counters to model a subtraction doubles fact.

- Count out 12 Counters. Divide them into two piles of 6 Counters each.
- Write $6 + 6 = 12$ and $12 - 6 = 6$ on the board. Then count 13 Counters.
- ▶ **Can I divide these Counters into two halves? Why or why not?**
  No; there is an odd number of Counters, so one set will have 6 Counters and the other will have 7 Counters; $6 + 7$ is not a doubles fact.
- Give pairs of students 20 Counters. Tell them to try to model subtraction doubles facts using 14, 15, 16, 17, 18, 19, and 20 Counters.

3 REFLECT

Think Critically

Review students' answers to the Reflect prompt at the bottom of **Student Workbook,** page 19, and then review the Engage activity.

Connect these problems to the **Subtraction Doubles** activity completed in class.

▶ **Why can't you write two different equations for each of the problems?**

Real-World Application

Write *10 dollars, 5 dollars, 50 cents,* and *10 cents* on the board. Use play money to show amounts that are half the value of each amount.

4 ASSESS

Informal Assessment

Use the online or print Student Record, **Assessment,** page 128, to record informal observations.

Subtraction Doubles	
Did the student	
☐ make important observations?	☐ provide insightful answers?
☐ extend or generalize learning?	☐ pose insightful questions?

Additional Practice

For additional practice, have students complete **Practice,** page 80.

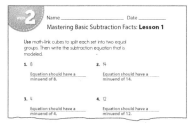

Practice, p. 80

Week 2 • Mastering Basic Subtraction Facts

Lesson 1

Key Idea
In doubles facts for subtraction, one of the doubles numbers is the difference.
$5 + 5 = 10 \qquad 8 + 8 = 16$
$10 - 5 = 5 \qquad 16 - 8 = 8$

Try This
Find the sum of each doubles fact. Then write a subtraction equation using the doubles fact.

1. $1 + 1 = \underline{\ 2\ }$
   $\underline{\ 2 - 1 = 1\ }$
2. $3 + 3 = \underline{\ 6\ }$
   $\underline{\ 6 - 3 = 3\ }$
3. $6 + 6 = \underline{\ 12\ }$
   $\underline{\ 12 - 6 = 6\ }$
4. $7 + 7 = \underline{\ 14\ }$
   $\underline{\ 14 - 7 = 7\ }$

Use math-link cubes to split each set into two equal groups. Then write the subtraction equation that is modeled.

5. 4 $\underline{\ 4 - 2 = 2\ }$
6. 8 $\underline{\ 8 - 4 = 4\ }$
7. 10 $\underline{\ 10 - 5 = 5\ }$
8. 20 $\underline{\ 20 - 10 = 10\ }$

18 Level D Unit 4 Subtraction

Practice

Use math-link cubes to split each set into two equal groups. Then write the subtraction equation that is modeled.

9. 6 $\underline{\ 6 - 3 = 3\ }$
10. 16 $\underline{\ 16 - 8 = 8\ }$
11. 10 $\underline{\ 10 - 5 = 5\ }$
12. 18 $\underline{\ 18 - 9 = 9\ }$

Write an addition equation and a subtraction equation using doubles facts to describe the following situations.

Half is 6.
13. Addition equation: $\underline{\ 6 + 6 = 12\ }$
14. Subtraction equation: $\underline{\ 12 - 6 = 6\ }$

Half is 9.
15. Addition equation: $\underline{\ 9 + 9 = 18\ }$
16. Subtraction equation: $\underline{\ 18 - 9 = 9\ }$

Reflect

In Problems 9–12, does each problem have only one correct answer? Explain.

Yes; the halves have to be equal, so there is only one correct answer.

Week 2 Mastering Basic Subtraction Facts • Lesson 1 19

Student Workbook, pp. 18–19

WEEK 2
Mastering Basic Subtraction Facts

Lesson 2

Objective
Students can use near-doubles facts to solve subtraction problems.

Standard
2.OA.2 Fluently add and subtract within 20 using mental strategies. By end of Grade 2, know from memory all sums of two one-digit numbers.

Creating Context
One difficulty that English Learners encounter in the classroom is the use of instructions and directions for completing assignments. Tell students that if they do not understand the instructions, they should ask someone for help before they begin. Review the written directions in this lesson by asking students to describe what the directions mean.

Materials
Additional Materials
math-link cubes, 40 per student

1 WARM UP

Prepare

▶ **Yesterday we used addition doubles facts to create subtraction doubles facts. Today we are going to work with near-doubles facts. Does anyone remember what a near-doubles fact is?** a fact that is close to a doubles fact, but one of the numbers is one number bigger or smaller

▶ **Write two addition near-doubles facts using 3.** Possible answers: $3 + 4 = 7$; $3 + 2 = 5$

▶ **Use these addition facts to write subtraction near-doubles facts. The rule is: If you take away one of the near-doubles numbers, the other number is the answer.** Possible answers: $7 - 3 = 4$; $5 - 2 = 3$

▶ **Write an addition near-doubles fact using 5.**

▶ **Write a subtraction near-doubles fact using 5.**

Just the Facts

Have students chorally call out the correct answer. Use questions such as the following:

▶ **What is 3 plus 1?** 4
▶ **What is 4 minus 1?** 3
▶ **What is 4 minus 3?** 1

2 ENGAGE

Develop: Subtraction Near-Doubles

"Today we are going to create subtraction near-doubles facts." Follow the instructions on the Activity Card **Subtraction Near-Doubles**. As students complete the activity, be sure to use the Questions to Ask.

Alternative Grouping

Individual: Partner with the student and complete the activity as written.

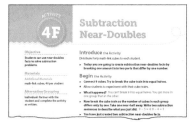

Activity Card 4F

Progress Monitoring

| If... students are having trouble, | Then... spend time explaining and reviewing the meaning of the word *difference*. |

Practice

Have students complete **Student Workbook,** pages 20–21. Guide students through the Key Idea example and the Try This exercises.

Interactive Differentiation

Consult the **Teacher Dashboard** for grouping suggestions. You can also use performance on the Engage activity to guide students.

Independent Practice

For additional practice, have students use the Sets Former Tool to model a near-doubles subtraction fact, such as $9 - 5 = 4$. Tell them to exchange the difference for the subtracted number, modeling the corresponding fact $(9 - 4 = 5)$. Repeat this activity with other near-doubles subtraction facts. Provide students with a list of near-doubles facts, such as the following:

- $11 - 6 = 5$ $11 - 5 = 6$
- $7 - 4 = 3$ $7 - 3 = 4$
- $13 - 7 = 6$ $13 - 6 = 7$

Supported Practice

For additional support, use math-link cubes with students to model near-doubles subtraction facts.

- Give each student 20 cubes. Tell students to model $11 - 5 = 6$. Students should connect 11 math-link cubes and then detach 5, showing that 6 are left.
- Have students model another near-doubles fact with the same 11 cubes. Students should reconnect the math-link cubes before detaching 6, showing that 5 are left.
- ▶ **Which number is the difference in a subtraction equation?** the number that is remaining; the number after the equal sign
- ▶ **Why do you think these are called *near-doubles facts*?** Answers may vary. Possible answer: The two addends are only one number apart. If the lesser number were one greater, or if the greater number were one less, then the two numbers would be the same. Two numbers that are the same are doubles, so these are nearly doubles, or near doubles.
- Have students model a variety of near-doubles subtraction facts with math-link cubes.

REFLECT

Think Critically

Review students' answers to the Reflect prompt at the bottom of **Student Workbook,** page 21, and then review the Engage activity.

Connect these problems to the **Subtraction Near-Doubles** activity completed in class. Ask students to write another equation for each of the problems.

▶ **What does this tell you about subtraction?**

Real-World Application

▶ **Have you ever had something that you wanted to share with a friend but you could not split it evenly between the two of you? Tell us about a time like this. Why could you not share evenly?**

▶ **What did you decide to do?**

ASSESS

Informal Assessment

Use the online or print Student Record, **Assessment,** page 128, to record informal observations.

Subtraction Near-Doubles	
Did the student	
☐ make important observations?	☐ provide insightful answers?
☐ extend or generalize learning?	☐ pose insightful questions?

Additional Practice

For additional practice, have students complete **Practice,** page 81.

Practice, p. 81

Week 2 • Mastering Basic Subtraction Facts

Lesson 2

Key Idea
In near-doubles facts for subtraction, one of the near-doubles numbers is the difference.
$5 + 4 = 9$ $9 - 5 = 4$ $9 - 4 = 5$

Try This
Find the sum of each near-doubles fact. Then write two subtraction equations using the near-doubles fact.

1. $1 + 2 = \underline{\ 3\ }$
   $\underline{\ 3 - 1 = 2\ }$
   $\underline{\ 3 - 2 = 1\ }$

2. $3 + 4 = \underline{\ 7\ }$
   $\underline{\ 7 - 4 = 3\ }$
   $\underline{\ 7 - 3 = 4\ }$

3. $7 + 8 = \underline{\ 15\ }$
   $\underline{\ 15 - 8 = 7\ }$
   $\underline{\ 15 - 7 = 8\ }$

4. $9 + 10 = \underline{\ 19\ }$
   $\underline{\ 19 - 10 = 9\ }$
   $\underline{\ 19 - 9 = 10\ }$

Use math-link cubes to split each set into two groups that differ by 1. Then write the subtraction equation that is modeled. **Sample answers are shown.**

5. 5 $\underline{\ 5 - 3 = 2\ }$

6. 11 $\underline{\ 11 - 6 = 5\ }$

20 Level D Unit 4 Subtraction

Practice

Use math-link cubes to split each set into two groups. One group should have one more cube than the other group. Then write a near-doubles fact describing what you did. **Sample answers are shown.**

7. 15 $\underline{\ 15 - 8 = 7\ }$

8. 13 $\underline{\ 13 - 6 = 7\ }$

9. 7 $\underline{\ 7 - 3 = 4\ }$

10. 9 $\underline{\ 9 - 4 = 5\ }$

11. 17 $\underline{\ 17 - 8 = 9\ }$

12. 19 $\underline{\ 19 - 10 = 9\ }$

13. 3 $\underline{\ 3 - 1 = 2\ }$

14. 21 $\underline{\ 21 - 10 = 11\ }$

Reflect

In Problems 5–14, does each problem have only one correct answer? Explain.

No; they are not halves, so you can have two answers.

Week 2 Mastering Basic Subtraction Facts • Lesson 2   21

Student Workbook, pp. 20–21

WEEK 2
Mastering Basic Subtraction Facts

Lesson 3

Objective
Students can use simple known facts to solve problems that involve greater numbers.

Standard
2.OA.2 Fluently add and subtract within 20 using mental strategies. By end of Grade 2, know from memory all sums of two one-digit numbers.

Creating Context
English Learners may find the academic use of some common vocabulary words confusing, such as the word *difference* in this lesson. Recommend to students that when they hear a word they are unsure about, they should try to assess the context of the situation they are studying. Here the word *difference* refers to the answer to a subtraction problem.

Materials
No materials needed.

Prepare Ahead
Write two columns of numbers on the board, and label the columns *A* and *B*. In column A, write the following numbers: 25, 26, 27, 28, 29, 35, 36, 37, 38, 39. In column B, write the following numbers: 10, 11, 12, 13, 14, 20, 21, 22, 23, 24.

1 WARM UP

Prepare

Write 48 − 44 = _____ on the board.

▶ **What is the answer to this problem?** 4 **How did you figure it out?** Answers may vary.

▶ Today I am going to show you another strategy you can use to solve problems with big numbers. We can think of 48 as 4 tens and 8 ones. We can also write this number as 40 + 8. We can think of 44 as 4 tens and 4 ones. We can also write this number as 40 + 4.

▶ To subtract these numbers, we can subtract the tens first. **What is 40 − 40?** 0

▶ Then we can subtract the ones. **What is 8 − 4?** 4

Demonstrate how to use this strategy to solve two more problems: 19 − 15 and 36 − 21.

Just the Facts

Have students clap out the number of the correct answer. Clap once as you say each word or number to establish a rhythm and tempo. Use doubles and near-doubles subtraction prompts such as the following:

▶ **Ten minus 5 is _____.** Students should clap 5 times.
▶ **Ten minus 4 is _____.** Students should clap 6 times.
▶ **Ten minus 6 is _____.** Students should clap 4 times.

2 ENGAGE

Develop: Differences around the Room

"Today we are going to create and solve subtraction problems with two-digit numbers." Follow the instructions on the Activity Card **Differences around the Room.** As students complete the activity, be sure to use the Questions to Ask.

Activity Card 4G

Alternative Grouping

Individual: Have the student choose numbers, and help him or her create and solve subtraction problems using those numbers.

Progress Monitoring	
If… a student is reluctant to participate in an activity,	**Then…** determine whether the reason is a social issue or a skills issue. Work on inclusive strategies if a student is not participating. Work individually on developing skills if the student cannot complete the task successfully.

Practice

Have students complete **Student Workbook,** pages 22–23. Guide students through the Key Idea example and the Try This exercises.

Interactive Differentiation

Consult the **Teacher Dashboard** for grouping suggestions. You can also use performance on the Engage activity to guide students.

Independent Practice

For additional practice with subtraction, have students use the Sets Former Tool to model subtraction problems. Provide students with a list of two-digit subtraction problems. The difficulty level of the problems should increase as students work down the list. For example, an early problem could be 17 − 6 = 11, a problem near the middle of the list could be 41 − 21 = 20, and a problem near the end could be 67 − 22 = 45. These problems should not involve regrouping.

Supported Practice

For additional support, use the Base 10 Blocks Tool with students to model two-digit subtraction.

• Briefly review how to use the tool and then tell students to model 15 − 3.
▶ **How many ones blocks should you drag to the bottom mat?** 3
▶ **How many tens flats and ones blocks are left?** There are 1 tens flat and 2 ones blocks left.
▶ **What number does that make?** 12

• Have students use the Base 10 Blocks Tool to model two-digit subtraction problems that become increasingly difficult. These problems should not involve regrouping.

3 REFLECT

Think Critically

Review students' answers to the Reflect prompt at the bottom of **Student Workbook,** page 23, and then review the Engage activity.

Discuss with students that different strategies are used for different problems.

▶ **What made you decide on the equations that you wrote?**

Real-World Application

▶ **When you have to complete a task that you think is too difficult, what do you do?**

Guide students to remember times in their lives when they have separated a task into smaller, more manageable tasks to complete.

4 ASSESS

Informal Assessment

Use the online or print Student Record, **Assessment,** page 128, to record informal observations.

Differences around the Room

Did the student
- ☐ pay attention to the contributions of others?
- ☐ contribute information and ideas?
- ☐ improve on a strategy?
- ☐ reflect on and check accuracy of work?

Additional Practice

For additional practice, have students complete **Practice,** page 82.

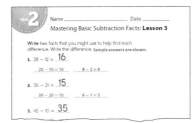

Practice, p. 82

Week 2 • Mastering Basic Subtraction Facts

Lesson 3

Key Idea
Use facts you know to help you solve facts you do not know.
$24 - 21 =$
Think of 24 as 20 and 4.
Think of 21 as 20 and 1.
Subtract $20 - 20 = 0$.   Subtract $4 - 1 = 3$.
So, $24 - 21 = 3$.

Try This
Use facts you know to help you find each difference. Show your work.

1. $15 - 12 = \underline{3}$ 2. $25 - 23 = \underline{2}$

Find each difference. Circle the two problems that have the same difference.

3. ⟨$35 - 15$⟩ $= \underline{20}$ 4. ⟨$44 - 42$⟩ $= \underline{2}$
   $40 - 10 = \underline{30}$ $64 - 60 = \underline{4}$
   ⟨$30 - 10$⟩ $= \underline{20}$ ⟨$4 - 2$⟩ $= \underline{2}$

22 Level D Unit 4 Subtraction

Practice
Write facts that you might use to help find each difference. Then write the difference. Sample answers are shown.

5. $26 - 21 = \underline{5}$ 6. $24 - 13 = \underline{11}$
   $20 - 20 = 0$ $6 - 1 = 5$ $20 - 10 = 10$ $4 - 3 = 1$

Write three subtraction problems that have the given answer. Sample answers are shown.

7. 6 $10 - 4 = 6$ $30 - 24 = 6$ $69 - 63 = 6$

8. 8 $9 - 1 = 8$ $38 - 30 = 8$ $59 - 51 = 8$

9. 7 $8 - 1 = 7$ $12 - 5 = 7$ $10 - 3 = 7$

10. 10 $20 - 10 = 10$ $12 - 2 = 10$ $15 - 5 = 10$

Reflect
Do the problems below have the same difference? Explain how to use facts you know to find whether the differences are the same.

$27 - 23 = \underline{4}$ $29 - 25 = \underline{4}$

Yes; use $20 - 20 = 0$ for each problem so that you are only finding the differences to $7 - 3$ and $9 - 5$. The differences are both the same: 4.

Week 2 Mastering Basic Subtraction Facts • Lesson 3   23

Student Workbook, pp. 22–23

WEEK 2
Mastering Basic Subtraction Facts

Lesson 4

Objective
Students can write and solve equations to determine the answers to word problems.

Standard
2.OA.2 Fluently add and subtract within 20 using mental strategies. By end of Grade 2, know from memory all sums of two one-digit numbers.

Creating Context
The key idea for this lesson is to be sure to answer the question that is being asked. Remind students to always go back and check the question before moving on to the next problem.

Materials
Program Materials
- How Many Are Left? 1 per pair
- Counters, 20 per student

1 WARM UP

Prepare

▶ **The Jones family had 2 dogs, 7 fish, and 1 parakeet. They gave 2 fish away. How many pets do they have left?**

Write the word problem on the board.

- Ask students to write a number story for the problem and to include labels in their story so we know what each number refers to. 2 dogs + 7 fish + 1 parakeet = 10 pets; 10 pets − 2 fish = 8 pets

Guide students through each step in this process by asking problem-solving questions, such as those included on Activity Card 4H.

Allow several students to answer these questions to ensure that all students understand the process.

Just the Facts

Have students hold up their fingers to show the answer. Use questions such as the following:

▶ If Jack has 5 stickers and gives 4 away, how many stickers does Jack have now? 1

▶ I had 4 crayons, but I lost 2 of them. How many do I have left? 2

▶ Today 5 children went to the park, and then 1 child went home. How many children were left? 4

2 ENGAGE

Develop: How Many Are Left?

"Today we are going to solve word problems that will require you to use addition and subtraction to find the answer." Follow the instructions on the Activity Card **How Many Are Left?** As students complete the activity, be sure to use the Questions to Ask.

Activity Card 4H

Alternative Grouping

Individual: Help the student create and solve subtraction problems, as needed.

Progress Monitoring

| **If...** students have trouble solving the word problems and writing the corresponding equations, | ▶ **Then...** have them use Counters to model the problems. Guide students through the problem-solving process. |

Practice

Have students complete **Student Workbook,** pages 24–25. Guide students through the Key Idea example and the Try This exercises.

Interactive Differentiation

Consult the **Teacher Dashboard** for grouping suggestions. You can also use performance on the Engage activity to guide students.

Independent Practice

For additional practice with solving and writing equations for word problems, students should use Word Problems with Tools 1: Find Result or Change. Students use tools provided to solve word problems (totals to 10).

Supported Practice

For additional support, use Word Problems with Tools 1: Find Result or Change with students. Have students take turns reading their word problems aloud. Ask questions such as the following:

▶ What does the story tell you?
▶ What numbers are in the story?
▶ What question does the problem ask?

3 REFLECT

Think Critically

Review students' answers to the Reflect prompt at the bottom of **Student Workbook,** page 25, and then review the Engage activity.

Have students share the different math stories they created. Challenge the rest of the class to solve them as they are being read.

Real-World Application

▶ Writers create their own stories. What is the difference between a math story and a story that you might write for science or language arts?

4 ASSESS

Informal Assessment

Use the online or print Student Record, **Assessment,** page 128, to record informal observations.

How Many Are Left?	
Did the student	
☐ make important observations?	☐ provide insightful answers?
☐ extend or generalize learning?	☐ pose insightful questions?

Additional Practice

For additional practice, have students complete **Practice,** page 83.

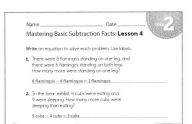

Practice, p. 83

Week 2 • Mastering Basic Subtraction Facts

Lesson 4

> **Key Idea**
> When solving a word problem, be sure to answer the question that is asked.

Try This

Write an equation to solve each math story. Use labels. Draw a picture if you need help.

1. Mrs. Fielder took her children to the zoo. She has 2 boys and 4 girls. How many more girls than boys does she have?

   $4 - 2 = 2$ more girls

2. Seven family members went to the zoo. Two of them saw the dolphins first, and the others saw the bears first. How many family members saw the bears first?

   $7 - 2 = 5$ family members saw the bears first

3. There were 6 different bear exhibits to see. In 3 of the exhibits, the bears were eating, and in 1 exhibit, they were playing. In the rest of the exhibits, the bears were sleeping. How many exhibits had sleeping bears?

   $6 - 3 - 1 = 2$ exhibits had sleeping bears

Practice

Write an equation to solve each problem. Use labels.

4. The gorilla exhibit had 12 gorillas. When the Fielder family arrived to see the gorillas, 7 were out playing. The rest of the gorillas were hiding in a cave. Two came outside. How many gorillas were still in the cave?

   $12 - 7 - 2 = 3$ gorillas were in the cave

5. Eight penguins were swimming, and 3 were waddling around on the ice. How many more penguins were swimming than waddling around on the ice?

   $8 - 3 = 5$ more penguins were swimming

Reflect

Write a math story about going to the zoo. Then write a subtraction equation that fits your story.

Answers will vary.

Student Workbook, pp. 24–25

WEEK 2
Mastering Basic Subtraction Facts

Lesson 5 Review

Objective
Students review skills learned this week and complete the weekly assessment and project.

Standard
2.OA.2 Fluently add and subtract within 20 using mental strategies. By end of Grade 2, know from memory all sums of two one-digit numbers.

Vocabulary
Review vocabulary introduced during the week.

Creating Context
When having students do mental math, suggest that English Learners draw what they are picturing so that you can check for understanding.

1 WARM UP

Prepare
- Name any strategy that you have used in the subtraction unit.
- When do you prefer to use each strategy?

2 ENGAGE

Practice
Have students complete *Student Workbook,* pages 26–27.

Week 2 • Mastering Basic Subtraction Facts

Lesson 5 Review

This week you looked at different subtraction strategies.

Lesson 1 Find the sum of each doubles fact. Then write a subtraction equation using the doubles fact.

1. $4 + 4 = \underline{8}$
   $8 - 4 = 4$

2. $11 + 11 = \underline{22}$
   $22 - 11 = 11$

3. $7 + 7 = \underline{14}$
   $14 - 7 = 7$

4. $8 + 8 = \underline{16}$
   $16 - 8 = 8$

Lesson 2 Find the sum of each near-doubles fact. Then write a subtraction equation using the near-doubles fact.

5. $6 + 5 = \underline{11}$
   $11 - 5 = 6$ or $11 - 6 = 5$

6. $10 + 9 = \underline{19}$
   $19 - 10 = 9$ or $19 - 9 = 10$

Lesson 3 Use facts you know to help you find each difference. Show your work.

7. $12 - 7 = \underline{5}$

8. $5 - 1 = \underline{4}$

26 Level D Unit 4 **Subtraction**

Lesson 4 Write an equation about each sentence.

9. The Guzman family had 4 hamsters, 2 cats, and 12 fish. They gave away 3 of the hamsters. How many pets do they have left?

   4 hamsters + 2 cats + 12 fish = 18 pets

   18 pets − 3 hamsters = 15 pets left

10. Danita runs a flower shop. She started the morning with 8 roses, 10 daisies, and 6 carnations. She sold 4 flowers. How many flowers does she have left?

    8 roses + 10 daisies + 6 carnations = 24 flowers

    24 flowers − 4 flowers = 20 flowers left

Reflect

Write and solve a math story that has 8 total objects and two amounts subtracted from the total.

Answers will vary. Possible answer: I had 8 stuffed animals. I gave away 2 and sold 3 more. 8 stuffed animals − 2 stuffed animals − 3 stuffed animals = 3 stuffed animals

Week 2 **Mastering Basic Subtraction Facts** • Lesson 5 27

Student Workbook, pp. 26–27

3 REFLECT

Think Critically

Review students' answers to the Reflect prompt at the bottom of *Student Workbook,* page 27.

Discuss the answer with the group to reinforce Week 2 concepts.

4 ASSESS

Formal Assessment

Students may take the weekly assessment online.

As an alternative, students may complete the weekly test on *Assessment,* pages 51–52. Record progress using the Student Assessment Record, *Assessment,* page 128.

Going Forward

Use the **Teacher Dashboard** to view results of the online assessments, to input the results of print student assessments, and to review progress before making decisions about next steps. Use the weekly test results and observations to determine the next steps for each student.

Retention	
Student displays good grasp of this week's concepts and skills.	**Building Blocks** Use the digital Building Blocks activity Word Problems with Tools 1: Find Result or Change with students.
Remediation	
Student is still struggling with the week's concepts and skills.	Use the Base-Ten Tool with students to model two-digit subtraction. • Briefly review how to use the tool and tell students to model 19 − 12. ▸ **How many tens blocks should you drag to the bottom mat?** 1 ▸ **How many ones blocks should you drag to the bottom mat?** 2 ▸ **How many tens flats and ones blocks are left?** There are no tens flats and 7 ones blocks left. ▸ **What number does that make?** 7 • Have students use the Base-Ten Tool to model additional two-digit subtraction problems. These problems should not involve regrouping.

Suggestions for Re-Evaluation: If a student has struggled without success for several weeks, use observations and test results to place the student at a level in which he or she can find success and build confidence to move forward.

Name _____ Date _____ WEEK 2
Mastering Basic Subtraction Facts

1. Circle the answer that shows how to divide 6 into halves.

   2 + 4   ⟨3 + 3⟩   1 + 5

2. 5 + 5 = __10__

3. 16 − 8 = __8__

4. Show how to divide 18 into halves.
   __9 + 9__

5. 7 + 7 = __14__

6. 5 + 6 = __11__

WEEK 2   Name _____ Date _____
Mastering Basic Subtraction Facts

7. 13 − 7 = __6__

8. __17__ − 9 = 8

9. Circle the answer that is the same as 25 − 20.

   40 − 20   25 − 25   ⟨40 − 35⟩

10. Circle the answer that is the same as 54 − 51.

    ⟨60 − 57⟩   25 − 20   45 − 33

11. Circle the two answers that will help you solve 27 − 14.

    ⟨7 − 4⟩   4 − 2   ⟨20 − 10⟩

12. Mark saw 8 cows and 3 sheep at a farm. Write an equation to show how many more cows he saw.

    8 cows − 3 sheep =
    5 more cows

Assessment, pp. 51–52

Project Preview

This week, students worked with doubles and near-doubles facts and wrote equations to describe two-step solutions. The project for this unit requires students to extend the knowledge they gained in Find the Math and what they have learned this week. For this week's project, students will use subtraction to determine how much inventory they have in their stores.

Project-Based Learning

Standards-driven Project-Based Learning is effective in building deep content understanding. Project-Based Learning increases long-term retention of concepts and has been shown to be more effective than traditional instruction. Completing a project to answer an essential question challenges students to apply and demonstrate mastery of concepts and skills by expressing understanding through discussion, research, and presentation.

Essential Question

WHEN would it be useful to write equations outside the classroom?

Project Evaluation Criteria

Review project evaluation criteria with students prior to beginning the project.

Exceeds Expectations
- ☐ Project result is explained and can be extended.
- ☐ Project result is explained in context and can be applied to other situations.
- ☐ Project result is explained using advanced mathematical vocabulary.
- ☐ Project result is described, and mathematics are used correctly and can be extended.
- ☐ Project result is explained and extended, and shows advanced knowledge of mathematical concepts and skills.

Meets Expectations
- ☐ Project result is explained.
- ☐ Project result is explained in context.
- ☐ Project result is explained using mathematical vocabulary.
- ☐ Project result is described, and mathematics are used correctly.
- ☐ Project result is explained, and shows satisfactory knowledge of mathematical concepts and skills.

Does Not Meet Expectations
- ☐ Project result is not explained.
- ☐ Project result is explained, but out of context.
- ☐ Project result is explained, but mathematical vocabulary is oversimplified.
- ☐ Project result is described, but mathematics are not used correctly.
- ☐ Project result is not explained and/or extended, or shows less than satisfactory knowledge of mathematical concepts and skills.

How Many Can You Sell?

Objective
Students can interpret and solve two-step subtraction word problems and write equations to describe each step.

Standard
2.OA.2 Fluently add and subtract within 20 using mental strategies. By end of Grade 2, know from memory all sums of two one-digit numbers.

Materials
Program Materials
Number Cards (6–12)

Prepare Ahead
Set aside Number Cards 6–12 from the deck for use in the project.

Best Practices
- Check for student understanding frequently.
- Select and provide the appropriate materials.
- Create adequate time lines for each project.

Introduce

Your clothing store sells pants for $15. There was a roof leak in your storage room, and 6 pairs of pants were stained.

- ▶ **What happens when you get a stain on your clothes?** Answers may vary. Possible answer: The clothes can't be used.
- • Point out that customers buying new clothes do not expect the clothes to have been treated for stains and then washed. That means that the store will not be able to sell the pants stained by the leaky roof.
- ▶ **How will you figure out how many pairs of pants you can still sell?** I will subtract 6 from the total number of pants.

Explore

- ▶ **Today you will write an equation to figure out how many pairs of pants you can still sell.**
- • Organize students into small groups. Give each group a set of Number Cards (6–12). Tell students to place the cards facedown in the center of the table or desk.
- ▶ **Use the Number Cards to figure out how many pants you have in your store.**
- • Have students take turns drawing a Number Card. Explain that the number on the card tells them how many pairs of pants were in the storage room.
- ▶ **Write the number on your worksheet. Then put all the cards back in the pile and shuffle them.**
- • Have students take turns drawing a second Number Card. Explain that this number tells them how many pairs of pants were on the racks in the front of the store.
- ▶ **Write the number down on your worksheet. Then write an equation that shows the total pairs of pants in the store.**
- ▶ **Complete *Student Workbook*, page 28, to find out how many pairs of pants you can sell.**

Wrap Up

- • Allow students time to write down their equations.
- • Solving some of the subtraction equations may require students to regroup. If students are not ready to regroup, have them count back 6 on a number line to solve the problem.
- • Make sure each student can explain how he or she determined each answer.
- • Discuss students' answers to the Reflect prompts at the bottom of ***Student Workbook*,** page 28.

Have students write a "damage report" for the 6 pairs of stained pants. Encourage students to be creative. Display student work if space permits.

If time permits, allow each student to check another student's equations.

Week 2 • Mastering Basic Subtraction Facts

Project
How Many Can You Sell?

Figure out how many pairs of pants you can still sell.

1. How many pairs of pants were in the storage room?
   Answers may vary.

2. How many pairs of pants were on the racks?
   Answers may vary.

3. Write and solve the addition equation that shows how many pairs of pants were in the store.
   Check students' work.

4. Write and solve the equation that shows how many pairs of pants you can sell if 6 pairs are stained.
   Answers may vary. The subtraction equation should have 6 as the subtrahend.

5. Draw a picture that represents your equation.
   Check students' work.

Reflect
What did you need to know? Explain if you used doubles or near-doubles facts to find the answer.

I needed to know how many pairs of pants I had in the store. Answers may vary; possible answer: Yes; I drew a 7 and then an 8, and 7 + 8 is a near-doubles fact. Because 7 + 7 = 14, I knew that 7 + 8 = 15.

28 Level D Unit 4 Subtraction

Student Workbook, p. 28

Teacher Reflect

- ☐ Did I adequately explain and discuss the Reflect questions with students?
- ☐ Did students use their time wisely and effectively?
- ☐ Did students finish all of the steps required by the project?

WEEK 3
Solving Subtraction Problems

Week at a Glance
This week, students continue **Number Worlds,** Level D, Subtraction, by exploring equations that involve subtraction.

Skills Focus
- Subtract a series of numbers and write equations to describe these operations.
- Understand the relationship between addition and subtraction.
- Solve comparison subtraction problems.
- Interpret and solve subtraction problems.

How Students Learn
Using real-life examples is an excellent way to help students make sense of new concepts. When students can practice using subtraction by removing objects or decreasing steps on a number line, the words and concept become more comprehensible.

English Learners ELL
For language support, use the **English Learner Support Guide,** pages 80–81, to preview lesson concepts and teach academic vocabulary.

Math at Home
Give one copy of the Letter to Home, page 21, to each student. Encourage students to share and complete the activity with their caregivers.

Weekly Planner

Lesson	Learning Objectives
1 pages 306–307	Students can perform a series of subtraction operations and can write equations to describe them.
2 pages 308–309	Students can solve subtraction problems and related addition problems.
3 pages 310–311	Students can solve subtraction comparison problems.
4 pages 312–313	Students can interpret and solve word problems and can write number stories to describe them.
5 pages 314–315	**Review and Assess** Students review skills learned this week and complete the weekly assessment and project.
Project pages 316–317	Students can use subtraction to solve problems.

304 Level D Unit 4 **Subtraction**

Key Standard for the Week

Domain: Number and Operations in Base Ten

Cluster: Use place value understanding and properties of operations to add and subtract.

2.NBT.5 Fluently add and subtract within 100 using strategies based on place value, properties of operations, and/or the relationship between addition and subtraction.

Materials		Technology
Program Materials • *Student Workbook,* pp. 29–31 • *Practice,* p. 84 • Activity Card 4I, **Going Fishing** • Fish Pond • Fish Pond Record Form • Number 1–6 Cube • Spinners		*Teacher Dashboard* Sets Former Tool
Program Materials • *Student Workbook,* pp. 32–33 • *Practice,* p. 85 • Activity Card 4J, **Break It and Make It**	Additional Materials math-link cubes*	*Teacher Dashboard* Sets Former Tool
Program Materials • *Student Workbook,* pp. 34–35 • *Practice,* p. 86 • Activity Card 4K, **How Many More?** • Counters	Additional Materials brown paper bags	*Teacher Dashboard* Building Blocks Word Problems with Tools 2
Program Materials • *Student Workbook,* pp. 36–37 • *Practice,* p. 87 • Activity Card 4L, **Subtraction Action** • Counters		*Teacher Dashboard*
Program Materials • *Student Workbook,* pp. 38–39 • Weekly Test, *Assessment,* pp. 53–54		Review previous activities.
Program Materials • *Student Workbook,* p. 40 • Number 1–6 Cube • Counters		

*Available from McGraw-Hill Education.

WEEK 3
Solving Subtraction Problems

Find the Math

Have students think about how they might use subtraction to solve real-world problems.

Use the following to begin a guided discussion:

▶ **When might you use subtraction in a store?** Answers may vary; possible answers: to see how much change you get after you pay for something; to see what the sale price of an item is

Have students complete *Student Workbook*, page 29.

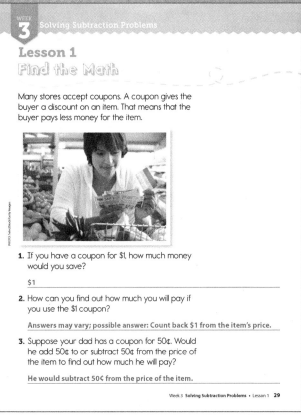

Student Workbook, p. 29

Lesson 1

Objective
Students can perform a series of subtraction operations and can write equations to describe them.

Standard
2.NBT.5 Fluently add and subtract within 100 using strategies based on place value, properties of operations, and/or the relationship between addition and subtraction.

Creating Context
Help English Learners make a poster that shows the use of a plus sign, a minus sign, and an equal sign.

Materials
Program Materials
- Fish Pond
- Fish Pond Record Form, 1 per pair
- Counters, 20
- Number 1–6 Cube, 1 per group
- Spinner, 1 per group

Lead students to recognize that the answer to the first equation is the first number in the second equation.

Repeat this process four times.

2 ENGAGE

Develop: Going Fishing

"Today we are going to learn about subtraction by going on a fishing trip with a partner." Follow the instructions on the Activity Card **Going Fishing**. As students complete the activity, be sure to use the Questions to Ask.

Activity Card 4I

Alternative Groupings

Pair: Have students play individually and complete the activity as written.

Individual: Instead of playing with two teams of two, have the student play against you individually and complete the activity as written.

Progress Monitoring	
If... one student is solving all of the equations,	▶ **Then...** have other students take turns writing and solving the equations.

Practice

Have students complete *Student Workbook*, pages 30–31. Guide students through the Key Idea example and the Try This exercises.

1 WARM UP

Prepare

Draw the first five lines of the Fish Pond Record Form on the board. Take out 20 Counters and a Number 1–6 Cube.

▶ **The number on the first line says we start with 20. I have twenty Counters on the table to show this amount.**

Ask a student to roll the Number Cube to determine how many you need to take away.

Write this number in the first equation and subtract that number of Counters from the set.

▶ **How many do I have left? How did you know?**

Write this number in the equation, and count the Counters to verify.

▶ **How many Counters do I have at the beginning of my next turn?**

306 Level D Unit 4 **Subtraction**

Interactive Differentiation

Consult the **Teacher Dashboard** for grouping suggestions. You can also use performance on the Engage activity to guide students.

Independent Practice

For additional practice with subtraction, have pairs of students use the Sets Former Tool. One student uses the Sets Former Tool to show a subtraction problem, and the other student writes the subtraction problem. For example, one student places 8 marbles in the top half of the mat. Then he or she drags 3 marbles into the bottom half. The other student writes 8 − 3 = 5 on a sheet of paper. Have students take turns using the Sets Former Tool and writing the subtraction problem.

Supported Practice

For additional support, use the Sets Former Tool with students.

- Write 7 − 4 = _____, and have students launch the Sets Former Tool.
- ▶ **How many marbles should you start with?** 7
- Have students place 7 marbles on the screen.
- ▶ **How many marbles should you subtract, or take away, from the 7 marbles?** 4
- Instruct students to drag 4 marbles to the bottom of the mat. Explain that the number of marbles left is the answer.
- ▶ **How many marbles are left?** 3
- Repeat the steps to help students model and solve 8 − 5 and 12 − 7.

3 REFLECT

Think Critically

Review students' answers to the Reflect prompt at the bottom of **Student Workbook,** page 31, and then review the Engage activity.

- ▶ **Why was Toby's computation incorrect?** He said 15 − 6 = 11.
- ▶ **Can you think of another way to solve the problem?**

4 ASSESS

Informal Assessment

Use the online or print Student Record, **Assessment,** page 128, to record informal observations.

Going Fishing
Did the student
- ☐ pay attention to the contributions of others?
- ☐ contribute information and ideas?
- ☐ improve on a strategy?
- ☐ reflect on and check accuracy of work?

Additional Practice

For additional practice, have students complete **Practice,** page 84.

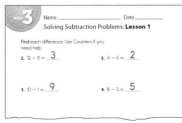

Practice, p. 84

Week 3 • Solving Subtraction Problems

Lesson 1

Key Idea
− is a minus sign.
It means to take away, or subtract.
= is an equal sign.
It means that both sides of the equation have the same value.

Try This
Circle either = or not = to tell if Column 1 equals Column 2.

	Column 1			Column 2
1.	4 − 2	⊜	not =	3 − 1
2.	4 + 3	=	⊝	10 − 1
3.	7 − 2	⊜	not =	5
4.	4 + 6	=	⊝	12 − 4

Practice
Find each difference. Use Counters if you need help.

5. 10 − 6 = **4** 6. 8 − 1 = **7**
7. 7 − 3 = **4** 8. 15 − 7 = **8**

Use Counters to find each difference.

9. 10 − 2 − 3 = **5** 10. 10 − 1 − 7 = **2**
11. 8 − 3 − 4 = **1** 12. 8 − 1 − 2 − 4 = **1**

Reflect
Toby and Matt tried to solve the subtraction problem below. Who answered it correctly? Explain how you know.

Matt is correct. Toby incorrectly solved 15 − 6 = 11. The difference should be 9.

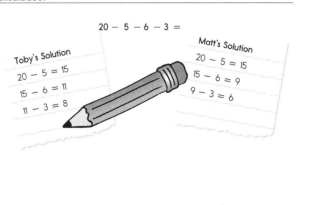

Student Workbook, pp. 30–31

WEEK 3
Solving Subtraction Problems

Lesson 2

Objective
Students can solve subtraction problems and related addition problems.

Standard
2.NBT.5 Fluently add and subtract within 100 using strategies based on place value, properties of operations, and/or the relationship between addition and subtraction.

Creating Context
Many English words have more than one meaning. Practice with English Learners using context to learn the meaning. For example, the word *check* in this lesson means "to look carefully at your work to be sure it is correct." We know also that someone might write a check or check into a hotel. Remind students they can also search for words in the dictionary.

Materials
Additional Materials
math-link cubes, 20 per student

1 WARM UP

Prepare
Give students 1 minute to write as many addition facts as they can. After the exercise, students should use math-link cubes to model each fact and verify that their sums are correct. Then give students 1 minute to write as many subtraction facts as they can. After the exercise, students should use math-link cubes to model each fact and verify that their differences are correct.

Just the Facts
Have students say the answer to a subtraction problem after hearing a related addition fact. Use questions such as the following:

- If $10 + 1 = 11$, what is $11 - 1$? 10
- If $5 + 2 = 7$, what is $7 - 2$? 5
- If $7 + 3 = 10$, what is $10 - 3$? 7

2 ENGAGE

Develop: Break It and Make It
"Today we are going to explore how addition and subtraction are related." Follow the instructions on the Activity Card **Break It and Make It**. As students complete the activity, be sure to use the Questions to Ask.

Alternative Grouping
Individual: Partner with the student and complete the activity as written.

Activity Card 4J

Progress Monitoring

| If... students always break the connecting cubes in half, | ▶ Then... instruct them to redo their work so their sentences are not all doubles facts. |

Practice
Have students complete **Student Workbook,** pages 32–33. Guide students through the Key Idea example and the Try This exercises.

Interactive Differentiation
Consult the **Teacher Dashboard** for grouping suggestions. You can also use performance on the Engage activity to guide students.

Independent Practice

For additional practice, have pairs of students use the Sets Former Tool. Explain that one student will write an addition equation and then use the Sets Former Tool to model it. The other student will write a related subtraction equation and then use the Sets Former Tool to model it. For example, if the first student writes $5 + 6 = 11$, the second student could write $11 - 5 = 6$. The Sets Former Tool model for the subtraction equation would show 5 marbles taken away from a group of 11, leaving 6 marbles.

Supported Practice

For additional support, use the Sets Former Tool to model related subtraction sentences for students.

- Tell students that you will model $15 + 5 = 20$. On the Sets Former Tool, count out 15 marbles. Add 5 marbles to the 15. Count the marbles to show students that there are 20 marbles.

  ▶ **I can check to see if my answer of 20 is correct by modeling a related subtraction equation.**

- Put 20 marbles on the top half of the mat. Drag 5 marbles onto the bottom mat. Point out to students that you are subtracting one of the numbers from the addition equation.

  ▶ **I just modeled $20 - 5 = 15$. Because taking 5 from 20 leaves 15, I know that 20 is the correct answer for $15 + 5$. How else could I check to see if my answer of 20 is correct?** You could subtract 15 from 20 to see if it equals 5.

- Put all 20 marbles back on the top half of the mat and then subtract 15 marbles. Count out the 5 remaining marbles on the top half of the mat. Point out that $20 - 15 = 5$ is another related subtraction equation for $15 + 5 = 20$.

- Have students use the Sets Former Tool to model the related subtraction equations for $12 + 2 = 14$.

3 REFLECT

Think Critically

Review students' answers to the Reflect prompt at the bottom of **Student Workbook,** page 33, and then review the Engage activity.

Discuss with students that the sum is always the greater number in an addition problem and that the first number of a subtraction equation is always the greater number.

▶ These facts are called a *fact family.* Why does that make sense?

Real-World Application

▶ Grocery stores model how addition and subtraction facts are opposite when they restock their shelves. Sam's Grocery Store started the day with 15 boxes of macaroni and cheese on the shelf. At the end of the day, only 4 boxes were left. How many boxes did they sell? 11

▶ The shelves were stocked again overnight, so the next morning 15 boxes were on the shelf. Write an addition equation that models the restocking of the shelf. $4 + 11 = 15$

4 ASSESS

Informal Assessment

Use the online or print Student Record, **Assessment,** page 128, to record informal observations.

Break It and Make It

Did the student
- ☐ pay attention to the contributions of others?
- ☐ contribute information and ideas?
- ☐ improve on a strategy?
- ☐ reflect on and check accuracy of work?

Additional Practice

For additional practice, have students complete **Practice,** page 85.

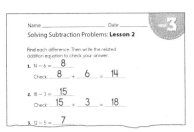

Practice, p. 85

Week 3 • Solving Subtraction Problems

Lesson 2

Key Idea
Subtraction is the opposite of addition.

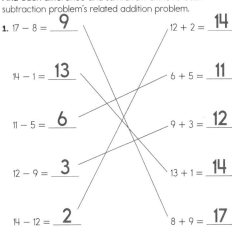

Try This
Find each difference and sum. Draw a line to each subtraction problem's related addition problem.

1. $17 - 8 = \underline{9}$ $12 + 2 = \underline{14}$

$14 - 1 = \underline{13}$ $6 + 5 = \underline{11}$

$11 - 5 = \underline{6}$ $9 + 3 = \underline{12}$

$12 - 9 = \underline{3}$ $13 + 1 = \underline{14}$

$14 - 12 = \underline{2}$ $8 + 9 = \underline{17}$

Practice
Find each difference. Write the related addition equation to check your answer.

2. $10 - 7 = \underline{3}$
   Check: $\underline{7} + \underline{3} = \underline{10}$

3. $9 - 2 = \underline{7}$
   Check: $\underline{2} + \underline{7} = \underline{9}$

4. $19 - 1 = \underline{18}$
   Check: $\underline{1} + \underline{18} = \underline{19}$

5. $11 - 3 = \underline{8}$
   Check: $\underline{3} + \underline{8} = \underline{11}$

Reflect
Use the numbers 5, 7, and 12 to write four related equations.

$5 + 7 = 12$ $7 + 5 = 12$
$12 - 7 = 5$ $12 - 5 = 7$

Student Workbook, pp. 32–33

WEEK 3
Solving Subtraction Problems

Lesson 3

Objective
Students can solve subtraction comparison problems.

Standard
2.NBT.5 Fluently add and subtract within 100 using strategies based on place value, properties of operations, and/or the relationship between addition and subtraction.

Creating Context
Modeling helps students learn new concepts. Encourage English Learners to discuss what they are doing in both English and in their primary language to reinforce comprehension.

Materials
Program Materials
Counters, 10 per student

Additional Materials
brown paper bag, 1 per student

1 WARM UP

Prepare
Draw a ten frame on the board, and place five *X*s in the top row and two *O*s in the bottom row.

▶ **How many *X*s are there? How many *O*s?** five; two

▶ **How many more *X*s are there than *O*s?** three

Let several students answer and explain their thinking. If necessary, tell students they can replace the word *more* with the word *extra* to make this question easier to think about.

▶ **We have two *O*s and five *X*s. If I want to find out how many extra *X*s there are, I can count up from two to five and use my fingers to keep track of how many extras there are: three, four, five.**

▶ **How many extra *X*s are there?** three

Repeat this process with two more problems: four *X*s and three *O*s; six *X*s and three *O*s.

Just the Facts
Have students hold up their fingers to show the answer. Use questions such as the following:

▶ **How many more is 5 than 3?** Students hold up 2 fingers.
▶ **How many more is 8 than 5?** Students hold up 3 fingers.
▶ **How many more is 9 than 2?** Students hold up 7 fingers.

2 ENGAGE

Develop: How Many More?
"Today we are going to practice solving 'how many more' problems." Follow the instructions on the Activity Card **How Many More?** As students complete the activity, be sure to use the Questions to Ask.

Activity Card 4K

Alternative Groupings
Whole Class: Pair students, and then complete the activity as written.

Small Group: Have two students in the group pull out handfuls of Counters. The whole group will then try to calculate how many more one student has than the other.

Progress Monitoring

| If... one student is dominating the game, | Then... have students wait 10 seconds before giving an answer. When 10 seconds pass, they both give an answer, and both students can earn a point if correct. |

Practice
Have students complete **Student Workbook,** pages 34–35. Guide students through the Key Idea example and the Try This exercises.

Interactive Differentiation
Consult the **Teacher Dashboard** for grouping suggestions. You can also use performance on the Engage activity to guide students.

Independent Practice

For additional practice solving basic subtraction problems, have students complete Word Problems with Tools 2: Find Result or Change, Counting +/−. Students use tools provided to solve word problems (totals to 20).

Supported Practice

For additional support, use the following word problem. Write the problem on the board.

- Javier has 8 crackers. He eats 3 crackers. How many crackers does Javier have left?

▶ **How many crackers did Javier start with?** 8
- Instruct students to count out 8 Counters.

▶ **How many crackers did Javier eat?** 3
- Instruct students to take away 3 of the 8 Counters. Tell students that the amount left will show how many crackers Javier has left. 5

▶ **How could you change this math story into a "how many more" problem?** Answers may vary. Possible answer: Instead of asking how many were left, you would ask how many more 8 is than 3. For example: Javier has 8 crackers and his friend has 3 crackers. How many more crackers does Javier have than his friend?

310 Level D Unit 4 **Subtraction**

3 REFLECT

Think Critically

Review students' answers to the Reflect prompt at the bottom of **Student Workbook,** page 35, and then review the Engage activity.

Discuss to reinforce that "how many more" problems are comparison problems with subtraction.

▶ **What other phrases do you think of when you are solving "how many more" problems?** Possible answer: how many extra

▶ **If I have three pencils and two pens, how many more pencils than pens do I have?** 1 **How did you solve that problem?** Answers may vary.

Real-World Application

▶ With every board game that you play, do you have different strategies to win?

▶ Do you have different strategies for the same game depending on who your opponent is?

▶ Is one strategy always the best one?

4 ASSESS

Informal Assessment

Use the online or print Student Record, **Assessment,** page 128, to record informal observations.

How Many More?
Did the student
- ☐ pay attention to the contributions of others?
- ☐ contribute information and ideas?
- ☐ improve on a strategy?
- ☐ reflect on and check accuracy of work?

Additional Practice

For additional practice, have students complete **Practice,** page 86.

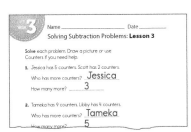

Practice, p. 86

Week 3 • Solving Subtraction Problems

Lesson 3

Key Idea
You can use addition or subtraction strategies to find how many more objects are in one group than are in another group.

Try This
Tell which color of counters there are more of and how many more there are.

1.

   There are more ___blue___ counters.

   How many more? __2__

   Circle the strategy you used to solve the problem.
   Student should circle one strategy.
   counting up from the smaller amount

   counting down from the larger amount

2. ⬭⬭⬭⬭⬭⬭⬭⬭⬭

   There are more ___yellow___ counters.

   How many more? __4__

   Circle the strategy you used to solve the problem.
   Student should circle one strategy.
   counting up from the smaller amount

   counting down from the larger amount

34 Level D Unit 4 Subtraction

Practice
Solve each problem. Draw a picture or use Counters if you need help.

3. Sheri has 3 counters. Patricia has 7 counters.

   Who has more counters? ___Patricia___

   How many more? __4__

4. Miguel has 10 counters. Alexis has 5 counters.

   Who has more counters? ___Miguel___

   How many more? __5__

5. Kim has 8 counters. Nan has 3 counters.

   Who has more counters? ___Kim___

   How many more? __5__

Reflect
There are 5 boys and 3 girls at Josh's party.
Are there more boys or more girls?
How many more?

There are 2 more boys at Josh's party.

Week 3 Solving Subtraction Problems • Lesson 3 35

Student Workbook, pp. 34–35

WEEK 3
Solving Subtraction Problems

Lesson 4

Objective
Students can interpret and solve subtraction word problems and can write number stories to describe them.

Standard
2.NBT.5 Fluently add and subtract within 100 using strategies based on place value, properties of operations, and/or the relationship between addition and subtraction.

Creating Context
Graphic organizers help English Learners develop higher-level thinking skills. Create a two-column chart to show the regular past-tense and irregular past-tense verbs used in this lesson. Help students identify verbs as regular or irregular, and write them on the chart.

Materials
Program Materials
Counters, 15 per pair

1 WARM UP

Prepare
Remind students that they can count down to solve subtraction problems.

Have students, as a group, continue a sequence, naming the next three numbers when counting backward.

- **10, 9, 8** 7, 6, 5
- **18, 17, 16** 15, 14, 13
- **30, 29, 28** 27, 26, 25
- **61, 60** 59, 58, 57

Repeat until students can do this with ease.

Then have students practice individually while the other students listen for errors.

Just the Facts
Have students chorally call out each answer. Use questions such as the following:

- **What is 9 minus 4?** 5
- **What is 13 minus 7?** 6
- **What is 12 minus 9?** 3

2 ENGAGE

Develop: Subtraction Action
"Today we are going to learn how to solve subtraction word problems." Follow the instructions on the Activity Card **Subtraction Action.** As students complete the activity, be sure to use the Questions to Ask.

Activity Card 4L

Alternative Grouping
Small Group: Have all group members take turns in each role. If there is an odd number of students, one student can watch and verify that the actor and the solver have not made errors.

Progress Monitoring

| If... students are not making the connection between subtraction and counting down, | ▶ Then... have them count backwards as each Counter is removed. |

Practice
Have students complete **Student Workbook,** pages 36–37. Guide students through the Key Idea example and the Try This exercises.

Interactive Differentiation
Consult the **Teacher Dashboard** for grouping suggestions. You can also use performance on the Engage activity to guide students.

Independent Practice
For additional practice, have students draw a picture to solve each math story. Use stories such as the following:

- **Janet had 7 pennies, but she lost 2 of them. How many pennies does Janet have now?** 5
- **Carol has 8 grapes and eats 3 of them. How many grapes does Carol have left?** 5
- **There are 10 angelfish at the pet store and 3 are sold. How many angelfish are left?** 7

Supported Practice
For additional support, use the following subtraction problem with students. Write the problem on the board:

- Kelly has 18 stickers. She gives 4 stickers to a friend. How many stickers does Kelly have left?
- **How many stickers did Kelly start with?** 18
- Instruct students to count out 18 Counters.
- **How many stickers did Kelly give to her friend?** 4
- Instruct students to take away, or subtract, 4 Counters. Tell students that the number of Counters remaining is how many stickers Kelly has left.
- **How many stickers does Kelly have left?** 14
- Have students draw a picture to solve the problem about Kelly. Explain that instead of using Counters to model the problem, students can draw 18 stickers and cross out 4 of them.
- Have students model with Counters and then draw a picture to solve more word problems as time permits.

312 Level D Unit 4 **Subtraction**

3 REFLECT

Think Critically

Review students' answers to the Reflect prompt at the bottom of **Student Workbook,** page 37, and then review the Engage activity.

▸ **Did your math stories have to be about the same thing?** no
▸ **Did your math stories have to have the same answer?** yes
▸ **How did you solve the problem?** Answers may vary. Possible answer: I counted to 6 and then counted back 2.

Real-World Application

Brainstorm real world situations that involve the concept of subtraction. Have each student describe a situation that occurred recently in which the remaining number of items had to be calculated.

4 ASSESS

Informal Assessment

Use the online or print Student Record, **Assessment,** page 128, to record informal observations.

Subtraction Action

Did the student
☐ pay attention to the contributions of others?
☐ contribute information and ideas?
☐ improve on a strategy?
☐ reflect on and check accuracy of work?

Additional Practice

For additional practice, have students complete **Practice,** page 87.

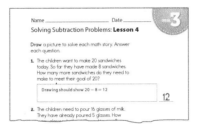
Practice, p. 87

Week 3 • Solving Subtraction Problems

Lesson 4

Key Idea
Act out the story or draw a picture to help solve word problems.

Try This
Draw a picture to solve each math story.

1. The Williams family went to the lake. They have 7 children in their family. Five of the children rode their bikes to the lake, and the rest of them walked. How many children walked to the lake?

   Students should draw a picture that shows 7 minus 5 is equal to 2.

2. When the 7 brothers and sisters arrived at the lake, all 4 boys used the boat to go fishing while the girls fished from the pier. How many girls are there?

   Students should draw a picture that shows 7 minus 4 is equal to 3.

3. The boys caught 10 fish and put them in a bucket. The boat started to shake, and the bucket tipped over. Six of the fish got out of the bucket. How many were left in the bucket?

   Students should draw a picture that shows 10 minus 6 is equal to 4.

36 Level D Unit 4 Subtraction

Practice
Draw a picture to solve each math story.

4. The girls made a goal to catch 12 fish by the end of the day. So far they have caught 6 fish. How many more fish do they have to catch to meet their goal of 12?

   Students should draw a picture that shows 12 minus 6 is equal to 6.

5. The boys made a goal to catch 18 fish by the end of the day. So far they have caught 8 fish. How many more fish do they have to catch to meet their goal of 18?

   Students should draw a picture that shows 18 minus 8 is equal to 10.

6. The Williams family caught a total of 20 fish. They decided to throw back 14 fish and keep the rest. How many did they keep?

   Students should draw a picture that shows 20 minus 14 is equal to 6.

Reflect
Solve the subtraction problem. Write a math story about the subtraction equation.

$6 - 2 = \underline{4}$

Answers may vary. Possible answer: Six monkeys were sitting in a tree. Two monkeys jumped out. How many monkeys were left in the tree?

Week 3 Solving Subtraction Problems • Lesson 4 37

***Student Workbook,* pp. 36–37**

WEEK 3
Solving Subtraction Problems

Lesson 5 Review

Objective
Students review skills learned this week and complete the weekly assessment and project.

Standard
2.NBT.5 Fluently add and subtract within 100 using strategies based on place value, properties of operations, and/or the relationship between addition and subtraction.

Vocabulary
Review vocabulary introduced during the week.

Creating Context
Help students review the words used to describe addition and subtraction in word problems by creating a chart. Have students match each of the following words or phrases to the correct operation.
- difference
- take away
- sum
- altogether
- how many more?

1 WARM UP

Prepare
Have students find each difference and share their strategies for solving the problems.
- $12 - 12 = 0$
- $18 - 2 - 7 = 9$
- $10 - 1 - 3 = 6$

▶ **Tell a math story for each of the problems you solved.**

2 ENGAGE

Practice
Have students complete **Student Workbook,** pages 38–39.

Week 3 • Solving Subtraction Problems

Lesson 5 Review

This week you continued working with subtraction. You modeled subtraction using Counters. You also explored subtracting a series of numbers, including equations that result in 0.

Lesson 1 Use Counters to find each difference.
1. $7 - 2 - 3 = \underline{2}$
2. $10 - 2 - 3 = \underline{5}$
3. $11 - 4 - 4 = \underline{3}$
4. $9 - 1 - 2 = \underline{6}$

Lesson 2 Find the difference. Write the related addition fact to check your answer.
5. $12 - 8 = \underline{4}$
   Check: $\underline{8} + \underline{4} = \underline{12}$
6. $17 - 4 = \underline{13}$
   Check: $\underline{4} + \underline{13} = \underline{17}$
7. $19 - 8 = \underline{11}$
   Check: $\underline{8} + \underline{11} = \underline{19}$
8. $14 - 9 = \underline{5}$
   Check: $\underline{9} + \underline{5} = \underline{14}$

38 Level D Unit 4 Subtraction

Lesson 3 Tell which color of counters there are more of and how many more.

9.
   There are more ___red___ counters.
   How many more? __1__

10.
    There are more ___green___ counters.
    How many more? __2__

Lesson 4 Write a subtraction equation about the math story.

11. On Friday night, 5 people went to the movies. Three people sat in Row J, and the others sat in Row K. How many people sat in Row K?

    $\underline{5\ people - 3\ people = 2\ people}$

Reflect
Write a math story about a group of children who have 17 objects. When some children leave, only 12 objects are left. Show your subtraction equation.

Stories will vary, but equations should show that $17 - 5 = 12$.

Week 3 Solving Subtraction Problems • Lesson 5 39

Student Workbook, pp. 38–39

314 Level D Unit 4 **Subtraction**

3 REFLECT

Think Critically

Review students' answers to the Reflect prompt at the bottom of **Student Workbook,** page 39.

Discuss the answer with the group to reinforce Week 3 concepts.

4 ASSESS

Formal Assessment

Students may take the weekly assessment online.

As an alternative, students may complete the weekly test on **Assessment,** pages 53–54. Record progress using the Student Assessment Record, **Assessment,** page 128.

Going Forward

Use the **Teacher Dashboard** to view results of the online assessments, to input the results of print student assessments, and to review progress before making decisions about next steps. Use the weekly test results and observations to determine the next steps for each student.

Retention	
Student displays good grasp of this week's concepts and skills.	Have students continue to use the Sets Former Tool to model subtraction problems. Challenge student pairs to write subtraction equations and model them using the Sets Former Tool.
Remediation	
Student is still struggling with the week's concepts and skills.	Have students work in pairs to complete Word Problems with Tools 2: Find Result or Change, Counting +/−.

Suggestions for Re-Evaluation: If a student has struggled without success for several weeks, use observations and test results to place the student at a level in which he or she can find success and build confidence to move forward.

Name_____ Date_____ WEEK 3
Solving Subtraction Problems

Circle the two problems that have the same answer.

1. (8 − 4)   (7 − 3)   10 − 5

2. (7 + 5)   9 − 6   (15 − 3)

Answer the questions below.

3. 14 − 6 = __8__

4. 10 − 7 − 2 = __1__

5. Cal has 18 pieces of paper. He gave 6 to his sister and 5 to his brother. How many pieces of paper does he have left? __7__

Level D Unit 4 Week 3 53

WEEK 3   Name_____ Date_____
Solving Subtraction Problems

6. Peg wants to run for 20 minutes. She already ran 12 minutes. How many more minutes must she run? __8__

7. Write an addition fact that goes with 11 − 7 = 4.

   $7 + 4 = 11$ or inverse

8. Write an addition fact that goes with 13 − 4 = 9.

   $9 + 4 = 13$ or inverse

9. Raj had 18 crackers. He gave 7 crackers to his sister. Write an equation to show how many crackers he had left.

   18 crackers − 7 crackers = 11 crackers

10. Lupe wants to read 15 pages of her science book today. She has already finished 6 pages. Write an equation to show how many pages she still has to read.

    15 pages − 6 pages = 9 pages

54 Level D Unit 4 Week 3

Assessment, pp. 53–54

Week 3 **Solving Subtraction Problems** • Lesson 5   315

Project Preview

This week, students focused on subtraction. They modeled subtraction using Counters and the Set Tool. The project for this unit requires students to extend the knowledge they gained in Find the Math and what they have learned this week. Students will use subtraction to determine how much inventory is left after a day's sales.

Project-Based Learning

Standards-driven Project-Based Learning is effective in building deep content understanding. Project-Based Learning increases long-term retention of concepts and has been shown to be more effective than traditional instruction. Completing a project to answer an essential question challenges students to apply and demonstrate mastery of concepts and skills by expressing understanding through discussion, research, and presentation.

Essential Question

WHEN would it be useful to write equations outside the classroom?

Project Evaluation Criteria

Review project evaluation criteria with students prior to beginning the project.

Exceeds Expectations
- ☐ Project result is explained and can be extended.
- ☐ Project result is explained in context and can be applied to other situations.
- ☐ Project result is explained using advanced mathematical vocabulary.
- ☐ Project result is described, and mathematics are used correctly and can be extended.
- ☐ Project result is explained and extended, and shows advanced knowledge of mathematical concepts and skills.

Meets Expectations
- ☐ Project result is explained.
- ☐ Project result is explained in context.
- ☐ Project result is explained using mathematical vocabulary.
- ☐ Project result is described, and mathematics are used correctly.
- ☐ Project result is explained, and shows satisfactory knowledge of mathematical concepts and skills.

Does Not Meet Expectations
- ☐ Project result is not explained.
- ☐ Project result is explained, but out of context.
- ☐ Project result is explained, but mathematical vocabulary is oversimplified.
- ☐ Project result is described, but mathematics are not used correctly.
- ☐ Project result is not explained and/or extended, or shows less than satisfactory knowledge of mathematical concepts and skills.

How Many Are Left?

Objective
Students can use subtraction to solve problems.

Standard
2.NBT.5 Fluently add and subtract within 100 using strategies based on place value, properties of operations, and/or the relationship between addition and subtraction.

Materials
Program Materials
- Number 1–6 Cube, 1 per group
- Counters, 20 per group

Best Practices
- Clearly enunciate instructions.
- Coach, demonstrate, and model.
- Select and provide the appropriate materials.

Introduce

Stores have to keep track of how much inventory they have.

▶ **Do you know what *inventory* means?** Answers may vary. Possible answer: *Inventory* is what a store sells. In a clothing store, the clothing items for sale are the inventory.

▶ **Why do you think it is important for stores to keep track of their inventory?** Answers may vary. Possible answers: to know when to order more; to know which item in their inventory is the most popular; to make sure nothing has been stolen

Explore

▶ **Today you will figure out how much inventory you have left at the end of the day. What kind of equation would you write to figure this out?** a subtraction equation

▶ **What information do you need to write this subtraction equation?** how much inventory I had at the beginning of the day and how many items were sold during the day

- Organize students into small groups. Give each group 20 Counters and a Number 1–6 Cube. Make sure each student has a copy of *Student Workbook,* page 40.

▶ **Complete *Student Workbook,* page 40, to find out how much inventory you have left.**

Wrap Up

- Allow students time to write their own problem and think of a scenario.
- Make sure each student can explain how he or she determined the amount of money the store made for the day.
- Check that students understand how to find the difference.
- If students struggle to find the answer, have them use Counters for help.
- Discuss students' answers to the Reflect prompt at the bottom of *Student Workbook,* page 40.

Have students make a list of the inventory they want to sell. Make sure students list the items in a single column, as they will add the prices next week. Save the inventory lists for next week.

If time permits, allow each student to check another student's subtraction solution by writing and solving the corresponding addition equation.

Week 3 • Solving Subtraction Problems

Project

How Many Are Left?

Read and follow the directions.

1. Circle one number between 15 and 20. Then circle one of the items of clothing. This is how many items you have at the beginning of the day.

   15   16   17   18   19   20   skirts   shirts   pants   shorts   scarves

2. Roll the Number Cube to find out how many of those items you sold today. Write the number on the line.

   Answers may vary.

3. Write a number equation that shows how many items are left at the end of the day.

   Check students' work.

4. Write your own subtraction story that fits the number equation you wrote.

   Check students' work.

Reflect

Explain how you can use Counters to solve the subtraction equation.

Answers may vary; possible answer: Count out the number of Counters that matches the number I circled. Then take away the number that I rolled from the pile of Counters.

Student Workbook, p. 40

Teacher Reflect

☐ Did students focus on the major concept of the activity?

☐ Did students tell or show the steps when they explained how to do something?

☐ Were students able to correctly answer the Reflect question?

WEEK 4: Subtraction Tools and Strategies

Week at a Glance

This week, students continue with **Number Worlds,** Level D, Subtraction, by exploring subtracting on a number line. Students use models on the number lines to write subtraction equations and to determine an unknown number in a subtraction sentence.

Skills Focus

- Model basic subtraction facts on a number line.
- Model subtraction equations that include a series of numbers being subtracted.
- Use the vertical format to solve subtraction problems.

How Students Learn

Encourage students to use what they know about addition facts and forward movement along a number line to help them construct relationships for subtraction facts. Students need opportunities to experiment with numbers and relationships between numbers. Students will eventually tie this relationship to backward movement on a number line and use this relationship to recall basic facts.

English Learners ELL

For language support, use the **English Learner Support Guide,** pages 82–83, to preview lesson concepts and teach academic vocabulary.

Math at Home

Give one copy of the Letter to Home, page 22, to each student. Encourage students to share and complete the activity with their caregivers.

Weekly Planner

Lesson	Learning Objectives
1 pages 320–321	Students can write equations to describe backward progression on a number line.
2 pages 322–323	Students can use number lines and counting back to solve subtraction problems.
3 pages 324–325	Students can use a number line and counting back to solve a series of subtraction problems.
4 pages 326–327	Students can solve subtraction problems with one- and two-digit numbers when displayed in a vertical format.
5 pages 328–329	**Review and Assess** Students review skills learned this week and complete the weekly assessment and project.
Project pages 330–331	Students can use subtraction to determine the correct amount of change a customer should receive.

318 Level D Unit 4 **Subtraction**

Key Standard for the Week

Domain: Number and Operations in Base Ten

Cluster: Use place value understanding and properties of operations to add and subtract.

2.NBT.5 Fluently add and subtract within 100 using strategies based on place value, properties of operations, and/or the relationship between addition and subtraction.

Materials		Technology
Program Materials • *Student Workbook,* pp. 41–43 • *Practice,* p. 88 • Activity Card 4M, **Counting Back** • Number Lines • Counters • Neighborhood Number Line (1–20) • Number Cards (1–10)	**Additional Materials** subtraction flash cards	*Teacher Dashboard*
Program Materials • *Student Workbook,* pp. 44–45 • *Practice,* p. 89 • Activity Card 4N, **Take It Back** • Number Lines • Number 1–6 Cubes • Number Cards (6–20)	**Additional Materials** • marker • masking tape • subtraction flash cards • colored pencils (for Variation)	*Teacher Dashboard*
Program Materials • *Student Workbook,* pp. 46–47 • *Practice,* p. 90 • Activity Card 4N, **Take It Back** • Number Lines • Number Cards (6–20) • Number 1–6 Cubes	**Additional Materials** • colored pencils • marker • masking tape • note cards	*Teacher Dashboard*
Program Materials • *Student Workbook,* pp. 48–49 • *Practice,* p. 91 • Activity Card 3P, **Vertical Operations** • Double-Digit Number Cards (Subtraction) • Place Value Mat • Neighborhood Number Line • Number 1–6 Cubes		*Teacher Dashboard* Building Blocks Word Problems with Tools 4
Program Materials • *Student Workbook,* pp. 50–51 • Weekly Test, *Assessment,* pp. 55–56		Review previous activities.
Program Materials • *Student Workbook,* p. 52 • Number Lines • Counters		

WEEK 4
Subtraction Tools and Strategies

Find the Math

In this week, students will use a number line to solve subtraction problems.

Use the following to begin a guided discussion:

▶ **How might a weather person use a number line in his or her job?** Answers may vary. Possible answer: A thermometer looks like a number line. The weather person might use it to tell about the difference between the temperature yesterday and the temperature today.

Have students complete *Student Workbook,* page 41.

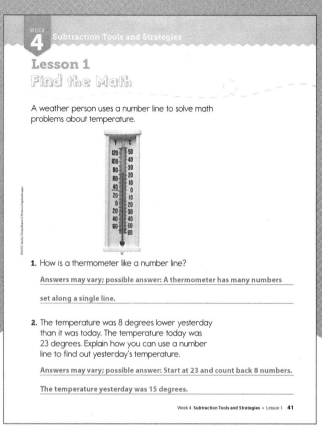

Student Workbook, p. 41

Lesson 1

Objective
Students can write equations to describe backward progression on a number line.

Standard
2.NBT.5 Fluently add and subtract within 100 using strategies based on place value, properties of operations, and/or the relationship between addition and subtraction.

Vocabulary
- minuend
- subtrahend

Creating Context
Creating direct experiences for English Learners can greatly enhance their comprehension. Taping a number line on the ground that they can hop or leap along is an excellent way to practice the skills in this lesson.

Materials
Program Materials
- Number Lines, 2 per student
- Counters, 20
- Neighborhood Number Line (1–20)
- Number Cards (1–10)

Additional Materials
subtraction flash cards

Prepare Ahead
Shuffle the Number Cards for each pair of students and place them facedown on the table.

1 WARM UP

Prepare

Place fifteen Counters on a table. Write "15 −" on the board. Explain to students that 15 is the *minuend,* or the number you will subtract from. Ask a student to take a few Counters from the pile and write that number (e.g., 4) in the equation. Tell students that this is the *subtrahend,* or the number being subtracted.

▶ **How can we figure out how many Counters are left on the table without counting them?**

If students don't suggest counting back by the number of Counters taken, suggest it as a strategy. Have students name the difference.

2 ENGAGE

Develop: Counting Back

"Today we are going to use questions to help solve subtraction problems." Follow the instructions on the Activity Card **Counting Back**. As students complete the activity, be sure to use the Questions to Ask.

Activity Card 4M

Alternative Grouping
Individual: Act as the student's partner and complete the activity as written.

Progress Monitoring

| **If…** students have trouble with the subtraction operation, | ▶ **Then…** have them start the game with ten Counters and use Number Cards 1–5. |

Practice

Have students complete *Student Workbook,* pages 42–43. Guide students through the Key Idea example and the Try This exercises.

Interactive Differentiation

Consult the *Teacher Dashboard* for grouping suggestions. You can also use performance on the Engage activity to guide students.

Independent Practice

For additional practice, provide each student with a stack of subtraction flash cards, twenty Counters, and two copies of Number Lines.

- Instruct students to model the problem on each card using Counters and a number line. Make sure students look at the minuend to determine the total number of Counters they need.

320 Level D Unit 4 **Subtraction**

- Have students write each problem and the answer on a dry-erase board or a blank piece of paper.

Supported Practice

For additional support, use Counters to help students model subtraction. Provide each student with twenty Counters and two Number Lines. Write a subtraction equation (such as 18 − 3 = ?) on the board.

▸ **We want to model this math problem with Counters. How many Counters should we start with?** 18

- Have students count out 18 Counters and place them on their desks.

▸ **Because this is a subtraction problem, we know we need to take some away. How many Counters should we take away?** 3

▸ **How many Counters do we have left?** 15

▸ **We can use a number line to help us answer the same math problem. Because we start with 18 Counters, place a Counter on each number on the number line from 1 to 18.**

▸ **We need to take 3 Counters away. Which 3 Counters should we take away to help us find the answer?** Start at 18 and work to the left. Take 3 Counters from the right end.

Point out that when students take away the Counters on the numbers 18, 17, and 16, they can easily see that there are 15 Counters left. Discuss the importance of taking away the three Counters at the right end of the number line rather than any other three Counters.

3 REFLECT

Think Critically

Review students' answers to the Reflect prompt at the bottom of **Student Workbook,** page 43, and then review the Engage activity.

4 ASSESS

Informal Assessment

Use the online or print Student Record, **Assessment,** page 128, to record informal observations.

Counting Back

Did the student

☐ make important observations? ☐ provide insightful answers?

☐ extend or generalize learning? ☐ pose insightful questions?

Additional Practice

For additional practice, have students complete **Practice,** page 88.

Practice, p. 88

Week 4 • Subtraction Tools and Strategies

Lesson 1

Key Idea
When subtracting more than one number, subtract one number at a time.

14 − 6 − 3 = ?

14 − 6 = 8 and 8 − 3 = 5

So, 14 − 6 − 3 = 5

Try This

Write and solve an equation for each set of cards.
Use the number line if you need help.

1.

   20 − 4 = 16

2. [14] − [3]

   14 − 3 = 11

3.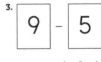

   9 − 5 = 4

4.

   15 − 5 = 10

Use Counters to help find each difference.

5. 16 − 8 − 4 = __4__   6. 14 − 8 − 5 = __1__
7. 8 − 2 − 3 = __3__   8. 17 − 3 − 5 = __9__

Practice

Write and solve an equation for each set of cards.

9. [14] − [8] − [2]

   14 − 8 = 6
   6 − 2 = 4
   or
   14 − 8 − 2 = 4

10. [16] − [4] − [5]

    16 − 4 = 12
    12 − 5 = 7
    or
    16 − 4 − 5 = 7

Reflect

Solve the problem. Show each step.
15 − 3 − 4 − 5 = ?

15 − 3 = 12

12 − 4 = 8

8 − 5 = 3

Student Workbook, pp. 42–43

WEEK 4
Subtraction Tools and Strategies

Lesson 2

Objective
Students can use number lines and counting back to solve subtraction problems.

Standard
2.NBT.5 Fluently add and subtract within 100 using strategies based on place value, properties of operations, and/or the relationship between addition and subtraction.

Vocabulary
- minuend
- subtrahend

Creating Context
It is important to review correctly completed problems with students. It reinforces the concepts for English Learners. It also helps students learn to correctly express themselves using math vocabulary.

Materials
Program Materials
- Number Lines, 2 per student
- Number 1–6 Cube, 1 per pair
- Number Cards (6–20), 1 set per pair

Additional Materials
- marker (dark-colored)
- masking tape
- subtraction flash cards

Prepare Ahead
- Using a long piece of masking tape and a dark-colored marker, create a large number line (0–20) on the classroom floor. Numbers should be far enough apart that students can stand on a single number, but close enough together that students can easily step from one number to the next.
- For **Take It Back**, remove Number Cards 1–5.

1 WARM UP

Prepare
Draw a 1–20 number line on the board. Write $12 - 4 = ?$ Circle 12 on the number line.

▶ **We're starting with 12 in this subtraction equation. Then we have to take something away. How much do we have to take away?**

Circle "−4" in the subtraction equation.

▶ **It says to take 4 away, so I'm going to jump backward 4 spaces. I'm starting at 12, so I jump over 11, 10, and 9, and land on 8. That's four spaces. We landed on 8—the number we have left after we take 4 away from 12. This is the *difference* between 4 and 12.**

Just the Facts
Have students use mental math to solve subtraction problems. Tell them to show you the answer by holding up the appropriate number of fingers. Use questions such as the following:

▶ **What is 14 − 8?** Students hold up 6 fingers.
▶ **What is 16 − 9?** Students hold up 7 fingers.
▶ **What is 19 − 9?** Students hold up 10 fingers.

2 ENGAGE

Develop: Take It Back
"Today we are going to use number lines to subtract." Follow the instructions on the Activity Card **Take It Back**. As students complete the activity, be sure to use the Questions to Ask.

Activity Card 4N

Alternative Grouping
Individual: Partner with the student and complete the activity as written.

Progress Monitoring
| If... students are struggling with using a number line on paper, | ▶ Then... use tape to make a number line on the floor. |

Practice
Have students complete **Student Workbook,** pages 44–45. Guide students through the Key Idea example and the Try This exercises.

Interactive Differentiation
Consult the **Teacher Dashboard** for grouping suggestions. You can also use performance on the Engage activity to guide students.

Independent Practice
For additional practice, provide students with a stack of subtraction flash cards and a copy of Number Lines. Instruct students to take turns standing on the giant number line on the number that corresponds to the first number on the flash card (the minuend). Students should then look at the second number (the subtrahend) to determine how many steps to the left to take. Once students find the difference, have them record the moves they made on a blank number line, circling the final answer.

Supported Practice
For additional support, use the giant number line to have students experience counting back on a number line.

- Write a subtraction equation (such as $15 - 9 = ?$) on the board.
- ▶ **We are going to use our giant number line to help us find the answer to the subtraction problem. Which number should I start on?** 15
- ▶ **Because this is a subtraction problem, which way should I move on the number line? How many steps should I take?** to the left; 9 steps
- ▶ **What number did I land on?** 6
- Have students record the problem and solution on a blank number line. Instruct them to draw a square around the starting number and then illustrate the steps along the number line. Have students circle their final answer and write the equation under the number line.
- Write a new subtraction problem on the board and choose a student to walk on the number line. Repeat as time permits.

REFLECT

Think Critically

Review students' answers to the Reflect prompt at the bottom of **Student Workbook,** page 45, and then review the Engage activity.

Discuss the meaning of each number.

▶ **Why is the largest number on the number line the beginning of an equation?** Possible answer: You can't subtract larger numbers from smaller numbers.

▶ **Can any of these numbers be placed in another position and still make a true equation?** The 4 and 10 can be switched.

Real-World Application

Tell a story about a situation that involves subtraction. The main character in the story should try to achieve the greatest difference or the least difference. Include strategies that the character used.

ASSESS

Informal Assessment

Use the online or print Student Record, **Assessment,** page 128, to record informal observations.

Take It Back
Did the student
- ☐ make important observations?
- ☐ provide insightful answers?
- ☐ extend or generalize learning?
- ☐ pose insightful questions?

Additional Practice

For additional practice, have students complete **Practice,** page 89.

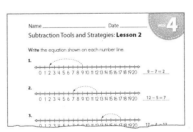

Practice, p. 89

Week 4 • Subtraction Tools and Strategies

Lesson 2

Key Idea
Subtraction is the opposite of addition.

[number line 0–10 with jump from 7 to 3]

Subtraction problems can be shown on a number line.
7 − 4 = 3
7 is where the jump starts.
4 is the distance of the jump.
3 is where the jump ends.

Try This
Show each subtraction problem on the number line. Write each difference.

1. [number line 1–21]
10 − 4 = **6**

2. [number line 1–21]
17 − 6 = **11**

44 Level D Unit 4 Subtraction

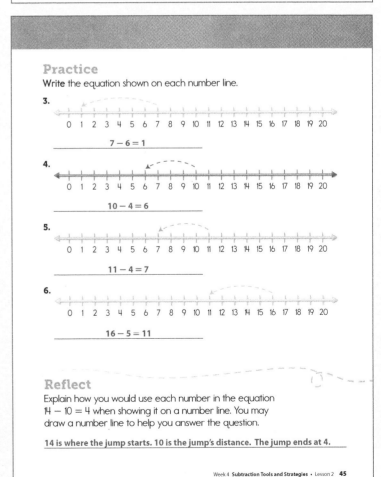

Practice
Write the equation shown on each number line.

3. [number line 0–20]
7 − 6 = 1

4. [number line 0–20]
10 − 4 = 6

5. [number line 0–20]
11 − 4 = 7

6. [number line 0–20]
16 − 5 = 11

Reflect
Explain how you would use each number in the equation 14 − 10 = 4 when showing it on a number line. You may draw a number line to help you answer the question.

14 is where the jump starts. 10 is the jump's distance. The jump ends at 4.

Week 4 Subtraction Tools and Strategies • Lesson 2 45

Student Workbook, pp. 44–45

Week 4 **Subtraction Tools and Strategies** • Lesson 2

WEEK 4
Subtraction Tools and Strategies

Lesson 3

Objective
Students can use a number line and counting back to solve a series of subtraction operations.

Standard
2.NBT.5 Fluently add and subtract within 100 using strategies based on place value, properties of operations, and/or the relationship between addition and subtraction.

Vocabulary
- minuend
- subtrahend

Creating Context
Tell students that in English we can turn some verbs into nouns that describe a person doing an action by adding -er to the end of the verb. Give examples, such as *jumper*.

Materials

Program Materials
- Number Lines, 3 copies per student
- Number 1–6 Cube, 1 per pair
- Number Cards (6–20), 1 set per pair

Additional Materials
- colored pencils, 3 per pair
- marker (dark-colored)
- masking tape
- note card, 1 per student

Prepare Ahead
- Write multistep subtraction problems on individual note cards, one per student. Use problems such as $25 - 3 - 6 = ?$ and $31 - 7 - 3 = ?$.
- Create a giant number line on the floor using masking tape and a dark-colored marker. Numbers should be far enough apart that students can stand on an individual number, but close enough together that students can easily step from one number to the next.
- For **Take It Way Back**, remove Number Cards 1–11.

1 WARM UP

Prepare
Draw a number line on the board. Review how to use the number line to solve subtraction problems.

▶ **How long is my jump if I start on 9 and finish at 6?** 3

▶ **How did you figure that out?** Possible answer: I counted the spaces between 6 and 9. If students answer that they counted the lines for numbers 6–9, discuss which lines and numbers the students counted. Remind them that they do not count the number (or line) they start on, but they should count the number they land on.

▶ **How long is my jump if I start on 15 and finish at 8?** 7

Just the Facts
Say a subtraction equation. Instruct students to jump up and down the number of times they would jump to the left on a number line to solve the problem. Use equations such as the following:

▶ $15 - 7 = ?$ Students should jump 7 times.

▶ $22 - 9 = ?$ Students should jump 9 times.

▶ $19 - 3 = ?$ Students should jump 3 times.

2 ENGAGE

Develop: Take It Back (Variation)
"Today we are going to use number lines to subtract more than one number." Follow the instructions on the Activity Card **Take It Back (Variation)**. As students complete the activity, be sure to use the Questions to Ask.

Activity Card 4N

Alternative Grouping
Individual: Partner with the student and complete the activity as written.

Progress Monitoring

If... students are getting confused when making multiple jumps,	Then... have the recorder illustrate each jump after rolling the Number 1–6 Cube.

Practice
Have students complete **Student Workbook,** pages 46–47. Guide students through the Key Idea example and the Try This exercises.

Interactive Differentiation
Consult the **Teacher Dashboard** for grouping suggestions. You can also use performance on the Engage activity to guide students.

Independent Practice
For additional practice, provide students with a subtraction problem note card and a copy of Number Lines. Students should take turns walking out the solution to the subtraction problem on the giant number line. Once students find the correct answer, have them illustrate the moves they made on a blank number line. Have students circle their final answer and then write the equation with the answer under the number line.

Supported Practice
For additional support, use the giant number line on the floor to help students model subtraction equations.

- Write a subtraction equation such as $23 - 5 - 4 = ?$ on the board.

▶ **We are going to use our giant number line to help us find the answer to the subtraction problem. Which number should I start on?** 23

▶ **What steps should I take to find the answer?** Move to the left 5 steps and then 4 more steps.

▶ **What number did I land on?** 14

- Have students record the moves you made on a blank number line. Instruct them to draw a square around the starting number and then illustrate the steps you took along the number line. Have students circle the final answer and then write the equation with the answer under the number line.

- Write a new subtraction problem on the board and choose a student to walk on the number line. Repeat as time permits.

324 Level D Unit 4 **Subtraction**

3 REFLECT

Think Critically

Review students' answers to the Reflect prompt at the bottom of **Student Workbook,** page 47, and then review the Engage activity.

Discuss the difference between adding and subtracting on a number line.

▶ Why do addition equations and subtraction equations move in opposite directions on the number line?

▶ Can we use addition to solve subtraction problems?

Real-World Application

Discuss with students that addition and subtraction are opposites, meaning you move in opposite directions on a number line.

▶ **Can you name other opposites?** Possible answers: small and large, tall and short, round and flat, high and low

4 ASSESS

Informal Assessment

Use the online or print Student Record, **Assessment,** page 128, to record informal observations.

Take It Back (Variation)
Did the student
- [ ] provide a clear explanation?
- [ ] choose appropriate strategies?
- [ ] communicate reasons and strategies?
- [ ] argue logically?

Additional Practice

For additional practice, have students complete **Practice,** page 90.

Practice, p. 90

Week 4 • Subtraction Tools and Strategies

Lesson 3

Key Idea

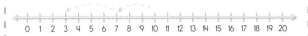

Number lines can be used to show a series of numbers subtracted from one number.

$10 - 3 - 4 = 3$

Try This

Show each equation on the number line.
Use a different-colored pencil for each jump.
Write each difference.

1. $10 - 6 - 1 = \underline{\ 3\ }$

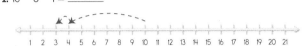

2. $8 - 3 - 4 = \underline{\ 1\ }$

46 Level D Unit 4 **Subtraction**

Practice

Write each equation shown.

3.
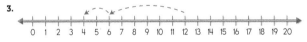
$12 - 6 - 2 = 4$

4.

$15 - 6 - 4 = 5$

5.

$8 - 1 - 1 = 6$

6.

$11 - 3 - 7 - 1 = 0$

Reflect

Look back at the number-line problems in this lesson. When subtracting numbers on a number line, in which direction did you move?

Answers may vary; possible answer: I moved left toward 0.

Week 4 **Subtraction Tools and Strategies** • Lesson 3   47

Student Workbook, pp. 46–47

WEEK 4
Subtraction Tools and Strategies

Lesson 4

Objective
Students can solve subtraction problems with one- and two-digit numbers when displayed in a vertical format.

Standard
2.NBT.5 Fluently add and subtract within 100 using strategies based on place value, properties of operations, and/or the relationship between addition and subtraction.

Vocabulary
- minuend
- subtrahend

Creating Context
Modeling new skills helps English Learners learn new concepts. Make sure to pause between each step in your instructions and check for understanding before you continue.

Materials
Program Materials
- Double-Digit Number Cards (Subtraction), 1 set per 6 students
- Place Value Mat, 1 per student
- Neighborhood Number Line
- Number 1–6 Cube, 1 per pair

1 WARM UP

Prepare
Write "$27 - 4 =$ _____" on the board as an equation and also in vertical format.

Point to the equation format. Ask students to solve the problem mentally and to describe their thinking aloud. Most students will use the count-back strategy.

Point to the vertical format, and tell students that we use a different strategy when the problem is written this way.

▶ **First you subtract the numbers in the ones column and write that number under the bar in the ones column. What is 7 minus 4?** 3

▶ **Next you subtract the numbers in the tens column and write that number under the bar in the tens column. What is 2 minus 0?** 2

▶ **The number under the bar is the answer.** 23

Write "$18 - 3$" and "$35 - 2$" in both formats on the board. Ask students to solve the problems aloud.

Just the Facts
Play a game of "Finger Flash." Tell students to "flash" the number of fingers that answers the question. Use questions such as the following:

▶ **What is $12 - 6$?** Students should show 6 fingers.
▶ **What is $9 - 5$?** Students should show 4 fingers.
▶ **What is $15 - 8$?** Students should show 7 fingers.

2 ENGAGE

Develop: Vertical Operations (Variation)

"Today we are going to do subtraction in the vertical format." Follow the instructions on the Activity Card **Vertical Operations (Variation)**. As students complete the activity, be sure to use the Questions to Ask.

Activity Card 3P

Alternative Grouping
Small Group: Complete the activity as written. Make sure all students take a turn recording and creating equations.

Progress Monitoring

| If... students have trouble correctly lining up the numbers in vertical format problems, | ▶ Then... give them graph paper, and demonstrate how to use the paper to help keep the places aligned. |

Practice
Have students complete **Student Workbook,** pages 48–49. Guide students through the Key Idea example and the Try This exercises.

Interactive Differentiation
Consult the **Teacher Dashboard** for grouping suggestions. You can also use performance on the Engage activity to guide students.

Independent Practice

For additional practice, have students complete Word Problems with Tools 4: Multidigit +/− to 100 and Multidigit Solver. As students read and work each word problem, have them record the appropriate equation in vertical form.

Supported Practice

For additional support, use the Place Value Mat to help students understand how to solve subtraction equations in vertical format.

- Write a subtraction equation such as $27 - 4 = ?$ on the board.
▶ **How many ones are in the number 27?** 7
▶ **How many tens?** 2
▶ **How many hundreds?** 0
- Have students record the minuend on the Place Value Mat, paying close attention to the digits and their place value.
▶ **How many are you going to take away from 27?** 4
▶ **How many ones does that number have?** 4
▶ **How many tens?** 0
▶ **How many hundreds?** 0
- Have students record the subtrahend on the Place Value Mat under the minuend. Guide students to draw a line under both numbers and then subtract to find the difference.
▶ **Which place value digits should we always subtract first?** the ones
- Repeat with other subtraction equations. You may want to model the numbers with base-ten blocks for students who need a more tactile experience.

326 Level D Unit 4 **Subtraction**

3 REFLECT

Think Critically

Review students' answers to the Reflect prompt at the bottom of **Student Workbook,** page 49, and then review the Engage activity.

Discuss the similarities and differences in vertical format problems and equations.

▶ **Which problems were easier to solve? Why?**

Real-World Application

▶ **Suppose that you are in charge of buying supplies for a big birthday party. You buy a package of 56 paper plates. After the party, you have 36 plates. How can you find the number of plates used by the partygoers?** subtract

▶ **How many plates were used? Write a number sentence.** 56 − 36 = 20

ASSESS

Informal Assessment

Use the online or print Student Record, **Assessment,** page 128, to record informal observations.

Vertical Operations (Variation)

Did the student
- ☐ provide a clear explanation?
- ☐ communicate reasons and strategies?
- ☐ choose appropriate strategies?
- ☐ argue logically?

Additional Practice

For additional practice, have students complete **Practice,** page 91.

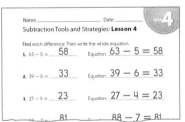

Practice, p. 91

Week 4 • Subtraction Tools and Strategies

Lesson 4

Key Idea
Subtraction equations can be shown two ways.

Horizontally:
38 − 4 = 34
Count back to find the answer to horizontal equations.
37, 36, 35, 34

Vertically:
38
− 4
34
Subtract the numbers in the ones column and the tens column to find the answer to vertical equations.

Try This
Find each difference.

1. 97
   − 3
   94

2. 28
   − 5
   23

3. 59
   − 7
   52

4. 74
   − 2
   72

5. 34
   − 1
   33

6. 67
   − 6
   61

7. 22 − 3 = **19**

8. 46 − 3 = **43**

48 Level D Unit 4 Subtraction

Practice
Find each difference. Then write the equation.

9. 84
   − 4
   80
   Equation: 84 − 4 = 80

10. 58
    − 7
    51
    Equation: 58 − 7 = 51

11. 24
    − 2
    22
    Equation: 24 − 2 = 22

12. 68
    − 5
    63
    Equation: 68 − 5 = 63

Reflect
How does addition help you with subtraction?

Possible answer: I can use addition to make sure I've subtracted correctly.

Week 4 Subtraction Tools and Strategies • Lesson 4 49

Student Workbook, pp. 48–49

WEEK 4
Subtraction Tools and Strategies

Lesson 5 Review

Objective
Students review skills learned this week and complete the weekly assessment and project.

Standard
2.NBT.5 Fluently add and subtract within 100 using strategies based on place value, properties of operations, and/or the relationship between addition and subtraction.

Vocabulary
Review vocabulary introduced during the week.

Creating Context
If English Learners have learned their basic mathematics strategies outside the United States, they might bring a more advanced mathematics experience, but not know how to express their questions and solutions in English. Be sure to talk through the solution of each math problem so English Learners can also develop their math vocabulary in English.

1 WARM UP

Prepare
Send five students to the board. Each student should create a number line, illustrate an equation similar to one from this week's lessons, and then select another student to write the equation associated with the illustration. If that student is struggling to write the equation, he or she can select another student to help complete the equation. Those remaining in their seats can verify that each equation is correct.

2 ENGAGE

Practice
Have students complete **Student Workbook,** pages 50–51.

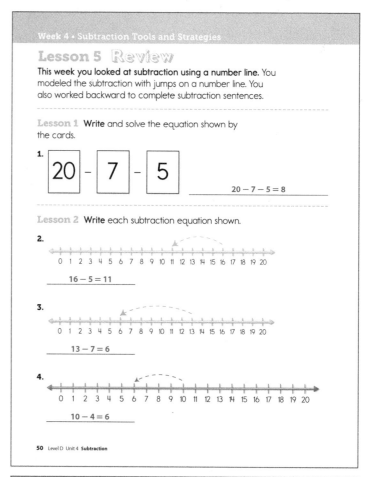

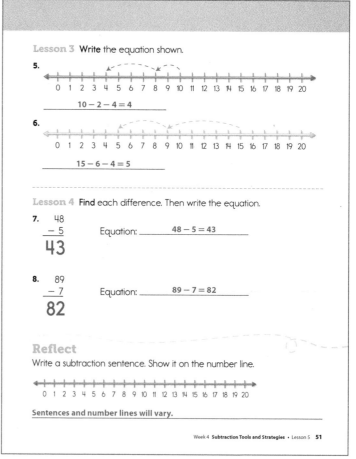

Student Workbook, pp. 50–51

328 Level D Unit 4 **Subtraction**

3 REFLECT

Think Critically

Review students' answers to the Reflect prompt at the bottom of **Student Workbook,** page 51.

Discuss the answer with the group to reinforce Week 4 concepts.

4 ASSESS

Formal Assessment ✓

Students may take the weekly assessment online.

As an alternative, students may complete the weekly test on **Assessment,** pages 55–56. Record progress using the Student Assessment Record, **Assessment,** page 128.

Going Forward

Use the **Teacher Dashboard** to view results of the online assessments, to input the results of print student assessments, and to review progress before making decisions about next steps. Use the weekly test results and observations to determine the next steps for each student.

Retention	
Student displays good grasp of this week's concepts and skills.	Students should continue to play **Vertical Subtraction.** If students are ready for the extra challenge, add even more subtrahends (more jumps backward).

Remediation	
Student is still struggling with the week's concepts and skills.	Use the Place Value Mat to help students solve vertically formatted subtraction equations. • Display a subtraction equation (such as 39 − 12 = ?) on the board. ▶ **How many ones are in the number 39?** 9 ▶ **How many tens?** 3 ▶ **How many hundreds?** 0 • Have students record the minuend on the Place Value Mat. ▶ **How many are you going to take away from 39? How many ones does that number have? How many tens? How many hundreds?** 12; 2; 1; 0 • Have students record the subtrahend on the Place Value Mat under the minuend. Guide students to draw a line under both numbers.

Suggestions for Re-Evaluation: If a student has struggled without success for several weeks, use observations and test results to place the student at a level in which he or she can find success and build confidence to move forward.

Name _____ Date _____
WEEK 4
Subtraction Tools and Strategies

Solve each problem.

1. $8 - 3 = \underline{5}$

2. $19 - 6 = \underline{13}$

3. $17 - 3 - 4 = \underline{10}$

4. $12 - 7 - 5 = \underline{0}$

Write an equation for the problem shown on the number line.

5.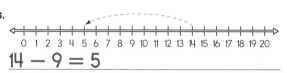
$14 - 9 = 5$

Level D Unit 4 Week 4 **55**

WEEK 4
Name _____ Date _____
Subtraction Tools and Strategies

6.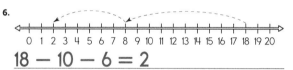
$18 - 10 - 6 = 2$

7.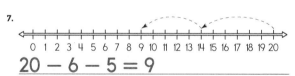
$20 - 6 - 5 = 9$

Write an equation to match each subtraction problem. Solve the equation.

8. 28
  −7
$28 - 7 = 21$

9. 89
  −5
$89 - 5 = 84$

10. 53
  −3
$53 - 3 = 50$

56 Level D Unit 4 Week 4

Assessment, pp. 55–56

Project Preview

This week, students learned how to use a number line to solve more complicated subtraction equations. The project for this unit requires students to extend the knowledge they gained in Find the Math and what they have learned this week. In today's project, students will use a number line to find the amount of change a customer should receive.

Project-Based Learning

Standards-driven Project-Based Learning is effective in building deep content understanding. Project-Based Learning increases long-term retention of concepts and has been shown to be more effective than traditional instruction. Completing a project to answer an essential question challenges students to apply and demonstrate mastery of concepts and skills by expressing understanding through discussion, research, and presentation.

Essential Question

WHEN would it be useful to write equations outside the classroom?

Project Evaluation Criteria

Review project evaluation criteria with students prior to beginning the project.

Exceeds Expectations
- ☐ Project result is explained and can be extended.
- ☐ Project result is explained in context and can be applied to other situations.
- ☐ Project result is explained using advanced mathematical vocabulary.
- ☐ Project result is described, and mathematics are used correctly and can be extended.
- ☐ Project result is explained and extended, and shows advanced knowledge of mathematical concepts and skills.

Meets Expectations
- ☐ Project result is explained.
- ☐ Project result is explained in context.
- ☐ Project result is explained using mathematical vocabulary.
- ☐ Project result is described, and mathematics are used correctly.
- ☐ Project result is explained, and shows satisfactory knowledge of mathematical concepts and skills.

Does Not Meet Expectations
- ☐ Project result is not explained.
- ☐ Project result is explained, but out of context.
- ☐ Project result is explained, but mathematical vocabulary is oversimplified.
- ☐ Project result is described, but mathematics are not used correctly.
- ☐ Project result is not explained and/or extended, or shows less than satisfactory knowledge of mathematical concepts and skills.

Subtracting for Change

Objective
Students can use subtraction to determine the correct amount of change a customer should receive.

Standard
2.NBT.5 Fluently add and subtract within 100 using strategies based on place value, properties of operations, and/or the relationship between addition and subtraction.

Materials
Program Materials
- Number Lines (4 per student)
- Counters

Best Practices
- Check for student understanding frequently.
- Coach, demonstrate, and model.
- Allow students to self-monitor.

Introduce

When you go into a store to buy a new toy, you might have more money than you need to pay for the toy.

- Explain that paper currency comes in amounts of $1, $5, $10, and $20 bills.
- ▶ **Let's say that a new toy car costs $3. You don't have three $1 bills, but you do have one $10 bill. Can you still buy the new toy? Why or why not?** Yes; $10 is more than $3.
- ▶ **What will happen when you give the cashier your $10 bill? Will he or she keep it all or give some of it back to you?** give some of it back
- Explain to students the concept of receiving change from a transaction.

Explore

- ▶ **Today you will use subtraction to find the amount of change a customer should receive after he or she pays for an item.**
- ▶ **You can find how much change to give the customer by writing a subtraction problem. You will subtract the price of the item from the amount of money the customer gives you.**
- ▶ **Look at Problem 1 on your workbook page. How much does a sun hat cost?** $4
- ▶ **What subtraction problem can you write to find out how much change to give?** $10 - 4 = ?$
- Model counting back on a number line to find the solution. Then have students count back on their own number lines. Finally, have students write the equation in vertical format:

    $$\begin{array}{r} 10 \\ -4 \\ \hline 6 \end{array}$$

- Guide students to use the same strategies as they continue solving the other subtraction problems.
- ▶ **Complete Student Workbook, page 52, to practice using strategies to solve subtraction problems.**

Wrap Up

- Encourage students to write the equation for each problem.
- Make sure students are using the number line and writing each equation in a vertical format.
- If students struggle to use one or both of the strategies, model how both methods will lead students to the same answer.
- Discuss students' answers to the Reflect prompts at the bottom of *Student Workbook,* page 52.

Have students use the inventory lists from Week 3. Have them add the price of each item. For this activity, encourage students to keep prices under $30.

If time permits, allow each student to check another student's answers.

Student Workbook, p. 52

Teacher Reflect

- ☐ Did I explain what students had to find, make, or do before they began their projects?
- ☐ Was I able to answer questions when students did not understand?
- ☐ Were students able to correctly answer the Reflect questions?

WEEK 5: Subtraction Word Problems within 100

Week at a Glance

This week students continue **Number Worlds,** Level D, Subtraction. This unit builds on the previous 4 weeks, and students will apply this knowledge to solve one- and two-step subtraction word problems with unknowns in all positions.

Skills Focus

- Students solve one- and two-step subtraction word problems within 100.
- Students write equations with unknown minuends, subtrahends, and differences.
- Students solve equations with unknown minuends, subtrahends, and differences.

How Students Learn

Encourage students to use what they know about addition facts to help them construct relationships for subtraction facts. Students need opportunities to experiment with numbers and relationships between numbers.

English Learners ELL

For language support, use the **English Learner Support Guide,** pages 84–85, to preview lesson concepts and teach academic vocabulary. **Number Worlds** Vocabulary Cards are listed as additional materials in many lessons and can be used to preteach and reinforce academic vocabulary.

Math at Home

Give one copy of the Letter to Home, page 23, to each student. Encourage students to share and complete the activity with their caregivers.

Weekly Planner

Lesson	Learning Objectives
1 pages 334–335	Students can solve one- and two-step subtraction word problems within 100 and with unknowns in all positions.
2 pages 336–337	Students can solve one- and two-step subtraction word problems within 100 and with unknowns in all positions.
3 pages 338–339	Students can solve one- and two-step subtraction word problems within 100 and with unknowns in all positions.
4 pages 340–341	Students can solve one- and two-step subtraction word problems within 100 and with unknowns in all positions.
5 pages 342–343	**Review and Assess** Students review skills learned this week and complete the weekly assessment and project.
Project pages 344–345	Students can use what they know about solving subtraction word problems to compose and solve their own subtraction story.

Key Standard for the Week

Domain: Operations and Algebraic Thinking

Cluster: Represent and solve problems involving addition and subtraction.

2.OA.1 Use addition and subtraction within 100 to solve one- and two-step word problems involving situations of adding to, taking from, putting together, taking apart, and comparing, with unknowns in all positions, e.g., by using drawings and equations with a symbol for the unknown number to represent the problem.

Materials		Technology
Program Materials • **Student Workbook,** pp. 53–55 • **Practice,** p. 92 • Activity Card 4O, **Parts of a Story** • Neighborhood Number Line • Counters	**Additional Materials** self-sticking notes	**Teacher Dashboard** Building Blocks Word Problems with Tools 2
Program Materials • **Student Workbook,** pp. 56–57 • **Practice,** p. 93 • Activity Card 4P, **Ask About It** • Neighborhood Number Line • Counters		**Teacher Dashboard**
Program Materials • **Student Workbook,** pp. 58–59 • **Practice,** p. 94 • Activity Card 4P, **Ask About It** • Neighborhood Number Line • Counters	**Additional Materials** base-ten blocks*	**Teacher Dashboard** Building Blocks Word Problems with Tools 4
Program Materials • **Student Workbook,** pp. 60–61 • **Practice,** p. 95 • Activity Card 4Q, **What's Missing?** • Neighborhood Number Line • Counters	**Additional Materials** note cards	**Teacher Dashboard** Building Blocks Word Problems with Tools 8
Program Materials • **Student Workbook,** pp. 62–63 • Weekly Test, **Assessment,** pp. 57–58		Review previous activities.
Program Materials **Student Workbook,** p. 64	**Additional Materials** department-store advertisements	

*Available from McGraw-Hill Education

WEEK 5

Subtraction Word Problems within 100

Find the Math

In this week, students will learn how to use information in math stories to solve problems.

Use the following to begin a guided discussion:

▶ **How can you describe what is happening in the picture?** Answers may vary. Possible answer: Some ducks are walking on wet ground. Some ducks have green heads and some have brown heads.

Have students complete *Student Workbook,* page 53.

Student Workbook, p. 53

Lesson 1

Objective
Students can solve one- and two-step subtraction word problems within 100 and with unknowns in all positions.

Standard
2.OA.1 Use addition and subtraction within 100 to solve one- and two-step word problems involving situations of adding to, taking from, putting together, taking apart, and comparing, with unknowns in all positions, e.g., by using drawings and equations with a symbol for the unknown number to represent the problem.

Creating Context
Help English Learners understand how to identify a subtraction word problem. Point out that when we subtract, we take away a part from a whole set and then a part is left over. As students read subtraction word problems, have them highlight clue words such as *take away, left over, lost,* and so on. Point out that these words are clues that the problem involves subtraction.

Materials
Program Materials
- Neighborhood Number Line
- Counters

Additional Materials
self-sticking notes, 3

1 WARM UP

Prepare
- Write a simple subtraction equation on the board, such as 10 − 4 = 6.
- ▶ **These numbers tell a story. Let's think of a story that we could tell using this subtraction equation.**
- Guide students to tell a simple subtraction story, such as *Ten children are eating lunch. Four children finish their lunch and go play on the swings. Six children are left.*
- Point out the different pieces of the subtraction equation in the story: 10 represents the whole set, 4 represents the part that was taken away, and 6 represents the part that is left.

2 ENGAGE

Develop: Parts of a Story
"Today we are going to ask questions to help us solve word problems about subtraction." Follow the instructions on the Activity Card **Parts of a Story.** As students complete the activity, be sure to use the Questions to Ask.

Activity Card 40

Alternative Grouping
Pair: Have one student write the subtraction equation while the other student models the story with a drawing, a number line, or Counters.

Progress Monitoring	
If... students have difficulty determining where to place each number from the story in the subtraction equation,	**Then...** remind them that the greatest number in a subtraction equation always comes first. You cannot take away more than you have.

Practice
Have students complete *Student Workbook,* pages 54–55. Guide students through the Key Idea example and the Try This exercises.

Interactive Differentiation
Consult the *Teacher Dashboard* for grouping suggestions. You can also use performance on the Engage activity to guide students.

Independent Practice
For additional practice, have students complete Word Problems with Tools 2: Find Results or Change, Counting +/− to practice solving subtraction word problems as well as to review how to solve addition word problems.

334 Level D Unit 4 **Subtraction**

Supported Practice

For additional support, write the following word problem on the board or on a large piece of paper.

- Aaron took 14 of his model cars to the park. He lost some of them in the sand. Now he has only 11 model cars. How many cars did he lose in the sand?
- On three self-sticking notes, write *whole set, amount taken away,* and *amount left.*

▶ **Which number tells us about the whole set?** The whole set is the amount of model cars Aaron started with. 14

▶ **Because we know that 14 is not the unknown number, I am going to cover up that number.**

- Cover the number 14 in the word problem with the self-sticking note labeled *whole set.*

▶ **What number tells us about the amount that is left?** The number 11 is the amount that is left.

- Cover the number 11 in the word problem with the self-sticking note labeled *amount left.*

▶ **What number tells us about the amount taken away?** The problem does not supply that number; it is an unknown.

▶ **If the unknown is the amount taken away, what should our subtraction equation look like?** 14 − _____ = 11

▶ **What can we take away from 14 to equal 11?** 3

▶ **Aaron lost 3 model cars in the sand.**

 # REFLECT

Think Critically

Review students' answers to the Reflect prompt at the bottom of **Student Workbook,** page 55, and then review the Engage activity.

▶ **What are the three parts of a subtraction equation?** the whole set, the part that was taken away, and the part that is left

ASSESS

Informal Assessment

Use the online or print Student Record, **Assessment,** page 128, to record informal observations.

Parts of a Story
Did the student
- ☐ respond accurately?
- ☐ respond quickly?
- ☐ respond with confidence?
- ☐ self-correct?

Additional Practice

For additional practice, have students complete **Practice,** page 92.

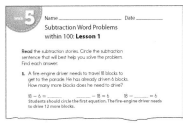

Practice, p. 92

Week 5 • Subtraction Word Problems within 100

Lesson 1

Key Idea
A subtraction equation has three parts: the whole set, the amount that is taken away from the whole set, and the amount that is left.

Look for the three parts of a subtraction equation to solve subtraction word problems.

| The Whole Set | − | Amount Taken Away | = | Amount Left |

Try This
Read the word problem. Answer the questions below it. If the answer to the question is the unknown value, write *unknown* in the blank.

1. Josh has a set of building tools. He left them out in the rain, and 5 of them got rusty. Now he has 6 tools left that are not rusty. How many tools are in Josh's set?

   Which number shows the amount of tools in the set? __unknown__
   Which number shows the amount of rusty tools? __5__
   Which number shows the amount of tools that are not rusty? __6__

2. Micah has 12 action figures. He shared some with his brother. He has 8 action figures left. How many action figures did Micah share with his brother?

   Which number shows the whole set of action figures? __12__
   Which number shows how many figures Micah shared with his brother? __unknown__
   Which number shows how many figures Micah has left? __8__

54 Level D Unit 4 Subtraction

Practice
Read the word problem. Answer the questions below it. If the answer to the question is the unknown value, write *unknown* in the blank.

3. Judy's coach asked her to run 10 laps. Judy has 7 laps left to run. How many laps has she run so far?
   Which number shows the whole set? __10__
   Which number shows the amount taken away? __unknown__
   Which number shows the amount left? __7__
   Write a subtraction equation to find the answer.

   10 − ___ = 7; She has run 3 laps so far.

4. In the morning, a bike shop had 13 bikes to fix. Some were fixed before lunch. After lunch, 4 bikes were left. How many bikes were fixed before lunch?
   Which number shows the whole set? __13__
   Which number shows the amount taken away? __unknown__
   Which number shows the amount left? __4__
   Write a subtraction equation to find the answer.

   13 − ___ = 4; 9 bikes were fixed before lunch.

Reflect
What does a subtraction equation tell us? How is it different from an addition equation?

Answers may vary; possible answer: A subtraction equation shows how a whole set is broken into separate parts. An addition equation shows how two or more groups are joined together to make a whole set.

Week 5 Subtraction Word Problems within 100 • Lesson 1 55

Student Workbook, pp. 54–55

WEEK 5
Subtraction Word Problems within 100

Lesson 2

Objective
Students can solve one- and two-step subtraction word problems within 100 and with unknowns in all positions.

Standard
2.OA.1 Use addition and subtraction within 100 to solve one- and two-step word problems involving situations of adding to, taking from, putting together, taking apart, and comparing, with unknowns in all positions, e.g., by using drawings and equations with a symbol for the unknown number to represent the problem.

Creating Context
Show English Learners how a whole set can be separated into two or more parts. Count out 12 Counters and place them together. Explain that these 12 Counters represent the whole set. Separate the Counters into two groups so that one group has 5 Counters and the other group has 7 Counters. Explain that the separate parts each represent a part of the whole set. When the two parts are put back together again, they form a whole set. Tell students that a subtraction equation often shows part of a set being taken away from the whole.

Materials
Program Materials
- Neighborhood Number Line
- Counters

Prepare Ahead
Prepare several subtraction stories you can share with students in which the subtrahend is the unknown value.

1 WARM UP

Prepare
- Review with students how to use a number line to solve a simple subtraction problem, such as $9 - 3$.
- Then lead students through solving a problem in which the subtrahend is the unknown value. Write $12 - \_\_\_\_ = 5$ on the board.
- ▶ **Look at the problem. How can we use a number line to find the number we subtracted from 12 to get 5?** Start at the number 12 on the number line. Count how many steps to the left we must take to get to 5.

Just the Facts
Present subtraction problems in which the subtrahend is the unknown value. Have students march in place the number of times they would step to the left on a number line to find the answer. Use equations such as the following:

- ▶ $9 - \_\_\_\_ = 6$ Students should march in place 3 times.
- ▶ $13 - \_\_\_\_ = 8$ Students should march in place 5 times.
- ▶ $25 - \_\_\_\_ = 21$ Students should march in place 4 times.

2 ENGAGE

Develop: Ask About It
"Today we are going to ask questions to help us solve subtraction word problems." Follow the instructions on the Activity Card **Ask About It.** As students complete the activity, be sure to use the Questions to Ask.

Activity Card 4P

Alternative Grouping
Pair: Have one student write out the subtraction equation while the other student models the story with a drawing, a number line, or Counters.

Progress Monitoring
| If... students have difficulty determining how to find the unknown value, | ▶ Then... have students label each piece of known data as either *the whole set* or *part of the whole set*. |

Practice
Have students complete **Student Workbook,** pages 56–57. Guide students through the Key Idea example and the Try This exercises.

Interactive Differentiation
Consult the **Teacher Dashboard** for grouping suggestions. You can also use performance on the Engage activity to guide students.

Independent Practice
For additional practice, write several subtraction equations on the board in which the subtrahend is the unknown value, such as $37 - \_\_\_\_ = 15$ or $65 - \_\_\_\_ = 41$. Have students compose a simple subtraction story for each equation. Make sure they understand that the amount taken away from the whole set is the unknown value. Have students solve for the unknown value by writing a subtraction equation in vertical format.

Supported Practice
For additional support, write the following word problem on the board: *There were 28 children swimming in the pool. Some children got out to eat a snack. Then there were 12 children in the pool. How many children got out of the pool to eat a snack?*

- ▶ **What do we know? What do we need to find out?** We know that there were 28 children in the pool. Some got out to eat a snack, and 12 stayed in the pool. We need to find out how many children are eating a snack.
- Have students draw a blank space or simple picture for the unknown value and then write the equation $28 - \_\_\_\_ = 12$.
- ▶ **How can we use a number line to help us solve the subtraction problem?** Answers may vary. Possible answer: We find the first number in the equation on the number line. We count back, or to the left, the number that was subtracted. The number we land on is the answer.
- ▶ **We can use those same steps to solve a subtraction problem like this one. Find the number 28 on the Neighborhood Number Line. We have to go back to get to the 12.**
- Count back along the number line to solve for the unknown value. 16
- Point out that when the value of the whole set is known and one of the parts of the set is known, students can subtract to find the value of the other part of the set. Model this concept with Counters if necessary.

336 Level D Unit 4 **Subtraction**

3 REFLECT

Think Critically

Review students' answers to the Reflect prompt at the bottom of **Student Workbook,** page 57, and then review the Engage activity.

▶ **When you read a word problem, what can you do to help yourself write the correct math equation?** Answers may vary. Possible answer: Look for a number that shows the whole set. That number comes first.

Real-World Application

Casey and her club are selling cookies for a fundraiser. They started with 24 dozen cookies, and after the sale they have 11 dozen cookies left.

▶ **How can you find out how many dozen cookies Casey's club sold?** Find the difference of 24 − 11. Casey's club sold 13 dozen cookies.

4 ASSESS

Informal Assessment

Use the online or print Student Record, **Assessment,** page 128, to record informal observations.

Ask About It
Did the student
☐ respond accurately?   ☐ respond with confidence?
☐ respond quickly?   ☐ self-correct?

Additional Practice

For additional practice, have students complete **Practice,** page 93.

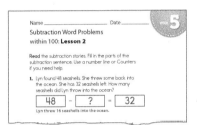

Practice, p. 93

Week 5 • Subtraction Word Problems within 100

Lesson 2

Key Idea
A subtraction story tells about something taken away from a whole set. Sometimes you know how much was taken away, and sometimes you do not. If you do not know how much was taken away, you can write a subtraction equation like the one below to help you solve the problem.

46 − ___ = 23

Try This
Read each word problem. Underline what you know. Circle what you need to find out.

1. Mike is allowed to play video games for 35 minutes after school. He has 12 minutes left to play. How many minutes has he already played video games?

   Write the subtraction equation.

   35 − ___ = 12

   Write an equation in vertical format to find the answer.

   35
   −12; 23 minutes
   ——
   23

2. Marissa bought 88 beads at the craft store. She used some of them to make a new necklace. She has 35 beads left. How many beads did she use to make the necklace?

   Write the subtraction equation.

   88 − ___ = 35

   Write an equation in vertical format to find the answer.

   88
   −35; 53 beads
   ——
   53

56 Level D Unit 4 Subtraction

Practice
Read each word problem. Think about what you know and what you need to find out. Use a blank in a math equation to show what you do not know.

3. Farmer Jones has 54 cows. Some of them have already been milked. He has 21 cows left to milk before the end of the day. How many cows has Farmer Jones already milked?
   Write the subtraction equation. 54 − ___ = 21
   Write a subtraction equation in vertical format to find the answer.
   54
   −21; 33 cows
   ——
   33

4. Mary Anne had 38 tomatoes in her garden. Bugs ate some of the tomatoes. She has 21 tomatoes left. How many tomatoes did the bugs eat?
   Write the subtraction equation. 38 − ___ = 21
   Write a subtraction equation in vertical format to find the answer.
   38
   −21; 17 tomatoes
   ——
   17

Reflect
To solve a subtraction equation such as 57 − ___ = 21, you can write 57 − 21 = ___. Use words such as *whole set* and *part* to explain why this works.

Answers may vary; possible answer: In this problem 57 is the whole set. I do not know one part, but I know that the other part is 21. I can take away the part I know from the whole set to find the other part because when I put both parts together, they make one whole set.

Week 5 Subtraction Word Problems within 100 • Lesson 2 57

Student Workbook, pp. 56–57

WEEK 5
Subtraction Word Problems within 100

Lesson 3

Objective
Students can solve one- and two-step subtraction word problems within 100 and with unknowns in all positions.

Standard
2.OA.1 Use addition and subtraction within 100 to solve one- and two-step word problems involving situations of adding to, taking from, putting together, taking apart, and comparing, with unknowns in all positions, e.g., by using drawings and equations with a symbol for the unknown number to represent the problem.

Creating Context
Help English Learners understand the purpose of asking questions to solve a problem. Think of a number between 1 and 20. Instruct students to ask yes/no questions such as *Is it a one- or two-digit number?* or *Is it larger than 10?* until students figure out the number. Encourage them to explain how each question they ask leads them closer to the answer.

Materials

Program Materials
- Neighborhood Number Line
- Counters

Additional Materials
base-ten blocks

Prepare Ahead
Prepare several subtraction stories you can share with students in which the minuend is the unknown value. Use larger two-digit numbers for these problems.

1 WARM UP

Prepare
- Write the problem _____ − 3 = 7 on the board.
▶ **What do we know about this subtraction equation?** We know that 3 was taken away from a number and 7 is left.
▶ **We do not know the number that tells us about the whole set, but we do know that one part has 3 and one part has 7. How can we find out the number that tells us about the whole set?** Add the two parts together.
- Use Counters to model joining two parts to make a whole set.

Just the Facts
Have students find sums. Speak rhythmically to encourage students to respond in unison. Use prompts such as the following:
▶ 5 + 9 = _____ 14
▶ 6 + 9 = _____ 15
▶ 7 + 4 = _____ 11

2 ENGAGE

Develop: Ask About It (Variation)
"Today we are going to ask questions to help us solve more subtraction word problems." Follow the instructions on the Activity Card **Ask About It (Variation)**. As students complete the activity, be sure to use the Questions to Ask.

Activity Card 4P

Alternative Grouping
Pair: Have one student write out the subtraction equation while the other student models the story with a drawing, a number line, or Counters.

Progress Monitoring

If... students assume that the greatest number within the word problem represents the whole,	▶ Then... ask, "Is there a number that shows how much was taken away? Is there a number that shows how much is left?"

Practice
Have students complete **Student Workbook,** pages 58–59. Guide students through the Key Idea example and the Try This exercises.

Interactive Differentiation
Consult the **Teacher Dashboard** for grouping suggestions. You can also use performance on the Engage activity to guide students.

Independent Practice

For additional practice, have students complete Word Problems with Tools 4: Multidigit +/− to 100 and Multidigit Solver. Encourage students to ask *What do I know?* and *What do I need to find out?* before solving each problem. Have them model how to solve the problem using a number line, Counters, or base-ten blocks. Students should then write the appropriate equation in vertical format and solve for the unknown.

Supported Practice

For additional support, use Counters to help students understand that they will sometimes add to solve a subtraction problem.

- Write the following subtraction story on the board: *I lost 12 pens. Now I only have 5 pens left. How many pens did I have before I lost some of them?*
▶ **What do we know? What do we need to find out?** We know you lost 12 pens and now you have 5. We do not know how many you had.
- Write the structure of a subtraction equation on the board, but leave the numbers blank.
▶ **What does the first number in a subtraction equation tell us? Do we know that number? What should we do?** It tells us about the whole set. We don't yet know that number. We should draw a question mark in the first blank.
- Have students supply the known numbers in the subtraction equation. ? − 12 = 5
- Count out 12 Counters and have a volunteer hold the Counters under the number 12 on the board. Count out 5 Counters and have another volunteer hold those counters under the number 5.
▶ **Do these Counters tell us about the whole set or about the parts that were separated from the whole set?** the parts that were separated

▶ **How can we find out how many pens I started with?** Put the two parts back together to make the whole set, or add the two parts to find the whole.

- Write $\begin{array}{r}12\\+\ 5\end{array}$ on the board.

▶ **How many pens did I start with? How do you know?** 17, because $12 + 5 = 17$

REFLECT

Think Critically

Review students' answers to the Reflect prompt at the bottom of **Student Workbook,** page 59, and then review the Engage activity.

▶ **Sometimes the greatest number in a subtraction word problem tells about the whole set and sometimes it does not. How can you tell the difference?** Answers may vary. Possible answer: Look for numbers that tell about the whole set, the amount that was taken away, and the amount that was left. If the number that shows the whole set is unknown, then the greatest number in the subtraction story tells about one of the parts.

Real-World Application

A store owner opened a box of new bike helmets. He put 6 helmets on display in the front of the store. He put 24 helmets on a rack.

▶ **How many bike helmets were in the box?** 30 bike helmets

ASSESS

Informal Assessment

Use the online or print Student Record, **Assessment,** page 128, to record informal observations.

Ask About It (Variation)
Did the student
☐ respond accurately?   ☐ respond with confidence?
☐ respond quickly?   ☐ self-correct?

Additional Practice

For additional practice, have students complete **Practice,** page 94.

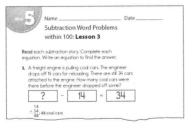

Practice, p. 94

Week 5 • Subtraction Word Problems within 100

Lesson 3

Key Idea
Subtraction stories tell us about something that was taken away. Sometimes we know how much there was at the beginning—the whole set—and sometimes we do not.
When we do not know how much was there before some was taken away, we can write a math equation, such as ___ − 12 = 37.

Try This

Read each word problem. Underline what you know. Circle what you need to find out.

1. Anita needs to write some thank-you notes. She has already written 14 of them and she has 15 left to write. How many thank-you notes did Anita need to write?

   Write the subtraction equation.
   ___ − 14 = 15

   Write an equation in vertical format to find the answer.
   $\begin{array}{r}14\\+15\\\hline 29\end{array}$; 29 thank-you notes

2. Raj practices the piano for the same amount of time each day. He has already practiced 17 minutes today, and he has 13 minutes left of practice time. How many minutes does Raj practice the piano each day?

   Write the subtraction equation.
   ___ − 17 = 13

   Write an equation in vertical format to find the answer.
   $\begin{array}{r}17\\+13\\\hline 30\end{array}$; 30 minutes

58 Level D Unit 4 Subtraction

Practice

Read each word problem. Think about what you know and what you need to find out. Use a blank to show what you do not know. Write an equation in vertical format to find the answer.

3. The servers have to refill all of a restaurant's ketchup bottles. So far, the servers have refilled 44 bottles. There are still 33 ketchup bottles left to refill. How many ketchup bottles need refilling?
   Write the subtraction equation. ___ − 44 = 33
   Find the answer. $\begin{array}{r}44\\+33\\\hline 77\end{array}$; 77 bottles

4. A group of children made some snowballs. They threw 24 of them at the neighbor's snow fort. They have 33 snowballs left. How many snowballs did the children start with?
   Write the subtraction equation. ___ − 24 = 33
   Find the answer. $\begin{array}{r}24\\+33\\\hline 57\end{array}$; 57 snowballs

Reflect

When you want to solve a subtraction equation such as ___ − 24 = 52, you can write 24 + 52 = ___.
Use words such as *whole set* and *part* to explain why this works.

Answers may vary; possible answer: I do not know how much is in the whole set, but I know how much was taken away and how much was left. These are parts of the whole set. If I put those two parts together, I can find out how much was in the whole set.

Week 5 Subtraction Word Problems within 100 • Lesson 3 **59**

Student Workbook, pp. 58–59

WEEK 5
Subtraction Word Problems within 100

Lesson 4

Objective
Students can solve one- and two-step subtraction word problems within 100 and with unknowns in all positions.

Standard CCSS
2.OA.1 Use addition and subtraction within 100 to solve one- and two-step word problems involving situations of adding to, taking from, putting together, taking apart, and comparing, with unknowns in all positions, e.g., by using drawings and equations with a symbol for the unknown number to represent the problem.

Creating Context
Help English Learners look for clues in subtraction word problems that will help them write an appropriate subtraction equation. Have students write down a subtraction story from a previous lesson. Tell them to highlight or circle words and phrases that give clues about what each number represents, such as *take away, put away, left, in all, to start with,* and so on.

Materials
Program Materials
- Neighborhood Number Line
- Counters

Additional Materials
note cards, 3 per student

Prepare Ahead
Prepare several subtraction stories to read to students, varying the placement of the unknown value. Use numbers up to 100. Use students' names within the word problems to help engage students in the lesson.

1 WARM UP

Prepare
- Review the parts of a subtraction equation. Point out that a part of the whole set is taken away, leaving the other part of the whole set.

▶ **What equation could you write to help you solve the problem 38 − _____ = 15? Why does that work?** 38 − 15; the whole set (38) is being separated into two parts, 15 and something else. I can take away either part from the whole set to find the other part.

▶ **What equation could you write to help you solve the problem _____ − 23 = 41? Why does that work?** 23 + 41; I know 23 is being taken away from the whole set, and I know there are 41 left. I know the two parts, so I can add them together to find the number that makes the whole set.

Just the Facts
Play a game of "Finger Flash." Tell students to "flash" the number of fingers that answers the question. Use questions such as the following:

▶ **What is 18 − 9?** Students should show 9 fingers.
▶ **What is 30 − 20?** Students should show 10 fingers.
▶ **What is 16 − 7?** Students should show 9 fingers.

2 ENGAGE

Develop: What's Missing?
"Today we are going to look for clues to help us solve subtraction word problems." Follow the instructions on the Activity Card **What's Missing?** As students complete the activity, be sure to use the Questions to Ask.

Activity Card 4Q

Alternative Grouping
Pair: Have one student write the equation in vertical format and solve while the other student models the problem on the Neighborhood Number Line or with Counters. Have partners compare their answers.

Progress Monitoring
If... students have difficulty determining which note card to hold up,

Then... have them write a subtraction equation on a piece of paper without the numbers. Reread the problem and have students fill in the blanks with the appropriate numbers. Students can then compare the subtraction equation they wrote with the blank note cards.

Practice
Have students complete **Student Workbook,** pages 60–61. Guide students through the Key Idea example and the Try This exercises.

Interactive Differentiation
Consult the **Teacher Dashboard** for grouping suggestions. You can also use performance on the Engage activity to guide students.

Independent Practice Building Blocks
For additional practice, have students complete Word Problems with Tools 8: Multidigit +/− (Adding Two Groups to 1000). Encourage students to ask *What do I know?* and *What do I need to find out?* before solving each problem.

Have students write an addition or subtraction equation in vertical form and solve for the unknown value.

Supported Practice Building Blocks
For additional support, guide students as they complete Word Problems with Tools 8: Multidigit +/− (Adding Two Groups to 1000).

- After listening to the first problem, have students read the problem again aloud. Point out that some of the problems will require addition and some will require subtraction.

▶ **What clues can you look for to decide whether you need to add or subtract?** Answers may vary. Possible answer: If two sets are joining together, then I will add. If something is being taken away, I know I need to subtract.

- Remind students to ask themselves *What do I know?* and *What do I need to find out?*
- Guide students to identify the number that represents the whole set, the amount that was taken away, and the amount that was left.
- Have students write and solve the appropriate equation in vertical format.

3 REFLECT

Think Critically

Review students' answers to the Reflect prompt at the bottom of **Student Workbook,** page 61, and then review the Engage activity.

▸ **What strategies do you use to help you solve subtraction stories?**
Answers may vary. Possible answers: I look for the numbers that represent the whole set, the amount that was taken away, and the amount that was left; I use a number line to help me understand the parts of the story; I model the numbers in the story with Counters.

Real-World Application

People use addition and subtraction to help them make decisions every day. Maybe you need to decide whether you should spend your allowance on a snow cone or wait until you earn more money so that you can buy a new baseball bat. Maybe you need to know if you will have enough stickers to share with everyone in your class if you give some of them to your best friend.

▸ **How have you used addition or subtraction to help you make decisions?**
Answers may vary. Possible answer: I have used subtraction to figure out if I have enough money to buy a movie ticket.

4 ASSESS

Informal Assessment

Use the online or print Student Record, **Assessment,** page 128, to record informal observations.

What's Missing?	
Did the student	
☐ respond accurately?	☐ respond with confidence?
☐ respond quickly?	☐ self-correct?

Additional Practice

For additional practice, have students complete **Practice,** page 95.

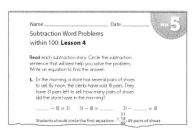

Practice, p. 95

Week 5 • Subtraction Word Problems within 100

Lesson 4

Key Idea
When you read a subtraction story, you need to find out what you do not know.

If you do not know the number that shows the whole set, you might write ___ − 32 = 12.

If you do not know the amount that was taken away, you might write 56 − ___ = 20.

If you do not know how much is left, you might write 74 − 33 = ___.

Try This
Write a math equation in vertical format to find what you do not know.

1. There are 38 children sledding down the hill. Some get cold and go inside. That leaves 16 children outside. How many went inside?

What do you need to find out?
how many children went inside

Write the subtraction equation.

38 − ___ = 16

Find the answer.
38
−16 ; 22 children
22

2. There are 49 dogs at the park. Some are puppies, and 36 are adult dogs. How many puppies are at the park?

What do you need to find out?
the number of puppies at the park

Write the subtraction equation.

49 − 36 = ___

Find the answer.
49
−36 ; 13 puppies
13

60 Level D Unit 4 Subtraction

Practice
Write a subtraction equation for each word problem. Then write a math equation in vertical format to find the answer.

3. A librarian has a cart full of books to put back on the shelf. He has put 12 books back already. He has 31 books left on the cart. How many books were on the cart? ___ − 12 = 31;   12
                                                                                                +31 ; 43 books
                                                                                                 43

4. Nathan found 77 pennies in his piggy bank. He spent 25 of them on a toy. How many pennies does he have left? 77 − 25 = ___;   77
                                                                                −25 ; 52 pennies
                                                                                 52

5. A pet store has 86 goldfish. At the end of the week, there are 63 goldfish left. How many goldfish did the pet store sell this week?
86 − ___ = 63;   86
                −63 ; 23 goldfish
                 23

Reflect
Subtraction stories are full of clues to help you write a correct subtraction equation. List some subtraction clue words that you might look for.

Answers may vary; possible answers: *gave away, spent, gone, left over*

Week 5 Subtraction Word Problems within 100 • Lesson 4   61

Student Workbook, pp. 60–61

WEEK 5
Subtraction Word Problems within 100

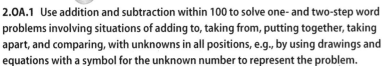

Objective
Students review skills learned this week and complete the weekly assessment and project.

Standard CCSS
2.OA.1 Use addition and subtraction within 100 to solve one- and two-step word problems involving situations of adding to, taking from, putting together, taking apart, and comparing, with unknowns in all positions, e.g., by using drawings and equations with a symbol for the unknown number to represent the problem.

Vocabulary
Review vocabulary introduced during the week.

Creating Context
Help English Learners understand how to identify a subtraction word problem and how to identify an addition word problem. Reread some of the word problems from the Word Problem activities in Building Blocks. Have students identify key words that indicate an addition problem, such as *in all*, *altogether*, and so on. Also have them identify key words that indicate a subtraction problem, such as *take away*, *left*, *how many more*, and so on.

1 WARM UP

Prepare

▶ **What are the three parts of a subtraction equation?** Possible answers: the whole set, the amount that was taken away, and the amount that is left; the minuend, the subtrahend, and the difference

▶ **What questions can you ask yourself to help you solve a subtraction story?** *What do I know?* and *What do I need to find out?*

2 ENGAGE

Practice
Have students complete **Student Workbook,** pages 62–63.

Week 5 • Subtraction Word Problems within 100

Lesson 5 Review

This week you learned about the parts of a subtraction story. You practiced asking questions to help you solve each problem, and you reviewed how to write a math equation in vertical format to find the answer.

Lesson 1 Match the numbers with the parts of a subtraction equation.

1. A treasure hunter finds 9 golden statues. He takes 5 of them home and leaves 4 of them hidden for another treasure hunter to find.

   takes 5 away — finds 9 statues — 4 are left
   [The Whole Set] − [Amount Taken Away] = [Amount Left]

Lesson 2 Read the word problem. Underline what you know. Circle what you need to find out.

2. The temperature yesterday was 47 degrees. The temperature went down during the night, and now it is only 35 degrees. How many degrees did the temperature go down during the night? Write a math equation in vertical format to find the answer.

   47
   −35 ; 12 degrees
   ___
   12

62 Level D Unit 4 Subtraction

Lesson 3 Read and solve each word problem.

3. Phil loves to ride roller coasters. So far, he has ridden 23 roller coasters on his list of roller coasters to ride. He has 12 roller coasters left to ride. How many roller coasters are on Phil's list?
   Write the subtraction equation. ___ − 23 = 12
   Write a math equation in vertical format to
   find the answer.   12
                    +23 ; 35 roller coasters
                    ___
                     35

Lesson 4 Write a math equation in vertical format to find the answer.

4. A total of 86 actors tried out for parts in a play. When the cast list was posted, 24 actors were picked for parts in the play. How many actors who tried out did not get a part?

   86
   −24 ; 62 actors
   ___
   62

Reflect
When you read a subtraction story, you need to look for all the parts of a subtraction equation in order to solve the problem. Explain the strategies you use or the clues you look for to solve a subtraction equation.

Answers may vary. Possible answer: First, I find out what I do not know. Then I use what I do know to solve the problem.

Week 5 Subtraction Word Problems within 100 • Lesson 5   63

Student Workbook, pp. 62–63

3 REFLECT

Think Critically

Review students' answers to the Reflect prompt at the bottom of **Student Workbook,** page 63.

Discuss the answer with the group to reinforce Week 5 concepts.

4 ASSESS

Formal Assessment

Students may take the weekly assessment online.

As an alternative, students may complete the weekly test on **Assessment,** pages 57–58. Record progress using the Student Assessment Record, **Assessment,** page 128.

Going Forward

Use the **Teacher Dashboard** to view results of the online assessments, to input the results of print student assessments, and to review progress before making decisions about next steps. Use the weekly test results and observations to determine the next steps for each student.

Retention	
Student displays good grasp of this week's concepts and skills.	Have students practice reading and solving subtraction stories. Instruct students to first identify the unknown value and then write an appropriate subtraction equation. Encourage students to write an equation in vertical format to find the answer.

Remediation	
Student is still struggling with the week's concepts and skills.	Use the Neighborhood Number Line to model how to solve a subtraction equation. Present subtraction equations such as $37 - 16 = \square$, $\square - 16 = 21$, and $37 - \square = 21$. Model how to count back and/or count on to find the unknown value.

Suggestions for Re-Evaluation: If a student has struggled without success for several weeks, use observations and test results to place the student at a level in which he or she can find success and build confidence to move forward.

Name _____ Date _____

Subtraction Word Problems within 100

1. Circle the subtraction sentence that matches this problem.

   A park has 16 statues. A group of 7 of them are beside the fountain in the middle of the park. How many statues are in other parts of the park?

   ___ − 7 = 16   16 − ___ = 7   (16 − 7 = ___)

2. Check the box for the unknown in this problem.

   Ruby is beginning a stamp collection. Her uncle gave her 6 stamps, and now she has 15 all together.

   ☐ the number of stamps her uncle gave her
   ☒ the number of stamps she had at the beginning
   ☐ the number of stamps she has now

Use this word problem to answer questions 3 and 4.
A herd of 31 deer were in a field. A car drove by and some of the deer ran away. Now there were only 12 deer in the field.

3. What do you need to find out in the problem?
   How many deer ran away?

4. Write a subtraction equation for the problem.
   31 − ? = 12 or equivalent

Subtraction Word Problems within 100

Use this word problem to answer questions 5 and 6.
A museum is showing 50 paintings by students. One gallery has 37 pictures, and the rest are in the hallway. How many are in the hallway?

5. Write a subtraction equation for the problem.
   50 − 37 = ?

6. Write the problem in the vertical format and find the answer.

   ```
     50
   − 37
   ----
     13
   ```

Use this word problem to answer questions 7 and 8.
A corn stalk is 23 inches tall. It will grow another 44 inches. How tall will the stalk be when it is fully grown?

7. Write a subtraction equation for the problem.
   ? − 23 = 44

8. Write the same problem in the vertical addition format and find the answer.

   ```
     44
   + 23
   ----
     67
   ```

Assessment, pp. 57–58

Project Preview

This week, students learned how to identify the parts of a subtraction equation within a word problem. The project for this unit requires students to use what they have learned during the week and apply that knowledge to a real-world situation. Students will use what they know to write and solve a subtraction word problem.

Project-Based Learning

Standards-driven Project-Based Learning is effective in building deep content understanding. Project-Based Learning increases long-term retention of concepts and has been shown to be more effective than traditional instruction. Completing a project to answer an essential question challenges students to apply and demonstrate mastery of concepts and skills by expressing understanding through discussion, research, and presentation.

Essential Question

WHEN would it be useful to write equations outside the classroom?

Project Evaluation Criteria

Review project evaluation criteria with students prior to beginning the project.

Exceeds Expectations
☐ Project result is explained and can be extended.
☐ Project result is explained in context and can be applied to other situations.
☐ Project result is explained using advanced mathematical vocabulary.
☐ Project result is described, and mathematics are used correctly and can be extended.
☐ Project result is explained and extended, and shows advanced knowledge of mathematical concepts and skills.

Meets Expectations
☐ Project result is explained.
☐ Project result is explained in context.
☐ Project result is explained using mathematical vocabulary.
☐ Project result is described, and mathematics are used correctly.
☐ Project result is explained, and shows satisfactory knowledge of mathematical concepts and skills.

Does Not Meet Expectations
☐ Project result is not explained.
☐ Project result is explained, but out of context.
☐ Project result is explained, but mathematical vocabulary is oversimplified.
☐ Project result is described, but mathematics are not used correctly.
☐ Project result is not explained and/or extended, or shows less than satisfactory knowledge of mathematical concepts and skills.

Write a Subtraction Story

Objective
Students can use what they know about solving subtraction word problems to compose and solve their own subtraction story.

Standard
2.OA.1 Use addition and subtraction within 100 to solve one- and two-step word problems involving situations of adding to, taking from, putting together, taking apart, and comparing, with unknowns in all positions, e.g., by using drawings and equations with a symbol for the unknown number to represent the problem.

Materials
Additional Materials
department-store advertisements

Prepare Ahead
Gather several department-store advertisements for students to use in the activity. Look for a variety of stores that sell food, clothing, toys, cars, or items that you know would interest your students.

Best Practices
- Select and provide the appropriate materials.
- Allow active learning with noise and movement.
- Provide project directions that are clear and brief.

Introduce

Your store already sells clothing. Many stores sell more than one type of item.

▸ What are some other things that you would like to sell in your store?

▸ What would you name your store?

Explore

▸ Today you will use what you learned this week about solving subtraction stories to write your own subtraction story.

- Lay out the store ads so that students can look through them.

▸ Find an ad that shows items like the ones you would like to sell in your store.

- Allow students time to browse through the store ads and choose one that best fits their interests.

▸ As you are looking at the items in the store ads, can you think of a subtraction story you could tell? What are some things you would include in a subtraction story? Answers may vary. Possible answers: the name of the store or the name of a customer, the name of an item for sale, the price of the item

▸ What are some questions you could ask in your subtraction story?
Answers may vary. Possible answers: How much money does the customer have left? How much did the item cost before it went on sale?

▸ Complete *Student Workbook,* page 64, to help you write a subtraction story.

Wrap Up

- Allow students adequate time to look through the store ads and choose an ad that interests them.
- Make sure students are using words and phrases that indicate a subtraction story rather than an addition story.
- Along with the ads, have students think about what they did in Weeks 1 through 4. They may find inspiration for their subtraction story in the sale tag, damage report, or inventory list they created.
- If students are struggling to write a story, have them first write a subtraction equation with an unknown. Have them identify whether the unknown represents the whole set, the part that was taken away, or the part that was left. Then have students write a question that asks about the unknown value. At that point students should be able to write an equation or two about the known parts of the situation.
- Discuss students' answers to the Reflect prompts at the bottom of *Student Workbook,* page 64.

Have students design and decorate a logo for their store.

If time permits, allow students to type their subtraction stories. Print the stories and save the stories and logos for use during next week's project.

Week 5 • Subtraction Word Problems within 100

Project
Write a Subtraction Story

Look at the department-store ads. Pretend that these items are for sale in your store.

1. What is your store's name? Which items will you sell?

   Answers may vary.

2. Choose one or two items you see in the ad. Write down the name of the item and how much it costs.

   Answers may vary.

3. Use your store's name and your chosen items in a subtraction story. Include two facts that the reader will know and a question for the reader to answer.

   Check students' work.

4. Write a subtraction equation to find the answer to your subtraction story. Make sure you include a symbol or simple drawing to show the unknown number.

   Check students' work.

Reflect
What did you have to think about as you were writing your subtraction story?

Answers may vary; possible answer: I had to think about the equation that I would write.

Student Workbook, p. 64

Teacher Reflect

☐ Did I supply the necessary materials?

☐ Did students organize their ideas?

☐ Did students' ideas logically relate to one another?

WEEK 6: Solving Subtraction Word Problems within 1,000

Week at a Glance
This week, students conclude **Number Worlds,** Level D, Subtraction, by solving subtraction word problems with regrouping. Students write and solve subtraction equations and check their answers using corresponding addition problems.

Skills Focus
- Subtract using concrete models.
- Write and solve subtraction equations that represent word problems.
- Understand the relationship between addition and subtraction.

How Students Learn
Encourage students to use what they know about addition facts to help them construct relationships for subtraction facts. Students need opportunities to experiment with numbers and relationships between numbers. After exploring subtraction with concrete manipulatives, students will learn to write and solve equations without using models, and extend their knowledge to relating addition equations to subtraction equations.

English Learners ELL
For language support, use the **English Learner Support Guide,** pages 86–87, to preview lesson concepts and teach academic vocabulary.

Math at Home
Give one copy of the Letter to Home, page 24, to each student. Encourage students to share and complete the activity with their caregivers.

Weekly Planner

Lesson	Learning Objectives
1 pages 348–349	Students can subtract within 1,000 using concrete models or drawings and understand the relationship between addition and subtraction.
2 pages 350–351	Students can subtract within 1,000 using concrete models or drawings and understand the relationship between addition and subtraction.
3 pages 352–353	Students can subtract within 1,000 using concrete models or drawings and understand the relationship between addition and subtraction.
4 pages 354–355	Students can subtract within 1,000 using concrete models or drawings and understand the relationship between addition and subtraction.
5 pages 356–357	**Review and Assess** Students review skills learned this week and complete the weekly assessment and project.
Project pages 358–359	Students can use what they know about solving subtraction problems to explain the parts of a subtraction equation to their peers.

Key Standard for the Week

Domain: Number and Operations in Base Ten

Cluster: Use place value understanding and properties of operations to add and subtract.

2.NBT.7 Add and subtract within 1000, using concrete models or drawings and strategies based on place value, properties of operations, and/or the relationship between addition and subtraction; relate the strategy to a written method. Understand that in adding or subtracting three-digit numbers, one adds or subtracts hundreds and hundreds, tens and tens, ones and ones; and sometimes it is necessary to compose or decompose tens or hundreds.

Materials		Technology
Program Materials • **Student Workbook,** pp. 65–67 • **Practice,** p. 96 • Activity Card 4R, **Race to Zero** • Number Construction Mat • Neighborhood Number Line • Number 1–6 Cube • Number Cards (20–99) • Spinner	**Additional Materials** • plastic straws • rubber bands	*Teacher Dashboard*
Program Materials • **Student Workbook,** pp. 68–69 • **Practice,** p. 97 • Activity Card 4S, **Neighborly Trading** • Number Construction Mat • Number Cards (10–99)	**Additional Materials** base-ten blocks*	*Teacher Dashboard* 🔘 Base 10 Blocks Tool
Program Materials • **Student Workbook,** pp. 70–71 • **Practice,** p. 98 • Activity Card 4T, **Going Shopping** • Number Construction Mat	**Additional Materials** • base-ten blocks* • department-store ads • dry-erase boards • note cards	*Teacher Dashboard* 🔘 Base 10 Blocks Tool
Program Materials • **Student Workbook,** pp. 72–73 • **Practice,** p. 99 • Activity Card 4T, **Going Shopping** • Counters	**Additional Materials** • department-store ads • dry-erase boards • note cards	*Teacher Dashboard*
Program Materials • **Student Workbook,** pp. 74–75 • Weekly Test, **Assessment,** pp. 59–60		Review previous activities.
Program Materials **Student Workbook,** p. 76	**Additional Materials** • poster paper • glue sticks • scissors	

*Available from McGraw-Hill Education

WEEK 6
Solving Subtraction Word Problems within 1,000

Find the Math
In this week, students will use models and equations in vertical format to solve subtraction problems with and without regrouping.

Use the following to begin a guided discussion:

▶ **What are some different ways you can show 65¢ with dimes and pennies?** Answers may vary; possible answers: 6 dimes and 5 pennies; 5 dimes and 15 pennies.

Have students complete *Student Workbook,* page 65.

Student Workbook, p. 65

Lesson 1

Objective
Students can subtract within 1,000 using concrete models or drawings and understand the relationship between addition and subtraction.

Standard
2.NBT.7 Add and subtract within 1000, using concrete models or drawings and strategies based on place value, properties of operations, and/or the relationship between addition and subtraction; relate the strategy to a written method. Understand that in adding or subtracting three-digit numbers, one adds or subtracts hundreds and hundreds, tens and tens, ones and ones; and sometimes it is necessary to compose or decompose tens or hundreds.

Creating Context
Help English Learners understand the concept of regrouping by making a list of things that come in groups, such as 12 eggs, 5 fingers, 4 wheels on a car, and so on. Point out that these items can be bundled together to form one group, or they can be separated and counted individually.

Materials
Program Materials
- Number Construction Mat, 1 per pair
- Neighborhood Number Line
- Number 1–6 Cube, 1 per pair
- Number Cards (20–99), 1 set per pair
- Spinner

Additional Materials
- plastic straws, 100 for each pair of students
- rubber bands, 10 for each pair of students

Prepare Ahead
Remove Number Cards 0–19 from the set.

1 WARM UP

Prepare
- Have a volunteer model the number 34 with straws. The student should hold three bundles of 10 and four individual straws.
- Show students how to regroup the straws by unwrapping one group of 10 straws and placing them with the individual straws. Have students count the straws to see that the total number does not change: $4 + 10 + 10 + 10 = 34$ and $14 + 10 + 10 = 34$.

2 ENGAGE

Develop: Race to Zero
"Today we are going to play a game to help us practice subtracting." Follow the instructions on the Activity Card **Race to Zero**. As students complete the activity, be sure to use the Questions to Ask.

Activity Card 4R

Alternative Groupings
Individual: Play the game with the student.

Small Group: Complete the activity as written. Group members will take turns rolling the Number Cube and taking away straws from the Number Construction Mat.

Progress Monitoring
| **If...** students have difficulty understanding when to regroup the straws, | ▶ **Then...** have students take away individual straws one at a time and then unwrap a bundle of ten when necessary. |

Practice
Have students complete *Student Workbook,* pages 66–67. Guide students through the Key Idea example and the Try This exercises.

Interactive Differentiation
Consult the **Teacher Dashboard** for grouping suggestions. You can also use performance on the Engage activity to guide students.

Independent Practice
For additional practice, write several subtraction equations on the board (two-digit minus one-digit), including some that require regrouping and some that do not. Have students work with a partner. Instruct them to first

discuss which equations require regrouping and which do not. Then have one student in the pair model the subtraction problem with straws on a Number Construction Mat. The other student in the pair should use the Neighborhood Number Line to subtract. Have partners compare answers and then switch roles to solve the next equation.

Supported Practice

For additional support, have several students use straws to model a number with the ones digit greater than 5, such as 28.

- Have a volunteer spin the Spinner.
- ▶ **The Spinner landed on the number 2, so we will take 2 away from 28. Do we take away from the bundles of ten or from the individual straws? How do you know?** We take away from the individual straws because the number 2 tells us about 2 ones, not 2 tens.
- Model for students how to take away 2 straws. Count the remaining straws by tens and then count on by ones.
- Have several students use straws to model a number with a small ones digit, such as 32. Have a volunteer spin the Spinner.
- ▶ **The Spinner landed on the number 4. We will take 4 away from 32. Let's start taking away individual straws: 1, 2. We don't have enough. Where can we get more straws?** from the bundles of ten
- Guide students as they unwrap one bundle of ten and add 10 straws to the 2 individual straws.
- Model how to take away 4 straws and then count to find the remaining number of straws.

  REFLECT

Think Critically

Review students' answers to the Reflect prompt at the bottom of **Student Workbook,** page 67, and then review the Engage activity.

▶ **What does it mean to regroup?** Answers may vary. Possible answer: to make new groups.

  ASSESS

Informal Assessment

Use the online or print Student Record, **Assessment,** page 128, to record informal observations.

Race to Zero
Did the student
- ☐ provide a clear explanation?
- ☐ communicate reasons and strategies?
- ☐ choose appropriate strategies?
- ☐ argue logically?

Additional Practice

For additional practice, have students complete **Practice,** page 96.

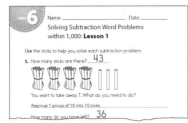
Practice, p. 96

Week 6 • Solving Subtraction Word Problems within 1,000

Lesson 1

Key Idea
You can show the number 20 with two groups of 10. You can also show the number 20 using one group of 10 sticks and 10 individual sticks. This is called regrouping.

The number stays the same, but the way that the sticks are grouped does not.

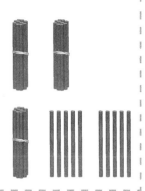

Try This
Find each answer using straw bundles.

1. 56
   − 5
   51

2. 32
   − 7
   25

3. 25
   − 8
   17

4. 31
   − 3
   28

5. 46
   − 7
   39

6. 24
   − 3
   21

66 Level D Unit 4 Subtraction

Practice
Find each answer using straw bundles.

7. 54
   − 9
   45

8. 42
   − 6
   36

9. 28
   − 8
   20

10. Belton had 21 inches of snow last year. Rock City had 4 in. of snow. How many more inches of snow fell in Belton than in Rock City?

   A. What do you know?
      Belton had 21 in. of snow last year and Rock City had 4 in.

   B. What do you need to find out?
      how many more inches of snow fell in Belton than in Rock City

   C. Write a subtraction equation.
      21 − 4 = ?

   D. Use straw bundles to help solve the problem.
      Belton had 17 more in. of snow than Rock City.

Reflect
Explain how you knew when you needed to regroup.

Answers may vary; possible answer: In Problem 1, when I was taking away 5 from 56, I did not need to regroup because there were 6 individual straws and I only needed to take 5 away. When I was taking 8 away from 25 in Problem 3, I knew I needed to regroup because 8 straws is more than 5 straws.

Week 6 Solving Subtraction Word Problems within 1,000 • Lesson 1   67

Student Workbook, pp. 66–67

WEEK 6

Solving Subtraction Word Problems within 1,000

Lesson 2

Objective
Students can subtract within 1,000 using concrete models or drawings and understand the relationship between addition and subtraction.

Standard
2.NBT.7 Add and subtract within 1000, using concrete models or drawings and strategies based on place value, properties of operations, and/or the relationship between addition and subtraction; relate the strategy to a written method. Understand that in adding or subtracting three-digit numbers, one adds or subtracts hundreds and hundreds, tens and tens, ones and ones; and sometimes it is necessary to compare or decompose tens or hundreds.

Creating Context
Help English Learners understand the concept of trading. Hold an item in your hand, such as a book, while the student holds another item, such as a box of markers. Say, "I will trade you my book for your markers." Act out the process of trading one item for another.

Materials

Program Materials
- Number Construction Mat, 1 per student
- Number Cards (10–99)

Additional Materials
base-ten blocks

Prepare Ahead
Remove the Number Cards 0–9 from the deck. They will not be used in the Independent and Supported Practice.

1 WARM UP

Prepare
Review with students how to make a trade to model regrouping.

- Have a volunteer model the number 31 with base-ten blocks.
- ▶ **I want to take away 7 base-ten unit blocks. What do I need to do?** Trade 1 base-ten rod for 10 base-ten blocks.
- Guide the volunteer to take 1 base-ten rod to the "bank" and trade it in for 10 base-ten blocks. Have the student count the blocks to show that the value of the blocks has not changed.
- ▶ **Now can I take away 7 base-ten blocks?** Yes **How do you know?** There are now 11 base-ten blocks, and 11 is greater than 7.

Just the Facts
Say a variety of simple subtraction equations that students may encounter when subtracting in the ones column. If there is a possible solution, have students show you the answer with their fingers. If the equation cannot be solved using positive whole numbers, have students cover their eyes. Use questions such as the following:

- ▶ **What is 9 — 3?** Students show 6 fingers.
- ▶ **What is 5 — 8?** Students cover their eyes.
- ▶ **What is 2 — 3?** Students cover their eyes.

2 ENGAGE

Develop: Neighborly Trading
"Today we are going to practice making trades to help us subtract." Follow the instructions on the Activity Card **Neighborly Trading**. As students complete the activity, be sure to use the Questions to Ask.

Alternative Grouping
Pair: Have one partner subtract using base-ten blocks while the other partner solves the subtraction equation. Then have students compare answers and switch roles.

Activity Card 4S

Progress Monitoring

| **If…** students have difficulty determining when they need to regroup, | ▶ **Then…** have students circle the greater number in the ones column. If the greater number is on the bottom, students know they need to regroup. |

Practice
Have students complete **Student Workbook,** pages 68–69. Guide students through the Key Idea example and the Try This exercises.

Interactive Differentiation
Consult the **Teacher Dashboard** for grouping suggestions. You can also use performance on the Engage activity to guide students.

Independent Practice
For additional practice, instruct students to draw two Number Cards. Students should place the greater number above the lesser number and write a subtraction equation to mirror the two cards. First have students use the Base 10 Blocks Tool to find the difference. Then have them solve the equation in vertical format. Remind students to regroup when necessary.

Supported Practice
For additional support, draw two Number Cards, such as 52 and 27, and place them in vertical format with the greater number on top.

- Have students use the Base 10 Blocks Tool to model the greater number.
- ▶ **We want to take away 27 blocks from the 52 blocks we already have. Where should we start?** Start by looking at the ones.
- ▶ **We need to take away 7 ones. Let's start counting: 1, 2. There are only 2 ones. We can't take away 7. What do we need to do?** We need to regroup.
- Guide students to select a base-ten rod and click the Break-Down button (hammer icon) to regroup the base-ten rod into 10 base-ten blocks.
- ▶ **Now how many ones do you have? Can you take away 7?** I have 12 ones, so I can take 7 away.
- Help students use the Base 10 Blocks Tool to complete the subtraction problem. Then model how to solve the problem on paper in vertical format.
- Have each student choose two Number Cards to solve a new subtraction equation. Instruct them to write the subtraction equation in vertical format. As students use the Base 10 Blocks Tool to find the difference, tell them to record each step on the vertical equation.

350 Level D Unit 4 **Subtraction**

3 REFLECT

Think Critically

Review students' answers to the Reflect prompt at the bottom of **Student Workbook,** page 69, and then review the Engage activity.

▶ **How do you know when you need to regroup?** Answers may vary. Possible answer: If the greater number in the ones column is on the bottom, I know I do not have enough to subtract. I need to regroup.

Real-World Application

Kaitlyn has a one-dollar bill. She wants to buy a sticker. Stickers cost 20¢.

▶ **What does Kaitlyn need to do to buy the sticker?** She needs to regroup her one-dollar bill by trading it for 10 dimes. Then she can spend 2 dimes on the sticker.

4 ASSESS

Informal Assessment

Use the online or print Student Record, **Assessment,** page 128, to record informal observations.

Neighborly Trading

Did the student
- [ ] provide a clear explanation?
- [ ] choose appropriate strategies?
- [ ] communicate reasons and strategies?
- [ ] argue logically?

Additional Practice

For additional practice, have students complete **Practice,** page 97.

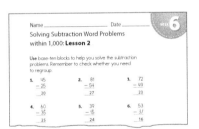

Practice, p. 97

Week 6 • Solving Subtraction Word Problems within 1,000

Lesson 2

Key Idea
Sometimes you need to regroup the number of tens and ones to subtract.

$$\begin{array}{r}\overset{1\,13}{\cancel{2}\cancel{3}}\\ -15\\ \hline 8\end{array}$$

Try This
Find each answer.

1. Finley can jump 62 centimeters on his skateboard. Micah can jump 43 cm. How many more centimeters can Finley jump than Micah?

   A. What do you know?

   Finley can jump 62 cm and Micah can jump 43 cm.

   B. What do you need to find out?

   how many more centimeters Finley can jump than Micah

   C. Write a subtraction equation.

   $62 - 43 = ?$

   D. Write the subtraction equation in vertical format. Find the answer.

   $\begin{array}{r}62\\ -43\\ \hline 19\end{array}$; Finley can jump 19 more cm.

68 Level D Unit 4 Subtraction

Practice
Find each answer.

2. There are 42 students in the art club. 15 students have finished painting pictures. The other students have not finished painting. How many students still need to finish their pictures?

   A. What do you know?

   There are 42 students total, and 15 have finished their pictures.

   B. What do you need to find out?

   the number of students who still need to finish their pictures

   C. Write a subtraction equation.

   $42 - ? = 15$

   D. Write a subtraction equation in vertical format. Find the answer.

   $\begin{array}{r}45\\ -15\\ \hline 27\end{array}$; 27 students still need to finish their pictures.

Reflect
Describe the steps you take to solve a subtraction problem. Make sure you explain how to regroup.

Answers may vary; possible answer: First, I look in the ones column. If the greater number is on the bottom, I know I need to regroup. I take 1 ten away from the tens column and break it apart into 10 ones. I add the 10 ones to the ones column. I subtract the ones first, then I subtract the tens.

Week 6 Solving Subtraction Word Problems within 1,000 • Lesson 2 69

Student Workbook, pp. 68–69

WEEK 6
Solving Subtraction Word Problems within 1,000

Lesson 3

Objective
Students can subtract within 1,000 using concrete models or drawings and understand the relationship between addition and subtraction.

Standard
2.NBT.7 Add and subtract within 1000, using concrete models or drawings and strategies based on place value, properties of operations, and/or the relationship between addition and subtraction; relate the strategy to a written method. Understand that in adding or subtracting three-digit numbers, one adds or subtracts hundreds and hundreds, tens and tens, ones and ones; and sometimes it is necessary to compose or decompose tens or hundreds.

Creating Context
Help English Learners understand the importance of using place value to write equations in vertical format. Show students a paragraph out of a book. Point out that all the words are aligned on the left side of the page. Explain to students that when writing an equation in vertical format, the numbers are aligned on the right side of the equation. Show students a Number Construction Mat. Demonstrate the columns for ones, tens, and hundreds.

Materials

Program Materials
 Number Construction Mat

Additional Materials
- base-ten blocks, 1 set per student
- department-store ads
- dry-erase boards
- note cards

Prepare Ahead
Gather several different department-store advertisements. The sale prices should be two- and three-digit numbers. On individual note cards, write different three-digit dollar amounts that are greater than the prices of the items in the ads. Also, prepare 15–20 two- and three-digit number cards, such as 580, 219, 27, and so on, for use in the Independent Practice.

1 WARM UP

Prepare
Review how to use place value to find the greater number.

- Write the numbers 347 and 377 on the board.
- ▶ **The hungry alligator (> or <) always wants to eat the greater number. Which place value should we look at first?** hundreds
- ▶ **Both numbers have 3 hundreds. Now what do we do?** Look at the tens place.
- ▶ **There are 4 tens and 7 tens. Which is greater? Which number does the hungry alligator want to eat?** Seven tens is greater than 4 tens. The hungry alligator wants to eat the number 377.

Just the Facts
Play a game of "Finger Flash" to help students practice basic subtraction.
- ▶ **What is 10 − 6?** Students show 4 fingers.
- ▶ **What is 13 − 5?** Students show 8 fingers.
- ▶ **What is 16 − 9?** Students show 7 fingers.

2 ENGAGE

Develop: Going Shopping
"Today we are going to pretend to go shopping to help us practice our subtraction skills." Follow the instructions on the Activity Card **Going Shopping**. As students complete the activity, be sure to use the Questions to Ask.

Activity Card 4T

Alternative Grouping
Pair: Have one partner use the base-ten blocks to subtract while the other partner solves the vertical subtraction equation. Have students compare answers and switch roles.

Progress Monitoring

| **If...** students have difficulty writing a vertical subtraction equation, | ▶ **Then...** have students write the equation on a Number Construction Mat. Guide students to place the ones digit in the ones column, the tens digit in the tens column, and the hundreds digit in the hundreds column. |

Practice
Have students complete **Student Workbook,** pages 70–71. Guide students through the Key Idea example and the Try This exercises.

Interactive Differentiation
Consult the **Teacher Dashboard** for grouping suggestions. You can also use performance on the Engage activity to guide students.

Independent Practice

For additional practice, have pairs of students work with the Base 10 Blocks Tool. Give each pair of students several of the two- and three-digit number cards. Have one partner draw a number card and write a vertical subtraction equation using 1,000 as the minuend and the number on the card as the subtrahend. Both students will solve the equation. One student will use the Base 10 Blocks Tool to make trades and then subtract while the other student solves the equation on paper. Have students switch roles and draw a new card. This becomes the subtrahend of a new equation that uses the difference from the previous equation as the minuend. Have students continue the activity until they can no longer subtract and get a positive number.

Supported Practice

For additional support, guide students through the process of subtracting across zeros.

- Write 205 − 76 in vertical format on the Number Construction Mat. Have students model 205 with the Base 10 Blocks Tool.
- ▶ **Look at the ones column. Can you take away 6 from 5? What do you need to do?** No; I need to go next door to the tens house to borrow a group of ten.
- ▶ **Look at Tom Ten's house. Does he have anything we can borrow?** no
- ▶ **Because Tom Ten does not have any tens, you will go to his neighbor, Harold Hundred, to borrow from him.**
- Model for students how to trade 1 base-ten hundred flat for 10 base-ten rods. Show students how to record the regrouping in the equation on the Number Construction Mat.

352 Level D Unit 4 **Subtraction**

- ▶ **Now how many tens does Tom have? Can Olivia One borrow from Tom now?** 10; yes
- ▶ **If Olivia borrows a group of ten from Tom, how many tens does Tom have now? How many ones does Olivia have?** 9; 15
- Guide students through the process of using the Base-Ten Tool to subtract 76 base-ten blocks. Then have students solve the vertical equation on paper.
- Continue the activity with additional equations that require subtracting across zeros.

3 REFLECT

Think Critically

Review students' answers to the Reflect prompt at the bottom of **Student Workbook,** page 71, and then review the Engage activity.

▶ **When you write a vertical subtraction problem, which number always goes on top?** the greater number, the minuend

Real-World Application

Andy has $20. He bought a book for $9.

▶ **How much money does Andy have left?** $11

4 ASSESS

Informal Assessment

Use the online or print Student Record, **Assessment,** page 128, to record informal observations.

Going Shopping

Did the student
- ☐ provide a clear explanation?
- ☐ communicate reasons and strategies?
- ☐ choose appropriate strategies?
- ☐ argue logically?

Additional Practice

For additional practice, have students complete **Practice,** page 98.

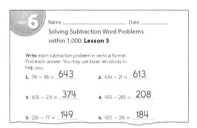

Practice, p. 98

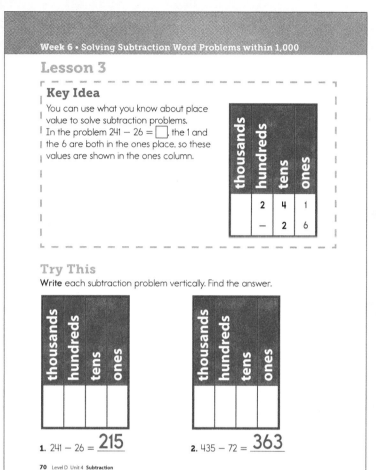

Student Workbook, pp. 70–71

Week 6 **Solving Subtraction Word Problems within 1,000** • Lesson 3 **353**

WEEK 6
Solving Subtraction Word Problems within 1,000

Lesson 4

Objective
Students can subtract within 1,000 using concrete models or drawings and understand the relationship between addition and subtraction.

Standard
2.NBT.7 Add and subtract within 1000, using concrete models or drawings and strategies based on place value, properties of operations, and/or the relationship between addition and subtraction; relate the strategy to a written method. Understand that in adding or subtracting three-digit numbers, one adds or subtracts hundreds and hundreds, tens and tens, ones and ones; and sometimes it is necessary to compose or decompose tens or hundreds.

Creating Context
Remind English Learners how addition and subtraction are related. Use Counters to make a group of 15. Take 7 Counters away to model the process of subtraction. Put the two groups back together to model addition.

Materials
Program Materials
Counters

Additional Materials
- department-store ads
- dry-erase boards
- note cards

Prepare Ahead
Gather several different department store advertisements. The sale prices should be two- and three-digit numbers. On individual note cards, write several different three-digit dollar amounts that are greater than the price of the items in the ads.

1 WARM UP

Prepare
▶ You already know that addition and subtraction are related. When people are related, we say they are in the same family. Addition and subtraction facts have families too. Addition and subtraction facts are in the same family when all the numbers are the same. They just come in a different order.

- Write three related numbers on the board, such as 2, 7, and 9. Have students use the numbers to write two addition and two subtraction equations.

Just the Facts
Say one addition fact and one subtraction fact. Have students cheer if they are related and boo if they are not. Use facts such as the following:

▶ $9 + 3 = 12$ and $12 - 3 = 9$ *cheer*
▶ $4 + 6 = 10$ and $10 - 3 = 7$ *boo*
▶ $8 + 5 = 13$ and $13 - 5 = 8$ *cheer*

2 ENGAGE

Develop: Going Shopping (Variation)
"Today we are going to go shopping again, but this time we are going to use addition to check to make sure our answers are correct." Follow the instructions on the Activity Card **Going Shopping (Variation)**. As students complete the activity, be sure to use the Questions to Ask.

Activity Card 4T

Progress Monitoring

| If... students have difficulty writing a related addition equation, | Then... tell them to reverse their subtraction equation. Have students rewrite the sum and subtrahend as an addition problem. |

Practice
Have students complete **Student Workbook,** pages 72–73. Guide students through the Key Idea example and the Try This exercises.

Interactive Differentiation
Consult the **Teacher Dashboard** for grouping suggestions. You can also use performance on the Engage activity to guide students.

Independent Practice
For additional practice, write several vertical subtraction equations on the board, including some with correct answers and some with incorrect answers. Have students check each answer by writing an addition equation. Instruct students to make corrections when necessary.

Supported Practice
For additional support, use Counters with students to model how to add to check subtraction equations.

- Give several students 60 Counters each. Instruct them to count out 52 Counters and then take 23 away.
▶ **How many Counters are left? Did you add or subtract?** 29; I subtracted.
▶ **You now have a group of 29 Counters and a group of 23 Counters. What will happen if you put the two groups back together again? Do you need to add or subtract?** I will have 52 Counters; I need to add.
- Have students solve $52 - 23$ by writing a vertical equation.
▶ **How can you check that the answer is correct?** Add the difference to 23. If the difference is correct, then the sum of the two numbers will be the minuend, 52.
- Engage students in a discussion about how addition and subtraction are related. Emphasize that subtraction starts with a whole group and divides it into two parts, and addition starts with those same two parts and joins them together to make a whole again.

3 REFLECT

Think Critically

Review students' answers to the Reflect prompt at the bottom of *Student Workbook,* page 73, and then review the Engage activity.

▶ **What parts of a subtraction problem do you use to write a related addition problem?** the difference and the subtrahend

Real-World Application

In the classroom, we use addition to check the answer to a subtraction problem because we want to make sure we solved the problem correctly.

▶ **Describe a situation outside the classroom when it would be a good idea to use addition to check a subtraction problem.** Answers may vary. Possible answer: Use addition to make sure you received the correct amount of change at the store.

4 ASSESS

Informal Assessment

Use the online or print Student Record, *Assessment,* page 128, to record informal observations.

Going Shopping (Variation)

Did the student

☐ provide a clear explanation?   ☐ choose appropriate strategies?

☐ communicate reasons and strategies?   ☐ argue logically?

Additional Practice

For additional practice, have students complete *Practice,* page 99.

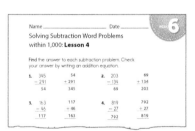

Practice, p. 99

Week 6 • Solving Subtraction Word Problems within 1,000

Lesson 4

Key Idea

Addition is the opposite of subtraction, so you can write an addition problem to check the answer to a subtraction problem.

$$\begin{array}{r}721\\-88\\\hline 633\end{array} \qquad \begin{array}{r}633\\+88\\\hline 721\end{array}$$

The sum matches the minuend, so the difference is correct.

Try This

Check the answers to each subtraction problem by writing and solving an addition equation. If the answer to the subtraction problem is incorrect, find the correct answer.

1. $\begin{array}{r}682\\-75\\\hline 607\end{array}$ $607 + 75 = 682$

2. $\begin{array}{r}345\\-87\\\hline 258\end{array}$ $258 + 87 = 345$

3. $\begin{array}{r}408\\-25\\\hline 423\end{array}$ 408 $423 + 25 = 448;$ $408 - 25 = 383$

Practice

Write each subtraction problem in vertical format. Check your answers by writing an addition equation.

4. $249 - 107 = $ __142; 142 + 107 = 249__

5. $418 - 234 = $ __184; 184 + 234 = 418__

72 Level D Unit 4 **Subtraction**

Complete the problem.

6. A bridge in Colorado is 486 feet tall. A bridge in Michigan is 199 ft tall. How much taller is the Colorado bridge than the Michigan bridge?

   A. What do you know?

   The Colorado bridge is 486 ft tall. The Michigan bridge is 199 ft tall.

   B. What do you need to find out?

   the difference in height between the two bridges

   C. Write a subtraction equation in vertical format and find the answer.

   $\begin{array}{r}486\\-199\\\hline 287\end{array}$; The Colorado bridge is 287 ft taller than the Michigan bridge.

   D. Write an addition equation to check your answer.

   $287 + 199 = 486$

   E. Was your answer correct? If not, why not?

   Answers may vary.

Reflect

Explain how you can write and solve an addition equation to check a subtraction equation answer.

Answers may vary; possible answer: When I subtract, I take a part away from a whole set and the other part is left. If I put the two parts back together again, they make the same whole set, so I know I subtracted correctly.

Week 6 Solving Subtraction Word Problems within 1,000 • Lesson 4 73

Student Workbook, pp. 72–73

WEEK 6
Solving Subtraction Word Problems within 1,000

Lesson 5 Review

Objective
Students review skills learned this week and complete the weekly assessment and project.

Standard
2.NBT.7 Add and subtract within 1000, using concrete models or drawings and strategies based on place value, properties of operations, and/or the relationship between addition and subtraction; relate the strategy to a written method. Understand that in adding or subtracting three-digit numbers, one adds or subtracts hundreds and hundreds, tens and tens, ones and ones; and sometimes it is necessary to compose or decompose tens or hundreds.

Creating Context
Help English Learners understand that unwrapping the straw bundles, trading the base-ten blocks, and borrowing from the tens house all demonstrate the process of regrouping. Have students list strategies they have used for addition and subtraction, such as counting Counters, stepping on a number line, learning doubles facts, and so on. Point out that more than one strategy can be used to arrive at the same answer.

1 WARM UP

Prepare

▶ **How do you know which number goes on top when you write a subtraction problem in vertical format?** The greater number always goes on top.

▶ **When you subtract, which place value do you always start with?** the ones

▶ **How do you know when you need to regroup?** Answers may vary. Possible answer: You need to regroup when the bottom number in the column is greater than the top number in that column.

2 ENGAGE

Practice
Have students complete **Student Workbook,** pages 74–75.

Week 6 • Solving Subtraction Word Problems within 1,000

Lesson 5 Review

This week you learned about subtraction equations. You learned that sometimes you need to regroup before you can subtract. You can regroup 1 hundred into 10 tens. You can regroup 1 ten into 10 ones.

Lesson 1 Find each answer.

1. 28 − 7 = **21**
2. 54 − 8 = **46**
3. 43 − 7 = **36**

4. Joel has 31 bugs in his collection. 8 are beetles. How many of them are not beetles?
   23 bugs are not beetles.

Lesson 2 Solve each equation. You may use base-ten blocks to help you.

5. 74 − 48 = **26**
6. 47 − 23 = **24**
7. 93 − 51 = **42**

8. Art has 53 coins on his dresser. He is saving 14 of them to use for snacks at the baseball game. He puts the rest into his coin jar. How many coins does Art put into his coin jar?
   Art puts 39 coins in the jar.

74 Level D Unit 4 Subtraction

Lesson 3 Find the answer.

9. Last year, 294 days had at least some sunshine. How many days last year had no sunshine? Note: There are 365 days in 1 year.
   71 days had no sunshine.

Lesson 4 Solve each subtraction problem. Write an addition equation to check your answer.

10. 804 − 27 = **777**   777 + 27 = 804
11. 358 − 64 = **294**   294 + 64 = 358

12. There are 326 birds and 245 mammals at the city zoo. How many more birds are there than mammals?
    There are 81 more birds.

Reflect
Why is it important to start in the ones column when you are subtracting? Explain your answer.

Answers may vary; possible answer: If there are not enough ones in the greater number to subtract, I need to regroup. If I started subtracting in the tens or hundreds column, I would not be able to correctly regroup the numbers in the minuend.

Week 6 Solving Subtraction Word Problems within 1,000 • Lesson 5 75

Student Workbook, pp. 74–75

3 REFLECT

Think Critically

Review students' answers to the Reflect prompt at the bottom of **Student Workbook,** page 75.

Discuss the answer with the group to reinforce Week 6 concepts.

4 ASSESS

Formal Assessment

Students may take the weekly assessment online.

As an alternative, students may complete the weekly test on **Assessment,** pages 59–60. Record progress using the Student Assessment Record, **Assessment,** page 128.

Going Forward

Use the **Teacher Dashboard** to view results of the online assessments, to input the results of print student assessments, and to review progress before making decisions about next steps. Use the weekly test results and observations to determine the next steps for each student.

Retention	
Student displays good grasp of this week's concepts and skills.	Have students choose two Number Cards (10–99) and write a vertical subtraction equation. Tell students to solve the equation, regrouping when necessary. Encourage students to check their answers with addition. Then supply students with a three-digit subtraction equation. Have them repeat the solve-and-check process.

Remediation	
Student is still struggling with the week's concepts and skills.	Have students choose two Number Cards (10–99) and write a vertical subtraction equation. Instruct students to circle the greater number in the ones column to determine whether regrouping is necessary. Have students use base-ten blocks or other concrete manipulatives to subtract. Then have students use the Base 10 Blocks Tool to model another two-digit subtraction equation. Finally, have students use the standard algorithm to solve a two-digit subtraction problem in vertical format.

Suggestions for Re-Evaluation: If a student has struggled without success for several weeks, use observations and test results to place the student at a level in which he or she can find success and build confidence to move forward.

Name _____ Date _____ WEEK 6

Solving Subtraction Word Problems within 1,000

Use this problem to answer questions 1–4.

The temperature was 28 degrees at 8:00 in the morning. By noon, it was up to 51 degrees. How much did the temperature rise?

1. What do you know?
   The temperature went from 28 to 51 degrees.

2. What do you need to find out?
   how many degrees did the temperature rise from 8:00 to noon

3. Write a subtraction equation. $51 - 28 = ?$

4. Write the subtraction equation in vertical format. Find the answer.

   $$\begin{array}{r} 51 \\ -28 \\ \hline 23 \end{array}$$

Level D Unit 4 Week 6   59

WEEK 6   Name _____ Date _____

Solving Subtraction Word Problems within 1,000

Use this problem to answer questions 5–9.

When the baseball game started, 731 people were in the stands. It started to rain and 154 people left. How many people stayed to watch the game?

5. What do you know?
   The game started with 731 people and 154 left.

6. What do you need to find out?
   How many people stayed

7. Write a subtraction equation.
   $731 - 154 = ?$

8. When you subtract, will you need to regroup? How do you know?
   Yes. There are fewer ones and tens in 731 than in 154

9. How many people stayed to watch the game? 577

60 Level D Unit 4 Week 6

Assessment, pp. 59–60

Project Preview

This week, students learned how to solve subtraction equations with and without regrouping. Students will use what they have learned during the whole unit to illustrate their own subtraction story and then explain the parts of a subtraction equation to their peers.

Project-Based Learning

Standards-driven Project-Based Learning is effective in building deep content understanding. Project-Based Learning increases long-term retention of concepts and has been shown to be more effective than traditional instruction. Completing a project to answer an essential question challenges students to apply and demonstrate mastery of concepts and skills by expressing understanding through discussion, research, and presentation.

Essential Question

WHEN would it be useful to write equations outside the classroom?

Project Evaluation Criteria

Review project evaluation criteria with students prior to beginning the project.

Exceeds Expectations
☐ Project result is explained and can be extended.
☐ Project result is explained in context and can be applied to other situations.
☐ Project result is explained using advanced mathematical vocabulary.
☐ Project result is described, and mathematics are used correctly and can be extended.
☐ Project result is explained and extended, and shows advanced knowledge of mathematical concepts and skills.

Meets Expectations
☐ Project result is explained.
☐ Project result is explained in context.
☐ Project result is explained using mathematical vocabulary.
☐ Project result is described, and mathematics are used correctly.
☐ Project result is explained, and shows satisfactory knowledge of mathematical concepts and skills.

Does Not Meet Expectations
☐ Project result is not explained.
☐ Project result is explained, but out of context.
☐ Project result is explained, but mathematical vocabulary is oversimplified.
☐ Project result is described, but mathematics are not used correctly.
☐ Project result is not explained and/or extended, or shows less than satisfactory knowledge of mathematical concepts and skills.

Draw a Picture

Objective

Students can use what they know about solving subtraction problems to explain the parts of a subtraction equation to their peers.

Standard

2.NBT.7 Add and subtract within 1000, using concrete models or drawings and strategies based on place value, properties of operations, and/or the relationship between addition and subtraction; relate the strategy to a written method. Understand that in adding or subtracting three-digit numbers, one adds or subtracts hundreds and hundreds, tens and tens, ones and ones; and sometimes it is necessary to compose or decompose tens or hundreds.

Materials

Additional Materials
- poster paper, 1 piece per student
- glue sticks
- scissors

Prepare Ahead

Students will need the subtraction stories and logos they created last week.

Best Practices

- Clearly enunciate instructions.
- Attend to the varying cognitive styles of individual students.
- Allow students to self-monitor.

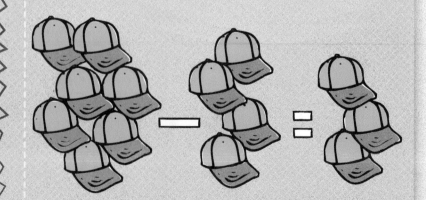

Introduce

▶ **A picture can be used to help you tell a story.**

▶ **Think about some of your favorite stories. How do the pictures help tell the story? How would the story be different without the pictures?**

Explore

▶ **Today you will illustrate your subtraction story.**

- Provide each student with a piece of poster paper.
- Instruct students to copy their subtraction story onto the bottom of the page. If students typed and printed their story last week, they can cut it out and glue it to the bottom of the poster paper.
- Students should glue the logo that they made for their store at the top of the poster paper. Have them write the name of their store next to the logo.
- Have students illustrate their story on the rest of the paper. Students can draw one large picture telling about the whole story, or several smaller pictures telling about the different parts of the story.
- Make sure students' illustrations show what is happening in the story and include numbers when necessary.

▶ **Complete Student Workbook, page 76, to help you draw a picture for your subtraction story.**

Wrap Up

- Have students present their illustrations and story problems to the whole class.
- Students should read the story and then have volunteers identify what they know and what they need to find out. Have students identify the whole set, the part that was taken away, and the part that was left.
- Tell a volunteer to write a subtraction equation on the board. Then have students individually solve the equation using a chosen strategy.
- Discuss students' answers to the Reflect prompt at the bottom of **Student Workbook,** page 76.

If time permits, allow each student to tell how the illustrations helped him or her to better understand the word problems.

Student Workbook, p. 76

Teacher Reflect

☐ Did students think about the questions they needed to answer in their presentation?

☐ Did students show knowledge of how their project related to the major concept?

☐ Did I explain what students had to find, make, or do before they began their projects?

UNIT 5: Geometry and Measurement

Unit at a Glance

This **Number Worlds** unit builds on prior knowledge of geometry and measurement. Students will identify various attributes of shapes and objects and measure time using digital and analog clocks. By the end of the unit, students should be able to describe shapes and cubes by their attributes, and measure time to five minutes on digital and analog clocks.

Skills Trace

Before Level D	Level D	After Level D
Level C Students can create maps based on their knowledge of the number sequence and spatial terms. They can read and understand basic bar graphs.	By the end of this unit, students should be able to describe shapes and cubes by their attributes, measure time, sort objects classify data, and use graphs to represent and analyze data.	**Moving on to Level E** Students will understand perimeter and area. They will measure volume and capacity, create graphs, and analyze the data shown on bar and picture graphs.

Learning Technology

The following activities are available online to support the learning goals in this unit.

Building Blocks

- Comparisons
- Geometry Snapshots 8
- Reptile Ruler
- Shape Shop 3

Digital Tools

- Geometry Sketch Tool
- Graphing and Spreadsheet Tool
- Pictograph Tool
- Shape Creator Tool
- Venn Diagram Tool

Unit Overview

Week	Focus
1	**Linear Measurement** • *Teacher Edition,* pp. 362–375 • *Activity Cards,* 5A, 5B • *Student Workbook,* pp. 5–16 • *English Learner Support Guide,* pp. 94–95 • *Assessment,* pp. 61–62
2	**Measurement Tools** • *Teacher Edition,* pp. 376–389, • *Activity Cards,* 5C, 5D, 5E, 5F • *Student Workbook,* pp. 17–28 • *English Learner Support Guide,* pp. 96–97 • *Assessment,* pp. 63–64
3	**Time Measurement to the Half Hour** • *Teacher Edition,* pp. 390–403 • *Activity Cards,* 5G, 5H • *Student Workbook,* pp. 29–40 • *English Learner Support Guide,* pp. 98–99 • *Assessment,* pp. 65–66
4	**Time Measurement to the Nearest Five Minutes** • *Teacher Edition,* pp. 404–417 • *Activity Cards,* 5I, 5J, 5K, 5L • *Student Workbook,* pp. 41–52 • *English Learner Support Guide,* pp. 100–101 • *Assessment,* pp. 67–68
5	**Attributes of Shapes** • *Teacher Edition,* pp. 418–431 • *Activity Cards,* 5M, 5N, 5O, 5P • *Student Workbook,* pp. 53–64 • *English Learner Support Guide,* pp. 102–103 • *Assessment,* pp. 69–70
6	**Graphs** • *Teacher Edition,* pp. 432–445 • *Activity Cards,* 5Q, 5R, 5S, 5T • *Student Workbook,* pp. 65–76 • *English Learner Support Guide,* pp. 104–105 • *Assessment,* pp. 71–72

Essential Question

HOW can I use my knowledge of length, time, and shapes to help someone complete a task?

In this unit, students will use the knowledge they gained this week, including linear measurement, to build a model of a park.

Learning Goals	CCSS Key Standards
Students can understand the concept of measurement and estimate length with nonstandard, customary, and metric units. **Project:** Students can continue to use estimation and nonstandard and standard units to find linear measures.	**Domain:** Measurement and Data **Cluster:** Measure and estimate lengths in standard units. **2.MD.3:** Estimate lengths using units of inches, feet, centimeters, and meters.
Students can choose the appropriate measuring tool and measure length using customary and metric units. **Project:** Students can use customary and metric tools to measure the length of equipment or features in the park or playground.	**Domain:** Measurement and Data **Cluster:** Measure and estimate lengths in standard units. **2.MD.1:** Measure the length of an object by selecting and using appropriate tools, such as rulers, yardsticks, meter sticks, and measuring tapes.
Students can match analog and digital displays of hour times, tell time to the half hour using analog and digital clocks and show time on an analog clock. **Project:** Students can tell time to the nearest half hour and to the nearest whole hour.	**Domain:** Measurement and Data **Cluster:** Work with time and money. **2.MD.7:** Tell and write time from analog and digital clocks to the nearest five minutes, using a.m. and p.m.
Students can tell and write time to the nearest five minutes and use a.m. and p.m. **Project:** Students can tell time to the nearest five minutes and to the nearest quarter hour.	**Domain:** Measurement and Data **Cluster:** Work with time and money. **2.MD.7:** Tell and write time from analog and digital clocks to the nearest five minutes, using a.m. and p.m.
Students can identify and name different types of triangles, quadrilaterals, and polygons with 3-8 sides. Students understand the attributes of a cube. **Project:** Students design a sandbox for the playground in the shape of a triangle, square, pentagon, or hexagon.	**Domain:** Geometry **Cluster:** Reason with shapes and their attributes. **2.G.1:** Recognize and draw shapes having specified attributes, such as a given number of angles or a given number of equal faces. Identify triangles, quadrilaterals, pentagons, hexagons, and cubes.
Students can draw a picture graph and bar graph and use picture graphs to analyze data. Students can create picture graphs from a set of data and use the graphs to analyze the data. **Project:** Students can create a picture graph from a tally chart.	**Domain:** Measurement and Data **Cluster:** Represent and Interpret Data **2.MD.10:** Draw a picture graph and a bar graph (with single-unit scale) to represent a data set with up to four categories. Solve simple put-together, take-apart, and compare problems using information presented in a bar graph.

CCSS Daily lesson activities emphasize using communication, logic, reasoning, modeling, tools, precision, structure, and patterns to solve problems. All student activities, reflections, and assessments require application of the **Common Core Standards for Mathematical Practice.**

WEEK 1: Linear Measurement

Week at a Glance
This week, students begin **Number Worlds,** Level D, Geometry and Measurement, by investigating linear measurement.

Skills Focus
- Measure length with nonstandard, customary, and metric units
- Estimate length using nonstandard, customary, and metric units

How Students Learn
When introducing students to linear measurement, it is important to make sure students begin measuring at zero. Students should understand that each space on the measuring tool represents one unit. When they measure distance, students are finding the space between two points.

English Learners ELL
For language support, use the **English Learner Support Guide,** pages 94–95, to preview lesson concepts and teach academic vocabulary.

Math at Home
Give one copy of the Letter to Home, page 25, to each student. Encourage students to share and complete the activity with their caregivers.

Weekly Planner

Lesson	Learning Objectives
1 pages 364–365	Students can understand the concept of measurement and can use various objects to measure.
2 pages 366–367	Students can measure and estimate length in inches.
3 pages 368–369	Students can measure and estimate length in feet and yards.
4 pages 370–371	Students can measure and estimate length in meters and centimeters.
5 pages 372–373	**Review and Assess** Students review skills learned this week and complete the weekly assessment.
Project pages 374–375	Students can continue to use estimation and nonstandard and standard units to find linear measures.

362 Level D Unit 5 **Geometry and Measurement**

Key Standard for the Week

Domain: Measurement and Data

Cluster: Measure and estimate lengths in standard units.

2.MD.3 Estimate lengths using units of inches, feet, centimeters, and meters.

Materials		Technology
Program Materials • **Student Workbook,** pp. 5–7 • **Practice,** p. 100 • Activity Card 5A, **Hands and Paper Clips**	**Additional Materials** • bottle of glue • eraser • paper clips • pencil • scissors	*Teacher Dashboard*
Program Materials • **Student Workbook,** pp. 8–9 • **Practice,** p. 101 • Activity Card 5B, **Greater Than, Less Than, or Equal To** • Lengths of Classroom Objects	**Additional Materials** • customary rulers* • various inch-long objects	*Teacher Dashboard* Building Blocks Comparisons
Program Materials • **Student Workbook,** pp. 10–11 • **Practice,** p. 102 • Activity Card 5B, **Greater Than, Less Than, or Equal To** • Lengths of Classroom Objects	**Additional Materials** • customary rulers* • various objects • yardsticks	*Teacher Dashboard*
Program Materials • **Student Workbook,** pp. 12–13 • **Practice,** p. 102 • Activity Card 5B, **Greater Than, Less Than, or Equal To** • Lengths of Classroom Objects	**Additional Materials** • metersticks • metric rulers* • various objects	*Teacher Dashboard*
Program Materials • **Student Workbook,** pp. 14–15 • Weekly Test, **Assessment,** pp. 61–62		Review previous activities.
Program Materials **Student Workbook,** p. 16 **Additional Materials** • poster paper • markers or crayons • paper for note-taking		

*Available from McGraw-Hill Education

WEEK 1
Linear Measurement

Find the Math

Introduce students to estimating length using non-standard, 1 inch, 1 foot, 1 yard, 1 centimeter, and 1 meter units.

Use the following to begin a guided discussion:

▶ **On a playground, what are some ways that you can find the length of objects or the distance between play areas?** Answers may vary. Possible answers: measurement by hands, the number of steps, the time it takes to go to different areas, the comparison of objects as more than or less than other objects

Have students complete *Student Workbook,* page 5.

Student Workbook, p. 5

Lesson 1

Objective
Students can understand the concept of measurement and can use various objects to measure.

Standard
2.MD.3 Estimate lengths using units of inches, feet, centimeters, and meters.

Creating Context
English spellings can be confusing to English Learners and to all students. Help English Learners practice pronouncing the word *measure*.

Materials
Additional Materials
- bottle of glue
- eraser
- paper clips, 2 per student
- pencil
- scissors, 1 per student

Preparing Ahead
Trace a student's handprint, and make two copies for each student.

1 WARM UP

Prepare
At the beginning of the lesson, give each student two handprints. Tell students they are going to use the handprints as a tool for measuring objects. Take time to carefully demonstrate how to measure an object, such as a long table, using the handprints. Emphasize the importance of marking the tip of one handprint (using a finger or pencil mark) so that you know where to put the heel of the handprint for the next measurement.

2 ENGAGE

Develop: Hands and Paper Clips
"Today we are going to measure objects around the classroom." Follow the instructions on the Activity Card **Hands and Paper Clips.** As students complete the activity, be sure to use the Questions to Ask.

Activity Card 5A

Alternative Groupings
Small Group: Assign each group member one or two objects to measure, and have them discuss their measurements as described on the Activity Card.

Progress Monitoring

| **If...** students are having trouble measuring, | ▶ **Then...** be sure they are correctly aligning the ends of their measuring tools. |

Practice
Have students complete *Student Workbook,* pages 6–7. Guide students through the Key Idea example and the Try This exercises.

Interactive Differentiation

Consult the **Teacher Dashboard** for grouping suggestions. You can also use performance on the Engage activity to guide students.

Independent Practice

For additional practice understanding the concept of measurement, have students measure classroom objects by thumb width. Tell students to trace a pencil, an eraser, and a bottle of glue on a piece of paper. Then have them measure each object's length and write the number of thumbs long.

Supported Practice

For additional support, use classroom objects and paper clip chains.

- Demonstrate a 5-unit paper clip chain to students. Ask them to name some objects in the classroom that they think are about the same length.
- Write students' guesses on the board. Choose about five objects to test with the paper clip chain, including objects that are significantly more than or less than 5 units long.
- Demonstrate how to use the paper clip chain to measure the objects.

▶ **Which objects are about the same length as the paper clip chain?** Answers may vary. Possible answers: board eraser, bottle of glue, box of crayons, small book, pencil, student's hand.

▶ **Which objects are longer than the paper clip chain?** Answers may vary. Possible answers: shoe, math book, writing paper, lunch box.

▶ **Which objects are shorter than the paper clip chain?** Answers may vary. Possible answers: pencil erasers, keys, beads.

3 REFLECT

Think Critically

Review students' answers to the Reflect prompt at the bottom of **Student Workbook** page 7, and then review the Engage activity.

Discuss to reinforce the idea that you can use any object to measure length.

▶ **Can you find an object in the classroom that is about 3 paper clips long? 5 paper clips long?**

4 ASSESS

Informal Assessment

Use the online or print Student Record, **Assessment,** page 128, to record informal observations.

Hands and Paper Clips	
Did the student	
☐ make important observations?	☐ provide insightful answers?
☐ extend or generalize learning?	☐ pose insightful questions?

Additional Practice

For additional practice, have students complete **Practice,** page 100.

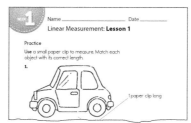

Practice, p. 100

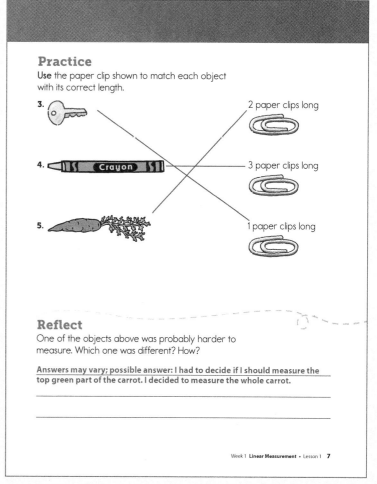

Student Workbook, pp. 6–7

Week 1 **Linear Measurement** • Lesson 1 **365**

WEEK 1
Linear Measurement

Lesson 2

Objective
Students can measure and estimate length in inches.

Standard
2.MD.3 Estimate lengths using units of inches, feet, centimeters, and meters.

Vocabulary
inch

Creating Context
English Learners might benefit from instruction of the meaning of the prefix *centi-*. Explain to students that the prefix *centi-* means "1/100", and that there are 100 centimeters in one meter.

Materials

Program Materials
- Lengths of Classroom Objects, 1 per student

Additional Materials
- customary rulers, 1 per student
- various inch-long objects

Preparing Ahead
Gather various small classroom objects and put them in a box. Include items such as a paper clips, teddy bear counters, crayons, books, staples, staplers, beads, and hole punch.

1 WARM UP

Prepare

▶ **About how long is an inch? Show me an object you think is about one inch long.**

Give students time to find two objects that are each about one inch long. After students have gathered a collection, compare all the objects. Have students agree on whether or not each object is about an inch in length. Have students verify their estimates by measuring each object.

Just the Facts

Have students demonstrate comparisons by holding their hands far apart to show "more than" and putting their hands together to show "less than." Use questions such as the following:

▶ **Is a pencil more than or less than 1 inch long?** hands far apart

▶ **Is a fingernail more than or less than 1 inch wide?** hands together

▶ **Is a vegetable more than or less than 1 inch long?** Answers may vary. Ask students to name the vegetables they used in their comparisons.

2 ENGAGE

Develop: Greater Than, Less Than, or Equal To

"Today we are going to measure using inches." Follow the instructions on the Activity Card **Greater Than, Less Than, or Equal To.** As students complete the activity, be sure to use the Questions to Ask.

Activity Card 5B

Alternative Groupings

Individual: Ask the student the Questions to Ask, and discuss the answers with him or her.

Progress Monitoring

If... students are unable to measure the objects with the ruler, ▶ **Then...** make sure they are placing one end of the object exactly at the 0 mark on the ruler.

Practice

Have students complete **Student Workbook,** pages 8–9. Guide students through the Key Idea example and the Try This exercises.

Interactive Differentiation

Consult the **Teacher Dashboard** for grouping suggestions. You can also use performance on the Engage activity to guide students.

Independent Practice

For additional practice comparing lengths, students should complete Comparisons.

Supported Practice

For additional support, demonstrate measuring various objects with a customary ruler.

- Point out the length of an inch on the ruler.
- Have students show a length of about one inch on one finger.
- Tell students that they will help you sort the objects in the box. The sorting groups will be *about 1 inch* and *more than 1 inch*.
- Choose an object that is about one-inch long. Place it in a designated area to begin sorting.

Ask questions such as the following:

▶ **Where should we place this paper clip?** about 1 inch
▶ **Where should we place this teddy bear counter?** about 1 inch
▶ **Where should we place this crayon?** more than 1 inch

3 REFLECT

Think Critically

Review students' answers to the Reflect prompt at the bottom of *Student Workbook* page 9, and then review the Engage activity.

Discuss that an *inch* is a small unit of measurement.

▸ **What part of your hand is about the length of one inch?**

▸ **What part of your hand is about the length of two inches?**

Real-World Application

▸ **Name some things that would be useful to measure in inches.**
Possible answers: haircuts, bicycle tubes, pants length, and so on

4 ASSESS

Informal Assessment

Use the online or print Student Record, *Assessment,* page 128, to record informal observations.

Greater Than, Less Than, or Equal To
Did the student
☐ make important observations?   ☐ provide insightful answers?
☐ extend or generalize learning?   ☐ pose insightful questions?

Additional Practice

For additional practice, have students complete **Practice,** p. 101.

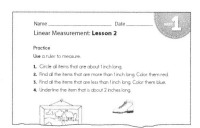
Practice, p. 101

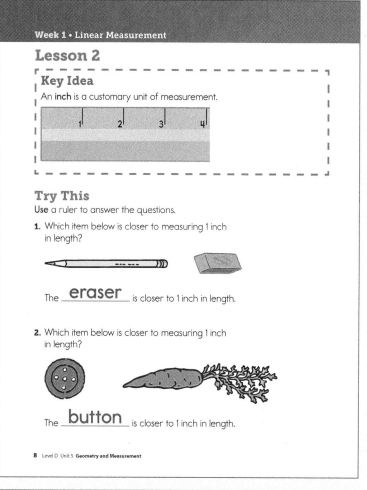

Week 1 • Linear Measurement

Lesson 2

Key Idea
An **inch** is a customary unit of measurement.

Try This
Use a ruler to answer the questions.

1. Which item below is closer to measuring 1 inch in length?

   The ___eraser___ is closer to 1 inch in length.

2. Which item below is closer to measuring 1 inch in length?

   The ___button___ is closer to 1 inch in length.

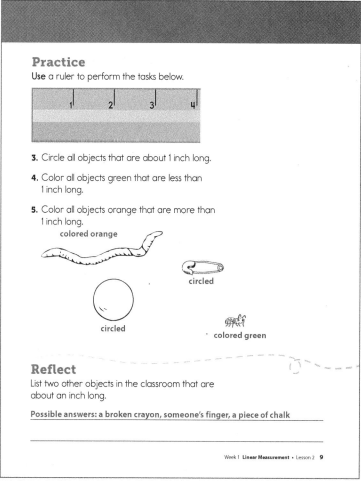

Practice
Use a ruler to perform the tasks below.

3. Circle all objects that are about 1 inch long.

4. Color all objects green that are less than 1 inch long.

5. Color all objects orange that are more than 1 inch long.

   colored orange

   circled

   circled   colored green

Reflect
List two other objects in the classroom that are about an inch long.

Possible answers: a broken crayon, someone's finger, a piece of chalk

Student Workbook, pp. 8–9

Week 1 **Linear Measurement** • Lesson 2 **367**

WEEK 1
Linear Measurement

Lesson 3

Objective
Students can measure and estimate length in feet and yards.

Standard
2.MD.3 Estimate lengths using units of inches, feet, centimeters, and meters.

Creating Context
Use a graphic organizer for higher-level thinking activities. Have students work with a two-column chart labeled *Agree* and *Disagree*. Have English Learners place some objects on a chart to show whether they agree or disagree that they are a foot long.

Materials
Program Materials
- Lengths of Classroom Objects, 2 per student

Additional Materials
- customary rulers, 1 per student
- various objects
- yardsticks, 1 per student

1 WARM UP

Prepare
▶ **Can you show me some objects that you think are about a foot long?**

Give students time to gather items around the classroom that are about a foot long. After they have gathered a collection, compare the objects. Have students decide whether or not each object is about a foot in length. Have students verify their estimates by measuring each object.

Just the Facts
Present students with addition facts to 20, using inches as the addition objects. Use a yardstick as a number line to check student answers. Use questions such as the following:

▶ **What is the length of 10 inches plus 5 inches?** 15 inches
▶ **What is the length of 12 inches plus 2 inches?** 14 inches
▶ **If I add the lengths of a 4-inch long ribbon and a 6-inch long ribbon, how many inches of ribbon do I have?** 10 inches

2 ENGAGE

Develop: Greater Than, Less Than, or Equal To

"Today we are going to measure using feet and yards." Follow the instructions on the Activity Card **Greater Than, Less Than, or Equal To** (Variation 1: Measuring by the Foot and the Yard). As students complete the activity, be sure to use the Questions to Ask.

Activity Card 5B

Individual: Ask the student the Questions to Ask, and discuss the answers with him or her.

Progress Monitoring

| If... a student cannot successfully identify objects that are about one foot long, | ▶ Then... show, describe, and list many more examples. |

Practice
Have students complete **Student Workbook,** pages 10–11. Guide students through the Key Idea example and the Try This exercises.

Interactive Differentiation
Consult the **Teacher Dashboard** for grouping suggestions. You can also use performance on the Engage activity to guide students.

Independent Practice
For additional practice understanding the concept of feet and yards, have students identify two classroom objects that are about 1 foot long and two classroom objects that are about 1 yard long. Tell them to draw pictures of the objects on cards and use the cards to sort by length.

Supported Practice
For additional support, use a customary ruler and a yardstick with students.

- Display a customary ruler, pointing out that it is marked in 12 inches.
- Display a yardstick and show that it is 36 inches in length.
- Tell students that you will name an object and ask them to tell whether it is closer to 1 foot or 1 yard.

Ask questions such as the following:

▶ **About how wide is a door?** about 1 yard
▶ **About how long is a math book?** about 1 foot
▶ **About how tall is a box of cereal?** about 1 foot

368 Level D Unit 5 **Geometry and Measurement**

3 REFLECT

Think Critically

Review students' answers to the Reflect prompt at the bottom of **Student Workbook** page 11, and then review the Engage activity.

Discuss the differences between an exact measurement and an estimated measurement.

▶ **When have you needed an estimated measurement?**

▶ **When have you needed an exact measurement?**

Real-World Application

▶ **What object might you measure in feet?**
Answers may vary. Possible answer: a person's height, the length of a desk, the width of a room

4 ASSESS

Informal Assessment

Use the online or print Student Record, **Assessment,** page 128, to record informal observations.

Greater Than, Less Than, or Equal To	
Did the student	
☐ make important observations?	☐ provide insightful answers?
☐ extend or generalize learning?	☐ pose insightful questions?

Additional Practice

For additional practice, have students complete **Practice,** page 102.

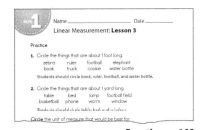
Practice, p. 102

Week 1 • Linear Measurement

Lesson 3

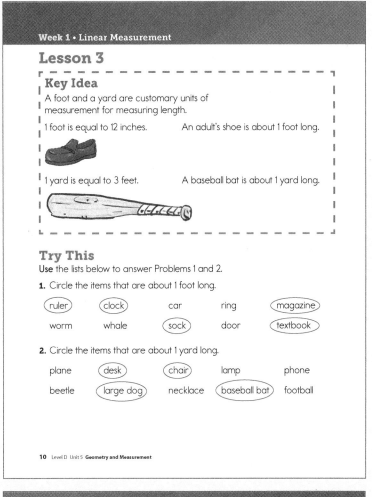

Key Idea
A foot and a yard are customary units of measurement for measuring length.

1 foot is equal to 12 inches.   An adult's shoe is about 1 foot long.

1 yard is equal to 3 feet.   A baseball bat is about 1 yard long.

Try This
Use the lists below to answer Problems 1 and 2.

1. Circle the items that are about 1 foot long.

   (ruler)   (clock)   car   ring   (magazine)
   worm   whale   (sock)   door   (textbook)

2. Circle the items that are about 1 yard long.

   plane   (desk)   (chair)   lamp   phone
   beetle   (large dog)   necklace   (baseball bat)   football

10 Level D Unit 5 Geometry and Measurement

Practice
Circle the unit of measurement that would be best for measuring the length of each object.

3. sofa	inches	(feet)	yards
4. ice skate	(inches)	feet	yards
5. phone	(inches)	feet	yards
6. parking lot	inches	feet	(yards)
7. child's height	inches	(feet)	yards
8. paper clip	(inches)	feet	yards
9. worm	(inches)	feet	yards
10. football field	inches	feet	(yards)

Reflect
Describe a situation in which an exact measurement is better than an estimate.

Answers will vary. Possible answer: when you are building something and the parts have to fit together

Week 1 Linear Measurement • Lesson 3 11

Student Workbook, pp. 10–11

WEEK 1
Linear Measurement

Lesson 4

Objective
Students can measure and estimate length in centimeters and meters.

Standard
2.MD.3 Estimate lengths using units of inches, feet, centimeters, and meters.

Vocabulary
- centimeter
- meter

Creating Context
Have students find out how tall each family member is using metric measurement. Then have them measure each person using feet and inches and make a chart with both measurements. Discuss these observations in a small group.

Materials
Program Materials
- Lengths of Classroom Objects, 2 per student

Additional Materials
- metersticks, 1 per student
- metric rulers, 1 per student
- various objects

 WARM UP

Prepare
▶ **About how long is a centimeter? A meter?**

Show students a centimeter and a meter on a meterstick. Have a variety of items at the front of the room on a desk or table. Some of the items should be about one centimeter in length, and some should be about one meter. Hold up any one of the items.

▶ **Is this item closer to one centimeter or one meter?** Answers may vary.

Just the Facts
Tell students that for each question, they will look up at you if the answer is yes and cover their eyes if the answer is *no*. Use questions such as the following:

▶ **Is 10 centimeters plus 10 centimeters 20 centimeters?** look up
▶ **Is 5 centimeters plus 7 centimeters 11 centimeters?**
cover eyes; 5 + 7 = 12 cm
▶ **Is 8 centimeters plus 6 centimeters 14 centimeters?** look up

2 ENGAGE

Develop: Greater Than, Less Than, or Equal To
"Today we are going to measure using centimeters and meters." Follow the instructions on the Activity Card, **Greater Than, Less Than, or Equal To** (Variation 2: Measuring by the Centimeter and the Meter). As students complete the activity, be sure to use the Questions to Ask.

Activity Card 5B

Alternative Groupings
Individual: Ask the student the Questions to Ask and discuss the answers with him or her.

Progress Monitoring
If... students are having difficulty measuring with a ruler,	▶ Then... review with them how to read a ruler and practice measuring line segments together.

Practice
Have students complete **Student Workbook,** pages 12–13. Guide students through the Key Idea example and the Try This exercises.

Interactive Differentiation
Consult the **Teacher Dashboard** for grouping suggestions. You can also use performance on the Engage activity to guide students.

Independent Practice
For additional practice understanding the concept of centimeters and meters, have students identify two classroom objects that are about 1 centimeter long and two classroom objects that are about 1 meter long. Tell them to draw pictures of the objects on cards and use the cards to sort by length.

Supported Practice
For additional support, use a metric ruler and a meter stick with students.

- Demonstrate a metric ruler. Point out that each centimeter is about the same as the width of a child's forefinger or an adult's pinky finger.
- Demonstrate a meter stick and tell students that it is 100 centimeters long.
- Tell students that you will name an object and ask them to tell whether it is closer to 1 centimeter or 1 meter.

Ask questions such as the following:

▶ **About how long is an eraser on the end of a pencil?**
about 1 centimeter
▶ **About how tall is a second-grader?** about 1 meter
▶ **About how wide is a shirt button?** about 1 centimeter

370 Level D Unit 5 **Geometry and Measurement**

3 REFLECT

Think Critically

Review students' answers to the Reflect prompt at the bottom of **Student Workbook** page 13, and then review the Engage activity.

Discuss that one meter is about the length of a baseball bat and that one centimeter is about the length of your thumbnail.

▶ Have you ever had to measure using metric units, other than in school?

▶ Name a place you might live if you measured using metric units all the time.

Real-World Application

After exploring the centimeter and the meter, have students list other items they know of that are about one centimeter or one meter long.

4 ASSESS

Informal Assessment

Use the online or print Student Record, **Assessment,** page 128, to record informal observations.

Greater Than, Less Than, or Equal To

Did the student

☐ make important observations? ☐ provide insightful answers?

☐ extend or generalize learning? ☐ pose insightful questions?

Additional Practice

For additional practice, have students complete **Practice,** page 103.

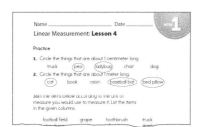

Practice, p. 103

Week 1 • Linear Measurement

Lesson 4

Key Idea
The **centimeter** and **meter** are metric units of measurement used to measure length.

1 meter is equal to 100 centimeters.

A small button is about 1 centimeter wide.

A wagon is about 1 meter long.

Try This
Use the lists below to answer Problems 1 and 2.

1. Circle the items that are about 1 centimeter long.

   (ant) stove car (pearl) (raisin)

2. Circle the items that are about 1 meter long.

   boat (desk) (television) light bulb cell phone

12 Level D Unit 5 Geometry and Measurement

Practice
Sort the items below according to the unit of measure that should be used to measure them. List the items in the given columns.

swimming pool	garden	pea	brush
marble	sidewalk	golf ball	walking path
computer screen	eyeglasses	fence	train

3. Centimeters

   pea

   brush

   marble

   golf ball

   computer screen

   eyeglasses

4. Meters

   swimming pool

   garden

   sidewalk

   walking path

   fence

   train

Reflect
Describe a time when an estimate is needed instead of an exact measurement.

Possible answer: If someone wants to know about how far he or she has walked, an estimate would be appropriate.

Week 1 Linear Measurement • Lesson 4 13

Student Workbook, pp. 12–13

WEEK 1
Linear Measurement

Lesson 5 Review

Objective
Students review skills learned this week and complete the weekly assessment.

Standard CCSS
2.MD.3 Estimate lengths using units of inches, feet, centimeters, and meters.

Vocabulary
Review vocabulary introduced during the week.

Creating Context
In English, we often use abbreviations for words that are repeated often in a context such as measuring. Help students make a table with the metric and customary measurements and their abbreviations. Remind students how important it is to place the abbreviation for the unit of measurement after the number so the reader knows which unit of measure was used.

1 WARM UP

Prepare
- Display items from the various activities of the week.
- Have students select items that are about one inch, one foot, one yard, one centimeter, and one meter in length.
- Have groups discuss, compare, and contrast these items.

2 ENGAGE

Practice
Have students complete **Student Workbook,** pages 14–15.

Week 1 • Linear Measurement

Lesson 5 Review
This week you explored different units of linear measurement.

Lesson 1 Measure the items below with a paper clip.

1. The bug is about ____1 paper clip____ long.

2. The chalk is about ____3 paper clips____ long.

Lesson 2 Use a ruler to measure.

3. Circle all objects that are about 1 inch long.

14 Level D Unit 5 Geometry and Measurement

Lesson 3 Circle the unit of measurement that would be best for measuring the length of each object listed.

4. Soccer field   inches   feet   (yards)
5. Bedspread   inches   (feet)   yards
6. Cell phone   (inches)   feet   yards

Lesson 4 Sort the items below according to the units of measure that should be used to measure them. List the items in the given columns.

football field   garden   comb
penny   driveway   baseball

7. Centimeters   8. Meters

comb   football field
penny   garden
baseball   driveway

Reflect
What metric unit of measure would you most likely use to find the length of a kitchen table?
____meter____

Week 1 **Linear Measurement** • Lesson 5 15

Student Workbook, pp. 14–15

372 Level D Unit 5 **Geometry and Measurement**

3 REFLECT

Think Critically

Review students' answers to the Reflect prompt at the bottom of *Student Workbook* page 15.

Discuss the answer with the group to reinforce Week 1 concepts

4 ASSESS

Formal Assessment

Students may take the weekly assessment online.

As an alternative, students may complete the weekly test on *Assessment,* pages 61–62. Record progress using the Student Assessment Record, *Assessment,* page 128.

Going Forward

Use the **Teacher Dashboard** to view results of the online assessments, to input the results of print student assessments, and to review progress before making decisions about next steps. Use the weekly test results and observations to determine the next steps for each student.

Retention	
Student displays good grasp of this week's concepts and skills.	Have students measure objects that require laying a measurement tool end-to-end more frequently, such as measuring the length of a hallway or a playground.

Remediation	
Student is still struggling with the week's concepts and skills.	Have students measure objects that are exactly the same length as the measuring tool. Then have students measure objects smaller than the measuring tool, which are the same length as a major increment on the tool, such as 10 inches on a 12-inch ruler. When students demonstrate competence with this skill, demonstrate how to measure objects greater than the length of the measuring tool.

Suggestions for Re-Evaluation: If a student has struggled without success for several weeks, use observations and test results to place the student at a level where they can find success and build confidence to move forward.

Name _____ Date _____
Linear Measurement — WEEK 1

Use the ruler at the bottom of the page to measure the length of each object. Write the length on the line beside the object.

1. __4 in.__
2. __2 in.__
3. __5 in.__
4. __1 in.__
5. __3 in.__

Level D Unit 5 Week 1 **61**

WEEK 1 — Name _____ Date _____
Linear Measurement

Fill in each space with inches, feet, or yards.

6. A football field is 100 __yards__ long.
7. A basketball goal is 10 __feet__ high.
8. A dollar bill is about 6 __inches__ long.

Decide whether each person would measure each thing with inches, feet, or yards. Each measure can only be used once.

9. A scientist would probably measure the length of a squirrel using __inches__
10. A worker would probably measure the width of a big highway in __yards__
11. A carpenter would probably measure the height of a room in __feet__

Circle the best answer to each question

12. About how long is a regular bed? (**2 meters**)   16 meters
13. About how long is a baseball bat? (**90 centimeters**)   25 centimeters
14. About how long is a school bus?   3 meters   (**13 meters**)
15. About how long is a grape? (**1 centimeter**)   8 centimeters

62 Level D Unit 5 Week 1

Assessment, pp. 61–62

Week 1 **Linear Measurement** • Lesson 5 **373**

Project Preview

This week, students learned to use nonstandard units to measure lengths. They were introduced to standard units to explore the concepts of 1 inch, 1 foot, 1 yard, 1 centimeter, and 1 meter. The project for this unit requires students to extend the knowledge they gained in Find the Math and what they have learned this week. They will use linear measurement to build a model of a park.

Project-Based Learning

Standards-driven Project-Based Learning is effective in building deep content understanding. Project-Based Learning increases long-term retention of concepts and has been shown to be more effective than traditional instruction. By completing a project to answer an essential question, students are challenged to apply and demonstrate mastery of concepts and skills by expressing understanding through discussion, research, and presentation.

Essential Question

HOW can I use my knowledge of length, time, and shapes to help someone complete a task?

Project Evaluation Criteria

Review project evaluation criteria with students prior to beginning the project.

Exceeds Expectations
☐ Project result is explained and can be extended.
☐ Project result is explained in context and can be applied to other situations.
☐ Project result is explained using advanced mathematical vocabulary.
☐ Project result is described, and mathematics are used correctly and can be extended.
☐ Project result is explained and extended, and shows advanced knowledge of mathematical concepts and skills.

Meets Expectations
☐ Project result is explained.
☐ Project result is explained in context.
☐ Project result is explained using mathematical vocabulary.
☐ Project result is described, and mathematics are used correctly.
☐ Project result is explained, and shows satisfactory knowledge of mathematical concepts and skills.

Does Not Meet Expectations
☐ Project result is not explained.
☐ Project result is explained, but out of context.
☐ Project result is explained, but mathematical vocabulary is oversimplified.
☐ Project result is described, but mathematics are not used correctly.
☐ Project result is not explained and/or extended, or shows less than satisfactory knowledge of mathematical concepts and skills.

Take a Walk in the Park

Objective
Students can continue to use estimation and nonstandard and standard units to find linear measures.

Standard
2.MD.3 Estimate lengths using units of inches, feet, centimeters, and others.

Materials
Additional Materials
- paper for note-taking
- poster paper
- markers or crayons

Prepare Ahead
Students will need access to a school playground or a park to complete the project. The ideal park will include a large swing set that has at least three swings. If a park or playground area is not available, you may use the wall of a school hallway, ribbon, and art paper to make a life-size picture of a swing set for students to use when estimating length by number of steps. The actual length of this representation should be between 25 and 30 feet long for a four-swing model. Each swing in the model should be about 18 inches wide. You will need to leave this life-size picture in place through next week, as students will need it again.

Best Practices
- Allow active learning with noise and movement.
- Meet individual student needs, and allow individualized solutions.
- Make decisions and contingency plans ahead of time.

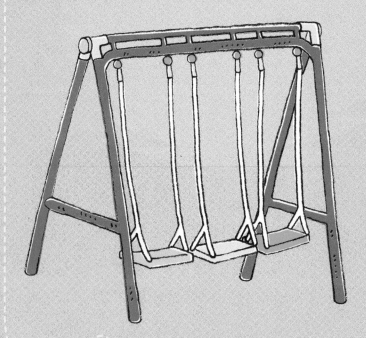

Introduce

Suppose that you were planning a new park or a school playground. What information would you need to know about the size of the park and what to include in the play areas?

- What are the major areas that you would like to have in the park?
- How much space would you need for each of these areas?
- Where could you go to find answers to these questions?

Explore

- Organize students into pairs. They will work with this same partner each week as they complete the unit project.
- **Today you will begin to make a map of your park or playground.**
- Bring students out to the playground or playground model. Have them work together in pairs to answer the following questions.
- **What object will you choose to measure first?**
- **What is another area you want for your park?**
- Give students a time limit, such as fifteen minutes, to decide which two playground areas they want to include and measure.
- Complete *Student Workbook,* **page 16, to find the length of the playground area and the distance between two areas.**

Wrap Up

- Make sure students can explain how they used steps to measure length.
- If students struggle to keep an accurate count of their steps, have them work together to measure and record the results.
- Discuss students' answers to the Reflect prompts at the bottom of *Student Workbook,* page 16.

Distribute to each pair of students a large piece of poster paper. Tell students that they will draw a map, or plan, of their playground on the poster. Instruct them to draw the two park areas they have decided to include, such as a swing set and a baseball diamond. Tell students to use only half of the poster for this activity because they must leave room for other parts of the playground.

- **You will use only half of the poster to draw two parts of the park.**
- **The areas do not have to be touching.**
- **Make sure that the larger playground object takes more poster space than the smaller one.**

If time permits, allow students to draw grass, students in the play areas, or other background items for the two areas they have drawn so far. Put students' posters aside for next week.

Week 1 • Linear Measurement

Project
Take a Walk in the Park

Walk to find the length of things in a park or playground. Then answer the questions.

1. What will you choose to measure first?

   Answers may vary. Possible answers: swing set; slide; baseball field; picnic area; tennis court

2. How many steps long is the object?

   Answers may vary. Possible answer: The swing set is twenty steps long.

3. You need to measure the distance between the first object and another object in the park or playground. What object will you choose next?

   Answers may vary. Possible answers: swing set; slide; baseball field; picnic area; tennis court

4. What is the number of steps between these two objects?

   Answers may vary. Possible answer: There are 25 steps between them.

Reflect

Is it better to measure a park by number of steps or number of inches? Why?

Answers may vary. Possible answer: It is faster and easier to measure a large area by the number of steps than the number of inches. Inches are small, and it would take a long time to measure a large distance in inches.

16 Level D Unit 5 Geometry and Measurement

Student Workbook, p. 16

Teacher Reflect

☐ Did students correctly use art, objects, graphs, or posters to explain their solutions?

☐ Did I explain what students had to find, make, or do before they began their projects?

☐ Did students tell or show the steps when they explained how to do something?

WEEK 2: Measurement Tools

Week at a Glance
This week, students continue **Number Worlds**, Level D, Geometry and Measurement, by using customary and metric measurement tools.

Skills Focus
- Choose the appropriate measuring tool.
- Measure length using customary units.
- Measure length using metric units.

How Students Learn
The United States is one of few countries who do not primarily use the metric system. In real-world applications, both systems are often used simultaneously. In this week of **Number Worlds** students also use both systems as they begin to learn *how* to measure.

English Learners ELL
For language support, use the **English Learner Support Guide,** pages 96–97, to preview lesson concepts and teach academic vocabulary.

Math at Home
Give one copy of the Letter to Home, page 26, to each student. Encourage students to share and complete the activity with their caregivers.

Weekly Planner

Lesson	Learning Objectives
1 pages 378–379	Students can choose the right tool: ruler, yardstick, meterstick, measuring tape.
2 pages 380–381	Students can use a customary ruler to measure.
3 pages 382–383	Students can use a metric ruler to measure.
4 pages 384–385	Students can use a yardstick and meterstick.
5 pages 386–387	**Review and Assess** Students review skills learned this week and complete the weekly assessment and project.
Project pages 388–389	Students can use customary and metric tools to measure the length of equipment or features in the park or playground.

Key Standard for the Week

Domain: Measurement and Data

Cluster: Measure and estimate lengths in standard units.

2.MD.1 Measure the length of an object by selecting and using appropriate tools, such as rulers, yardsticks, meter sticks, and measuring tapes.

Materials		Technology
Program Materials • *Student Workbook,* pp. 17–19 • *Practice,* p. 104 • Activity Card 5C, **Pick a Measurement Tool** • Pick a Measurement Tool Recording Sheet	**Additional Materials** • measuring tape* • meterstick • ruler (customary or metric depending on scale used)* • yardstick	*Teacher Dashboard* Building Blocks Reptile Ruler
Program Materials • *Student Workbook,* pp. 20–21 • *Practice,* p. 105 • Activity Card 5D, **Customary Length** • Customary Length Recording Sheet	**Additional Materials** • box of objects • rulers (customary)*	*Teacher Dashboard*
Program Materials • *Student Workbook,* pp. 22–23 • *Practice,* p. 106 • Activity Card 5E, **Metric Length** • Metric Length Recording Sheet	**Additional Materials** • box of objects • math-link cubes • rulers (metric)*	*Teacher Dashboard*
Program Materials • *Student Workbook,* pp. 24–25 • *Practice,* p. 107 • Activity Card 5F, **Yardsticks and Metersticks**	**Additional Materials** • index cards • metersticks • yardstick	*Teacher Dashboard*
Program Materials • *Student Workbook,* pp. 26–27 • Weekly Test, *Assessment,* pp. 63–64		Review previous activities.
Program Materials *Student Workbook,* p. 28	**Additional Materials** • customary ruler • measuring tape • meterstick • metric ruler • poster paper from previous week • yardstick	

*Available from McGraw-Hill Education

WEEK 2
Measurement Tools

Find the Math

In this week, introduce students to choosing the right tool for measuring length: ruler, yardstick, meterstick, measuring tape.

Use the following to begin a guided discussion:

▶ **In the classroom, what tools can you use to measure the length of objects?** Answers may vary. Possible answer: Use a yardstick to measure big objects and a ruler to measure small objects.

Have students complete *Student Workbook,* page 17.

Student Workbook, p. 17

Lesson 1

Objective
Students can choose the right tool: ruler, yardstick, meterstick, measuring tape.

Standard
2.MD.1 Measure the length of an object by selecting and using appropriate tools, such as rulers, yardsticks, meter sticks, and measuring tapes.

Creating Context
Help English Learners identify the measurement tools they need to learn. Display a poster showing a large, labeled picture of each tool. Point out each tool students will use while saying its name.

Materials
Program Materials
- Pick a Measurement Tool Recording Sheet, 1 per student

Additional Materials
- measuring tape
- meterstick
- ruler (customary or metric depending on scale used)
- yardstick

Prepare Ahead
Select and label twelve to fifteen classroom objects before students begin the activity. For example, label the whiteboard, "Whiteboard."

1 WARM UP

Prepare
- Have students recall how they the estimated the lengths of objects on a playground in Week 1.
- Display a customary ruler, a yardstick, a meterstick, and a measuring tape. Display the names of the tools. Have students name objects in the classroom that are about the same length as each tool.

2 ENGAGE

Develop: Pick a Measurement Tool

"Today we are going to choose the right tool to measure an object." Follow the instructions on the Activity Card **Pick a Measurement Tool.** As students complete the activity, be sure to use the Questions to Ask.

Activity Card 5C

Alternative Groupings

Individual: Give each student only one measurement tool. Tell students to find objects in the classroom that they can accurately measure with that tool. When students have finished the activity, have them meet to compare and discuss the results noted on their recording sheets.

Progress Monitoring

| **If...** students are having difficulty deciding which measurement tool to use, | ▶ **Then...** remind them to compare the size of the tool to the size of the object. The tool closest to the object's size is usually the best to use. |

Practice

Have students complete *Student Workbook,* pages 18–19. Guide students through the Key Idea example and the Try This exercises.

378 Level D Unit 5 **Geometry and Measurement**

Interactive Differentiation

Consult the **Teacher Dashboard** for grouping suggestions. You can also use performance on the Engage activity to guide students.

Independent Practice

For additional practice with understanding the concept of using a measurement tool have students use Reptile Ruler.

Supported Practice

For additional support, use the appropriate tool to demonstrate how to measure other familiar objects. Say the names of the objects (some are listed below) and have volunteers answer with the name of the most appropriate tool to measure each.

▶ **paperclip, eraser, activity bucket, book, small waste can, crayon, marker, shoe, pencil, pen, notebook paper:** Answers may vary. Possible answer: ruler.

▶ **closet door, classroom door, bookcase, desk, work table, large trash can, wall clock:** Answers may vary. Possible answers: yardstick, meterstick.

▶ **How do you decide which measurement tool to use?** Answers may vary. Possible answer: The size of the object tells you which tool to use. If the object is small, use a ruler. If the object is large, use a yardstick, meterstick, or measuring tape.

3 REFLECT

Think Critically

Review students' answers to the Reflect prompt at the bottom of **Student Workbook** page 19, and then review the Engage activity.

▶ **How do you choose a tool to measure objects?** Answers may vary. Possible answer: Look at the size of the object. If it is smaller than a ruler, choose a ruler. If the object is larger than a ruler, choose another tool that fits the size better.

4 ASSESS

Informal Assessment

Use the online or print Student Record, **Assessment,** page 128, to record informal observations.

Pick a Measurement Tool

Did the student

☐ pay attention to the contributions of others?   ☐ improve on a strategy?

☐ contribute information and ideas?   ☐ reflect on and check accuracy of work?

Additional Practice

For additional practice, have students complete **Practice,** page 104.

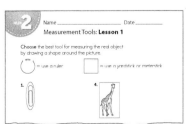

Practice, p. 104

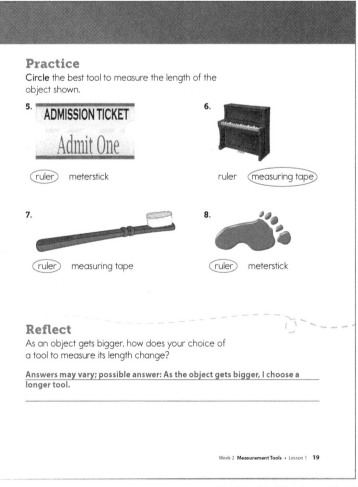

Student Workbook, pp. 18–19

Week 2 **Measurement Tools** • Lesson 1   **379**

WEEK 2
Measurement Tools

Lesson 2

Objective
Students can use a customary ruler to measure.

Standard
2.MD.1 Measure the length of an object by selecting and using appropriate tools such as rulers, yardsticks, metersticks, and measuring tapes.

Creating Context
English Learners may have used metric units in the past. Help English Learners practice naming measurements using the U.S. customary units, inches and feet. Have students fold a sheet of paper in half lengthwise and label the columns *Inches* and *Feet*. Have students draw pictures of some common objects in each column, with smaller objects drawn under *Inches* and objects longer than one foot drawn under *Feet*.

Materials
Program Materials
- Customary Length Recording Sheet

Additional Materials
- box of objects for each group that includes some of the following items: pencil, pen, paperclip, eraser, crayons, note card, marker, book, paintbrush, glue bottle, scissors, toy car, Counters, large paperclip, short piece of string, and empty CD case. If students already have some of these items, they can be omitted from the box.
- rulers (customary)

Prepare Ahead
Assemble a box of objects for each group that includes some of the items on the Additional Materials list.

1 WARM UP
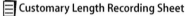

Prepare
- Discuss with students how they estimated the lengths of objects to be more than or less than 1-inch long in Week 1.
- Have students work with a partner to make a list of objects that are about 1-inch long.

Just the Facts
To increase fluency, have students work with doubles facts and customary measurements. Tell them to nod "yes" if the addition fact is correct. If the addition fact is incorrect, tell them to cover their eyes. Use prompts such as the following:

▶ **3 inches plus 3 inches is 6 inches.** nod "yes"
▶ **4 inches plus 4 inches is 10 inches.** cover eyes
▶ **2 feet plus 2 feet is 4 feet.** nod "yes"

2 ENGAGE

Develop: Customary Length
"Today we are going to use a customary ruler to measure." Follow the instructions on the Activity Card **Customary Length**. As students complete the activity, be sure to use the Questions to Ask.

Activity Card 5D

Alternative Groupings
Whole Class: Complete as written. Assign one student the job of recording the measurements. Have as many students as possible take a turn measuring each object.

Progress Monitoring	
If… students are unable to measure an object with a ruler,	▶ **Then…** make sure they are placing one end of the object exactly at the 0 point of the ruler.

Practice
Have students complete **Student Workbook,** pages 20–21. Guide students through the Key Idea example and the Try This exercises.

Interactive Differentiation
Consult the **Teacher Dashboard** for grouping suggestions. You can also use performance on the Engage activity to guide students.

Independent Practice
For additional practice, cut lengths of yarn appropriate for measuring with a ruler, a yardstick, or measuring tape. Have students measure each length of yarn and write the measurements down on a piece of paper.

Supported Practice
For additional support, measure students' feet.
- Lay a ruler or yardstick on the floor.
- Have a student place his or her heel at the zero mark on the ruler.
- Show students how to count the inches along the length of the foot up to the toes.
- Explain the concept of the "nearest inch." Show students how to determine which inch mark is closest to the actual measurement.

Make sure students understand the process by asking questions such as the following:

▶ **What would happen if you put the back of your heel on the 1 inch mark instead of the 0 inch mark?** Answers may vary. Possible answer: Your foot would measure one inch longer than it really is.

380 Level D Unit 5 **Geometry and Measurement**

3 REFLECT

Think Critically

Review students' answers to the Reflect prompt at the bottom of **Student Workbook** page 21, and then review the Engage activity.

▶ **How do you use a ruler to find the length of your pencil?** Answers may vary. Possible answer: Line up the point of the pencil with the zero mark on the left end of the ruler. Be sure the ruler is straight across the pencil. Then count the inches forward along the length of the pencil. See which number on the ruler is closest to the eraser end of the pencil.

Real-World Application

You are designing a scrapbook page that includes photographs of your vacation.

▶ **How would you measure the sides of a photograph using a ruler?** Answers may vary. Possible answer: For each side of the photograph, line up one corner with the zero mark on the ruler. Count the inches forward along the length of that side. Identify the number on the ruler that is closest to the other corner of that side.

4 ASSESS

Informal Assessment

Use the online or print Student Record, **Assessment,** page 128, to record informal observations.

Customary Length	
Did the student	
☐ pay attention to the contributions of others?	☐ improve on a strategy?
☐ contribute information and ideas?	☐ reflect on and check accuracy of work?

Additional Practice

For additional practice, have students complete **Practice,** page 105.

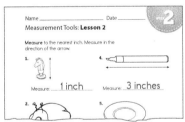

Practice, p. 105

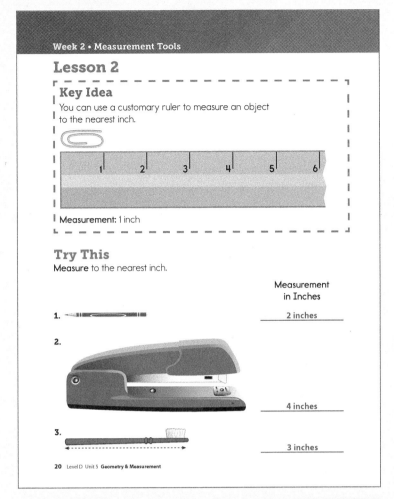

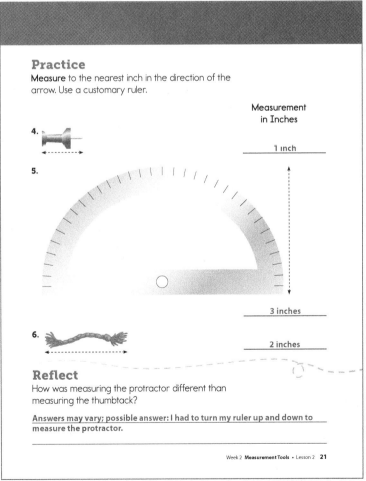

Student Workbook, pp. 20–21

Week 2 **Measurement Tools** • Lesson 2 **381**

WEEK 2
Measurement Tools

Lesson 3

Objective
Students can use a metric ruler to measure.

Standard
2.MD.1 Measure the length of an object by selecting and using appropriate tools such as rulers, yardsticks, meter sticks, and measuring tapes.

Vocabulary
- centimeter
- meter

Creating Context
English Learners may have used metric units in the past. Help English Learners practice naming measurements using centimeters and meters. Have students fold a sheet of paper lengthwise and label the columns *Centimeters* and *Meters*. Have students draw pictures of common objects in each column. Smaller objects should be drawn under *Centimeters* and objects longer than 1 meter drawn under *Meters*.

Materials
Program Materials
 Metric Length Recording Sheet, 1 per group

Additional Materials
- box of objects for each group that includes some of the following items: pencil, pen, paperclip, eraser, crayons, note card, marker, book, paintbrush, glue bottle, scissors, toy car, Counters, large paperclip, short piece of string, and empty CD case. If students already have some of these items, they can be omitted from the box.
- math-link cubes
- rulers (metric)

Prepare Ahead
Assemble a box of objects for each group using items shown on the Additional Materials list. If students already have some of these items, they can be omitted from the box.

1 WARM UP

Prepare
- Remind students that in Week 1 they identified objects that are about one centimeter long and objects that are about one meter long.
- Review the difference between customary and metric units of length with the class.

Just the Facts
Review "plus ten" facts with students. Show the appropriate number of fingers as you say each addend. Remind students to say the unit when they answer. Use questions such as the following:

▶ **What is 3 centimeters plus 10 centimeters?** 13 centimeters
▶ **What is 4 meters plus 10 meters?** 14 meters
▶ **What is 10 meters plus 6 meters?** 16 meters

382 Level D Unit 5 **Geometry and Measurement**

2 ENGAGE

Develop: Metric Length
"Today we are going to measure objects using a metric ruler." Follow the instructions on the Activity Card **Metric Length**. As students complete the activity, be sure to use the Questions to Ask.

Activity Card 5E

Alternative Groupings
Whole Class: Complete as written. Assign one student to record measurements on the Metric Length Recording Sheet.

Progress Monitoring

If... students are struggling with the terms for customary or metric units of measure,	▶ **Then...** have them make a flashcard for each unit of measure, which lists its definition and shows an illustration or drawing of an object labeled with its measurement in the corresponding units.

Practice
Have students complete **Student Workbook,** pages 22–23. Guide students through the Key Idea example and the Try This exercises.

Interactive Differentiation
Consult the **Teacher Dashboard** for grouping suggestions. You can also use performance on the Engage activity to guide students.

Independent Practice
For additional practice, have students measure different combinations of one to ten math-link cubes. Tell them to measure each object to the nearest centimeter. Have students write these measurements on a sheet of paper.

Supported Practice
For additional support, demonstrate measurement errors with various small objects and a metric ruler.

- Lay the metric ruler along an object, but do not align the end of the object with the zero mark on the ruler. Tell students that the object measures the number of centimeters marked on the ruler. Have students identify your error. Answers may vary. Possible answer: You have to align the end of the object with the zero mark of the metric ruler. Otherwise the object measurement is too long.

- Align the non-zero end of the metric ruler with the end of the object. Tell students that the object measures the number of centimeters marked on the ruler. Have them explain your error. Answers may vary. Possible answer: You have to align the end of the object with the zero mark on the metric ruler or the measurement will be wrong.

3 REFLECT

Think Critically

Review students' answers to the Reflect prompt at the bottom of **Student Workbook** page 23, and then review the Engage activity.

▶ **How would you measure the length of a pen using a metric ruler?** Answers may vary. Possible answer: Line up the point of the pen with the zero mark on the metric ruler. Count the centimeters along the length of the pen. Identify which number on the ruler is closest to other end of the pen.

Real-World Application

Your cousin in Germany wants to know how long your pet guinea pig is.

▶ **How would you use a metric ruler to measure your guinea pig in centimeters?** Answers may vary. Possible answer: Lay the metric ruler next to the guinea pig with the zero mark by the animal's nose. Count the centimeters along the ruler until you reach the other end of the guinea pig.

4 ASSESS

Informal Assessment

Use the online or print Student Record, **Assessment,** page 128, to record informal observations.

Metric Length
Did the student
- □ pay attention to the contributions of others?
- □ contribute information and ideas?
- □ improve on a strategy?
- □ reflect on and check accuracy of work?

Additional Practice

For additional practice, have students complete **Practice,** page 106.

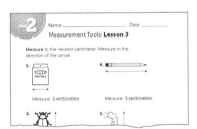

Practice, p. 106

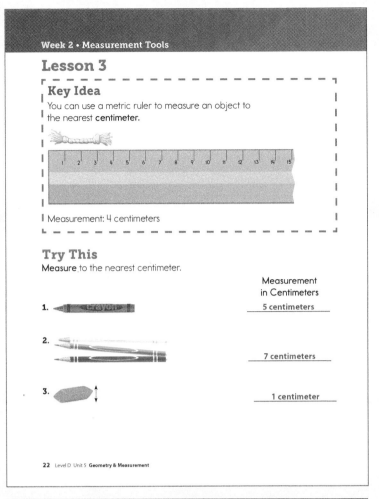

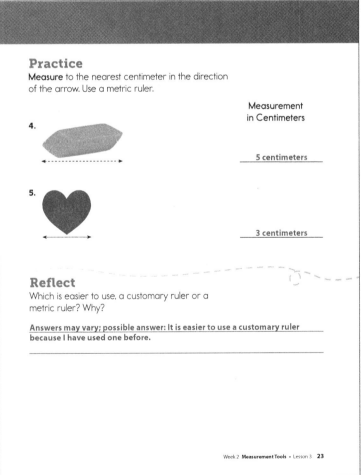

Student Workbook, pp. 22–23

WEEK 2
Measurement Tools

Lesson 4

Objective
Students can use a yardstick and meterstick.

Standard
2.MD.1 Measure the length of an object by selecting and using appropriate tools such as rulers, yardsticks, meter sticks, and measuring tapes.

Creating Context
Review the U.S. customary and metric units of measure: *inch, foot, yard; centimeter,* and *meter.* Encourage students to learn to recognize and correctly pronounce these words. The word *measure* can also be challenging for students. Make sure they can recognize the word *measure* in print and understand it when it is spoken.

Materials
Additional Materials
- index cards, 1 per student
- meterstick
- yardstick

Prepare Ahead
Before distributing the index cards to students, alternately write *yardstick* or *meterstick* on the back of the cards.

1 WARM UP

Prepare
To introduce the lesson, ask students questions such as the following:

▶ **What units are on a yardstick?** inch, foot

▶ **What units are on a meterstick?** centimeters

▶ **Which is longer, a yardstick or a meterstick?** Answers may vary. Possible answer: A meterstick is slightly longer than a yardstick.

Just the Facts
Practice subtracting ten with customary and metric measurements. Have students clap the number of units in the answer. Use questions such as the following:

▶ **What is 11 feet minus 10 feet?** Students clap once.

▶ **What is 13 meters minus 10 meters?** Students clap 3 times.

▶ **What is 15 inches minus 10 inches?** Students clap 5 times.

2 ENGAGE

Develop: Yardsticks and Metersticks
"Today we are going to use a yardstick and a meterstick." Follow the instructions on the Activity Card **Yardsticks and Metersticks.** As students complete the activity, be sure to use the Questions to Ask.

Alternative Groupings
Small Group: Have the group decide on an estimate together before measuring the object.

Activity Card 5F

Progress Monitoring

| **If...** students are having difficulty measuring with a meterstick, | ▶ **Then...** have them measure smaller objects in centimeters and gradually measure longer and longer objects. |

Practice
Have students complete **Student Workbook,** pages 24–25. Guide students through the Key Idea example and the Try This exercises.

Interactive Differentiation
Consult the **Teacher Dashboard** for grouping suggestions. You can also use performance on the Engage activity to guide students.

Independent Practice
For additional practice with yardsticks and metersticks, have students measure the same object with each tool. After students have measured several objects, ask them what they have observed. About how long is each unit of measure?

Supported Practice
For additional support, use a meterstick to review how to measure an object in centimeters.

- Show students how to line up an object with the zero mark on the meterstick.
- Demonstrate how to count the centimeters forward along the length of the object. Also show students how to read the number mark on the meterstick that is closest to the length of the object.
- Help students measure the length of their arms on a meterstick. Display the measurements on the board next to each student's name.
- ▶ **What is the relationship between your height and the length of your arm?** Answers may vary. Possible answer: Taller students have longer arms, and shorter students have shorter arms.

384 Level D Unit 5 **Geometry and Measurement**

REFLECT

Think Critically

Review students' answers to the Reflect prompt at the bottom of **Student Workbook** page 25, and then review the Engage activity.

▶ **How would you measure the height of a desk using a meterstick?** Answers may vary. Possible answer: Align the zero mark on the meterstick with one leg of the desk. Make sure it is straight up and down. Then count the centimeters forward from the floor upwards along the leg to the height of the desk, or see which number on the meterstick is closest to height of the desk.

Real-World Application

Your brother wants to put three posters on the wall of his room, and he wants the posters to be the same distance apart.

▶ **How would you measure the distance between posters using a yardstick?** Answers may vary. Possible answer: Hang up the first two posters. Line up the right edge of the first poster with the zero mark on the yardstick. Then see which number on the yardstick is closest to the left edge of the next poster. Hang up the third poster, with the left edge the same distance from the right edge of the second poster.

ASSESS

Informal Assessment

Use the online or print Student Record, **Assessment,** page 128, to record informal observations.

Yardsticks and Metersticks

Did the student

☐ pay attention to the contributions of others? ☐ improve on a strategy?

☐ contribute information and ideas? ☐ reflect on and check accuracy of work?

Additional Practice

For additional practice, have students complete **Practice,** page 107.

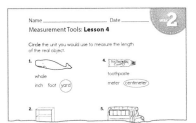

Practice, p. 107

Week 2 • Measurement Tools

Lesson 4

Key Idea
You can use a yardstick or a meterstick to measure long objects.
A baseball bat is about 1 yard or 1 meter long.

Try This
Find each object shown. Use a yardstick to measure each object's height to the nearest foot. Then use a meterstick to measure the object again to the nearest centimeter. **Answers may vary. Make sure students are using the appropriate tool correctly.**

Find the Object	Measurement in Feet	Measurement in Centimeters
1. chair	_____	_____
2. desk	_____	_____
3. window	_____	_____

24 Level D Unit 5 Geometry & Measurement

Practice
Find each object in the classroom. Use a yardstick to measure each object's length to the nearest foot. Then use a meterstick to measure it again to the nearest centimeter. **Answers may vary. Check students' measurements for reasonableness.**

Find the Object	Measurement in Feet	Measurement in Centimeters
4. row of books	_____	_____
5. bookcase	_____	_____
6. door	_____	_____

Reflect
Suppose you measured the door in inches with the yardstick. How would your measurement be different?

Answers may vary; possible answer: The door would be many more inches tall than it is feet tall. It would not be as many inches tall as it is centimeters tall.

Week 2 **Measurement Tools** • Lesson 4 25

Student Workbook, pp. 24–25

WEEK 2
Measurement Tools

Lesson 5 Review

Objective
Students review skills learned this week and complete the weekly assessment and project.

Standard
2.MD.1 Measure the length of an object by selecting and using appropriate tools such as rulers, yardsticks, meter sticks, and measuring tapes.

Vocabulary
Review vocabulary introduced during the week.

Creating Context
Have English Learners make a chart to review the vocabulary and new concepts in the unit. The chart will provide students with a way to review for assessments. Have them create a chart similar to the sample shown.

Word	Definition	Sentence
measure	to find the size of something	I use a ruler, a yardstick, or a meterstick to <u>measure</u>.

1 WARM UP

Prepare

▶ **How do you use a measuring tape to measure the height of your seat to the nearest inch?** Answers may vary. Possible answer: Place the zero mark of the measuring tape on the floor and measure to the bottom of the seat. The height of the seat is the number on the measuring tape at the seat.

▶ **How do you use a metric ruler to measure the width of your pencil box to the nearest centimeter?** Place the end of the metric ruler on one edge of the pencil box and see which number of centimeters on the metric ruler is closest to the edge of the pencil box.

2 ENGAGE

Practice
Have students complete **Student Workbook,** pages 26–27.

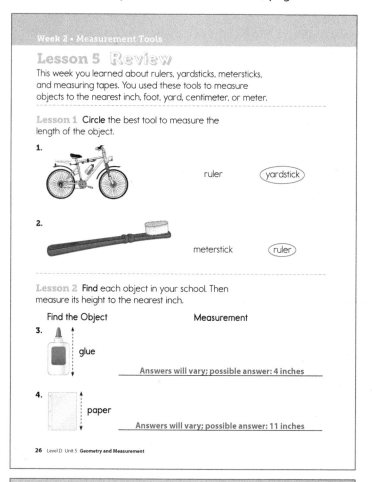

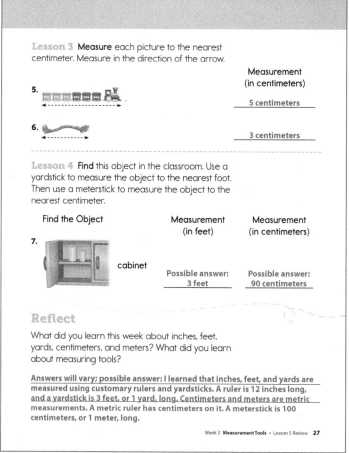

Student Workbook, pp. 26–27

386 Level D Unit 5 **Geometry and Measurement**

3 REFLECT

Think Critically

Review students' answers to the Reflect prompt at the bottom of *Student Workbook* page 27.

Discuss the answer with the group to reinforce Week 2 concepts.

4 ASSESS

Formal Assessment

Students may take the weekly assessment online.

As an alternative, students may complete the weekly test on *Assessment*, pages 63–64. Record progress using the Student Assessment Record, *Assessment*, page 128.

Going Forward

Use the *Teacher Dashboard* to view results of the online assessments, to input the results of print student assessments, and to review progress before making decisions about next steps. Use the weekly test results and observations to determine the next steps for each student.

Retention	
Student displays good grasp of this week's concepts and skills.	Give students two classroom objects, including one that is longer than 12 inches. Have students write how they would measure the objects using a customary ruler, metric ruler, yardstick, or meterstick.

Remediation	
Student is still struggling with the week's concepts and skills.	Partner students who need remediation with students who do not. Give the partners two classroom objects, including one that is longer than 12 inches. Have students discuss how to measure the objects using a customary ruler, metric ruler, yardstick, or meterstick. Then have them make and record their measurements.

Suggestions for Re-Evaluation: If a student has struggled without success for several weeks, use observations and test results to place the student at a level where they can find success and build confidence to move forward.

Assessment, pp. 63–64

Week 2 **Measurement Tools** • Lesson 5

Project Preview

This week, students learned how to use a customary ruler, metric ruler, yardstick, and meterstick. The project for this week requires students to extend the knowledge they gained in Find the Math and make measurements with customary and metric measuring tools as they plan a playground.

Project-Based Learning

Standards-driven Project-Based Learning is effective in building deep content understanding. Project-Based Learning increases long-term retention of concepts and has been shown to be more effective than traditional instruction. By completing a project to answer an essential question, students are challenged to apply and demonstrate mastery of concepts and skills by expressing understanding through discussion, research, and presentation.

Essential Question

HOW can I use my knowledge of length, time, and shapes to help someone complete a task?

Project Evaluation Criteria

Review project evaluation criteria with students prior to beginning the project.

Exceeds Expectations
☐ Project result is explained and can be extended.
☐ Project result is explained in context and can be applied to other situations.
☐ Project result is explained using advanced mathematical vocabulary.
☐ Project result is described, and mathematics are used correctly and can be extended.
☐ Project result is explained and extended, and shows advanced knowledge of mathematical concepts and skills.

Meets Expectations
☐ Project result is explained.
☐ Project result is explained in context.
☐ Project result is explained using mathematical vocabulary.
☐ Project result is described, and mathematics are used correctly.
☐ Project result is explained, and shows satisfactory knowledge of mathematical concepts and skills.

Does Not Meet Expectations
☐ Project result is not explained.
☐ Project result is explained, but out of context.
☐ Project result is explained, but mathematical vocabulary is oversimplified.
☐ Project result is described, but mathematics are not used correctly.
☐ Project result is not explained and/or extended, or shows less than satisfactory knowledge of mathematical concepts and skills.

Measure the Playground

Objective
Students can use customary and metric tools to measure the length of equipment or features in the park or playground.

Standard
2.MD.1 Measure the length of an object by selecting and using appropriate tools such as rulers, yardsticks, meter sticks, and measuring tapes.

Materials
Additional Materials
- customary ruler
- measuring tape
- meterstick
- metric ruler
- yardstick
- poster paper from previous week

Prepare Ahead
Try to have two or three of each measurement tool available. Measuring tapes will be the most useful for measuring the distance between pieces of equipment, so it will be helpful to have multiple measuring tapes on hand.

Best Practices
- Allow active learning with noise and movement.
- Organize materials ahead of time.
- Create adequate time lines for each project.

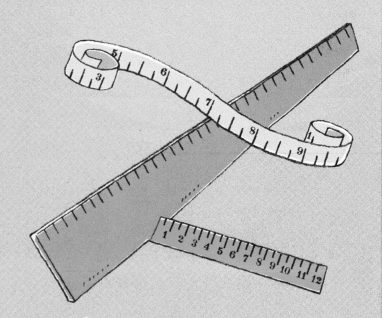

Introduce

Last week, you used nonstandard units to take measurements. This week you need to remeasure the same lengths and distances using rulers, yardsticks, metersticks, and measuring tapes.

▶ **What objects will you measure?** Answers may vary. Possible answers: teeter-totter and slide

▶ **Let's brainstorm what measuring tools you should use to measure the distance between the objects and the lengths of the equipment.**
Answers may vary. Possible answers: measuring tape, yardstick, meterstick

Explore

- Review how to use the measurement tools students have learned about this week.
▶ **Today you will work with your partner to measure the playground equipment that you measured last week.**
▶ **You will choose a measurement tool and measure the playground equipment. You can measure how wide the equipment is and how tall it is.**
- Allow pairs of students to choose a ruler, yardstick, meterstick, or measuring tape. Have students measure the same playground equipment they measured last week. They should also measure other pieces of playground equipment to gain experience with the measuring tools.
- Make sure students are taking turns making and recording the measurements on **Student Workbook,** page 28.
- Move students through each measuring tool as quickly as possible. Supervise the trading of rulers, yardsticks, metersticks, and measuring tapes among the pairs of students.
▶ **Complete Student Workbook, page 28, to list all the equipment you measured.**

Wrap Up

- Make sure all students can explain why they chose a certain tool and how they made the measurements.
- Discuss students' answers to the Reflect prompts at the bottom of **Student Workbook,** page 28.

Return the poster paper maps to pairs of students. Tell them to write the measurements of the equipment or playground feature they drew last week next to their pictures. When students have finished, put the posters away for next week.

If time permits, have students talk about the most important thing they learned.

Student Workbook, p. 28

Teacher Reflect

☐ Did I clearly explain how to organize the activity?

☐ Did students talk about the most important thing they learned?

☐ Did students show knowledge of how their project related to the major concept?

WEEK 3
Time Measurement to the Half Hour

Week at a Glance
This week students continue **Number Worlds,** Level D, Geometry and Measurement, by exploring different ways to measure time.

Skills Focus
- Match analog and digital displays of hour times.
- Tell time to the half hour using analog and digital clocks.
- Show times on an analog clock.

How Students Learn
Take advantage of opportunities throughout the day to help students develop concepts of time and the way it is measured. Guide students to notice and understand the patterns of minutes, hours, days, weeks, and months. Use different concrete models and manipulatives to make connections between analog and digital clocks.

English Learners ELL
For language support, use the **English Learner Support Guide,** pages 98–99, to preview lesson concepts and teach academic vocabulary.

Math at Home
Give one copy of the Letter to Home, page 27, to each student. Encourage students to share and complete the activity with their caregivers.

Weekly Planner

Lesson	Learning Objectives
1 pages 392–393	Students can tell time to the hour.
2 pages 394–395	Students can match analog and digital displays of hour times.
3 pages 396–397	Students can use analog and digital clocks to tell time to the hour and half hour.
4 pages 398–399	Students can show times on an analog clock to the half and whole hour.
5 pages 400–401	**Review and Assess** Students review skills learned this week and complete the weekly assessment.
Project pages 402–403	Students can tell time to the nearest half hour and to the nearest whole hour.

390 Level D Unit 5 **Geometry and Measurement**

Key Standard for the Week

Domain: Measurement and Data
Cluster: Work with time and money.
2.MD.7 Tell and write time from analog and digital clocks to the nearest five minutes, using a.m. and p.m.

Materials		Technology
Program Materials • **Student Workbook,** pp. 29–31 • **Practice,** p. 108 • Activity Card 5G, **Clock Concentration** • Clock Concentration Cards • Number Cube (7–12)	**Additional Materials** manipulative clock*	*Teacher Dashboard*
Program Materials • **Student Workbook,** pp. 32–33 • **Practice,** p. 109 • Activity Card 5G, **Clock Concentration** • Clock Concentration Cards • Digital Time Cards • Number Cubes (7–12)	**Additional Materials** manipulative clock*	*Teacher Dashboard*
Program Materials • **Student Workbook,** pp. 34–35 • **Practice,** p. 110 • Activity Card 5H, **What Time Is It?** • Clock Faces • Digital Time Cards	**Additional Materials** manipulative clock*	*Teacher Dashboard*
Program Materials • **Student Workbook,** pp. 36–37 • **Practice,** p. 111 • Activity Card 5H, **What Time Is It?** • Clock Faces • Digital Time Cards	**Additional Materials** manipulative clock*	*Teacher Dashboard*
Program Materials • **Student Workbook,** pp. 38–39 • **Assessment,** pp. 65–66		Review previous activities.
Program Materials **Student Workbook,** p. 40	**Additional Materials** • 2 long tubes, sticks, or dowels • 12 large notecards with the numbers 1 to 12 written on them • yardstick	

*Available from McGraw-Hill Education

WEEK 3
Time Measurement to the Half Hour

Find the Math

In this week, introduce students to clocks and telling time to the hour and to the half hour.

Use the following to begin a guided discussion:

▶ **How can you tell the time on the clock with the student?** Answers may vary. Possible answer: Look at the hour the student's arms are pointing to on the clock.

Have students complete *Student Workbook,* page 29.

Student Workbook, p. 29

Lesson 1

Objective
Students can tell time to the hour.

Standard
2.MD.7 Tell and write time from analog and digital clocks to the nearest five minutes, using a.m. and p.m.

Creating Context
Discuss with English Learners the meanings of *force, hands,* and *second,* and look for others that might have more than one meaning.

Materials
Program Materials
- Clock Concentration Cards, 1 set per student pair
- Number Cube (7–12)

Additional Materials
manipulative clock

 WARM UP

Prepare
Display a manipulative clock so that all students can see it. Discuss the clock with students, asking questions such as the following:

▶ **What numbers does the clock show?** 1–12

▶ **How many hands are there?** 2

▶ **Is there anything that would help you tell the hands apart?**
Possible answer: One hand is shorter than the other hand.

Explain that the shorter hand is called the *hour hand* and the longer hand is called the *minute hand*. Tell students that an *hour* is a measure of time equal to 60 minutes. A *minute* is a measure of time equal to 60 seconds. Use a timer or stopwatch to demonstrate the length of one second and the length of one minute.

2 ENGAGE

Develop: Clock Concentration

"Today we are going to learn how to tell time to the hour." Follow the instructions on the Activity Card **Clock Concentration.** As students complete the activity, be sure to use the Questions to Ask.

Activity Card 5G

Alternative Groupings

Small Group: Give groups two sets of cards and have them complete the activity as written.

Individual: Place the analog Clock Concentration Cards faceup in rows. Have the student draw a digital Clock Concentration Card from a facedown pile and find its matching analog Clock Concentration Card.

Progress Monitoring

| **If...** students have difficulty reading the time on an analog clock, | ▶ **Then...** remind them to look at the hour hand to write the hour as it would appear on a digital clock. |

Practice
Have students complete *Student Workbook,* pages 30–31. Guide students through the Key Idea example and the Try This exercises.

392 Level D Unit 5 **Geometry and Measurement**

Interactive Differentiation

Consult the **Teacher Dashboard** for grouping suggestions. You can also use performance on the Engage activity to guide students.

Independent Practice

For additional practice, give student pairs one set of Clock Concentration Cards. Have students place the Clock Cards facedown in a stack and spread out the time cards faceup. One partner will draw a Clock Card and show it to the other. The partner will find the time in the array of time cards.

Supported Practice

For additional support, use the manipulative clock to reinforce telling time. Roll a Number Cube (7–12) and then demonstrate the hour on the clock.

- Write times to the hour on the board and have students say the time aloud. Ask questions such as the following:
▶ **Where should the minute hand and the hour hand point to show 7 o'clock?** minute hand on *12*, hour hand on *7*

3 REFLECT

Think Critically

Review students' answers to the Reflect prompt at the bottom of **Student Workbook** page 31, and then review the Engage activity.

Have students talk about which kind of clock they like better.

▶ **Can you show me (say a time) on the clock?**

Real-World Application

The hours of the day are a.m. hours or p.m. hours. The a.m. hours are the hours between midnight and noon, and the p.m. hours are the hours between noon and midnight. Write the meaning of a.m. and p.m. on the board for students to refer to.

▶ **Are you more likely to be asleep at 3:00 a.m. or 3:00 p.m.?** 3:00 a.m.
▶ **Are you more likely to be at school at 1:00 a.m. or 1:00 p.m.?** 1:00 p.m.

4 ASSESS

Informal Assessment

Use the online or print Student Record, **Assessment,** page 128, to record informal observations.

Clock Concentration

Did the student
☐ respond accurately?   ☐ respond with confidence?
☐ respond quickly?   ☐ self-correct?

Additional Practice

For additional practice, have students complete **Practice,** page 108.

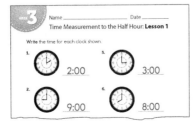

Practice, p. 108

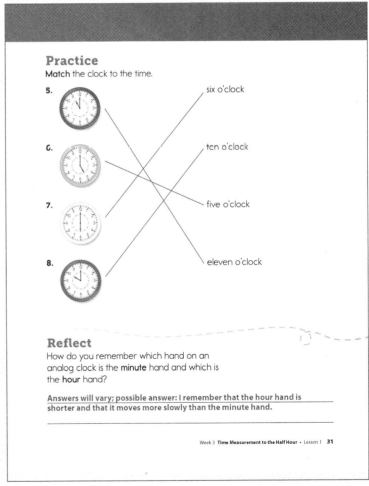

Student Workbook, pp. 30–31

Week 3 **Time Measurement to the Half Hour** • Lesson 1   **393**

WEEK 3
Time Measurement to the Half Hour

Lesson 2

Objective
Students can match analog and digital displays of hour times.

Standard
2.MD.7 Tell and write time from analog and digital clocks to the nearest five minutes, using a.m. and p.m.

Creating Context
Write *hour* and *our* on the board, and explain the difference between these two homophones. Encourage students to ask when they are uncertain of the meaning of a word or phrase.

Materials

Program Materials
- Clock Concentration Cards, 1 set per student pair
- Digital Time Cards, 1 set per student pair
- Number Cubes (7–12), 1 per student pair

Additional Materials
manipulative clock

1 WARM UP

Prepare

Distribute a copy of the Digital Time Cards to each student, and have students cut them apart.

Review digital hour times with students by showing students hour times on an analog clock and asking the following questions:

▶ What time does this clock show?

▶ How would you write this time? Raise the card showing how you would write the time.

If students struggle when answering the questions, demonstrate how to tell and write the time. Repeat the procedure using several different times.

Just the Facts

Play a game of "Fingers Up." Display times to the hour on a manipulative clock. Have students respond by holding up as many fingers as the hour time they see on the clock. If the clock reads 11:00 or 12:00, have students say the time aloud in unison. Use questions such as the following:

▶ **What time is this?** Display 12:00. Students say *twelve*.

▶ **What time is this?** Display 3:00. Students hold up three fingers.

▶ **What time is this?** Display 11:00. Students say *eleven*.

2 ENGAGE

Develop: Clock Concentration

"Today we are going to continue telling time." Follow the instructions on the Activity Card **Clock Concentration.** As students complete the activity, be sure to use the Questions to Ask.

Activity Card 5G

Alternative Groupings

Small Group: Give groups two sets of cards and have them complete the activity as written.

Individual: Place the analog Clock Concentration Cards faceup in rows. Have the student draw a digital Clock Concentration Card from a facedown pile and find its matching analog Clock Concentration Card.

Progress Monitoring

| If... students can readily tell time to the hour, | ▶ Then... consider introducing times to the half-hour. |

Practice

Have students complete **Student Workbook,** pages 32–33. Guide students through the Key Idea example and the Try This exercises.

Interactive Differentiation

Consult the **Teacher Dashboard** for grouping suggestions. You can also use performance on the Engage activity to guide students.

Independent Practice

For additional practice, give pairs of students Clock Concentration Cards and Digital Time Cards. Have students place the Analog Clock Cards facedown in a stack and spread out the Digital Clock Cards faceup. One partner will draw an Analog Clock Card and show it to the other student. The partner will find the matching digital clock in the array of Digital Clock Cards. Then have students write the time in analog notation. They can repeat the activity for the whole set of cards.

Supported Practice

For additional support, use the Digital Time Cards and a manipulative clock to demonstrate times on the hour. Remind students that an analog clock will have the minute hand pointing to the *12* at the beginning of each hour. The digital clock will show two zeros to the right of the colon at the beginning of the hour.

- Give each student a set of Digital Time Cards.
- Show a time on the manipulative clock. Have students hold up a digital time card that matches the time on the analog clock.
- Ask questions such as the following:

▶ **Which numbers should the minute hand and the hour hand point to at 4 o'clock?** minute hand on 12, hour hand on 4

▶ **Which numbers do you see on a digital clock to show 4 o'clock?** four and two zeros

394 Level D Unit 5 **Geometry and Measurement**

3 REFLECT

Think Critically

Review students' answers to the Reflect prompt at the bottom of **Student Workbook** page 33, and then review the Engage activity.

Discuss advantages and disadvantages of digital clocks. Ask students why it is still important to learn to tell time using both types of clocks.

▶ **How would you explain telling time to someone who has never done it before?**

Real-World Application

Being on time is important, and we need clocks to help us do so. Discuss with students different activities that begin at a certain time. For example:

▶ **What time does school start?**

▶ **What time do you catch the bus?**

▶ **What time do you go to bed?**

Discuss with students the ending times for each activity.

4 ASSESS

Informal Assessment

Use the online or print Student Record, **Assessment,** page 128, to record informal observations.

Clock Concentration
Did the student
- ☐ respond accurately?
- ☐ respond quickly?
- ☐ respond with confidence?
- ☐ self-correct?

Additional Practice

For additional practice, have students complete **Practice,** page 109.

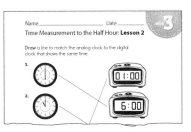

Practice, p. 109

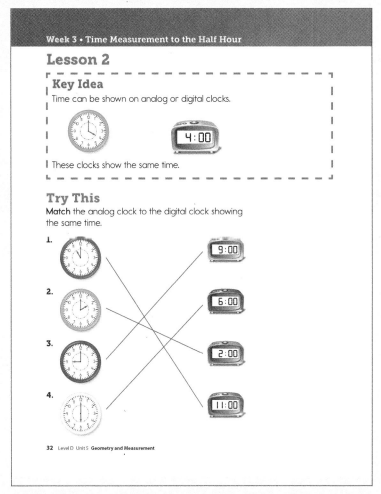

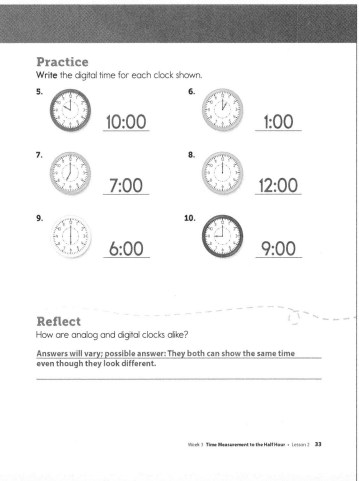

Student Workbook, pp. 32–33

Week 3 **Time Measurement to the Half Hour** • Lesson 2 **395**

WEEK 3
Time Measurement to the Half Hour

Lesson 3

Objective
Students can use analog and digital clocks to tell time to the hour and half hour.

Standard
2.MD.7 Tell and write time from analog and digital clocks to the nearest five minutes, using a.m. and p.m.

Vocabulary
- hour
- minute

Creating Context
Discuss with English Learners the meaning of *face*, *hands*, and *second*, and look for other words that might have more than one meaning.

Materials
Program Materials
- Clock Faces
- Digital Time Cards

Additional Materials
manipulative clock, 1 for each pair of students

1 WARM UP

Prepare
Show students an analog clock with the time set at 2:30.

▶ **Who can tell me what time this is?** Half past two; 2:30

- Allow several students to answer and explain their reasoning.
- Point out that there are two ways of naming this time. We can say it is "half past two" because the minute hand has moved half way around the dial. It is pointing straight down, towards the 6. The hour hand has also moved half way towards the 3. It is between the 2 and the 3 but it hasn't reached the 3 yet so we say the hour time is 2. When we add the minute time "half past" and the hour time "two", we get "half past two".
- We can also say the time is "2:30" because the minute hand has moved 30 spaces away from the 12, which shows zero minutes on a clock. Ask students to count the minute spaces with you as you point to each one to verify that there are 30 spaces, or 30 minutes, between the 12 and 6 positions on the dial.

Just the Facts
Play a game of "Face the Facts," using analog and digital clocks to tell time. Have students answer each question by looking up to show "yes" and covering their eyes to show "no." Use questions such as the following:

▶ **Is this time 4:00?** Display 4:00 on an analog clock.
  Students look up.

▶ **Is this time 6:00?** Display 8:00 using a digital time card.
  Students cover their eyes.

2 ENGAGE

Develop: What Time Is It?
"Today we are going to use the minute hand as well as the hour hand on a clock to tell time for hours and half hours." Follow the instructions on the Activity Card **What Time Is It?** As students complete the activity, be sure to use the Questions to Ask.

Activity Card 5H

Alternative Groupings
Small Group: After students record the time, have them discuss their records within the group.

Progress Monitoring	
If... students have difficulty reading half-hour times on an analog clock,	**Then...** have them look at the two numbers the hour hand is between and choose the smaller number for the hour time. Next, have them look at the minute hand. If it is pointing straight down, toward the 6, they should say "30 minutes" for the minute time.

Practice
Have students complete **Student Workbook**, pages 34–35. Guide students through the Key Idea example and the Try This exercises.

Interactive Differentiation
Consult the **Teacher Dashboard** for grouping suggestions. You can also use performance on the Engage activity to guide students.

Independent Practice
For additional practice, pair students with partners who have a better understanding of time concepts. Provide pairs with a manipulative clock. Have one student display times to the hour or to the half hour. Then have the other partner tell the time. Make sure students switch roles.

Supported Practice
For additional support, use the Digital Time Cards to demonstrate times on the clock, both on the hour and on the half hour. Remind students of the positions of the minute and hour hands on an analog clock for times on the hour and half hour. Also remind students that a digital clock will display 30 to the right of the colon when the time is on the half hour.

- Display a Digital Time Card and have students tell the time aloud.
- Display a time on the manipulative clock, and have students tell the time aloud. Do this for times that are on the hour and on the half hour. Then ask questions such as the following:

▶ **Where should the minute hand and the hour hand point to show 8:30 on an analog clock?** minute hand on 6, hour hand between 8 and 9

▶ **What do you notice about all half hour times on a digital clock?**
  Answers may vary. Possible answer: The minutes read 30 because 30 minutes have passed since the beginning of the hour.

396 Level D Unit 5 **Geometry and Measurement**

3 REFLECT

Think Critically

Review students' answers to the Reflect prompt at the bottom of **Student Workbook** page 35, and then review the Engage activity.

Have students volunteer their responses and talk about which kind of clock they like better.

▶ **Can you show me** (say a time) **on the clock?**

▶ **Would you rather read an analog or digital clock?**

Real-World Application

Remind students that the hours of the day are a.m. hours or p.m. hours. Write the meaning of a.m. and p.m. on the board for students to refer to as you ask questions, such as the following:

▶ **Would a circus more likely start at 7:30 a.m. or 7:30 p.m.?** 7:30 p.m.

▶ **Are you more likely to eat a snack at 2:30 a.m. or 2:30 p.m.?** 2:30 p.m..

▶ **Are you more likely to see the moon at 3:30 a.m. or 3:30 p.m.?** 3:30 a.m.

4 ASSESS

Informal Assessment

Use the online or print Student Record, **Assessment,** page 128, to record informal observations.

What Time Is It?	
Did the student	
☐ respond accurately?	☐ respond with confidence?
☐ respond quickly?	☐ self-correct?

Additional Practice

For additional practice, have students complete **Practice,** page 110.

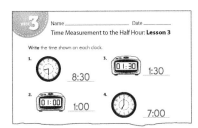

Practice, p. 110

Week 3 • Time Measurement to the Half Hour

Lesson 3

Key Idea
Clocks can be used to tell time to the nearest hour or half hour.

Say: "six o'clock" Say: "three o'clock" Say: "twelve o'clock"
Write: 6:00 Write: 3:00 Write: 12:00

Say: "six thirty" Say: "three thirty" Say: "twelve thirty"
Write: 6:30 Write: 3:30 Write: 12:30

Try This
Use the clocks above to answer each question.

1. Where is the long hand on the clock pointing for each time in the top row?
   to the twelve

2. Where is the short hand on the clock pointing for each time in the top row?
   to number that tells the hour

3. Where would the long hand be pointing according to the clocks in the bottom row?
   to the six

34 Level D Unit 5 Geometry and Measurement

Practice
Write the time shown on each clock.

4. 7:30   5. 2:00

6. 5:30   7. 1:30

8. 6:00   9. 4:00

Reflect
How is telling time to the half hour different than telling time to the whole hour on clocks with hands?

Answers may vary; possible answer: The hour hand is halfway between two numbers when you tell time to the half hour, and pointing to the number if you tell time to the whole hour. The minute hand is always on the 6 for time to the half hour, and always pointing to the 12 for time to the whole hour.

Week 3 Time Measurement to the Half Hour • Lesson 3 35

Student Workbook, pp. 34–35

WEEK 3
Time Measurement to the Half Hour

Lesson 4

Objective
Students can show times on an analog clock to the half and whole hour.

Standard
2.MD.7 Tell and write time from analog and digital clocks to the nearest five minutes, using a.m. or p.m.

Creating Context
Time is a central part of culture in the United States. Americans place value on being on time, sticking to a time schedule by starting on time, and not keeping others waiting. Discuss with English Learners how these expectations impact students, workers, and other adults when they don't adhere to the same concepts of time.

Materials
Program Materials
- Clock Faces
- Digital Time Cards

Additional Materials
manipulative clock

1 WARM UP

Prepare
Review how to tell time to the half hour and the hour with students by showing students half-hour and hour times on an analog clock and asking the following questions:

▶ **What time does this clock show?**

▶ **How would you write this time?**

Just the Facts
Play a game of "Heads Up," using analog clocks to tell time. Have students answer each question by holding their heads up to show "yes" and putting their heads down to show "no." Use questions such as the following:

▶ **Is this time 12:30?** Display 12:30 on a clock. heads up

▶ **Is this time 5:30?** Display 4:30 on a clock. heads down

▶ **Is this time 9:30?** Display 9:30 on a clock. heads up

2 ENGAGE

Develop: What Time Is It? (Variation)
"Today we are going to continue practicing telling and writing hour times and half-hour times." Follow the instructions on the Activity Card **What Time Is It?** (Variation: Show the Time). As students complete the activity, be sure to use the Questions to Ask.

Activity Card 5H

Alternative Groupings
Small Group: After students record the time, have them discuss their records within the group.

Progress Monitoring	
If... students struggle with placing the hour hand on times to the half hour,	▶ **Then...** point out that at 1:30, you are halfway between one o'clock and two o'clock.

Practice
Have students complete **Student Workbook,** pages 36–37. Guide students through the Key Idea example and the Try This exercises.

Interactive Differentiation
Consult the **Teacher Dashboard** for grouping suggestions. You can also use performance on the Engage activity to guide students.

Independent Practice
For additional practice provide students with blank copies of Clock Faces. Have them draw 2:00, 4:00, 6:00, …, 12:00 on the clock faces. Also have them draw 1:30, 3:30, 5:30, …, 11:30 on the clock faces.

Supported Practice
For additional support, provide students with blank copies of Clock Faces. When you name a time to the hour or half hour, students should draw the time on the clock face. Tell them to write the time using digital notation as well. Before your begin, remind students of the following information.

▶ **The long hand is the minute hand. It points to the twelve on the hour and to the six on the half hour.**

▶ **The short hand is the hour hand. It points right at a number on the hour and between two numbers on the half hour.**

▶ **When you write the time, write two zeros after the colon when the time is on the hour. Write *30* after the colon when the time is on the half hour.**

Draw the answer on the board and write the time below it after you have given students a chance to respond to each time that you name.

398 Level D Unit 5 **Geometry and Measurement**

3 REFLECT

Think Critically

Review students' answers to the Reflect prompt at the bottom of **Student Workbook** page 37, and then review the Engage activity.

Discuss how telling time to the half hour and the hour differs from what they did today.

▶ **What was the hardest thing you learned today?** Answers may vary. Possible answer: I learned to say the next hour when the hour hand was between two numbers

Real-World Application

Start time is an important fact that we use every day to structure our lives and to make sure we show up when we are supposed to. Discuss with students activities in their lives that are dependent on start time.s

▶ **When does the sun rise?** Answers may vary. Possible answer: 6:30
▶ **When does your recess start?** Answers may vary. Possible answer: 11:30
▶ **When does your favorite television show start?** Answers may vary. Possible answer: 8:30

4 ASSESS

Informal Assessment

Use the online or print Student Record, **Assessment,** page 128, to record informal observations.

What Time Is It? (Variation)	
Did the student	
☐ respond accurately?	☐ respond with confidence?
☐ respond quickly?	☐ self-correct?

Additional Practice

For additional practice, have students complete Practice, page 111.

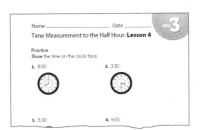

Practice, p. 111

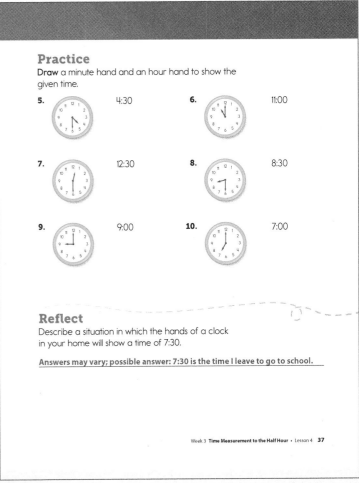

Student Workbook, pp. 36–37

Week 3 **Time Measurement to the Half Hour** • Lesson 4   **399**

WEEK 3
Time Measurement to the Half Hour

Lesson 5 Review

Objective
Students review skills learned this week and complete the weekly assessment and project.

Standard
2.MD.7 Tell and write time from analog and digital clocks to the nearest five minutes, using a.m. and p.m.

Vocabulary
Review vocabulary introduced during the week.

Creating Context
Remind students of the difference between the homophones *hour* and *our*. Remind students to ask when they are uncertain of the meaning of a word or phrase.

1 WARM UP

Prepare

▶ **Tell me something you do every day and when you begin.**
Answers may vary. Possible response: I brush my teeth each morning at 7:30 a.m.

2 ENGAGE

Practice
Have students complete **Student Workbook,** pages 38–39..

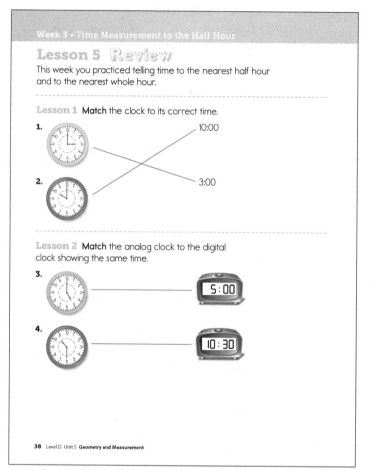

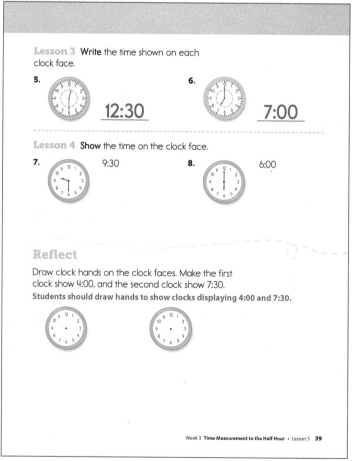

Student Workbook, pp. 38–39

400 Level D Unit 5 **Geometry and Measurement**

3 REFLECT

Think Critically

Review students' answers to the Reflect prompts at the bottom of **Student Workbook** page 39.

Discuss the answer with the group to reinforce Week 3 concepts.

4 ASSESS

Formal Assessment

Students may take the weekly assessment online.

As an alternative, students may complete the weekly test on **Assessment,** pages 65–66. Record progress using the Student Assessment Record, **Assessment,** page 128.

Going Forward

Use the **Teacher Dashboard** to view results of the online assessments, to input the results of print student assessments, and to review progress before making decisions about next steps. Use the weekly test results and observations to determine the next steps for each student.

Retention	
Student displays good grasp of this week's concepts and skills.	Give students two analog clock faces, one that shows 5:00 and one that shows 8:30. Have students describe how to tell the time on each clock. Then give them two digital clock times. Have students show the times on a manipulative clock.

Remediation	
Student is still struggling with the week's concepts and skills.	Partner students who need remediation with students who do not. Give the partners two clock faces, one that shows 8:00 and one that shows 2:30. Have students discuss how to tell the time on each clock. Then have them give the times verbally and record the times. Some students may benefit from reading books about telling time.

Suggestions for Re-Evaluation: If a student has struggled without success for several weeks, use observations and test results to place the student at a level in which he or she can find success and build confidence to move forward.

Assessment, pp. 65–66

Week 3 **Time Measurement to the Half Hour** • Lesson 5

Project Preview

This week, students learned how to tell time to the nearest half hour and to the nearest whole hour. The project for this week requires students to extend the knowledge they gained in Find the Math and demonstrate that they can tell time and show time using a human analog clock.

Project-Based Learning

Standards-driven Project-Based Learning is effective in building deep content understanding. Project-Based Learning increases long-term retention of concepts and has been shown to be more effective than traditional instruction. By completing a project to answer an essential question, students are challenged to apply and demonstrate mastery of concepts and skills by expressing understanding through discussion, research, and presentation.

Essential Question

HOW can I use my knowledge of length, time, and shapes to help someone complete a task?

Project Evaluation Criteria

Review project evaluation criteria with students prior to beginning the project.

Exceeds Expectations
☐ Project result is explained and can be extended.
☐ Project result is explained in context and can be applied to other situations.
☐ Project result is explained using advanced mathematical vocabulary.
☐ Project result is described, and mathematics are used correctly and can be extended.
☐ Project result is explained and extended, and shows advanced knowledge of mathematical concepts and skills.

Meets Expectations
☐ Project result is explained.
☐ Project result is explained in context.
☐ Project result is explained using mathematical vocabulary.
☐ Project result is described, and mathematics are used correctly.
☐ Project result is explained, and shows satisfactory knowledge of mathematical concepts and skills.

Does Not Meet Expectations
☐ Project result is not explained.
☐ Project result is explained, but out of context.
☐ Project result is explained, but mathematical vocabulary is oversimplified.
☐ Project result is described, but mathematics are not used correctly.
☐ Project result is not explained and/or extended, or shows less than satisfactory knowledge of mathematical concepts and skills.

Make a Human Clock

Objective
Students can tell time to the nearest half hour and to the nearest whole hour.

Standard
2.MD.7 Tell and write time from analog and digital clocks to the nearest five minutes, using a.m. and p.m.

Materials
Additional Materials
- 2 long wrapping paper tubes, sticks, or dowels; one should be shorter than the other
- 12 large notecards with the numbers 1 to 12 written on them
- yardstick

Prepare Ahead
To complete the project, students will need access to a large, empty space such as a playground, gym, or wide hallway.

Best Practices
- Provide project directions that are clear and brief.
- Allow active learning with noise and movement.
- Allow students to self-monitor.

Introduce

Imagine that you and your classmates want to form a human clock on a playground. You need to decide how to construct the clock and how to find a reasonable size for it. People will stand in for the numbers on the clock face.

▶ **How many students will it take to make the clock?** 12 students

▶ **How large should the clock be? What do you think is a reasonable distance from one side of the clock to the other?** Answers may vary. Possible answer: It should be at least ten feet wide.

Explore

- Depending on the number of students in the class, you will need to form two groups of at least twelve students or one group of at least twelve students. Provide student groups with number cards from 1 to 12. Give each of twelve students a number card and have them stand in a circle.

▶ **Now that you are in a circle, you need to put yourselves in order from 1 to 12.**

- Make sure students are in clockwise numerical order. Tell them to hold their number cards so that everyone can read them.
- If the group has more than twelve students, have students without number cards measure the diameter of the circle. Help students measure through the center of the circle with the yardstick. Make sure all students are the same distance from one another.

▶ **Now you need to think of six times to show on the human clock, with three times on the half hour.**

- Accept suggestions for six times from students. Have them write down the times in their **Student Workbooks. Depending on the number of students in the group, one student may serve as the recorder for the entire group.**
- Put the wrapping paper tubes or dowels in the center of the circle. If a student is available, give him or her the job of moving the "minute" and "hour" hands to the correct positions as another student calls out the times on the list. If only twelve students are in the group, you can have students with notecards step out of the clock periodically to adjust the minute and hour hands.
- Remind students to check off each time as they display it on their human clock.

▶ **Complete Student Workbook, page 40, to list the times your group made.**

Wrap Up

- Make sure each student can explain the position of the minute hand and the position of the hour hand to make a certain time.
- If groups struggle to complete the project worksheet, reduce the number of times students must show.
- Discuss students' answers to the Reflect prompts at the bottom of **Student Workbook,** page 40.

At the end of the activity, tell students that next week they will add the clock design to their playground map. Retain their maps for use in Week 4.

If time permits, have students show additional times on the human clock.

Student Workbook, p. 40

Teacher Reflect
- ☐ Did students use their time wisely and effectively?
- ☐ Was I able to explain questions when students did not understand?
- ☐ Did students focus on the major concept of the activity?

WEEK 4
Time Measurement to the Nearest Five Minutes

Week at a Glance
This week students continue **Number Worlds,** Level D, Geometry and Measurement, by exploring different ways to measure time.

Skills Focus
- Use analog and digital clocks.
- Tell time to the hour, half hour, quarter hour, and five minutes.
- Show time to the hour, half hour, quarter hour, and five minutes.

How Students Learn
Students should continue to notice and understand the patterns of minutes, hours, days, weeks, and months in relationship to activities they do every day. The use of concrete models and manipulatives builds student understanding as they continue to learn about fractional parts of an hour.

English Learners ELL
For language support, use the **English Learner Support Guide,** pages 100–101, to preview lesson concepts and teach academic vocabulary.

Math at Home
Give one copy of the Letter to Home, page 28, to each student. Encourage students to share and complete the activity with their caregivers.

Weekly Planner

Lesson	Learning Objectives
1 pages 406–407	Students can tell and write time to the nearest 15 minutes.
2 pages 408–409	Students can tell and write time using the terms *quarter past*, *half past*, and *quarter 'til*.
3 pages 410–411	Students can tell and write time to the nearest five minutes.
4 pages 412–413	Students can use a.m. and p.m.
5 pages 414–415	**Review and Assess** Students review skills learned this week and complete the weekly assessment.
Project pages 416–417	Students can tell time to the nearest five minutes and to the nearest quarter hour.

404 Level D Unit 5 **Geometry and Measurement**

Key Standard for the Week

Domain: Measurement and Data

Cluster: Work with time and money.

2.MD.7 Tell and write time from analog and digital clocks to the nearest five minutes, using a.m. and p.m.

Materials		Technology
Program Materials • *Student Workbook,* pp. 41–43 • *Practice,* p. 112 • Activity Card 5I, **What is the Time?** • Clock Faces	**Additional Materials** • manipulative clock* • Vocabulary Card 22, *hour* • Vocabulary Card 22, *minute*	*Teacher Dashboard*
Program Materials • *Student Workbook,* pp. 44–45 • *Practice,* p. 113 • Activity Card 5J, **Words 'n Time** • Clock Faces	**Additional Materials** • index cards (3 × 5) • manipulative clock* • Vocabulary Card 22, *hour* • Vocabulary Card 26, *minute*	*Teacher Dashboard*
Program Materials • *Student Workbook,* pp. 46–47 • *Practice,* p. 114 • Activity Card 5K, **Telling Time on the Fives** • Clock Faces	**Additional Materials** • manipulative clock* • Vocabulary Card 22, *hour* • Vocabulary Card 26, *minute*	*Teacher Dashboard*
Program Materials • *Student Workbook,* pp. 48–49 • *Practice,* p. 115 • Activity Card 5L, **A.M. and P.M.**	**Additional Materials** • manipulative clock* • Vocabulary Card 22, *hour* • Vocabulary Card 26, *minute*	*Teacher Dashboard*
Program Materials • *Student Workbook,* pp. 50–51 • Weekly Test, *Assessment,* pp. 67–68		Review previous activities.
Program Materials *Student Workbook,* p. 52	Additional Materials • brads • construction paper	

*Available from McGraw-Hill Education

WEEK 4
Time Measurement to the Nearest Five Minutes

Find the Math

In this week, introduce students to telling time to the quarter hour and to the nearest five minutes.

Use the following to begin a guided discussion:

▶ **When you look at a clock, how can you tell what time it is?** Answers may vary. Possible answer: Look at where the hour and minute hands are on the clock.

Have students complete **Student Workbook,** page 41.

Student Workbook, p. 41

Lesson 1

Objective
Students can tell and write time to the nearest 15 minutes.

Standard
2.MD.7 Tell and write time from analog and digital clocks to the nearest five minutes, using a.m. and p.m.

Vocabulary
- hour
- minute

Creating Context
Review the elements of an analog clock with English Learners. Remind them that there are 60 minutes in an hour, but analog clocks only show twelve numbers, one for every 5 minutes. Make sure students can recognize the numbers 15, 30, and 45, both when the numbers are spoken aloud and when they are written out.

Materials
Program Materials
 Clock Faces

Additional Materials
- manipulative clock
- Vocabulary Card 22, *hour*
- Vocabulary Card 26, *minute*

1 WARM UP

Prepare
- Review and demonstrate how to tell time on the half hour and on the hour. Show students half-hour and hour times on the manipulative clock and ask the following questions:

▶ **What time does the clock show?**
▶ **How would you write this time?**

2 ENGAGE

Develop: What is the Time?
"Today we are going to learn how to tell and write time to the nearest 15 minutes." Follow the instructions on the Activity Card **What Is the Time?** As students complete the activity, be sure to use the Questions to Ask.

Activity Card 5I

Alternative Groupings
Individual: Complete as written.

Progress Monitoring

| **If...** students struggle with telling time on an analog clock to the nearest 15 minutes, | ▶ **Then...** remind them that each section on the clock represents 5 minutes. Students can skip count by fives from 12 to the minute hand to count the number of minutes that have passed since the hour began. |

Practice
Have students complete **Student Workbook,** pages 42–43. Guide students through the Key Idea example and the Try This exercises.

406 Level D Unit 5 **Geometry and Measurement**

Interactive Differentiation

Consult the **Teacher Dashboard** for grouping suggestions. You can also use performance on the Engage activity to guide students.

Independent Practice

For additional practice, pair students with partners that have a better understanding of time concepts. Provide each student with a Clock Faces worksheet. Have one partner draw the minute and hour hands on clock faces to show times on the quarter hour, such as 8:00, 8:15, 8:30, and 8:45. Then have the partner say the times aloud. Have students switch roles.

Supported Practice

For additional support, let individual students take turns showing a time on the manipulative clock. You may need to assist them in positioning the hour hand appropriately. Then work with the rest of the class to determine the time to the nearest 15 minutes. Ask questions, such as the following:

▶ **Where is the hour hand pointing?**

▶ **Where is the minute hand pointing? Is it closest to the 3, 6, 9, or 12?**

Help students come to a consensus about the time to the nearest 15 minutes. Then allow another student to show a time on the manipulative clock.

3 REFLECT

Think Critically

Review students' answers to the Reflect prompt at the bottom of **Student Workbook,** page 43, and then review the Engage activity.

▶ **Is a clock that shows 6:30 showing the time to the nearest 15 minutes? Why or why not?** Answers may vary. Possible answer: Yes; when a clock shows 6:30, the minute hand is on the 6. That means that 15 minutes have passed twice since six o'clock.

4 ASSESS

Informal Assessment

Use the online or print Student Record, **Assessment,** page 128, to record informal observations.

What Is the Time?	
Did the student	
☐ make important observations?	☐ provide insightful answers?
☐ extend or generalize learning?	☐ pose insightful questions?

Additional Practice

For additional practice, have students complete **Practice,** page 112.

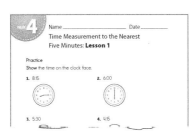

Practice, p. 112

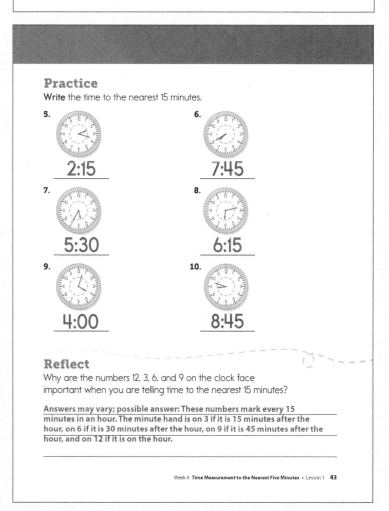

Student Workbook, pp. 42–43

Week 4 **Time Measurement to the Nearest Five Minutes** • Lesson 1 **407**

WEEK 4
Time Measurement to the Nearest Five Minutes

Lesson 2

Objective
Students can tell and write time using the terms *quarter past*, *half past*, and *quarter 'til*.

Standard
2.MD.7 Tell and write time from analog and digital clocks to the nearest five minutes, using a.m. and p.m.

Vocabulary
- hour
- minute

Creating Context
Help English Learners understand how the phrases *quarter past*, *half past*, and *quarter 'til* are related to times on the quarter hour. Have students make index cards showing the terms with examples. For instance, on the index card for *quarter past*, they should write 3:15, 5:15, 12:15, and so on. English Learners can also include diagrams of clocks showing the example times.

Materials
Program Materials
 Clock Faces, two for each student

Additional Materials
- index cards (3 × 5), 12 per student
- manipulative clock
- Vocabulary Card 22, *hour*
- Vocabulary Card 26, *minute*

1 WARM UP

Prepare
Write the terms *quarter past*, *half past*, and *quarter 'til* on the board. Then draw a clock face and show the numbers that divide it into quarters: 3, 6, 9, and 12.

▶ **How many quarters are in a dollar?** 4

- Explain that when a whole is divided into four parts, the parts are called *quarters*. Divide the clock face on the board into quarters with two perpendicular lines. Tell students that each part of the circle is one quarter of the whole. Then shade the right half.

▶ **How much of the clock face is shaded?** half

- Point out that two quarters make one half. Ask a volunteer to explain how this is related to the quarter hours on a clock. Answers may vary. Possible answer: When two quarters of an hour have passed, one half hour has passed.

Just the Facts
To increase fluency, have students add units of five minutes. Tell students to clap their hands if the addition fact is correct. If it is incorrect, tell them to cover their eyes. Use prompts, such as the following:

▶ **5 minutes + 5 minutes + 5 minutes = 15 minutes** clap hands
▶ **5 minutes + 5 minutes + 5 minutes + 5 minutes = 25 minutes** cover eyes

2 ENGAGE

Develop: Words 'n Time
"Today we are going to tell and write time using the terms *quarter past*, *half past*, and *quarter 'til*." Follow the instructions on the Activity Card **Words 'n Time**. As students complete the activity, be sure to use the Questions to Ask.

Activity Card 5J

Alternative Groupings
Whole Class: Complete as written using twelve index cards with times written in different ways. Have all students write the times on the worksheet.

Small Group: Have group members put all the index cards face down in one draw pile. Shuffle the cards. Students will take six cards from the pile and draw the times on their Clock Faces worksheets.

Progress Monitoring
| **If...** students struggle telling time using the terms *quarter past*, *half past*, and *quarter 'til*, | ▶ **Then...** have them work with a manipulative clock to show times to the quarter hour. |

Practice
Have students complete **Student Workbook,** pages 44–45. Guide students through the Key Idea example and the Try This exercises.

Interactive Differentiation
Consult the **Teacher Dashboard** for grouping suggestions. You can also use performance on the Engage activity to guide students.

Independent Practice
For additional practice, have one partner use a sentence frame to say, "My clock shows [a quarter past] [2]." Have the other partner repeat the sentence and draw the time on a clock face. Then have students repeat the exercise with other sentence frames using the terms *quarter past*, *half past*, and *quarter 'til*.

Supported Practice
For additional support, give students an index card for each term: *quarter past*, *half past*, and *quarter 'til*. Use a manipulative clock to show various times on the quarter hour. Have students hold up the index card that corresponds to the time. Ask questions, such as the following:

▶ **Where should the minute hand and the hour hand point to show a quarter 'til eight?** minute hand on 9, hour hand between 7 and 8
▶ **What number does the minute hand point to on a clock to show half past three?** 6

3 REFLECT

Think Critically

Review students' answers to the Reflect prompt at the bottom of **Student Workbook** page 45, and then review the Engage activity.

▶ **How can using the term *half past* help you draw 8:30 on a clock face?**
Answers may vary. Possible answer: The word *half* reminds me that 8:30 is halfway around the hour. Six is halfway around the clock face, so I know that the minute hand will be on the 6.

Real-World Application

Your parents tell you to set your alarm clock for a quarter past 6 to get up in time for a field trip.

▶ **What time should you set your alarm clock for?** 6:15

4 ASSESS

Informal Assessment

Use the online or print Student Record, **Assessment,** page 128, to record informal observations.

> **Words 'n Time**
> Did the student
> ☐ make important observations?   ☐ provide insightful answers?
> ☐ extend or generalize learning?   ☐ pose insightful questions?

Additional Practice

For additional practice, have students complete **Practice,** page 113.

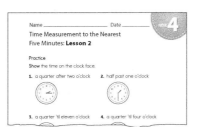

Practice, p. 113

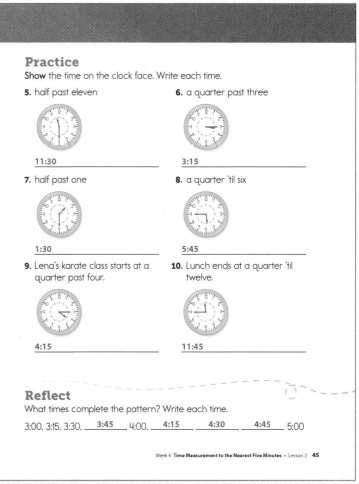

Student Workbook, pp. 44–45

Week 4 **Time Measurement to the Nearest Five Minutes** • Lesson 2 **409**

WEEK 4
Time Measurement to the Nearest Five Minutes

Lesson 3

Objective
Students can tell and write time to the nearest 5 minutes.

Standard
2.MD.7 Tell and write time from analog and digital clocks to the nearest five minutes, using a.m. and p.m.

Vocabulary
- hour
- minute

Creating Context
Help English Learners read times in five-minute intervals. For example, 1:05 is read "one oh five," 1:15 is "one fifteen," and so on. Relate the times on a digital clock to the position of the minute hand on an analog clock. Have students make flash cards with the time on one side of the card and the equivalent terms in their own language on the other side. If needed, students can draw a picture of the time on a clock.

Materials

Program Materials
- Clock Faces, several copies per student

Additional Materials
- manipulative clock
- Vocabulary Card 22, *hour*
- Vocabulary Card 26, *minute*

1 WARM UP

Prepare
- Have students count aloud to 60 by fives. Remind them that there are 60 minutes in one hour.
- Review how to tell time to the nearest 15 minutes.
- Display a time to the nearest 15 minutes on a manipulative clock.
- ▶ **What time does this clock show? What time would a digital clock show?**

Have students draw the times on their clock faces worksheet and write the times in digital notation. If students have difficulty answering these questions, repeat the procedure with several different times.

Just the Facts
Tell students to put their hands on top of their heads. Explain that if the answer to the question is "yes," they should raise their hands. If the answer to the question is "no," they should keep their hands on their heads. Use questions such as the following:

- ▶ **Does 10 minutes plus 10 minutes equal 20 minutes?** Students should raise their hands.
- ▶ **Does 15 minutes plus 15 minutes equal 45 minutes?** Students should keep their hands on their heads.
- ▶ **Does 5 minutes plus 5 minutes equal 55 minutes?** Students should keep their hands on their heads.

2 ENGAGE

Develop: Telling Time on the Fives

"Today we are going to tell and write time to the nearest 5 minutes." Follow the instructions on the Activity Card **Telling Time on the Fives.** As students complete the activity, be sure to use the Questions to Ask.

Activity Card 5K

Alternative Groupings

Whole Class: Choose twelve students to form the circle. Then complete the same way as written.

Small Group: Instead of forming a circle, students in the group will take turns counting by five-minute intervals until they reach 60 minutes. Then complete the same way as written.

Progress Monitoring

If... students struggle telling time to the nearest five minutes,	▶ Then... give them a blank analog clock face. Have them use their fingers to count clockwise by fives until they reach the number that the minute hand is pointing to.

Practice
Have students complete **Student Workbook,** pages 46–47. Guide students through the Key Idea example and the Try This exercises.

Interactive Differentiation
Consult the **Teacher Dashboard** for grouping suggestions. You can also use performance on the Engage activity to guide students.

Independent Practice
For additional practice, display 4:35 on a manipulative clock and write 4:35 on the board. Tell students that whenever the minute hand points to a number on the clock, the number after the colon in digital notation will end in a 5 or a 0. Then give pairs of students a copy of the Clock Faces worksheet. One partner will use a sentence frame to state a time to the nearest five minutes; for example, "My clock shows [3] [20]." The partner will repeat the sentence and draw the hands on a clock face. Tell students to switch roles and continue the activity until they have filled the worksheet.

Supported Practice
For additional support, display 4:35 on a manipulative clock and write 4:35 on the board. Tell students that whenever the minute hand points to a number on the clock, the number after the colon in digital notation will end in a 5 or a 0.

- Write the following times on index cards: 7:50, 10:45, 2:25, 4:15, 5:30, and 6:00. Distribute copies of the clock face worksheet to the class.
- Have a student choose an index card and show it to the class. Then have all students draw the time on a clock face worksheet.
- If time permits, write more times to the nearest five minutes on the board. Then ask questions, such as the following:
- ▶ **Where should the minute hand and the hour hand point to show 7:50?** The minute hand should point to 10; the hour hand to just before the 8.

3 REFLECT

Think Critically

Review students' answers to the Reflect prompt at the bottom of **Student Workbook** page 47, and then review the Engage activity.

▶ **Why do you think clocks have numbers 1 through 12 on their faces instead of 1 through 60?** Answers may vary. Possible answers: It is easier to read a clock with only twelve numbers on the face; knowing the time to the exact minute is not as important as knowing the time to the nearest five minutes; there are twelve hours in the day, not sixty.

Real-World Application

You just replaced the battery in an analog clock and have to set the time. Your brother says it is 3:35.

▶ **Where do you set the minute and hour hands to make the clock read 3:35?** Set the minute hand on 7 and the hour hand between 3 and 4.

4 ASSESS

Informal Assessment

Use the online or print Student Record, **Assessment,** page 128, to record informal observations.

Telling Time on the Fives	
Did the student	
☐ make important observations?	☐ provide insightful answers?
☐ extend or generalize learning?	☐ pose insightful questions?

Additional Practice

For additional practice, have students complete **Practice,** page 114.

Practice, p. 114

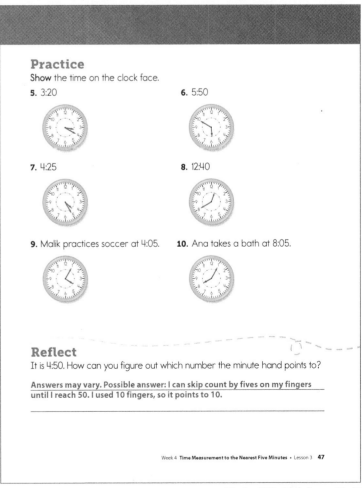

Student Workbook, pp. 46–47

Week 4 **Time Measurement to the Nearest Five Minutes** • Lesson 3 **411**

WEEK 4
Time Measurement to the Nearest Five Minutes

Lesson 4

Objective
Students can use a.m. and p.m.

Standard
2.MD.7 Tell and write time from analog and digital clocks to the nearest five minutes, using a.m. and p.m.

Vocabulary
- hour
- minute

Creating Context
English Learners may be unfamiliar with a.m. and p.m. In some non–English-speaking countries, the clock is divided into 24 hours, and A.M. and P.M. are not used. Spanish-speakers usually qualify a time with "in the morning/afternoon/evening" rather than a.m. and p.m. It may be helpful to tell English Learners that a.m. and p.m. stand for the Latin phrases *ante meridiem* and *post meridiem*, which mean "before noon" and "after noon."

Materials
Additional Materials
- manipulative clock
- Vocabulary Card 22, *hour*
- Vocabulary Card 26, *minute*

1 WARM UP

Prepare
- Remind students that a day has 24 hours and that a new day starts at midnight. Then tell them the following:
▶ **The 24 hours in a day are divided into two 12-hour parts. The first 12 hours in the day are called *a.m.* and the second 12 hours are called *p.m.***
- Explain that the twelve a.m. hours are between midnight and noon, and the twelve p.m. hours are between noon and midnight.
- Have students think of activities they regularly do during the day, such as getting up or eating dinner. As a class determine whether each activity happens at an a.m. time or a p.m. time.

Just the Facts
To increase fluency, review addition and subtraction with multiples of 5. Tell students to answer with a number and the unit of minutes. Use prompts, such as the following:
▶ 5 minutes plus 10 minutes equals _____. 15 minutes
▶ 20 minutes minus 10 minutes equals _____. 30 minutes
▶ 15 minutes plus 10 minutes equals _____. 25 minutes

2 ENGAGE

Develop: A.M. and P.M.
"Today we are going to use a.m. and p.m. to tell time." Follow the instructions on the Activity Card **A.M. and P.M.** As students complete the activity, be sure to use the Questions to Ask.

Activity Card 5L

Alternative Groupings
Whole Class: Choose twelve students to form the circle. Then complete the same way as written.

Small Group: Instead of forming a circle, students in the group will take turns counting off the hours of the day using a.m. and p.m. Then complete the same way as written.

Progress Monitoring

| If... students have difficulty identifying whether a time is a.m. or p.m., | ▶ Then... it may help to give students the mnemonic "about morning," and "past morning." |

Practice
Have students complete **Student Workbook,** pages 48–49. Guide students through the Key Idea example and the Try This exercises.

Interactive Differentiation
Consult the **Teacher Dashboard** for grouping suggestions. You can also use performance on the Engage activity to guide students.

Independent Practice
For additional practice, have students write the time they do each activity listed below and tell whether it is a.m. or p.m.
- wake up
- brush teeth before school
- do homework
- watch TV during the week
- get dressed for school
- get home from school

Supported Practice
For additional support, use the manipulative clock to demonstrate times on the clock to the nearest five minutes. Have volunteers give an example of what they may be doing at that time and identify the time as a.m. or p.m. Then ask questions, such as the following:
▶ **Where should the minute hand and the hour hand point to show 1:30 in the afternoon?** minute hand on 6, hour hand between 1 and 2
Is 1:30 in the afternoon *a.m.* or *p.m.*? p.m.
▶ **How many hours past midnight is 4:00 a.m.?** 4

412 Level D Unit 5 **Geometry and Measurement**

3 REFLECT

Think Critically
Review students' answers to the Reflect prompt at the bottom of **Student Workbook** page 49, and then review the Engage activity.

▶ **How would you explain a.m. or p.m. to a new student in the class?**

Real-World Application
Your father tells you that you have a dentist appointment tomorrow at 8:30 a.m.

▶ **How do you know whether your dentist appointment is in the morning or the evening?** Answers may vary. Possible answer: I know it is in the morning because *a.m.* means "before noon."

4 ASSESS

Informal Assessment
Use the online or print Student Record, **Assessment,** page 128, to record informal observations.

A.M. and P.M.
Did the student
- ☐ make important observations?
- ☐ provide insightful answers?
- ☐ extend or generalize learning?
- ☐ pose insightful questions?

Additional Practice
For additional practice, have students complete **Practice,** page 115.

Practice, p. 115

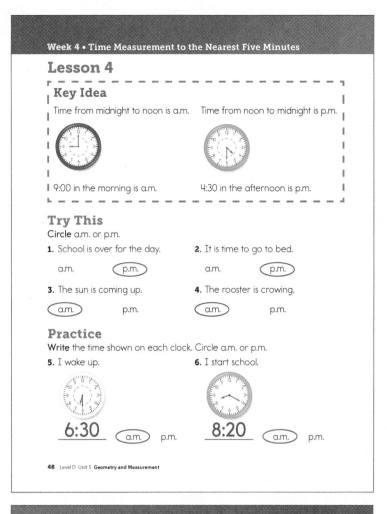

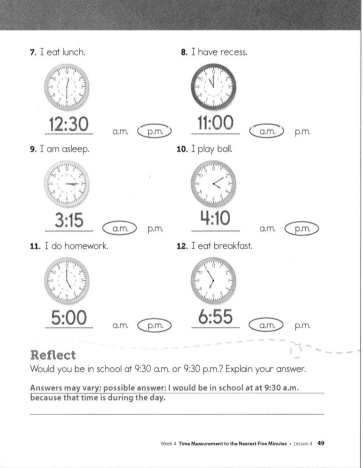

Student Workbook, pp. 48–49

WEEK 4
Time Measurement to the Nearest Five Minutes

Lesson 5 Review

Objective
Students review skills learned this week and complete the weekly assessment and project.

Standard
2.MD.7 Tell and write time from analog and digital clocks to the nearest five minutes, using a.m. and p.m.

Vocabulary
Review vocabulary introduced during the week.

Creating Context
Help English Learners assemble a paper or cardstock analog clock. Have them draw a large circle and add the numbers to the clock, starting with 3, 6, 9, and 12. Remind students that 3, 6, 9, and 12 are the numbers that mark the quarter hours. Provide students with two arrows, one long and one short, to represent the minute hand and the hour hand. Help students fasten the arrows to the center of the clock with a brad. Encourage them to use the clock as they review the week's lessons.

1 WARM UP

Prepare
Review how to tell time to the hour and to the nearest five minutes. Show 3:05, 5:20, and 10:40 on an analog clock and ask the following questions:

▶ **What time does this clock show?** 3:05, 5:20, 10:40

▶ **Explain how would you read the time on this clock.** Show 6:50 on a clock. The clock reads 6:50 because the minute hand is on the 10 and the hour hand is almost at the 7.

▶ **If the sun were coming up, would it be 6:50 a.m. or 6:50 p.m.? How do you know?** Answers may vary. Possible answer: 6:50 a.m. If the sun is coming up, it is before noon, which is a.m.

2 ENGAGE

Practice
Have students complete **Student Workbook,** pages 50–51.

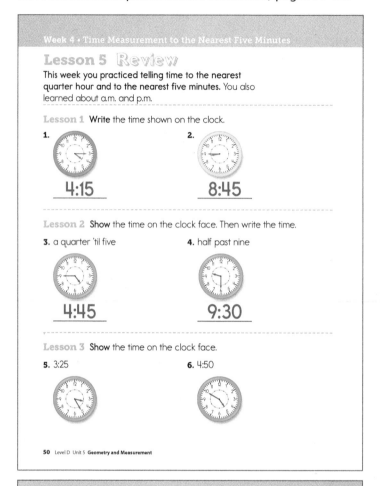

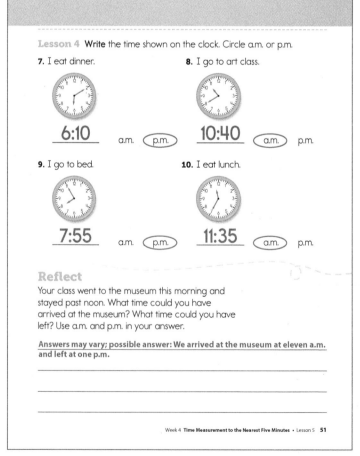

Student Workbook, pp. 50–51

3 REFLECT

Think Critically
Review students' answers to the Reflect prompts at the bottom of **Student Workbook** page 51. Discuss the answer with the group to reinforce Week 4 concepts.

4 ASSESS

Formal Assessment
Students may take the weekly assessment online.

As an alternative, students may complete the weekly test on **Assessment,** pages 67–68. Record progress using the Student Assessment Record, **Assessment,** page 128.

Going Forward
Use the **Teacher Dashboard** to view results of the online assessments, to input the results of print student assessments, and to review progress before making decisions about next steps. Use the weekly test results and observations to determine the next steps for each student.

Retention	
Student displays good grasp of this week's concepts and skills.	Give students two clock faces, one that shows 4:55 and one that shows 9:30. Tell them that one clock shows a time in the afternoon, and the other clock shows a morning time. Have them explain the positions of the minute hand and hour hand and describe how to tell the time on each clock, including a.m. or p.m. Tell students to name another way to say 9:30 (half past nine).
Remediation	
Student is still struggling with the week's concepts and skills.	Have students set the manipulative clock to a time to the nearest five minutes. Tell them what time the clock shows, and write the time on the board. Explain where the time falls in the day. For example, the clock shows 3:35. Tell students that it is 3:35 a.m. That time is three hours and 35 minutes past noon, so it is in the afternoon. Let each student take a turn.

Suggestions for Re-Evaluation: If a student has struggled without success for several weeks, use observations and test results to place the student at a level in which he or she can find success and build confidence to move forward.

Assessment, pp. 67–68

Project Preview

This week, students learned how to tell time to the nearest five minutes. The project for this lesson week requires students to extend the knowledge they gained in Find the Math and demonstrate that they can build a clock.

Project-Based Learning

Standards-driven Project-Based Learning is effective in building deep content understanding. Project-Based Learning increases long-term retention of concepts and has been shown to be more effective than traditional instruction. By completing a project to answer an essential question, students are challenged to apply and demonstrate mastery of concepts and skills by expressing understanding through discussion, research, and presentation.

Essential Question

HOW can I use my knowledge of length, time, and shapes to help someone complete a task?

Project Evaluation Criteria

Review project evaluation criteria with students prior to beginning the project.

Exceeds Expectations
☐ Project result is explained and can be extended.
☐ Project result is explained in context and can be applied to other situations.
☐ Project result is explained using advanced mathematical vocabulary.
☐ Project result is described, and mathematics are used correctly and can be extended.
☐ Project result is explained and extended, and shows advanced knowledge of mathematical concepts and skills.

Meets Expectations
☐ Project result is explained.
☐ Project result is explained in context.
☐ Project result is explained using mathematical vocabulary.
☐ Project result is described, and mathematics are used correctly.
☐ Project result is explained, and shows satisfactory knowledge of mathematical concepts and skills.

Does Not Meet Expectations
☐ Project result is not explained.
☐ Project result is explained, but out of context.
☐ Project result is explained, but mathematical vocabulary is oversimplified.
☐ Project result is described, but mathematics are not used correctly.
☐ Project result is not explained and/or extended, or shows less than satisfactory knowledge of mathematical concepts and skills.

Build a Clock

Objective
Students can tell time to the nearest five minutes and to the nearest quarter hour.

Standard
2.MD.7 Tell and write time from analog and digital clocks to the nearest five minutes, using a.m. and p.m.

Materials
Additional Materials
- brads
- construction paper

Prepare Ahead
Cut out large circles from construction paper and thin arrows in two sizes, one for the hour hand and one for the minute hand. Alternatively, you can use paper plates for clock faces.

Best Practices
- Provide meaning and organization to lessons and concepts.
- Allow active learning with noise and movement.
- Create adequate time lines for each project.

Introduce

Today you are going to add a clock to your playground maps. This will not be like a clock on the wall, it will be a clock on the ground. People will be able to stand in a circle on the playground and make a human clock. You will work with your partner to make a model of the clock and add the clock to your map.

▶ **Which numbers should go on the clock?** 1 through 12

Explore

- Review how to read time on a clock to the nearest five minutes and to the nearest quarter hour.
- Have students find their partners from the previous weeks and return the maps they created.
- ▶ **Now you need to decide how to add a human clock to your playground design.**
- Help students design their human clocks by asking questions such as the following:
- ▶ **Do you want to put stepping stones in a circle in the grass for the numbers? Or do you want to paint a circle on concrete to show the clock?**
- ▶ **What will the hands of the clock be made of?**
- Have students write a description of the human clock on a sheet of paper. Remind them to include a size measurement for the clock.
- Give each pair of students a circle, two different-sized arrows, and a brad with which to attach the arrows to the clock. Have students divide the circle into fourths and mark the center of the circle. Make sure students number the clock in a clockwise fashion with the 12 at the top. Then have them use the brad to fasten the arrows to the center of the circle.
- ▶ **Take turns showing times to the nearest quarter hour and to the nearest five minutes on your model clock.**
- Give students time to take turns using the model clock.
- ▶ **Complete Student Workbook, page 52, to list the times you and your partner showed on the model clock.**

Wrap Up

- Make sure each student can explain the position of the minute hand and the position of the hour hand to tell time to the nearest five minutes.
- Discuss students' answers to the Reflect prompts at the bottom of *Student Workbook,* page 52.

Have students add their clock designs to their maps.

▶ **Where will you add the human clock? Look at your map so far and decide where it will fit best. Remember how big a human clock is in real life as you think about where to put it.**

Make sure students write the numbers clockwise around the clock face on their maps. If possible, attach the model clock and the description of the playground clock to the back of the poster.

If time permits, allow students to add additional details and color to their playground maps.

Student Workbook, p. 52

Teacher Reflect

☐ Did students tell or show the steps when they explained how to do something?

☐ Did I adequately explain and discuss the Reflect questions with students?

☐ Did students focus on the major concept of the activity?

WEEK 5
Attributes of Shapes

Week at a Glance
This week students begin **Number Worlds,** Level D, Geometry and Measurement, by identifying attributes of shapes. Students will recognize and name different triangles and quadrilaterals. They will also be able to name polygons by looking at the number of sides.

Skills Focus
- Identify and name different types of triangles.
- Describe the properties of the special quadrilaterals.
- Sort and identify polygons by category.
- Recognize and name polygons with 3–8 sides.

How Students Learn
Students benefit most from "doing" geometry. Their visualization and thinking skills develop with hands-on experiences in sorting, drawing, tracing, and measuring geometric shapes. As students begin to analyze and compare the properties of shapes, they make conjectures about relationships among shapes and develop mathematical arguments to support their conjectures.

English Learners ELL
For language support, use the **English Learner Support Guide,** pages 102–103, to preview lesson concepts and teach academic vocabulary.

Math at Home
Give one copy of the Letter to Home, page 29, to each student. Encourage students to share and complete the activity with their caregivers.

Weekly Planner

Lesson	Learning Objectives
1 pages 420–421	Students can classify triangles by the measure of their angles.
2 pages 422–423	Students can identify and draw polygons given the number of sides.
3 pages 424–425	Students can classify special quadrilaterals by the number of congruent sides, parallel sides, and right angles.
4 pages 426–427	Students can identify the attributes of a cube.
5 pages 428–429	**Review and Assess** Students review skills learned this week and complete the weekly assessment.
Project pages 430–431	Students design a sandbox for the playground in the shape of a triangle, square, pentagon, or hexagon.

418 Level D Unit 5 **Geometry and Measurement**

Key Standard for the Week

Domain: Geometry

Cluster: Reason with shapes and their attributes.

2.G.1 Recognize and draw shapes having specified attributes, such as a given number of angles or a given number of equal faces. Identify triangles, quadrilaterals, pentagons, hexagons, and cubes.

Materials		Technology
Program Materials • *Student Workbook,* pp. 53–55 • *Practice,* p. 116 • Activity Card 5M, **Naming Triangles by Angles** • Naming Triangles by Angles	**Additional Materials** index cards • Vocabulary card 2, *angle* • Vocabulary card 47, *triangle*	*Teacher Dashboard* Shape Creator Tool
Program Materials • *Student Workbook,* pp. 56–57 • *Practice,* p. 117 • Activity Card 5N, **Naming Polygons** • Naming Polygons	**Additional Materials** pattern blocks*	*Teacher Dashboard* Geometry Sketch Tool
Program Materials • *Student Workbook,* pp. 58–59 • *Practice,* p. 118 • Activity Card 5O, **Special Quadrilaterals** • Special Quadrilaterals	**Additional Materials** • index cards • rulers*	*Teacher Dashboard* Building Blocks Shape Shop 3 Shape Creator Tool
Program Materials • *Student Workbook,* pp. 60–61 • *Practice,* p. 119 • Activity Card 5P, **Cubes** • Cubes • Number Cube (1–6)	**Additional Materials** • scissors • tape • Vocabulary Card 7, *cube* • Vocabulary Card 15, *face*	*Teacher Dashboard* Building Blocks Geometry Snapshots 8
Program Materials • *Student Workbook,* pp. 62–63 • Weekly Test, *Assessment,* pp. 69–70		Review previous activities.
Program Materials *Student Workbook,* p. 64	**Additional Materials** • construction paper • ruler* • scissors • tape or gluestick	

*Available from McGraw-Hill Education

WEEK 5
Attributes of Shapes

Find the Math

In this week, students learn about geometric shapes and their attributes.

Use the following to begin a guided discussion:

▶ **Have you ever seen road signs like these before? What do you think each road sign means?** Answers may vary. Possible answers: The sign on the far left means that you must yield to other drivers; the next sign means that you must stop; the square sign means that you can't turn left; the next sign means that you must watch out for people walking on the road; the curvy arrow sign means that the road is not straight.

Have students complete **Student Workbook,** page 53.

Student Workbook, p. 53

Lesson 1

Objective
Students can classify triangles by the measure of their angles.

Standard
2.G.1 Recognize and draw shapes having specified attributes, such as a given number of angles or a given number of equal faces. Identify triangles, quadrilaterals, pentagons, hexagons, and cubes.

Vocabulary
- angle
- acute angle
- acute triangle
- obtuse angle
- obtuse triangle
- right angle
- right triangle
- triangle

Creating Context
Identify affixes that help show word meanings. Brainstorm examples of prefixes, such as *tri-*, meaning "three."

Materials
Program Materials
- Naming Triangles by Angles, 1 copy per student pair

Additional Materials
index cards, 1 per student
- Vocabulary card 2, *angle*
- Vocabulary card 47, *triangle*

1 WARM UP

Prepare

Draw a right angle, an obtuse angle, and an acute angle on the board. Indicate each angle as you discuss it.

▶ **This is a right angle. The measure of a right angle is 90 degrees. The corner of a piece of paper is an example of a right angle.**

▶ **This is an obtuse angle. An obtuse angle is any angle that has a measure greater than 90 degrees.**

▶ **This is an acute angle. An acute angle is any angle that has a measure less than 90 degrees.**

After you have shown students all three types of angles, have them draw examples of each type of angle on their own.

2 ENGAGE

Develop: Naming Triangles by Angles

"Today we are going to identify and form different shapes." Follow the instructions on the Activity Card **Naming Triangles by Angles.** As students complete the activity, be sure to use the Questions to Ask.

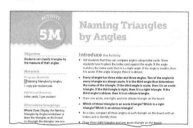

Activity Card 5M

Alternative Groupings

Whole Class: Display the Naming Triangles by Angle worksheet or draw the triangles on the board. Go through the triangles one at a time, having each student determine the type of triangle.

Individual: Have the student complete the worksheet and then draw three more triangles. The student will ask you to identify each triangle by their angles.

Progress Monitoring

| **If...** students are having trouble determining if an angle is acute, right, or obtuse, | ▶ **Then...** have students compare the angle with corners of an index card. |

Practice

Have students complete **Student Workbook,** pages 54–55. Guide students through the Key Idea example and the Try This exercises.

Interactive Differentiation

Consult the **Teacher Dashboard** for grouping suggestions. You can also use performance on the Engage activity to guide students.

Independent Practice

For additional practice with classifying triangles by angle measures, students should use the Shape Creator Tool. Have students set the appropriate properties in the first three categories only (Number of Angles/Sides, Equal Angles/Sides, and Size of Angles) in order to form right triangles, acute triangles, and obtuse triangles. Guide students to see that a right triangle can only have one right angle and two acute angles, an acute triangle can have no right angles or obtuse angles, and an obtuse triangle can have only one obtuse angle.

Supported Practice

For additional support, have students use the Naming Triangles by Angle worksheet as a guide.

- Have groups of students draw right, acute, or obtuse triangles on index cards.
- Have students put all the cards into a pile in the middle of the group. Then they should sort the triangles by angle type. Tell students to make one pile of acute triangles, one pile of right triangles, and one pile of obtuse triangles. Have students say the name of the triangle when they pick up a card.

▸ **Which type of triangle has all angles smaller than the corner of an index card?** acute triangle

▸ **Which type of triangle has one angle equal to the corner of an index card?** right triangle

▸ **Which type of triangle has one angle larger than the corner of an index card?** obtuse triangle

3 REFLECT

Think Critically

Review students' answers to the Reflect prompt at the bottom of **Student Workbook** page 55, and then review the Engage activity.

▸ **Is there more than one way to draw an acute angle? What about an obtuse angle?** yes; yes

▸ **Is there more than one way to draw a right angle?** no

4 ASSESS

Informal Assessment

Use the online or print Student Record, **Assessment,** page 128, to record informal observations.

Naming Triangles by Angles
Did the student
☐ make important observations?   ☐ provide insightful answers?
☐ extend or generalize learning?   ☐ pose insightful questions?

Additional Practice

For additional practice, have students complete **Practice,** page 116.

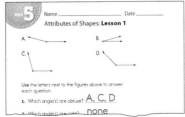
Practice, p. 116

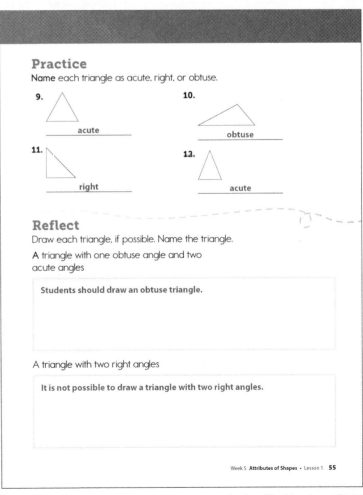

Student Workbook, pp. 54–55

Week 5 **Attributes of Shapes** • Lesson 1   **421**

WEEK 5
Attributes of Shapes

Lesson 2

Objective
Students can identify and draw polygons given the number of sides.

Standard
2.G.1 Recognize and draw shapes having specified attributes, such as a given number of angles or a given number of equal faces. Identify triangles, quadrilaterals, pentagons, hexagons, and cubes.

Vocabulary
- hexagon
- pentagon
- polygon
- quadrilateral

Creating Context
Help English Learners practice naming polygons. Have students work with a partner. Tell them to fold a sheet of paper into thirds and label the columns with *Number of Sides*, *Name*, and *Examples*. Have students write 3, 4, 5, and 6 in the "Number of Sides" column. In the next column have them write the name of the polygon with the corresponding number of sides. In the "Examples" column, have students draw pictures of the polygons.

Materials
Program Materials
Naming Polygons, 1 per student

Additional Materials
pattern blocks

Prepare Ahead
If necessary, prepare the Naming Polygons worksheet for classroom display.

1 WARM UP

Prepare
- Have students draw a diagram of a square, rectangle, and triangle.
- Have students color with a red crayon the shapes that have 4 sides. Have them color with a green crayon the shape with 3 sides.
- Have students draw additional shapes with 4 sides and 3 sides and color them as before. Then have them check their work in pairs.

Just the Facts
To increase computational fluency, tell students to nod up-and-down if the subtraction fact is correct. If the subtraction fact is incorrect, tell them to shake their heads "no". Use prompts such as the following:

▶ $10 - 5 = 5$ nod
▶ $9 - 8 = 2$ shake head "no"
▶ $12 - 4 = 8$ nod

2 ENGAGE

Develop: Naming Polygons
"Today we are going to name and draw polygons given the number of sides." Follow the instructions on the Activity Card **Naming Polygons**. As students complete the activity, be sure to use the Questions to Ask.

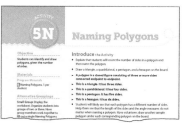

Activity Card 5N

Alternative Groupings
Small Group: Display the worksheet. Organize students into groups of two or three. Have group members work together to fill out a single Naming Polygons worksheet. Then complete the activity as written.

Progress Monitoring	
If... students have difficulty naming the polygons,	▶ **Then...** point out that *tri-* means "three," as in *tricycle*; *quad-* means "four;" *pent-* means "five;" and *hex-* means "six."

Practice
Have students complete **Student Workbook,** pages 56–57. Guide students through the Key Idea example and the Try This exercises.

Interactive Differentiation
Consult the **Teacher Dashboard** for grouping suggestions. You can also use performance on the Engage activity to guide students.

Independent Practice
For additional practice identifying shapes, students should use the Geometry Sketch Tool to draw triangles, quadrilaterals, pentagons, and hexagons. Have students discuss the attributes of the shapes with a partner.

Supported Practice
For additional support, have students use pattern blocks.

- Organize students into small groups. Give each group a generous number of pattern blocks.
- Each student should take a variety of pattern blocks. Tell students to separate their own pattern blocks into triangles, quadrilaterals, pentagons, and hexagons.
- Have students work together as a group to combine individual piles into four separate piles of triangles, quadrilaterals, pentagons, and hexagons.
- Check each group's piles. Then discuss the attributes of each shape with students.

3 REFLECT

Think Critically

Review students' answers to the Reflect prompt at the bottom of **Student Workbook** page 57, and then review the Engage activity.

▶ **Why can't a polygon have two sides?** Answers may vary. Possible answer: Two sides cannot form a closed shape, which is one of the attributes of a polygon.

Real-World Application

Your family needs to order a new glass top for a six-sided end table.

▶ **What shape of glass should they order?** hexagon

4 ASSESS

Informal Assessment

Use the online or print Student Record, **Assessment,** page 128, to record informal observations.

Naming Polygons
Did the student
☐ make important observations?   ☐ provide insightful answers?
☐ extend or generalize learning?   ☐ pose insightful questions?

Additional Practice

For additional practice, have students complete **Practice,** page 117.

Practice, p. 117

Week 5 • Attributes of Shapes

Lesson 2

Key Idea

If a polygon has 3 sides, then it is a **triangle**.

If a polygon has 4 sides, then it is a **quadrilateral**.

If a polygon has 5 sides, then it is a **pentagon**.

If a polygon has 6 sides, then it is a **hexagon**.

Try This

Answer the questions below using the letter next to each polygon.

A.   B.   C.   D.   E.

F.   G.   H.   I.   J.

1. Which polygons are hexagons? __B, J__
2. Which polygons are quadrilaterals? __A, F, H, I__
3. Which polygons have 5 sides? __C, D__
4. Which polygons have 3 sides? __E, G__
5. Which polygons are pentagons? __C, D__

56 Level D Unit 5 Geometry and Measurement

Practice

6. Draw three examples of each polygon. Answers will vary. Check student's work.

Triangle	Quadrilateral	Pentagon	Hexagon

Write the number of sides.

7. __4__ sides

8. __6__ sides

9. __3__ sides

10. __5__ sides

Reflect

How would you draw a polygon with 5 sides?

Answers may vary; possible answer: I would draw one side, add another side to each end of the first side, and then draw the other two sides to connect the polygon.

Week 5 Attributes of Shapes • Lesson 2 57

Student Workbook, pp. 56–57

Week 5 **Attributes of Shapes** • Lesson 2   **423**

WEEK 5
Attributes of Shapes

Lesson 3

Objective
Students can classify special quadrilaterals by the number of congruent sides, parallel sides, and right angles.

Standard
2.G.1 Recognize and draw shapes having specified attributes, such as a given number of angles or a given number of equal faces. Identify triangles, quadrilaterals, pentagons, hexagons, and cubes.

Vocabulary
- congruent
- parallel lines
- parallelogram
- rectangle
- rhombus
- square
- trapezoid

Creating Context
In English, words often have more than one meaning. For example, when you are instructed to make a table, *table* means "an organized display of information," not a piece of furniture.

Materials
Program Materials
- Special Quadrilaterals

Additional Materials
- index cards, 1 per student
- rulers, 1 per student

Prepare Ahead
Prepare the Special Quadrilaterals worksheet for classroom display or copy the table onto the board for students to see. Complete the table as you go along.

1 WARM UP

Prepare
Draw two parallel lines on the board. Mark them with arrows to show they are parallel.

▶ **If I extend these lines, will they ever touch?** no

▶ **These are called *parallel lines*. We mark them with arrows to show they are parallel. If there are two sets of parallel lines, we mark the second set with two arrows.**

Draw a rectangle on the board with two 12- inch sides and 2 nine-inch sides. Measure each side with a ruler.

▶ **Are any of these lines the same length?** yes

▶ **The lines that are the same length in this rectangle are called *congruent lines*.**

- Have a student volunteer draw a square with 12-inch congruent sides.

▶ **Which sides are congruent? How many pairs of congruent sides are there?** all sides; 2

Just the Facts
Play a game of "Face the Facts." Have students nod to show a correct answer and shake their heads "no" to show an incorrect answer. Use prompts such as the following:

▶ $8 + 4 = 12$ nod

▶ $5 + 9 = 13$ shake head "no"

2 ENGAGE

Develop: Special Quadrilaterals

"Today we are going to identify polygons according to their properties." Follow the instructions on the Activity Card **Special Quadrilaterals**. As students complete the activity, be sure to use the Questions to Ask.

Activity Card 50

Alternative Groupings

Whole Class: Allow student volunteers to draw special quadrilaterals on the board, marking any parallel or congruent sides.

Pair: Allow students to draw their own special quadrilaterals and trade them with one another. They must name the quadrilaterals.

Progress Monitoring

| **If...** students are having trouble determining if two sides of a shape are parallel, | ▶ **Then...** have them extend both sides using a ruler to see if the sides ever touch. |

Practice
Have students complete **Student Workbook,** pages 58–59. Guide students through the Key Idea example and the Try This exercises.

Interactive Differentiation
Consult the **Teacher Dashboard** for grouping suggestions. You can also use performance on the Engage activity to guide students.

Independent Practice

For additional practice, pair students to complete Shape Shop 3. This Building Blocks activity will help students identify shapes by their attributes. Have students use the examples they generate to complete the table shown below.

How many...	Parallelogram	Rectangle	Rhombus	Square	Trapezoid
right angles?	0 or 4	4	0 or 4	4	0, 1, or 2
pairs of congruent opposite sides?	2	2	2	2	0
pairs of parallel sides?	2	2	2	2	1

Supported Practice

For additional support, students should use the Shape Creator Tool. Have students set appropriate properties as described in each part below to generate various quadrilaterals and answer the questions. For each question, have students choose from these shape names: parallelogram, rectangle, rhombus, square, or trapezoid.

▶ **Check the Number of Angles/Sides box and set it to 4. Check the Equal Angles/Sides box and set it to exactly 2 sides equal. What kinds of shapes might have these properties?** parallelogram, rectangle, trapezoid

▶ **Check the Number of Angles/Sides box and set it to 4. Check the Equal Angles/Sides box and set it to exactly 3 sides equal. What kinds of shapes might have these properties?** trapezoid

424 Level D Unit 5 **Geometry and Measurement**

3 REFLECT

Think Critically

Review students' answers to the Reflect prompt at the bottom of *Student Workbook* page 59, and then review the Engage activity.

▶ **How would you describe a category that includes squares, rectangles, and parallelograms?** quadrilaterals that have two pairs of parallel sides and two pairs of congruent sides.

▶ **What about a category that includes just squares and rectangles?** quadrilaterals that have two pairs of parallel sides, two pairs of congruent sides, and four right angles.

Real-World Application

Find five different shapes in the classroom. Describe the characteristics of the shapes.

▶ **Do the shapes have any right angles?**
▶ **Are there any acute or obtuse angles?**
▶ **How many sides are parallel?**
▶ **Are any of the sides congruent?**

4 ASSESS

Informal Assessment

Use the online or print Student Record, **Assessment,** page 128, to record informal observations.

Special Quadrilaterals	
Did the student	
☐ make important observations?	☐ provide insightful answers?
☐ extend or generalize learning?	☐ pose insightful questions?

Additional Practice

For additional practice, have students complete **Practice,** page 118.

Practice, p. 118

Week 5 • Attributes of Shapes

Lesson 3

Key Idea
Quadrilaterals have four sides. There are five special quadrilaterals.

Parallelogram Rectangle Rhombus Square Trapezoid

Try This
Write the letter of the quadrilateral that fits the description. Some descriptions may match more than one quadrilateral.

A. Parallelogram B. Rectangle C. Rhombus D. Square E. Trapezoid

1. A quadrilateral with no right angles. __A, C, E__
2. A quadrilateral with four congruent sides. __C, D__
3. A quadrilateral with two pairs of parallel sides. __A, B, C, D__
4. A quadrilateral with four right angles. __B, D__
5. A quadrilateral with one pair of parallel sides. __E__
6. A quadrilateral with two pairs of congruent sides. __A, B, C, D__
7. A quadrilateral with four right angles and four congruent sides. __D__

Practice

Name the quadrilateral(s).

8. A quadrilateral with four right angles and two pairs of congruent sides
   __rectangle and square__

9. A quadrilateral with no right angles and two pairs of parallel sides
   __parallelogram and rhombus__

10. A quadrilateral with four congruent sides and no right angles
    __rhombus__

11. A quadrilateral with one pair of parallel sides
    __trapezoid__

Reflect

What do all the shapes in this lesson have in common?

All of the shapes in this lesson are four-sided figures, so all are quadrilaterals.

What are the differences between a parallelogram and a square? How are they alike?

Parallelograms do not have to have right angles, but a square does. Parallelograms have two pairs of parallel sides, but not all sides have to be congruent. A square has four congruent sides and four right angles. They are alike because they both have two pairs of parallel sides.

Student Workbook, pp. 58–59

WEEK 5
Attributes of Shapes

Lesson 4

Objective
Students can identify the attributes of a cube.

Standard
2.G.1 Recognize and draw shapes having specified attributes, such as a given number of angles or a given number of equal faces. Identify triangles, quadrilaterals, pentagons, hexagons, and cubes.

Vocabulary
- cube
- face
- net

Creating Context
English Learners may have difficulty distinguishing the common meanings of *face* and *edge* from their mathematical meanings. Explain that, in math, a *face* is the flat surface of a solid figure and an *edge* is the line between the faces. Have students make an index card that shows a sketch of a cube. Students should label a face, an edge, and a corner on the cube.

Materials

Program Materials
- Cubes, 1 per student
- Number Cube (1–6)

Additional Materials
- scissors, 1 per student
- tape
- Vocabulary Card 7, *cube*
- Vocabulary Card 15, *face*

Prepare Ahead
Copy Cubes on cardstock, 1 per student

1 WARM UP

Prepare
- Have students draw a box that has all squares as its sides. Have them describe their boxes. If students need help, give them a number cube and have them draw it.

Just the Facts
To increase computational fluency, play a game of "Heads Up." Have students answer each question by raising their heads up to show a correct answer and lowering their heads down to show an incorrect answer. Use prompts such as the following:

▶ 12 − 7 = 5   heads up

▶ 18 − 9 = 8   heads down

▶ 11 − 5 = 6   heads up

2 ENGAGE

Develop: Cubes
"Today we are going to identify the attributes of a cube." Follow the instructions on the Activity Card **Cubes.** As students complete the activity, be sure to use the Questions to Ask.

Activity Card 5P

Alternative Groupings
Small Group: Have one student number the squares on the net of the cube. Have another student assemble and tape the cube. Have a third student point to the faces and edges of the cube. Then complete the activity as written.

Progress Monitoring	
If… students have difficulty identifying the faces of the cube they made,	▶ **Then…** have them point to each face of a number cube.

Practice
Have students complete **Student Workbook,** pages 60–61. Guide students through the Key Idea example and the Try This exercises.

Interactive Differentiation
Consult the **Teacher Dashboard** for grouping suggestions. You can also use performance on the Engage activity to guide students.

Independent Practice

For additional practice, have students complete Geometry Snapshots 8. Students can use this Building Blocks activity to create a cube. Have students write down the attributes of a cube.

Supported Practice

For additional support, have students draw cubes on isometric dot paper or on plain paper. Tell them to use a different color for the vertical edges than for the horizontal edges.

▶ **Draw a smiley face onto each face of the cube that you can see. How many faces are there? How many edges are there?** Check students' drawings: 6; 12.

▶ **What shape is each face, or side, of the cube?** square

▶ **How are all cubes alike?** Answers may vary. Possible answers: Their faces are squares; they have six faces.

426 Level D Unit 5 **Geometry and Measurement**

REFLECT

Think Critically

Review students' answers to the Reflect prompt at the bottom of **Student Workbook** page 61, and then review the Engage activity.

▶ **How would you explain what a cube is to a new student?**
Answers may vary. Possible answer: I would explain that it is an object with six rectangular sides.

Real-World Application

You have a cube-shaped seat in your room. You want to paint each side of the cube a different color.

▶ **How many different colors of paint do you need?** 6

ASSESS

Informal Assessment

Use the online or print Student Record, **Assessment,** page 128, to record informal observations.

Cubes
Did the student
☐ make important observations?   ☐ provide insightful answers?
☐ extend or generalize learning?   ☐ pose insightful questions?

Additional Practice

For additional practice, have students complete **Practice,** page 119.

Practice, p. 119

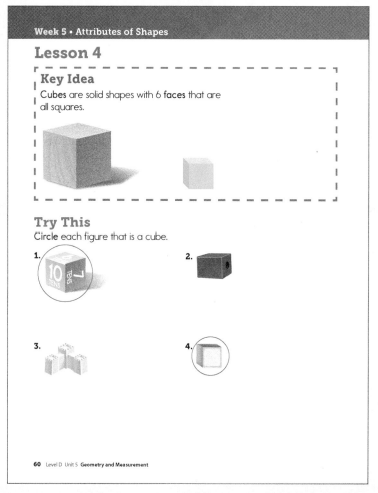

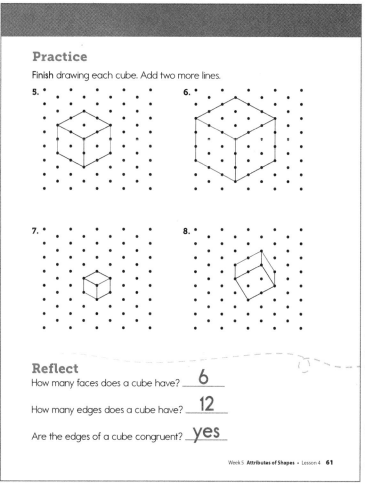

Student Workbook, pp. 60–61

WEEK 5
Attributes of Shapes

Lesson 5 Review

Objective
Students review skills learned this week and complete the weekly assessment and project.

Standard CCSS
2.G.1 Recognize and draw shapes having specified attributes, such as a given number of angles or a given number of equal faces. Identify triangles, quadrilaterals, pentagons, hexagons, and cubes.

Vocabulary
Review vocabulary introduced during the week.

Creating Context
Have English Learners make a note card for each shape they learned this week. Have them include the name of the shape, a drawing of the shape, and the attributes of the shape. For triangles, have them make separate cards for acute triangles, right triangles, and obtuse triangles. For quadrilaterals, have them make separate cards for squares, rectangles, trapezoids, rhombuses, and parallelograms.

1 WARM UP

Prepare
Review how to classify triangles and quadrilaterals and how to name polygons by the number of sides. Then ask the following questions:

▶ **Which polygon has three sides and one right angle?**
right triangle

▶ **Which polygon has six sides?** hexagon

▶ **Which polygon has four sides and one pair of parallel sides?**
trapezoid

2 ENGAGE

Practice
Have students complete **Student Workbook,** pages 62–63.

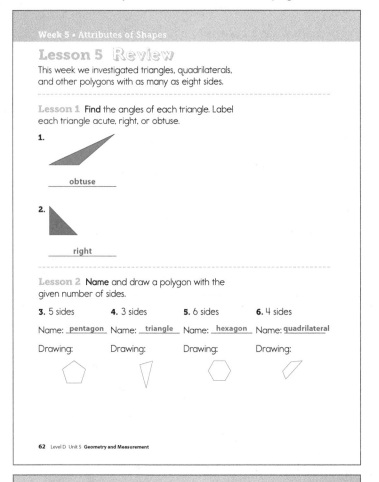

Student Workbook, pp. 62–63

428 Level D Unit 5 **Geometry and Measurement**

3 REFLECT

Think Critically

Review students' answers to the Reflect prompts at the bottom of **Student Workbook** page 63.

Discuss the answer with the group to reinforce Week 5 concepts.

4 ASSESS

Formal Assessment

Students may take the weekly assessment online.

As an alternative, students may complete the weekly test on **Assessment,** pages 69–70. Record progress using the Student Assessment Record, **Assessment,** page 128.

Going Forward

Use the **Teacher Dashboard** to view results of the online assessments, to input the results of print student assessments, and to review progress before making decisions about next steps. Use the weekly test results and observations to determine the next steps for each student.

Retention	
Student displays good grasp of this week's concepts and skills.	Draw shapes learned this week on the board and have students copy the shapes onto paper. Have them list the attributes of the shapes and then compare their papers with a partner.
Remediation	
Student is still struggling with the week's concepts and skills.	Partner students who need remediation with students who do not. Have the partners draw shapes learned this week onto their papers. Have them list the attributes of the shapes and then compare their papers to make sure they included all attributes.

Suggestions for Re-Evaluation: If a student has struggled without success for several weeks, use observations and test results to place the student at a level in which he or she can find success and build confidence to move forward.

Assessment, pp. 69–70

Week 5 **Attributes of Shapes** • Lesson 5   **429**

Project Preview

This week, students found the attributes of geometric shapes, including angle measurements and congruent or parallel sides. The project for this unit requires students to extend the knowledge they gained in Find the Math and what they have learned this week to design a sandbox for their playgrounds.

Project-Based Learning

Standards-driven Project-Based Learning is effective in building deep content understanding. Project-Based Learning increases long-term retention of concepts and has been shown to be more effective than traditional instruction. By completing a project to answer an essential question, students are challenged to apply and demonstrate mastery of concepts and skills by expressing understanding through discussion, research, and presentation.

Essential Question

HOW can I use my knowledge of length, time, and shapes to help someone complete a task?

Project Evaluation Criteria

Review project evaluation criteria with students prior to beginning the project.

Exceeds Expectations
☐ Project result is explained and can be extended.
☐ Project result is explained in context and can be applied to other situations.
☐ Project result is explained using advanced mathematical vocabulary.
☐ Project result is described, and mathematics are used correctly and can be extended.
☐ Project result is explained and extended, and shows advanced knowledge of mathematical concepts and skills.

Meets Expectations
☐ Project result is explained.
☐ Project result is explained in context.
☐ Project result is explained using mathematical vocabulary.
☐ Project result is described, and mathematics are used correctly.
☐ Project result is explained, and shows satisfactory knowledge of mathematical concepts and skills.

Does Not Meet Expectations
☐ Project result is not explained.
☐ Project result is explained, but out of context.
☐ Project result is explained, but mathematical vocabulary is oversimplified.
☐ Project result is described, but mathematics are not used correctly.
☐ Project result is not explained and/or extended, or shows less than satisfactory knowledge of mathematical concepts and skills.

Build a Shapely Sandbox

Objective
Students design a sandbox for the playground in the shape of a triangle, square, pentagon, or hexagon.

Standard
2.G.1 Recognize and draw shapes having specified attributes, such as a given number of angles or a given number of equal faces. Identify triangles, quadrilaterals, pentagons, hexagons, and cubes.

Materials
Additional Materials
- construction paper, 1 sheet per pair of students
- ruler, 1 per pair of students
- scissors, 1 pair per pair of students
- tape or glue stick

Best Practices
- Create opportunities for students to experience success.
- Allow active learning with noise and movement.
- Provide meaning and organization to the lessons and concepts.

Introduce

Suppose that you want to add a sandbox to the playground. You need to choose a polygon for the shape of the sandbox. The sandbox design will have to fit on the playground map.

▶ **How big do you think a playground sandbox should be?**
Answers may vary. Possible answer: It should be made from a square with sides that are six-feet long.

▶ **Where do you think is the best place for the sandbox?**
Answers may vary. Possible answer: It should be next to the slide.

Explore

- Organize students into their usual project pairs. Provide each pair of students with construction paper, a ruler, and a pair of scissors.
▶ **Today you will work together to design a sandbox that is in the shape of a triangle, square, pentagon, or hexagon.**
- Have students brainstorm which shape they will use for the sandbox.
- After students have made a decision regarding the shape of the sandbox, demonstrate how to use a ruler to draw a triangle, square, pentagon, and hexagon.
- Tell students to cut the sandbox shape out of the construction paper. Encourage them to use the ruler to draw straight lines for the sides. Remind students that they will add their cutouts to their playground maps, so they need to make sure the sandbox shape will fit on the map.
- Tell students to think of the attributes of the polygon they chose. Ask questions such as the following:
▶ **How many sides does the shape have?**
▶ **Are the sides parallel?**
▶ **Are the sides congruent?**
▶ **Are the angles acute, obtuse, or right?**
▶ **Complete *Student Workbook*, page 64, to list the attributes of the shape.**

Wrap Up

- Give students time to decide on other aspects of the sandbox, such as how deep it is.
- Make sure each student can describe the attributes of the shape that was used for his or her sandbox.
- Discuss students' answers to the Reflect prompts at the bottom of ***Student Workbook,*** page 64.

Have students tape or glue the sandbox shape to their playground maps. At this point, the map is complete. It will include a couple of pieces of playground equipment, a human clock, and a sandbox.

If time permits, allow students to talk about their favorite part of their playground design.

Student Workbook, p. 64

Teacher Reflect

- ☐ Did students show knowledge of how their project related to the major concept?
- ☐ Did I adequately explain and discuss the Reflect questions with students?
- ☐ Did students use their time wisely and effectively?

WEEK 6 Graphs

Week at a Glance
This week, students conclude **Number Worlds**, Level D, Geometry and Measurement. Students will build data sets and graph and analyze data.

Skills Focus
- Sort by attributes and build data sets.
- Draw picture graphs and bar graphs to represent data sets.
- Solve problems using information presented in a graph.

How Students Learn
Certain ways of showing information, such as graphs and charts, work better with some types of data than others. Students should understand that collected data need to be classified and organized depending on the original question the data are intended to answer. Students will benefit from classroom discussions about the different ways data can be grouped and classified.

English Learners ELL
For language support, use the **English Learner Support Guide**, pages 104–105, to preview lesson concepts and teach academic vocabulary.

Weekly Planner

Lesson	Learning Objectives
1 pages 434–435	Students can sort zoo animals based on various attributes.
2 pages 436–437	Students can transfer data to a math-link cubes graph and then to a bar graph, and draw a picture and a bar graph.
3 pages 438–439	Students can use picture graphs to analyze data.
4 pages 440–441	Students can create picture graphs from a set of data and can use the graphs to analyze the data.
5 pages 442–443	**Review and Assess** Students review skills learned this week and complete the weekly assessment.
Project pages 444–445	Students can create a picture graph from a tally chart.

Math at Home
Give one copy of the Letter to Home, page 30, to each student. Encourage students to share and complete the activity with their caregivers.

Level D Unit 5 **Geometry and Measurement**

Key Standard for the Week

Domain: Measurement and Data

Cluster: Represent and Interpret Data

2.MD.10 Draw a picture graph and a bar graph (with single-unit scale) to represent a data set with up to four categories. Solve simple put-together, take-apart, and compare problems using information presented in a bar graph.

Materials		Technology
Program Materials • *Student Workbook*, pp. 65–67 • *Practice*, p. 120 • Activity Card 5Q, **Sort a Zoo** • Zoo Animals	**Additional Materials** • chart paper • scissors • tape	*Teacher Dashboard* 🌐 Venn Diagram
Program Materials • *Student Workbook*, pp. 68–69 • *Practice*, p. 121 • Activity Card 5R, **Math-Link Cubes and Bar Graphs** • Sample Graphs • Zoo Animal Chart	**Additional Materials** • math-link cubes* • grid paper • rulers* • Vocabulary Card 4, *bar graph*	*Teacher Dashboard* 🌐 Graphing and Spreadsheet
Program Materials • *Student Workbook*, pp. 70–71 • *Practice*, p. 122 • Activity Card 5S, **Dog Walk**	**Additional Materials** newspapers, magazines, or websites that show picture graphs	*Teacher Dashboard* 🌐 Pictograph
Program Materials • *Student Workbook*, pp. 72–73 • *Practice*, p. 123 • Activity Card 5T, **Make a Picture Graph**	**Additional Materials** • math-link cubes* • picture graph from Lesson 3	*Teacher Dashboard* 🌐 Pictograph
Program Materials • *Student Workbook*, pp. 74–75 • Weekly Test, *Assessment*, pp. 71–72 • Dog Walk Picture Graph		Review previous activities.
Program Materials *Student Workbook*, p. 76	**Additional Materials** self-stick notes	

*Available from McGraw-Hill Education

WEEK 6
Graphs

Find the Math

In this week, introduce students to creating picture graphs and bar graphs to represent data.

Use the following to begin a guided discussion:

▶ **What food is your favorite lunch?** Answers may vary. Possible answer: pizza

Have students complete **Student Workbook,** page 65.

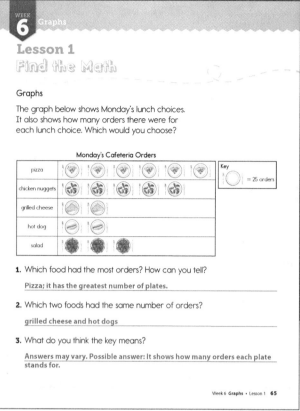

Student Workbook, pp. 65

Lesson 1

Objective
Students can sort zoo animals based on various attributes.

Standard
2.MD.10 Draw a picture graph and a bar graph (with single-unit scale) to represent a data set with up to four categories. Solve simple put-together, take-apart, and compare problems using information presented in a bar graph.

Creating Context
An excellent organizer for English Learners to begin to sort and classify is one in which they use their five senses to describe objects or activities. Make a worksheet or class chart with icons that represents each of the five senses as a reminder.

Materials
Program Materials
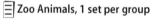 Zoo Animals, 1 set per group

Additional Materials
- chart paper
- scissors, 1 pair per student
- tape, 1 roll per group

1 WARM UP

Prepare
Display pictures of a tiger, shark, walrus, and lion.

▶ **Where might you see these animals?** at the zoo

▶ **If you were going to group or sort these animals, how might you do it?**

Have students practice sorting the animals into categories they choose. Possible answers: land animals, and water animals

2 ENGAGE

Develop: Sort a Zoo
"Today you are going to sort pictures of animals." Follow the instructions on the Activity Card **Sort a Zoo.** As students complete the activity, be sure to use the Questions to Ask.

Alternative Groupings
Whole Class: Find more pictures of animals so that each student has at least three animals to place in the correct category.

Activity Card 5Q

Progress Monitoring	
If… students have trouble brainstorming groups,	▶ **Then…** have them think of how animals are arranged or sorted.

Practice
Have students complete **Student Workbook,** pages 66–67. Guide students through the Key Idea example and the Try This exercises.

Interactive Differentiation
Consult the **Teacher Dashboard** for grouping suggestions. You can also use performance on the Engage activity to guide students.

Independent Practice

For additional practice in sorting items, have students use the Venn Diagram. Tell them to choose two or three circles for the Venn diagram, click the Select Data Set button, and choose attribute blocks. Then students can sort the polygons by shape, size, or color. After students have practiced with the Venn Diagram, provide them with three categories they can use to sort animals. For example, write *Small, Medium,* and *Large* on the board and

434 Level D Unit 5 **Geometry and Measurement**

under the headings write *mouse, raccoon,* and *hippo,* respectively. Tell students to fold a sheet of paper into thirds, write the categories at the top, and under each heading list all the animals they can think of in that category.

Supported Practice

For additional support, provide students with three categories they can use to sort animals. For example, write *Two Legs, Four Legs,* and *No Legs* on the board and under the headings write *parrot, cat,* and *worm.* Have students work in groups to list all the animals they can think of in each category. Tell students to fold a sheet of paper into thirds and write the categories at the top. They will list the animals under each heading.

▶ **How many animals did you think of that have no legs? What are they?** Answers may vary. Possible answer: 2: snake, shark

▶ **In which category does a lion fit?** Four legs

▶ **Are there any birds with four legs?** no

3 REFLECT

Think Critically

Review students' answers to the Reflect prompt at the bottom of **Student Workbook page** 67, and then review the Engage activity.

Discuss to reinforce the idea that sometimes the same animal could be a part of two or more groups based on its attributes and habits.

▶ **How are animals grouped in a pet store?**

▶ **Could you group them differently? How?**

4 ASSESS

Informal Assessment

Use the online or print Student Record, **Assessment,** page 128, to record informal observations.

Sort a Zoo
Did the student
☐ make important observations?   ☐ provide insightful answers?
☐ extend or generalize learning?   ☐ pose insightful questions?

Additional Practice

For additional practice, have students complete **Practice,** page 120.

Practice, p. 120

Week 6 • Graphs

Lesson 1

Key Idea
Animals can be sorted.

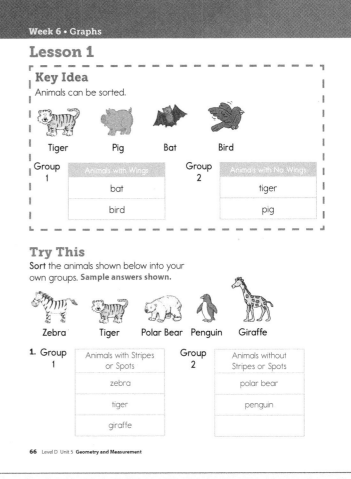

Try This
Sort the animals shown below into your own groups. **Sample answers shown.**

Practice
Look at the sorted animals below. Label each group with a category that fits every animal.

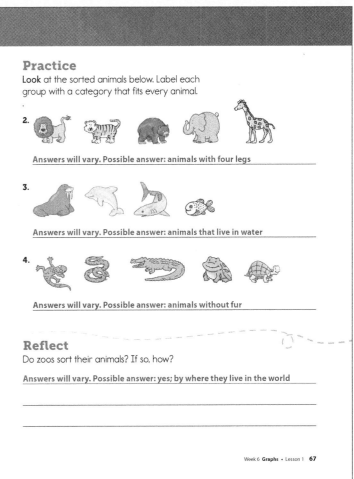

Reflect
Do zoos sort their animals? If so, how?

Answers will vary. Possible answer: yes; by where they live in the world

Student Workbook, pp. 66–67

Week 6 **Graphs** • Lesson 1   **435**

WEEK 6
Graphs

Lesson 2

Objective
Students can transfer data to a math-link cubes graph and then to a bar graph, and draw a picture graph and a bar graph.

Standard
2.MD.10 Draw a picture graph and a bar graph (with single-unit scale) to represent a data set with up to four categories. Solve simple put together, take-apart, and compare problems using information presented in a bar graph.

Vocabulary
- bar graph
- data

Creating Context
Use pictures and small plastic figures of animals to help English Learners categorize animals in the zoo. For example, give students a plastic bird, elephant, zebra, and lion. Have English Learners separate the figures into animals with two legs and animals with four legs. Then tell students to replace each figure with one math-link cube. The results will be one math-link cube for two-legged animals and three math-link cubes for four-legged animals. Have students align the cubes to echo the shape of a bar graph.

Materials

Program Materials
- Sample Graphs
- Zoo Animal charts from Lesson 1

Additional Materials
- math-link cubes of various colors
- grid paper
- rulers, 1 per group
- Vocabulary Card 4, *bar graph*

Prepare Ahead
Label each sheet of grid paper with the headings from one of the charts created in Lesson 1. Write the headings on the grid paper so that later they can become the *x*-axis labels for a bar graph. Match the labeled grid paper with its Zoo Animal chart. Prepare Sample Graphs for classroom display.

1 WARM UP

Prepare
Display Sample Graphs. Discuss the sample graphs on the page. Explain that the two graphs communicate the same information, but in a slightly different format. Help students see the connection between the math-link cubes graphs at the top of the page and the bar graph at the bottom of the page.

Just the Facts
To increase computational fluency, tell students to clap if the addition fact is correct. If the addition fact is incorrect, tell them to remain silent. Use prompts such as the following:

- ▶ 10 + 5 = 15   clap
- ▶ 9 + 8 = 20   remain silent
- ▶ 7 + 5 = 12   clap

2 ENGAGE

Develop: Math-Link Cubes and Bar Graphs

"Today we are going to use data to make a math-link cubes graph and then a bar graph." Follow the instructions on the Activity Card **Math-Link Cubes and Bar Graphs.** As students complete the activity, be sure to use the Questions to Ask.

Activity Card 5R

Alternative Groupings
Individual: Complete the activity as written.

Progress Monitoring

| If... students have trouble thinking of the parts of the graph that are missing, | ▶ Then... revisit the sample graphs used in the Warm Up activity. |

Practice
Have students complete **Student Workbook,** pages 68–69. Guide students through the Key Idea example and the Try This exercises.

Interactive Differentiation
Consult the **Teacher Dashboard** for grouping suggestions. You can also use performance on the Engage activity to guide students.

Independent Practice
For additional practice, have students use the Graphing and Spreadsheet Tool to make bar graphs. Show students how to use the features of the Graphing and Spreadsheet Tool. Then tell students to create a bar graph on the computer similar to the bar graphs they drew during the activity.

Supported Practice
For additional support, demonstrate how to translate a math-link cube graph into a bar graph. Use the data in the table below.

Neighborhood Dog Walkers	
Dog Walker	Number of Dogs Walked
Maura	5
Julio	1
Amity	3

▶ **If I were going to use math-link cubes to show the number of dogs walked, how many cubes would I need in all? How do you know?**
You would need nine cubes because the three people walked a total of nine dogs.

- Demonstrate how to draw the labels on a sheet of paper (*Maura, Julio, Amity*), and stack the appropriate number of math-link cubes above each name.

▶ **How can I turn this math-link cube graph into a bar graph?**
Answers may vary. Possible answer: Draw five squares above *Maura,* one square above *Julio,* and three squares above *Amity.*

▶ **What other information do I need to add to the graph to make it a true bar graph?** A bar graph needs a title and labels. You need to label the horizontal line *Dog Walker* and the vertical line *Number of Dogs Walked.* You also need to add the numbers to the vertical line so that someone looking at the graph knows how many dogs each square represents.

436 Level D Unit 5 **Geometry and Measurement**

REFLECT

Think Critically

Review students' answers to the Reflect prompt at the bottom of *Student Workbook* page 69, and then review the Engage activity.

▶ **Does a bar graph tell you which animals go into each category? Explain.** No. A bar graph only tells you how many animals are in each category, not which animals they are.

Real-World Application

A company wants to know which of its five games is the most popular.

▶ **How can a bar graph help the company see how popular its games are?** Answers may vary. Possible answer: If the company puts the number of each game sold on a bar graph, it can see which game was the most popular and which was the least popular. The most popular game will have the tallest bar.

ASSESS

Informal Assessment

Use the online or print Student Record, **Assessment,** page 128, to record informal observations.

Math-Link Cubes and Bar Graphs	
Did the student	
☐ apply learning to new situations?	☐ contribute answers?
☐ contribute concepts?	☐ connect mathematics to the real world?

Additional Practice

For additional practice, have students complete **Practice,** page 121.

Practice, p. 121

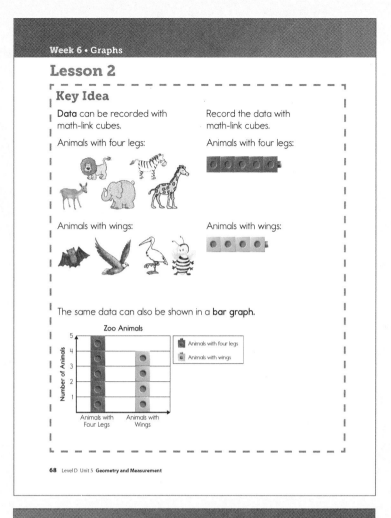

Student Workbook, pp. 68–69

Week 6 **Graphs** • Lesson 2   **437**

WEEK 6
Graphs

Lesson 3

Objective
Students can use picture graphs to analyze data.

Standard
2.MD.10 Draw a picture graph and a bar graph (with single-unit scale) to represent a data set with up to four categories. Solve simple put-together, take-apart, and compare problems using information presented in a bar graph.

Vocabulary
picture graph

Creating Context
The word *picture graph* begins with the syllable *pict-*, which is the same beginning as the word *picture*. Help English Learners look for this root in other words to see how they are related. Looking for similarity in words can help English Learners accelerate their acquisition of English academic language.

Materials
Program Materials

Additional Materials
newspapers, magazines, or websites that show picture graphs

Prepare Ahead
Prepare the Dog Walk Picture Graph on the activity card for classroom display.

 WARM UP

Prepare
Draw a row of 10 apples on the board. Underneath this, write a key stating that each apple picture equals 1 apple.

▶ **How many apples does this row represent?** 10

Change the key to state that each apple picture equals 2.

▶ **How many apples does this row represent now?** 20 (If students have trouble answering, write a 2 under each apple, and ask students to skip count until they have counted all the twos.)

Just the Facts
To build computational fluency, have students play "Heads Up." If the answer is correct, have them put their heads up. If it is not correct, have them put their heads down. Use prompts such as the following:

▶ $12 - 4 = 8$ heads up

▶ $9 + 6 = 15$ heads up

▶ $7 + 7 = 16$ heads down

2 ENGAGE

Develop: Dog Walk
"Today you are going to learn about picture graphs." Follow the instructions on the Activity Card **Dog Walk**. As students complete the activity, be sure to use the Questions to Ask.

Activity Card 5S

Alternative Groupings
Individual: Have the student answer the questions on his or her own.

Progress Monitoring

| If… students struggle with interpreting the picture graph, | ▶ Then… help them see how many dogs are represented in each row and write these numbers at the ends of the rows. |

Practice
Have students complete **Student Workbook,** pages 70–71. Guide students through the Key Idea example and the Try This exercises.

Interactive Differentiation
Consult the **Teacher Dashboard** for grouping suggestions. You can also use performance on the Engage activity to guide students.

Independent Practice
For additional practice, have students use the Pictograph to make a picture graph from the data in the table below. Use an *X* to represent two items. Have students add the labels and title to complete the graph.

Supplies in my Desk	
Crayons	10
Pencils	6
Pens	2
Key: *X* = 2 items	

Supported Practice
For additional support, create a picture graph based on the shirt colors of students in the class. Draw the data table below on the board. In the left column, write the shirt colors of students in the class. Have students go to the board and add tally marks in the right column next to their shirt colors.

Shirt Colors in the Class	
Shirt Colors	Number of Students

▶ **How many students have each color shirt?**

- Have students find the total number of students with each shirt color
- Write the totals in each row.

▶ **Now we will make a picture graph.**

- Guide students through the steps of making a picture graph from the data table. Make sure students remember the labels. Help students think of a way to use a key in the picture graph. An example of a key for this data would be one color dot symbol for every two students.

3 REFLECT

Think Critically

Review students' answers to the Reflect prompt at the bottom of *Student Workbook* page 71, and then review the Engage activity.

Discuss with students how a picture graph could mean many different things depending on the key of the graph. For example, each picture could represent 3 items or 300 items. The key makes a big difference.

▶ **Can you tell me something you learned about picture graphs?**
Answers may vary. Possible answer: I learned they are an easier way to show large numbers.

▶ **How do you find out how much a picture or an object in a picture graph is worth?** Possible answer: You look at the key

Real-World Application

Search for some different examples of picture graphs in the real world, such as those found in newspapers or magazines, and share them with students. Ask students to identify characteristics of the graphs, such as the key.

4 ASSESS

Informal Assessment

Use the online or print Student Record, *Assessment,* page 128, to record informal observations.

Dog Walk
Did the student
☐ make important observations?   ☐ provide insightful answers?
☐ extend or generalize learning?   ☐ pose insightful questions?

Additional Practice

For additional practice, have students complete *Practice,* page 122.

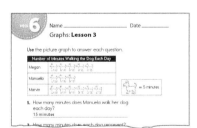
Practice, p. 122

Week 6 • Graphs

Lesson 3

Key Idea
A **picture graph** uses pictures to represent data. The key of a picture graph shows how many each picture represents.

Warriors Basketball Team

		Key
Mary	🏀🏀🏀🏀🏀	🏀 = 2 points
Shaun	🏀🏀🏀	
Ricky	🏀🏀🏀🏀🏀🏀	

According to this picture graph, Mary scored 10 points.

Try This
Use the picture graph to answer each question.

For how long did you practice the trumpet?

		Key
Bobby	🎺🎺🎺	🎺 = 30 minutes
Brittany	🎺🎺🎺🎺🎺	
Sandy	🎺🎺🎺	

1. What length of time does each trumpet represent?
30 minutes

---

2. Who spent the longest time practicing the trumpet?
Brittany

3. How long did Sandy spend practicing?
90 minutes, or 1 hour and 30 minutes

Practice
Use the picture graph to answer each question.

Car Wash Fund-Raiser

		Key
Third Graders	🚗🚗🚗🚗🚗	🚗 = 5 cars washed
Fourth Graders	🚗🚗🚗	
Fifth Graders	🚗🚗🚗🚗🚗	

4. How many car washes does each car represent?
5 cars

5. How many cars did the fifth graders wash?
25 cars

Reflect
Why is it important to include a key in a picture graph? Explain.

Sample answer: The key tells how many each picture represents. Without the key, you cannot interpret the picture graph.

Student Workbook, pp. 70–71

WEEK 6
Graphs

Lesson 4

Objective
Students can create picture graphs from a set of data and can use the graphs to analyze the data.

Standard
2.MD.10 Draw a picture graph and a bar graph (with single-unit scale) to represent a data set with up to four categories. Solve simple put-together, take-apart, and compare problems using information presented in a bar graph.

Vocabulary
picture graph

Creating Context
English Learners know that when we want to compare and use superlatives we usually use *-er* and *-est* at the end of adjectives. When an adjective has three or more syllables, we do not add these, but place the words *more/most, less/least* before them.

Materials
Program Materials

Additional Materials
- math-link cubes, 5 per student
- picture graph from Lesson 3

Prepare Ahead
Prepare the Number of Student Volunteers Table from the activity card for classroom display.

1 WARM UP

Prepare
Have students use math-link cubes to represent the pictures in a picture graph. Skip count to find how many cubes should be used for each of the following keys.

- Key: Each cube represents 2 yards. Gregory raked 6 yards. **Count by twos until you reach 6. How many cubes do you need?** 3 cubes
- Key: Each cube represents 5 seashells. Carmen found 25 seashells. **Count by fives until you reach 25. How many cubes do you need?** 5 cubes

Just the Facts
To build computational fluency, have students practice addition doubles facts. Ask the questions rhythmically to encourage students to respond in unison. Use questions such as the following:

▶ **What is 6 + 6?** 12

▶ **What is 8 + 8?** 16

▶ **What is 20 + 20?** 40

2 ENGAGE

Develop: Make a Picture Graph
"Today you are going to make picture graphs." Follow the instructions on the Activity Card **Make a Picture Graph.** As students complete the activity, be sure to use the Questions to Ask.

Activity Card 5T

Alternative Groupings
Individual: Have the student answer the questions on his or her own.

Progress Monitoring

| **If...** students have trouble finding a common multiple to use in the key, | ▶ **Then...** encourage them to use a number line or other manipulatives. |

Practice
Have students complete **Workbook,** pages 72–73. Guide students through the Key Idea example and the Try This exercises.

Interactive Differentiation
Consult the **Teacher Dashboard** for grouping suggestions. You can also use performance on the Engage activity to guide students.

Independent Practice

For additional practice, have students use the Pictograph and the data in the table below to make a picture graph. Use a circle to represent five coins. Have students add the labels and title to complete the picture graph.

Coins in Piggy Bank	
Quarters	25
Dimes	10
Nickels	15

Key: ● = 5 coins

Supported Practice

For additional support, create a picture graph based on the shoe sizes of students in the class. Draw the data table below on the board. In the left column, write the shoe sizes of students in the class. Have students go to the board and add tally marks in the right column next to their shoe size.

Shoe Sizes in the Class	
Shoe Size	Number of Students

- **How many students wear each size shoe?** Answers may vary.
- Have students find the total number of students with each shoe size. Write the totals in each row.
- **Now we will make a picture graph.**
- Guide students through the steps of making a picture graph from the data table. Make sure students remember the labels. Help students think of a way to use a key in the picture graph. An example of a key for this data would be one footprint symbol for every three students.
- **What can you learn from this picture graph?** Answers may vary. Possible answers: You can learn which shoe size is the most common in the class; you can learn what the smallest and largest shoe sizes are.

3 REFLECT

Think Critically

Review students' answers to the Reflect prompt at the bottom of **Student Workbook** page 73, and then review the Engage activity.

Be sure students understand the relationship between the key and the number of pictures that must be drawn.

- **Can you explain what a picture graph is?** Possible answer: A picture graph is a graph that uses pictures to represent the data you collect.
- **If each math-link cube represents 5 points, can you show me how many to connect to make 40 points?** Possible answer: Students should connect 8 cubes.

Real-World Application

Revisit the picture graphs you shared with students in Lesson 3 from newspapers, magazines, and the Internet. Instruct students to choose one of the graphs and create a tally sheet from which this graph could have been built.

4 ASSESS

Informal Assessment

Use the online or print Student Record, **Assessment,** page 128, to record informal observations.

Make a Picture graph
Did the student
☐ apply learning to new situations? ☐ contribute answers?
☐ contribute concepts? ☐ connect mathematics to the real world?

Additional Practice

For additional practice, have students complete **Practice,** page 123.

Practice, p. 123

Week 6 • Graphs

Lesson 4

Key Idea
When you make a picture graph, you must include a key.

Try This
Make a picture graph by following the steps below.

Favorite Yogurt Flavor		
Flavor	Tally	Number
Strawberry	||||| |||||	10
Vanilla	|||||		6
Lemon	||||| ||||| ||	12

Step 1 Write the flavors in the left-hand column.

Step 2 Write the key so each picture represents two votes.

Step 3 Divide the number of votes for each flavor by 2. Draw yogurt cups to match each flavor.

Students should draw a picture in the Key box.

1.

Favorite Yogurt Flavor	
Strawberry	Students should draw five pictures of the item used in the key.
Vanilla	Students should draw three pictures of the item used in the key.
Lemon	Students should draw six pictures of the item used in the key.

Key
= 2 votes

72 Level D Unit 5 **Geometry and Measurement**

Practice

Use your picture graph from the previous page to answer each question.

2. What is the most popular flavor?

   lemon

3. What is the least popular flavor?

   vanilla

4. How many people voted for strawberry as their favorite flavor?

   10 people

5. Suppose 8 people voted for lemon. How many pictures would you draw to show this?

   4 pictures

6. Suppose 9 people voted for peach. Four pictures equal 8 votes. How could you show the ninth vote?

   Answers may vary. Possible answer: Show half of a yogurt cup.

Reflect

Draw a row that could be combined with the picture graph on the previous page to show that peach yogurt received 9 votes.

Peach	Students should draw four full pictures and one half picture.

Week 6 **Graphs** • Lesson 4 73

Student Workbook, pp. 72–73

WEEK 6
Graphs

Lesson 5 Review

Objective
Students review skills learned this week and complete the weekly assessment and project.

Standard
2.MD.10 Draw a picture graph and a bar graph (with single-unit scale) to represent a data set with up to four categories. Solve simple put-together, take-apart, and compare problems using information presented in a bar graph.

Vocabulary
Review vocabulary introduced during the week.

Creating Context
Have English Learners think of attributes of animals that may be used to categorize them. For example, in this lesson animals have been categorized by the number of legs and whether or not they fly. Have students create a large table or chart with each attribute as the column head. Then have students write the name of the animal in each column that applies.

1 WARM UP

Prepare
Review how to interpret a picture graph and a bar graph and how to create a picture graph or bar graph from a set of data. Ask the following questions:

- **How do you know how many things each picture graph symbol represents?** You use the key.
- **How can you use a set of data to draw a bar graph?** Answers may vary. Possible answer: Draw and label the categories for the bar graph, and give the graph a title. Include numbers along one side of the graph so that people can look at the height of a bar and know its value. For each category of data, draw a bar whose height equals the data value for that category.

2 ENGAGE

Practice
Have students complete **Student Workbook,** pages 74–75.

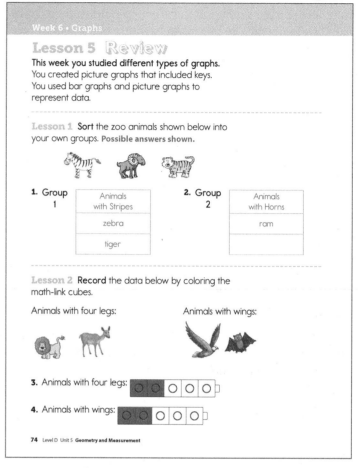

Student Workbook, pp. 74–75

442 Level D Unit 5 **Geometry and Measurement**

3 REFLECT

Think Critically

Review students' answers to the Reflect prompts at the bottom of **Student Workbook** page 75.

Discuss the answer with the group to reinforce Week 5 concepts.

4 ASSESS

Formal Assessment

Students may take the weekly assessment online.

As an alternative, students may complete the weekly test on **Assessment,** pages 71–72. Record progress using the Student Assessment Record, **Assessment,** page 128.

Going Forward

Use the **Teacher Dashboard** to view results of the online assessments, to input the results of print student assessments, and to review progress before making decisions about next steps. Use the weekly test results and observations to determine the next steps for each student.

Retention	
Student displays good grasp of this week's concepts and skills.	On the board, list the names of sports and have students generate several categories into which the sports can be placed. Have students copy the sport names and categories onto paper. Tell students to group the sports by categories and then compare their papers with a partner. Have students create a math-link cube graph and a bar graph corresponding to the numbers of sports in the categories.
Remediation	
Student is still struggling with the week's concepts and skills.	Partner students who need remediation with students who do not. On the board, list the names of animals and several categories into which the animals can be placed. Have students copy the animal names and categories onto paper. Tell partners to help each other group the animals by categories. Then have the partners create a math-link cube graph and a bar graph corresponding to the numbers of animals in the categories.

Suggestions for Re-Evaluation: If a student has struggled without success for several weeks, use observations and test results to place the student at a level in which he or she can find success and build confidence to move forward.

Assessment, pp. 71–72

Project Preview

This week, students sorted animals by various attributes, then created a picture graph and a bar graph. They also created picture graphs from a set of data and used the graphs to analyze the data. Today students will apply their knowledge about graphs to their playground-design maps and conclude their projects.

Project-Based Learning

Standards-driven Project-Based Learning is effective in building deep content understanding. Project-Based Learning increases long-term retention of concepts and has been shown to be more effective than traditional instruction. By completing a project to answer an essential question, students are challenged to apply and demonstrate mastery of concepts and skills by expressing understanding through discussion, research, and presentation.

Essential Question

HOW can I use my knowledge of length, time, and shapes to help someone complete a task?

Project Evaluation Criteria

Review project evaluation criteria with students prior to beginning the project.

Exceeds Expectations
☐ Project result is explained and can be extended.
☐ Project result is explained in context and can be applied to other situations.
☐ Project result is explained using advanced mathematical vocabulary.
☐ Project result is described, and mathematics are used correctly and can be extended.
☐ Project result is explained and extended, and shows advanced knowledge of mathematical concepts and skills.

Meets Expectations
☐ Project result is explained.
☐ Project result is explained in context.
☐ Project result is explained using mathematical vocabulary.
☐ Project result is described, and mathematics are used correctly.
☐ Project result is explained, and shows satisfactory knowledge of mathematical concepts and skills.

Does Not Meet Expectations
☐ Project result is not explained.
☐ Project result is explained, but out of context.
☐ Project result is explained, but mathematical vocabulary is oversimplified.
☐ Project result is described, but mathematics are not used correctly.
☐ Project result is not explained and/or extended, or shows less than satisfactory knowledge of mathematical concepts and skills.

Graph the Playground

Objective
Students can create a picture graph from a tally chart.

Standard
2.MD.10 Draw a picture graph and a bar graph (with single-unit scale) to represent a data set with up to four categories. Solve simple put together, take-apart, and compare problems using information presented in a bar graph.

Materials
Additional Materials
self-stick notes, 1 per student pair

Prepare Ahead
Draw a large tally chart on the board, such as the one on *Student Workbook*, page 76.

Best Practices
- Pace the instruction appropriately for student needs.
- Coach, demonstrate, and model.
- Create adequate time lines for each project.

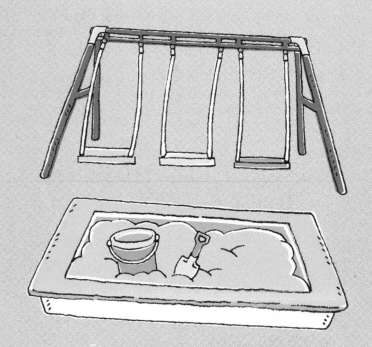

Introduce

▶ **It is time to display the playground maps you created. We will make a class tally chart of the shapes you used in your sandbox designs. Then you will use the tally chart to make a picture graph.**

▶ **What is a tally chart?** Answers may vary. Possible answer: It is a chart that helps you keep track of a group of items in categories. The tally marks help you show how many items there are in each category.

▶ **What would be a good picture symbol for the sandbox?** Answers may vary. Possible answer: a square

Explore

- Provide each pair of students with a self-stick note. Draw a tally chart on the board with *Triangle, Square, Pentagon,* and *Hexagon* as row headings.

▶ **Today we will make a class tally chart of the shapes you used for the sandbox.**

- Have one student from each pair come to the board and put a self-stick note on the chart in the appropriate row.

▶ **How many tally marks are in each row?** Answers will vary.

- Record the total for each row in the *Number* column on the right.

▶ **With your partner, decide how many tally marks (self-stick notes) will equal one symbol in your picture graph. Add that number to the *Key* box on the student workbook page.**

▶ **Choose a symbol to represent the sandbox in the picture graph.**

▶ **Complete *Student Workbook*, page 76, to make a picture graph of your tally chart.**

- Have students work together to complete their picture graphs.

Wrap Up

▶ **Describe how the number of triangle-shaped sandboxes is shown on the class tally chart.** Answers may vary. Possible answer: There is one sticky note for each triangle-shaped sandbox.

▶ **The tally chart and the picture graph look rather similar. How are they different?** Answers may vary. Possible answer: They are different because the picture graph shows you more quickly which polygons were most and least popular. With the tally chart, you have to count the tally marks or read the number.

- Discuss students' answers to the Reflect prompts at the bottom of *Student Workbook*, page 76.

Have each student pair present their playground map to the class. Permit students to ask the presenters questions about their design, but encourage positive comments.

If time permits, allow each student to tell about the most important thing they learned.

Student Workbook, p. 76

Teacher Reflect

☐ Did students talk about the most important thing they learned?

☐ Were students able to answer my questions about their solutions?

☐ Did students correctly use art, objects, graphs, or posters to explain their solutions?

Appendix

Table of Contents

Appendix A
Blackline Masters

Ten Frame A1
Ten Frame Flash Cards A2
Ten Frame Flash Cards A3
Ten Frame Flash Cards A4
The Neighborhood A5
Introduction to the Hotel A6
Room Service Delivery Slips A7
Number Cards (0–9) A8
Number Construction Mat A9
Place-Value Mat A10
Dot Set Cards A11
Number Cards (11–20) A12
Party! A13
What Number am I? A14
Monster Cards A15
1–100 Chart A16
Money Construction Mat A17
Withdrawal Slip A18
Deposit Slip A19
Count and Compare A20
Addition Table A21
Grab and Add A22
Dragon Quest Record Form A23
Double-Digit Number Cards (Addition) A24
Double-Digit Number Cards (Addition) A25
Story Time! Recording Sheets A26
Story Time! Recording Sheets A27
Story Time! Recording Sheets A28
Problems with Unknowns Recording Sheets A29
Problems with Unknowns Recording Sheets A30
Problems with Unknowns Recording Sheets A31

Appendix

Two-Step Word Problems Recording Sheet ... A32
Hotel Mystery Recording Sheets ... A33
Hotel Mystery Recording Sheets ... A34
Hotel Mystery Recording Sheets ... A35
Add Them Up Recording Sheet ... A36
Adding Hundreds Recording Sheet ... A37
How Many Are Left? ... A38
Fish Pond ... A39
Fish Pond Record Form ... A40
Number Lines ... A41
Double-Digit Number Cards (Subtraction) ... A42
Double-Digit Number Cards (Subtraction) ... A43
Length of Classroom Objects ... A44
Pick a Measurement Tool ... A45
Customary Length Recording Sheet ... A46
Metric Length Recording Sheet ... A47
Clock Concentration Cards (clock faces) ... A48
Clock Concentration Cards (times) ... A49
Digital Time Cards ... A50
Clock Faces ... A51
Naming Triangles by Angles ... A52
Naming Polygons ... A53
Special Quadrilaterals ... A54
Cubes ... A55
Zoo Animals ... A56
Sample Graphs ... A57
Number Cards (1–10) ... A58

Appendix

Appendix B
About Math Intervention B1
Program Research B2
Content Strands B6
Key Standards B7
Scope and Sequence B17

Appendix C
Glossary C1

Ten Frame

Ten Frame Flash Cards

A2   Level D   **Blackline Masters**

Ten Frame Flash Cards

Ten Frame Flash Cards

The Neighborhood

Activity

Tell students they will help you assemble the Neighborhood Number Line.

- Have students count with you as you count the ten houses on the first section. Ask students to predict what number the next house will be. Repeat this procedure with the next section.

- Before showing the third section, mention that the houses come in blocks of ten, and ask the students to predict what the last house on the new block will be. Reinforce justifications which suggest that students are focusing on the emerging number patterns of the neighborhood blocks.

- Have the students predict the first and last house numbers of each new block before showing the sections and counting the houses.

- When all ten blocks are in place, engage the students in some counting activities, such as counting the houses as a group, or counting the houses individually, or skip counting by tens.

Questions to Ask

Ask students to explain their predictions using questions such as the following:

▶ Why do you think that will be the next house?

As you add each block to the Neighborhood Number Line, ask questions to encourage students to think about the length of the number sequence.

▶ Is the number line getting longer or shorter when I put up new blocks?

▶ Are we getting closer to or farther away from house number 1?

▶ What would happen if we removed a block?

Introduction to the Hotel

Program Materials
Hotel Game Board

Activity

Tell students they are going to pretend to be the servers at a hotel, and their job will be to deliver room-service orders.

- Have students find the kitchen (Door 0), and explain that they will take turns receiving a delivery in the kitchen, bringing it to the appropriate room, and returning for another order.
- Explain that if someone receives an order for a ground-floor room, he or she will need to exit the kitchen on the right and move down the hallway one door at a time to the room.
- Have students read aloud the numbers on the ground-floor doors. If students say that Room 01 is Room 10, ask them in which position, the ones or the tens, the number 1 is, and in which position the 0 is.
- Have volunteers demonstrate a delivery on the ground floor.

Explain that if someone receives an order for a room above the ground floor, he or she will need to go up the stairs and then through the hallway to the destination.

- Point out the up and down arrows on the kitchen door, and explain that behind the door is a flight of ten steps that leads to the level above.
- Trace a route from Door 0 to Door 90. Stop at each level to ask what number is on the door at that level, how many flights you have climbed, how many steps are in each flight, and how many steps you have climbed so far.
- Have volunteers trace and count by tens from the kitchen at Door 0 up to Door 90 and then back down to the kitchen.
- Have volunteers explain how they would make a delivery to a room above the ground floor, and have them demonstrate.

A6 Level D **Blackline Masters**

Room Service Delivery Slips

Room Service Delivery	Room Service Delivery
Floor/Tokens = 0 1 2 3 4 5 6 7 8 9	Floor/Tokens = 0 1 2 3 4 5 6 7 8 9
Room Service Delivery Floor/Tokens = 0 1 2 3 4 5 6 7 8 9	Room Service Delivery Floor/Tokens = 0 1 2 3 4 5 6 7 8 9
Room Service Delivery Floor/Tokens = 0 1 2 3 4 5 6 7 8 9	Room Service Delivery Floor/Tokens = 0 1 2 3 4 5 6 7 8 9
Room Service Delivery Floor/Tokens = 0 1 2 3 4 5 6 7 8 9	Room Service Delivery Floor/Tokens = 0 1 2 3 4 5 6 7 8 9

Number Cards (0–9)

0	1	2	3	4
5	6	7	8	9

Number Construction Mat

Hundreds	Tens	Ones

Place-Value Mat

Tens Waiting Area	Ones Waiting Area

A10 Level D **Blackline Masters**

Dot Set Cards

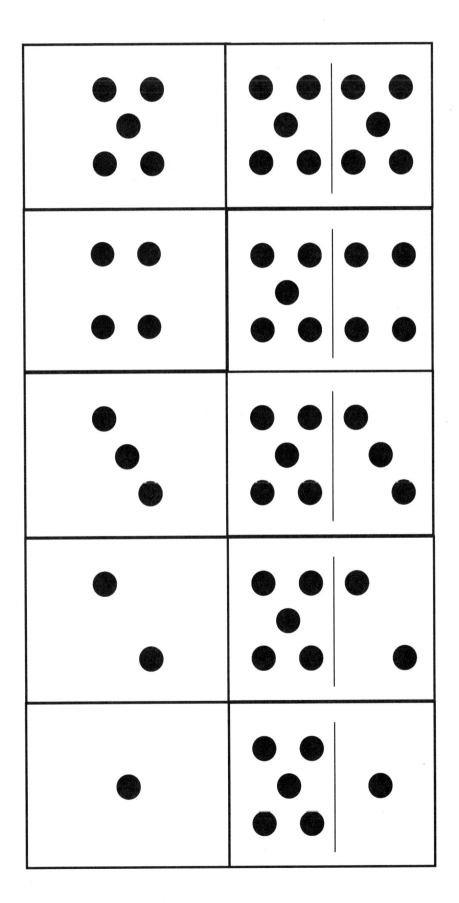

Party!

Name _____

What Number Am I?

Clue Set A

- I am close to the middle of the number line.
- I am somewhere between 40 and 50.
- One of my neighbors is 42.
- I come 2 numbers before 45.
- I am 2 more than 41.
- If you are on house 40 and you go forward 3 more houses, you will land on me.
- I am 4 blocks and 3 houses away from the beginning of the number line.

Clue Set B

- I am close to the end of the number line.
- I am somewhere between 85 and 95.
- I am the last house on my block.
- One of my neighbors is 91.
- I come 4 houses before 94.
- If you are on house 81 and you go forward 9 more houses, you will land on me.

Clue Set C

- I come somewhere in the first 2 blocks of the number line.
- I come after 10.
- One of my neighbors is 14.
- I come 10 numbers before 23.
- If you are on house 9 and move forward 4 houses, you will land on me.
- I am 1 block and 3 houses from the beginning of the number line.

Monster Cards

Blackline Masters Level D A15

1–100 Chart

1	2	3	4	5	6	7	8	9	10
11	12	13	14	15	16	17	18	19	20
21	22	23	24	25	26	27	28	29	30
31	32	33	34	35	36	37	38	39	40
41	42	43	44	45	46	47	48	49	50
51	52	53	54	55	56	57	58	59	60
61	62	63	64	65	66	67	68	69	70
71	72	73	74	75	76	77	78	79	80
81	82	83	84	85	86	87	88	89	90
91	92	93	94	95	96	97	98	99	100

Money Construction Mat

$10 Bills	$1 Bills	Dimes	Pennies

Withdrawal Slip

Withdrawal
Name: _____   Date: _____
Amount [_____] cents
Signature: _____

Withdrawal
Name: _____   Date: _____
Amount [_____] cents
Signature: _____

Withdrawal
Name: _____   Date: _____
Amount [_____] cents
Signature: _____

Deposit Slip

Deposit

Name:_____   Date:_____

Amount [¢]

Signature:_____

Deposit

Name:_____   Date:_____

Amount [¢]

Signature:_____

Deposit

Name:_____   Date:_____

Amount [¢]

Signature:_____

Count and Compare

Name _____

A20   Level D **Blackline Masters**

Addition Table   Name _____

+	0	1	2	3	4	5	6	7	8	9	10
0	0	1	2	3	4	5	6	7	8	9	10
1	1	2	3	4	5	6	7	8	9	10	11
2	2	3	4	5	6	7	8	9	10	11	12
3	3	4	5	6	7	8	9	10	11	12	13
4	4	5	6	7	8	9	10	11	12	13	14
5	5	6	7	8	9	10	11	12	13	14	15
6	6	7	8	9	10	11	12	13	14	15	16
7	7	8	9	10	11	12	13	14	15	16	17
8	8	9	10	11	12	13	14	15	16	17	18
9	9	10	11	12	13	14	15	16	17	18	19
10	10	11	12	13	14	15	16	17	18	19	20

Copyright © McGraw-Hill Education. Permission is granted to reproduce for classroom use.

Grab and Add   Name _____

1. _____   _____   _____   _____
   Grab 1 + Grab 2 + Grab 3 = Sum

2. _____   _____   _____   _____
   Grab 1 + Grab 3 + Grab 2 = Sum

3. _____   _____   _____   _____
   Grab 2 + Grab 1 + Grab 3 = Sum

4. _____   _____   _____   _____
   Grab 2 + Grab 3 + Grab 1 = Sum

5. _____   _____   _____   _____
   Grab 3 + Grab 1 + Grab 2 = Sum

6. _____   _____   _____   _____
   Grab 3 + Grab 2 + Grab 1 = Sum

Dragon Quest
Record Form

Name _____

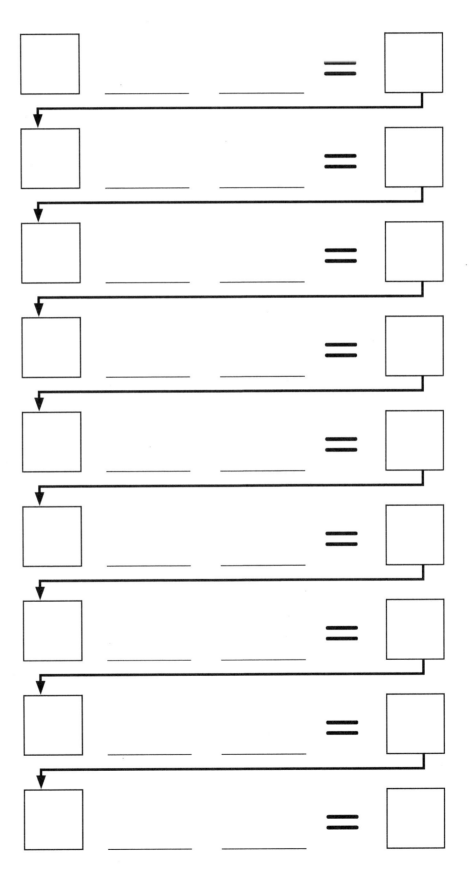

Blackline Masters Level D **A23**

Double-Digit Number Cards (Addition)

21	22	30	32
33	41	42	51

Double-Digit Number Cards (Addition)

53	62	63	72
73	81	82	93

Story Time! Recording Sheets

Directions
1. Choose three Number Cards.
2. Arrange the cards as part of an equation
3. Write a story to go with the equation and write the solution.

Sample: Number Cards:

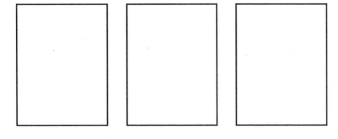

Equation: 43 + 6 = ☐

Story: There were 43 chickens at the farm. The farmer brought home 6 more chickens. How many total chickens were at the farm?

Solution: There were 49 chickens at the farm.

1. Number Cards:

☐ ☐ ☐

Equation: _____

Story: _____

_____

Solution: _____

2. Number Cards:

☐ ☐ ☐

Equation: _____

Story: _____

_____

Solution: _____

3. Number Cards:

☐ ☐ ☐

Equation: _____

Story: _____

_____

Solution: _____

4. Number Cards:

☐ ☐ ☐

Equation: _____

Story: _____

_____

Solution: _____

5. Number Cards:

☐ ☐ ☐

Equation: _____

Story: _____

_____

Solution: _____

Problems with Unknowns Recording Sheets

Directions
1. Choose a Number Card.
2. Roll the Number Cube. Don't show anyone the number.
3. Add the amount on the cube to the amount on the card.
4. Write an equation, story, and solution.

Sample:

Number Card:

35

Equation: $35 + \square = 39$

Story: Karen had 35 marbles. She added more to her collection. Now she has 39 marbles. How many marbles were in the bag?

Solution: There were 4 marbles in the bag.

1. Number Card:

Equation: _____

Story: _____

_____

Solution: _____

Blackline Masters Level D **A29**

2. Number Card:

Equation: _____

Story: _____

_____

Solution: _____

3. Number Card:

Equation: _____

Story: _____

_____

Solution: _____

4. Number Card:

[ ]

Equation: _____

Story: _____

_____

Solution: _____

5. Number Card:

[ ]

Equation: _____

Story: _____

_____

Solution: _____

Two-Step Word Problems Recording Sheet

Directions:
1. Write a word problem.
2. Write the equations that go with the word problem and write the solution.

Sample

Story: Kent had 24 quarters in his collection. On Saturday he found 4 more. Then his dad gave him 5 more quarters. How many quarters does Kent now have in his collection?

Solution: $24 + 4 = 28$

$28 + 5 = \square$

Kent has 33 quarters in his collection.

1. Story: _____

_____

Solution: _____   _____

_____

2. Story: _____

_____

Solution: _____   _____

_____

3. Story: _____

_____

Solution: _____   _____

_____

Hotel Mystery Recording Sheets

Directions:
1. Choose a Number Card.
2. Roll the Number Cube twice.
3. Write a story to go with the equation, leaving one number out.
4. Write the solution.

Sample: Addends:

Equation: 34 + 5 + ☐ = 41

Story: Mrs. Vickers buys 34 new books at the store. Her son gives her 5 new books and her daughter gives her some books. Now Mrs. Vickers has 41 books. How many books did her daughter give her?

Solution: Her daughter gave her 2 books.

1. Addends:

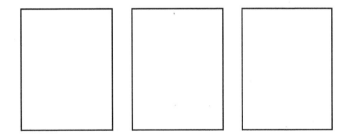

Equation: _____

Story: _____

_____

Solution: _____

2. Addends:

☐ ☐ ☐

Equation: _____

Story: _____

_____

Solution: _____

3. Addends:

☐ ☐ ☐

Equation: _____

Story: _____

_____

Solution: _____

4. Addends:

☐ ☐ ☐

Equation: _____

Story: _____

_____

Solution: _____

5. Addends:

☐ ☐ ☐

Equation: _____

Story: _____

_____

Solution: _____

Add Them Up Recording Sheet

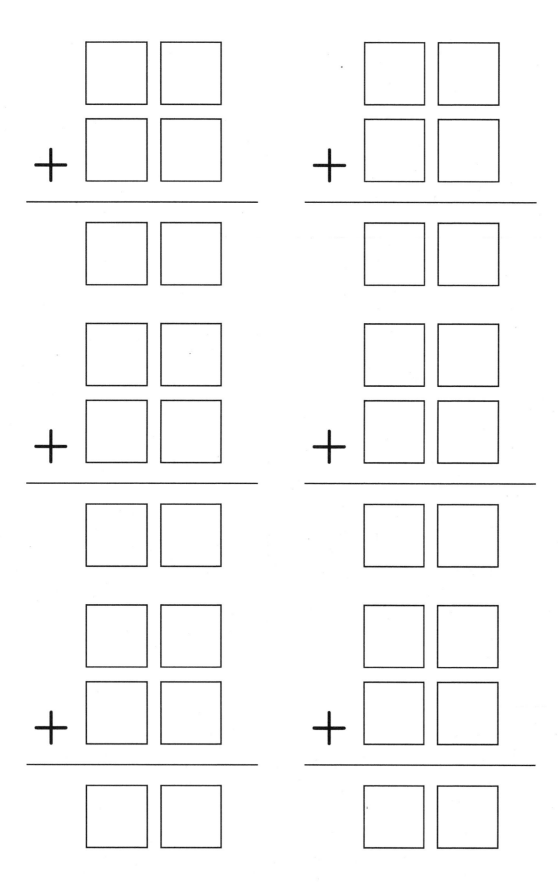

Adding Hundreds Recording Sheet

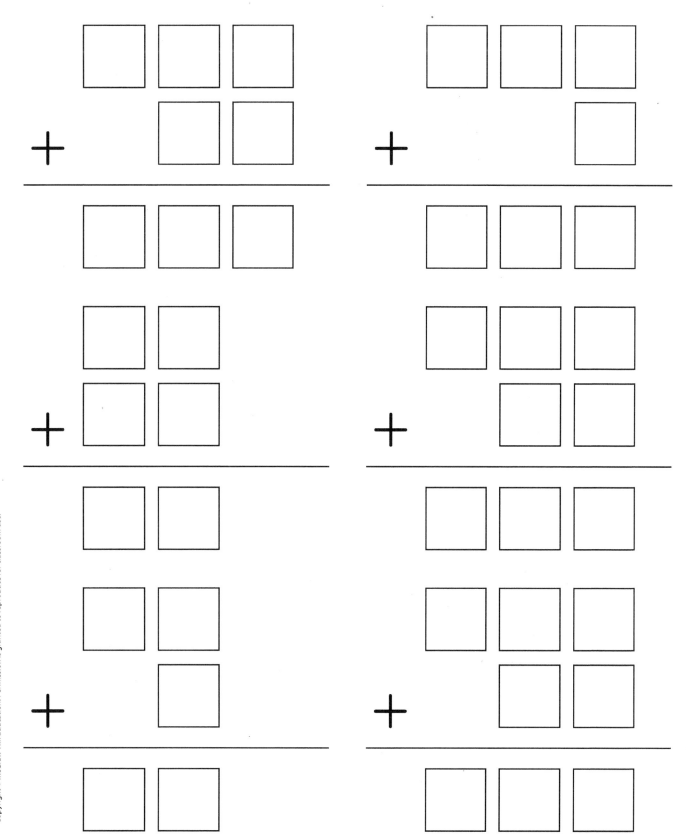

How Many Are Left?   Name _____

1. The pizza shop has only 20 pizzas left. Two people having pizza parties called to order pizzas. One person wants 11 pizzas, and the other person wants 8 pizzas. Are there enough pizzas for both people? Will there be any left? How many?

....

2. One loaf of bread will make 15 sandwiches. I need to make enough sandwiches for 8 people to have 2 sandwiches each. How many loaves of bread do I need to buy?

....

3. When I left home this morning, I had $5. A friend gave me $3. Then I went to the store and spent $5 on groceries. How much money do I have left?

Fish Pond

Name _____

Blackline Masters Level D **A39**

Fish Pond Record Form   Name _____

| How many fish are left in the pond? |

20 − ____ = ____

____ − ____ = ____

____ − ____ = ____

____ − ____ = ____

____ − ____ = ____

____ − ____ = ____

____ − ____ = ____

____ − ____ = ____

Number Lines

Name _____

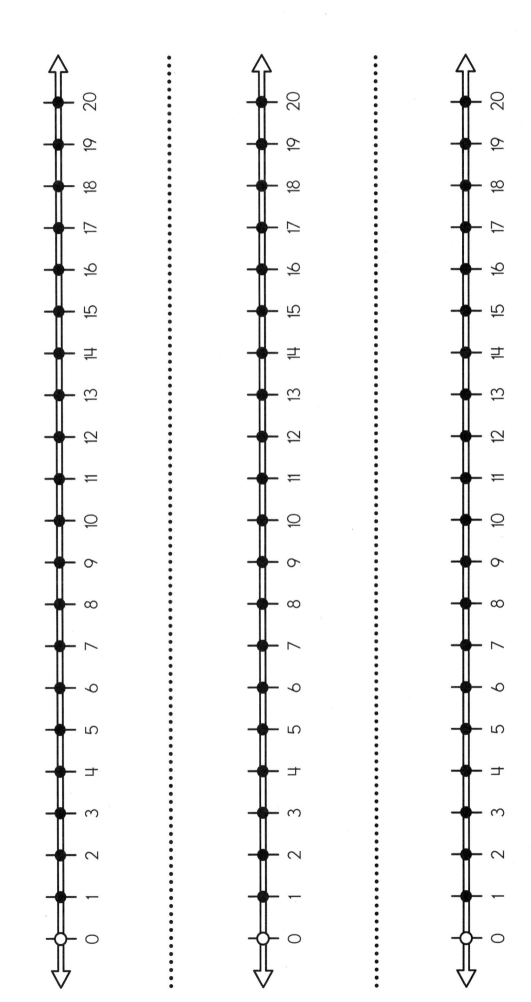

Blackline Masters Level D   **A41**

Double-Digit Number Cards (Subtraction)

26	27	28	29
36	37	38	39

A42　Level D　Blackline Masters

Double-Digit Number Cards (Subtraction)

46	47	58	59
67	68	77	88

Length of Classroom Objects   Name _____

Less than 1 ___	About 1 ___	Greater than 1 ___

Pick a Measurement Tool Recording Sheet

Object	Measurement Tool (yardstick, meterstick, measuring tape, ruler)

Customary Length Recording Sheet

Object	Length in Inches

Metric Length Recording Sheet

Object	Length in Centimeters

Clock Concentration Cards

Clock Concentration Cards

1:00	2:00	3:00	4:00
5:00	6:00	7:00	8:00
9:00	10:00	11:00	12:00

Digital Time Cards

A50   Level D **Blackline Masters**

Clock Faces

Naming Triangles by Angles

1.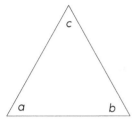

   a = _____ angle
   b = _____ angle
   c = _____ angle
   This is a(n) _____ triangle.

2.

   d = _____ angle
   e = _____ angle
   f = _____ angle
   This is a(n) _____ triangle.

3.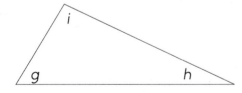

   g = _____ angle
   h = _____ angle
   i = _____ angle
   This is a(n) _____ triangle.

4.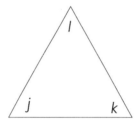

   j = _____ angle
   k = _____ angle
   l = _____ angle
   This is a(n) _____ triangle.

5.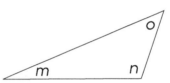

   m = _____ angle
   n = _____ angle
   o = _____ angle
   This is a(n) _____ triangle.

6.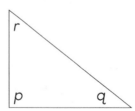

   p = _____ angle
   q = _____ angle
   r = _____ angle
   This is a(n) _____ triangle.

Naming Polygons

1.

_____ Number of sides
_____ Name of polygon

2.

_____ Number of sides
_____ Name of polygon

3.

_____ Number of sides
_____ Name of polygon

4.

_____ Number of sides
_____ Name of polygon

5.

_____ Number of sides
_____ Name of polygon

6.

_____ Number of sides
_____ Name of polygon

7.

_____ Number of sides
_____ Name of polygon

8.

_____ Number of sides
_____ Name of polygon

Special Quadrilaterals

	Parallelogram	Rectangle	Rhombus	Square	Trapezoid
Right angles					
Pairs of congruent sides					
Pairs of parallel sides					

Cubes

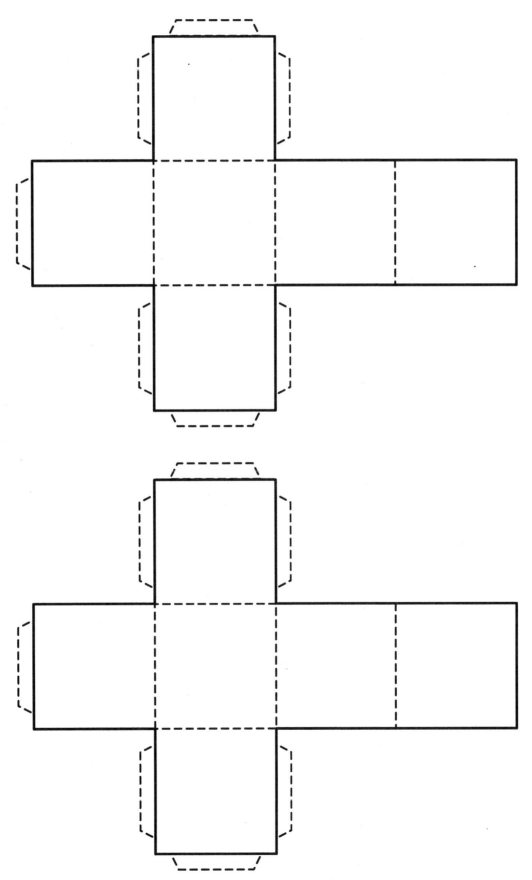

Blackline Masters Level D **A55**

Zoo Animals

Sample Graphs

Math-Link Cubes Graph

Animals that live on land:

Animals that live in the sea:

Bar Graph

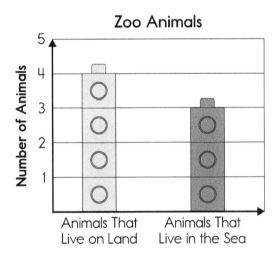

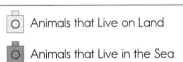 Animals that Live on Land
Animals that Live in the Sea

About Mathematics Intervention

What is an intervention?

An intervention is any instructional or practice activity designed to help students who are not making adequate progress.

- For struggling students, this requires an acceleration of development over a sufficient period of time.
- If the problem is small, the intervention might be brief.
- If the problem or lag in development is large, the intervention may last for weeks

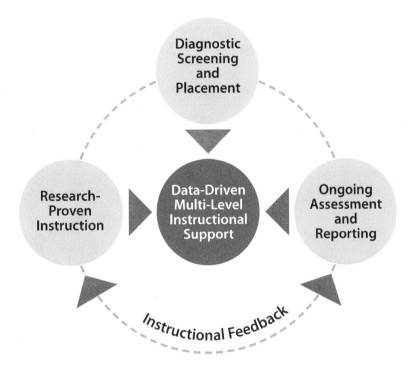

How is *Number Worlds* mathematics intervention different?

Number Worlds offers:

- Instruction built on the Common Core State Standards, with each lesson week focusing on a key standard to ensure mastery learning.
- Flexible implementation option to support grouping and targeted instruction.
- Engaging resources, including games, digital math tools, and hands on manipulatives to support application and practice of skills and strategies.
- Assessment and reporting for placing students, differentiating instruction, and measuring progress toward learning goals and standards.

Number Worlds is designed for both Tier 2 and Tier 3 students within a multi-level instructional support system. Tier 2 students may spend a brief time in the program mastering a key standard and then quickly reintegrate back into the core instructional program.

Students in Tier 3 will most likely need to complete the entire *Number Worlds* curriculum at their learning levels. Tier 3 requires intensive intervention for students with low skills and a sustained lack of adequate progress within Tiers 1 and 2. Teaching at this level is more intensive and includes more explicit instruction that is designed to meet the individual needs of struggling students. Group size is smaller, and the duration of daily instruction is longer.

Building Number Sense with *Number Worlds:*

A Mathematics Program for Young Children

What is number sense? We all know number sense when we see it but, if asked to define what it is and what it consists of, most of us, including the teachers among us, would have a much more difficult time. Yet this is precisely what we need to know to teach number sense effectively. Consider the answers three kindergarten children provide when asked the following question from the Number Knowledge Test (Griffin & Case, 1997): Which is bigger: seven or nine?"

Brie responds quickly, saying "Nine." When asked how she figured it out, she says, "Well, you go, 'seven' (pause) 'eight', 'nine' (putting up two fingers while saying the last two numbers). That means nine has two more than seven. So it's bigger."

Leah says, hesitantly, "Nine?" When asked how she figured it out, she says, "Because nine's a big number."

Caitlin looks genuinely perplexed, as if the question was not a sensible thing to ask, and says, "I don't know."

Kindergarten teachers will immediately recognize that Brie's answer provides evidence of a well-developed number sense for this age level and Leah's answer, a more fragile and less-developed number sense. The knowledge that lies behind this "sense" may be much less apparent. What knowledge does Brie have that enables her to come up with the answer in the first place and to demonstrate good number sense in the process?

1. Knowledge that underlies number sense

Research conducted with the Number Knowledge Test and several other cognitive developmental measures (see Griffin, 2002; Griffin & Case, 1997 for a summary of this research) suggests that the following understandings lie at the heart of the number sense that 5-year-olds like Brie are able to demonstrate on this problem. They know (a) that numbers indicate quantity and therefore, that numbers, themselves, have magnitude; (b) that the word "bigger" or "more" is sensible in this context; (c) that the numbers 7 and 9, like every other number from 1 to 10, occupy fixed positions in the counting sequence; (d) that 7 comes before 9 when you are counting up; (e) that numbers that come later in the sequence— that are higher up— indicate larger quantities and therefore, that 9 is bigger (or more) than 7.

Brie provided evidence of an additional component of number sense in the explanation she provided for her answer. By using the Count-On strategy to show that nine comes two numbers after seven and by suggesting that this means "it has two more than seven," Brie demonstrated that she also knows (f) that each counting number up in the sequence corresponds precisely to an increase of one unit in the size of a set. This understanding, possibly more than any of the others listed above, enables children to use the counting numbers alone, without the need for real objects, to solve quantitative problems involving the joining of two sets. In so doing, it transforms mathematics from something that can only be done out there (e.g., by manipulating real objects) to something that can be done in their own heads, and under their own control.

This set of understandings, the core of *number sense,* forms a knowledge network that Case and Griffin (1990), see also Griffin and Case (1997), have called a *central conceptual structure for number.* Research conducted by these investigators has shown that this structure is central in at least two ways (see Griffin, Case, & Siegler, 1994). First, it enables children to make sense of a broad range of quantitative problems across contexts and to answer questions, for example, about two times on a clock (Which is longer?), two positions on a path (Which is farther?), and two sets of coins (Which is worth more?). Second, it provides the foundation on which children's learning of more complex number concepts, such as those involving double-digit numbers, is built. For this reason, this network of knowledge is an important set of understandings that should be taught in the preschool years, to all children who do not spontaneously acquire them.

2. How can this knowledge be taught?

Number Worlds, a mathematics program for young children (formerly called *Rightstart*), was specifically developed to teach this knowledge and to provide a test for the cognitive developmental theory (i.e., Central Conceptual Structure theory; see Case & Griffin, 1990) on which the program was based. Originally developed for kindergarten, the program (see Griffin & Case, 1995) was expanded to teach a broader range of understandings when research findings provided strong evidence that (a) children who were exposed to the program acquired the knowledge it was designed to teach (i.e., the central conceptual structure for number), and (b) the theoretical postulates on which the program was based were valid (see Griffin & Case, 1996; Griffin, Case, &

Capodilupo, 1995; Griffin et al., 1994). Programs for grades one and two were developed to teach the more complex central conceptual structures that underlie base-ten understandings (see Griffin, 1997, 1998) and a program for preschool was developed (see Griffin, 2000) to teach the "precursor" understandings that lay the foundation for the development of the central conceptual structure for number.

Because the four levels of the program are based on a well-developed theory of cognitive development, they provide a finely graded sequence of activities (and associated knowledge objectives) that recapitulate the natural developmental progression for the age range of 3–9 years, and allow each child to enter the program at a point that is appropriate for his or her own development. To progress through the program to teach 20 or more children at any one time, every effort has been made in the construction of the **Number Worlds** program to make it as easy as possible for teachers to accommodate the developmental needs of individual children (or groups of children) in their classroom. Five instructional principles that lie at the heart of the program are described below and are used to illustrate several features of the program that have already been mentioned and several that have not yet been introduced.

2.1. Principle 1: Build upon children's current knowledge

Each new idea that is presented to children must connect to their existing knowledge if it is going to make any sense at all. Children must also be allowed to use their existing knowledge to construct new knowledge that is within reach—that is one step beyond where they are now—and a set of bridging contexts and other instructional supports should be in place to enable them to do so.

In the examples of children's thinking presented earlier, three different levels of knowledge are apparent. Brie appears to have acquired the knowledge network that underlies number sense and to be ready, therefore, to move on to the next developmental level: to connect this set of understandings to the written numerals (i.e., the formal symbols) associated with each counting word. Leah appears to have some understanding of some of the components of this network (i.e., that a number has magnitude) and to be ready to use this understanding as a base to acquire the remaining understandings (e.g., that a number's magnitude and its position in the counting sequence are directly related). Caitlin demonstrated little understanding of any element of this knowledge network and she might benefit, therefore, from exposure to activities that will help her acquire the "precursor" knowledge needed to build this network, namely knowledge of counting (e.g., the one-to-one correspondence rule) and knowledge of quantity (e.g., an intuitive understanding of relative amount). Although all three children are in kindergarten, each child appears to be at a different point in the developmental trajectory and to require a different set of learning opportunities; ones that will enable each child to use her existing knowledge to construct new knowledge at the next level up.

To meet these individual needs, teachers need (a) a way to assess children's current knowledge, (b) activities that are multi-leveled so children with different entering knowledge can all benefit from exposure to them, and (c) activities that are carefully sequenced and that span several developmental levels so children with different entering knowledge can be exposed to activities that are appropriate for their level of understanding. These are all available in the **Number Worlds** program and are illustrated in various sections of this paper.

2.2. Principle 2: Follow the natural developmental progression when selecting new knowledge to be taught

Researchers who have investigated the manner in which children construct number knowledge between the ages of 3 and 9 years have identified a common progression that most, if not all, children follow (see Griffin, 2002; Griffin and Case, 1997 for a summary of this research). As suggested earlier, by the age of 4 years, most children have constructed two "precursor" knowledge networks—knowledge of counting and knowledge of quantity—that are separate in this stage and that provide the base for the next developmental stage. Sometime in kindergarten, children become able to integrate these knowledge networks—to connect the world of counting numbers to the world of quantity—and to construct the central conceptual understandings that were described earlier. Around the age of 6 or 7 years, children connect this integrated knowledge network to the world of formal symbols and, by the age of 8 or 9 years, most children become capable of expanding this knowledge network to deal with double-digit numbers and the base-ten system. A mathematics program that provides opportunities for children to use their current knowledge to construct new knowledge that is a natural next step, and that fits their spontaneous development, will have the best chance of helping children make maximum progress in their mathematics learning and development.

Because there are limits in development on the complexity of information children can handle at any particular age/stage (see Case, 1992), it makes no sense to attempt to speed up the developmental process by accelerating children through the curriculum. However, for children who are at an age when they should have acquired the developmental milestones but for some reason haven't, exposure to a curriculum that will give them ample opportunities to do so makes tremendous sense. It will enable them to catch up to their peers and thus, to benefit from the formal mathematics instruction that is provided in school. Children who are developing normally also benefit from opportunities to broaden and deepen the knowledge networks they are constructing, to strengthen these understandings, and to use them in a variety of contexts.

2.3. Principle 3: Teach computational fluency as well as conceptual understanding

Because computational fluency and conceptual understanding have been found to go hand in hand in children's mathematical development (see Griffin, 2003; Griffin et al., 1994), opportunities to acquire computational fluency, as well as conceptual understanding, are built into every **Number Worlds** activity. This is nicely illustrated in the following activities, drawn from different levels of the program.

In The Mouse and the Cookie Jar Game (created for the preschool program and designed to give 3- to 4-year-olds an intuitive understanding of subtraction), children are given a certain number of counting chips (with each child receiving the same number but a different color) and told to pretend their chips are cookies. They are asked to count their cookies and, making sure they remember how many they have and what their color is, to deposit them in the cookie jar for safe keeping. While the children sleep, a little mouse comes along and takes one (or two) cookies from the jar. The problem that is then posed to the children is "How can we figure out whose cookie(s) the mouse took?"

Although children quickly learn that emptying the jar and counting the set of cookies that bears their own color is a useful strategy to use to solve this problem, it takes considerably longer for many children to realize that, if they now have four cookies (and originally had five), it means that they have one fewer and the mouse has probably taken one of their cookies. Children explore this problem by counting and recounting the remaining sets, comparing them to each other (e.g., by aligning them) to see who has the most or least, and ultimately coming up with a prediction. When a prediction is made, children search the mouse's hole to see whose cookie had been taken and to verify or revise their prediction. As well as providing opportunities to perfect their counting skills, this activity gives children concrete opportunities to experience simple quantity transformations and to discover how the counting numbers can be used to predict and explain differences in amount.

The Dragon Quest Game that was developed for the Grade 1 program teaches a much more sophisticated set of understandings. Children are introduced to Phase 1 activity by being told a story about a fire-breathing dragon that has been terrorizing the village where children live. The children playing the game are heroes who have been chosen to seek out the dragon and put out his fire. To extinguish this dragon's fire (as opposed to the other, more powerful dragons they will encounter in later phases) a hero will need at least 10 pails of water. If a hero enters into the dragon's area with less than 10 pails of water, he or she will become the dragon's prisoner and can only be rescued by one of the other players.

To play the game, children take turns rolling a die and moving their playing piece along the colored game board. If they land on a well pile (indicated by a star), they can pick a card from the face-down deck of cards, which illustrate, with images and symbols (e.g., +4) a certain number of pails of water. Children are encouraged to add up their pails of water as they receive them and they are allowed to use a variety of strategies to do so, ranging from mental math (which is encouraged) to the use of tokens to keep track of the quantity accumulated. The first child to reach the dragon's lair with at least 10 pails of water can put out the dragon's fire and free any teammates who have become prisoners.

2.5. Principle 5: Expose children to the major way number is represented and talked about in developed societies

As children play this game and talk about their progress, they have ample opportunity to connect numbers to several different quantity representations (e.g., dot patterns on the die; distance of their pawn along the path; sets of buckets illustrated on the cards; written numerals also provided on the cards) and to acquire an appreciation of numerical magnitude across these contexts. With repeated play, they also become capable of performing a series of successive addition operations in their heads and of expanding the well pile. When they are required to submit formal proof to the mayor of the village that they have amassed sufficient pails of water to put out the dragon's fire before they are allowed to do so, they become capable of writing a series of formal expressions to record the number of pails received and spilled over the course of the game. In contexts such as these children receive ample opportunity to use the formal symbol system in increasingly efficient ways to make sense of quantitative problems they encounter in the course of their own activity.

2.4. Principle 4: Provide plenty of opportunity for hands-on exploration, problem-solving, and communication

Like the Dragon Quest Game that was just described, many of the activities created for the **Number Worlds** program are set in a game format that provides plenty of opportunity for hands-on exploration of number concepts, for problem-solving and for communication. Communication is explicitly encouraged in a set of question prompts that are included with each small group game (e.g., How far are you now? How many more buckets do you need to put out the dragon's fire? How do you know?) as well as in a more general set of dialogue prompts that are included in the teacher's guide. Opportunities for children to discuss what they learned during game play each day, to share their knowledge with their peers, and to make their reasoning explicit are also provided in a Wrap-Up session that is included at the end of each math lesson.

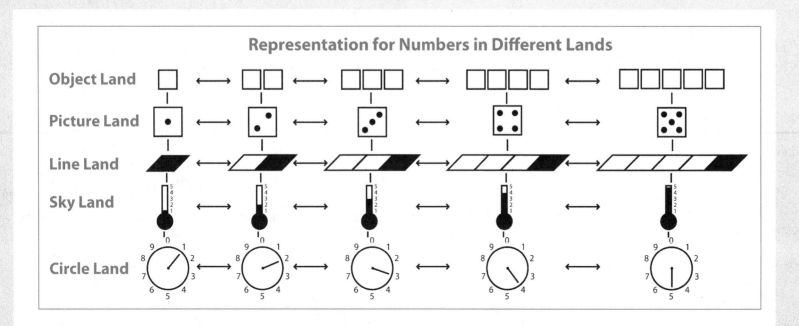

Finally, in the whole group games and activities that were developed for the Warm-Up portion of each math lesson, children are given ample opportunity to count (e.g., up from 1 and down from 10) and to solve mental math problems, in a variety of contexts. In addition to developing computational fluency, these activities expose children to the language of mathematics and give them practice using it. Although this is valuable for all children, it is especially useful for ESL children, who may know how to count in their native language but not yet in English. Allowing children to take turns in these activities and to perform individually gives teachers opportunities to assess each child's current level of functioning, important for instructional planning, and gives children opportunities to learn from each other.

Number is represented in our culture in five major ways: as a group of objects, a dot-set pattern, a position on a line, a position on a scale (e.g., a thermometer), and a point on a dial. In each of these contexts, number is also talked about in different ways, with a larger number (and quantity) described as "more" in the world of dot-sets, as "further along" in the world of paths and lines, as "higher up" in the world of scale measures, and as "further around" in the world of dials. Children who are familiar with these forms of representation and the language used to talk about number in these contexts have a much easier time making sense of the number problems they encounter inside and outside of school.

In the Number Worlds program, children are systematically exposed to these forms of representations as they explore five different "lands." Learning activities developed for each land share a particular form of number representation while they simultaneously address specific knowledge goals for each grade level. Many of the games, like Dragon Quest, also expose children to multiple representations of number in one activity so children can gradually come to see the ways they are equivalent.

3. Discussion

Children who have been exposed to the **Number Worlds** program do very well on number questions like the one presented in the introduction and on the Number Knowledge Test (Griffin & Case, 1997) from which this question was drawn. In several evaluation studies conducted with children from low-income communities, children who received the **Number Worlds** program made significant gains in conceptual knowledge of number and in number sense, when compared to matched-control groups who received readiness training of a different sort. These gains enabled them to start their formal schooling in grade one on an equal footing with their more advantaged peers, to perform as well as groups of children from China and Japan on a computation test administered at the end of grade one, and to keep pace with their more advantaged peers (an even outperform them on some measures) as they progressed through the first few years of formal schooling (Griffin & Case, 1997).

Teachers also report positive gains from using the **Number Worlds** program and from exposure to the instructional principles on which it is based. Although all teachers acknowledge that implementing the program and putting the principles into action is not an easy task, many claim that their teaching of all subjects has been transformed in the process. They now facilitate discussion rather than dominating it; they pay much more attention to what children say and do; and they now allow children to take more responsibility for their own learning, with positive and surprising results. Above all, they now look forward to teaching math and they and their students are eager to do more of it.

Griffin, S. "Building Number Sense with Number Worlds: A mathematics program for young children." *In Early Childhood Research Quarterly 19 (2004) 173–180*. Elsevier.

References

Case, R. (1992). *The mind's staircase: Exploring the conceptual underpinnings of children's thought and knowledge*. Hillsdale, NJ: Erlbaum.

Case, R., & Griffin, S. (1990). Child cognitive development: The role of central conceptual structures in the development of scientific and social thought. In E.A. Hauert (Ed.), *Developmental psychology: Cognitive, perceptuo-motor, and neurological perspectives* (pp. 193–230). North-Holland: Elsevier.

Griffin, S. (1997). *Number Worlds: Grade one level*. Durham, NH: Number Worlds Alliance Inc.

Griffin, S. (1998). *Number Worlds: Grade two level*. Durham, NH: Number Worlds Alliance Inc.

Griffin, S. (2000). *Number Worlds: Preschool level*. Durham, NH: Number Worlds Alliance Inc.

Griffin, S. (2002). The development of math competence in the preschool and early school years: Cognitive foundations and instructional strategies. In J. Royer (Ed.), *Mathematical cognition* (pp. 1–32). Greenwich, CT: Information Age Publishing.

Griffin, S. (2003). Laying the foundation for computational fluency in early childhood. *Teaching children mathematics,* 9, 306–309.

Griffin, S. (1997). *Number Worlds: Kindergarten level*. Durham, NH: Number Worlds Alliance Inc.

Griffin, S., & Case, R. (1996). Evaluating the breadth and depth of training effects when central conceptual structures are taught. *Society for research in child development monographs,* 59, 90–113.

Griffin, S., & Case, R. (1997). Re-thinking the primary school math curriculum: An approach based on cognitive science. *Issues in Education, 3,* 1–49.

Griffin, S., Case, R., & Siegler, R. (1994). Rightstart: Providing the central conceptual prerequisites for first formal learning of arithmetic to students at-risk for school failure. In K. McGilly (Ed.), *Classroom lessons: Integrating cognitive theory and classroom practice* (pp. 24-49). Cambridge, MA: Bradford Books MIT Press.

Griffin, S., Case, R., & Capodilupo, A. (1995). Teaching for understanding: The importance of central conceptual structures in the elementary mathematics curriculum. In A. McKeough, I. Lupert, & A. Marini (Eds.), *Teaching for transfer: Fostering generalization in learning* (pp. 121–151). Hillsdale, NJ: Erlbaum.

Content Strands

Counting and Cardinality

Know number names, master the counting sequence, use counting to tell the number of objects, compare numbers.

Number Worlds and Counting and Cardinality

Recognition of numbers and their meaning is the foundation for number sense. Developing number sense is a primary goal of **Number Worlds** at every level. Numbers are presented in a variety of representations and integrated in many contexts so that students develop a thorough understanding of numbers.

Number and Operations in Base Ten

Understand place value, use the properties of operations, and perform multi-digit arithmetic with whole numbers and decimal numbers.

Number Worlds and Number and Operations in Base Ten

Mastery of a Base-Ten number system allows students to see patterns and relationships between whole numbers and operations and decimal numbers and operations. An understanding of numbers and operations in Base Ten lays a foundation for future work with rational numbers.

Number and Operations—Fractions

Understand fractions as numbers. Compare and order fractions, and perform basic operations—addition, subtraction, multiplication, and division—with fractions.

Number Worlds and Number and Operations—Fractions

Students understand rational numbers and the relationship of fractions to decimal numbers. They build a greater understanding of addition, subtraction, multiplication, and division of fractions, along with using the multiplication of fractions in proportions and as a model for areas of regular quadrilaterals.

Operations and Algebraic Thinking

Understand addition and subtraction and their relationship to each other, understand multiplication and division and their relationship to each other, and recognize and analyze patterns and relationships.

Number Worlds and Operations and Algebraic Thinking

Arithmetic, one of the oldest branches of mathematics, arises from the most foundational of mathematical operations: counting. Whatever other skills and understandings children acquire, they must have the ability to calculate a precise answer when necessary. Once taught, arithmetic skills are also integrated into other topics, such as measurement, data analysis, and geometry.

The algebra readiness instruction that begins in the Pre-K level is designed to prepare students for future work in algebra by exposing them to algebraic thinking, including looking for patterns, using variables, working with functions, using integers and exponents, and being aware that mathematics is far more than just arithmetic.

The Number System

Apply number sense and operations to rational numbers.

Ratios and Proportional Relationships

Understand ratio concepts and analyze proportional relationships

Number Worlds and Ratios and Proportional Relationships

Students can expand their understanding of arithmetic to calculate ratios, proportions, and percentages. A solid foundation in the meaning of fractions will allow students to express ratios as fractions.

Expressions and Equations

Apply understandings of arithmetic to algebraic expressions and equations, solve one-variable equations and inequalities.

Functions

Define, evaluate, and compare functions, use to model relationships between quantities.

Measurement and Data

Classify and count objects, measure and convert measurements, work with time and money, represent and interpret data and geometric measurements: angle measurements, perimeter, area, and volume.

Number Worlds and Measurement and Data

Measurement allows a real-life application of numbers and number sense. Understanding conversions between small and larger amounts, as well as relationships utilizing money and time, are skills developed throughout the Number Worlds program. Students work with graphs beginning in the Pre-K and each level of the program emphasizes understanding what data in show in those graphs.

Geometry

Analyze, compare, create, and compose shapes, reason with shapes and their attributes, draw and identify lines and angles, classify shapes by properties, and graph on the coordinate plane.

Number Worlds and Geometry

Geometry is the branch of mathematics that deals with the properties of space, in two-dimensions and in three-dimensions, including representation, location, and direction. Shapes can be identified and classified by their properties, they can be decomposed and composed into new shapes, and they can be transformed by rotations, slides, and flips.

Statistics and Probability

Use random sampling for a population, investigate chance, and evaluate probability models.

Key Standards by Lesson Week

Key Common Core State Standards

Each lesson week in Number Worlds focuses on a key Common Core Standard. Expert reviewers identified these standards as most critical for student success in mastering math concepts.

Level A

Week	Lesson	Domain	Cluster	Key Standard
1	Counting Objects	Counting and Cardinality	Know number names and the count sequence.	K.CC.1
2	Counting and Sorting Objects	Measurement and Data	Classify objects and count the number of objects in each category.	K.MD.3
3	More Counting and Sorting Objects	Counting and Cardinality	Compare numbers.	K.CC.6
4	Sequencing Sets	Counting and Cardinality	Know number names and the count sequence.	K.CC.3
5	Shapes	Geometry	Identify and describe shapes (squares, circles, triangles, rectangles, hexagons, cubes, cones, cylinders, and spheres).	K.G.2
6	More Shapes	Geometry	Identify and describe shapes (squares, circles, triangles, rectangles, hexagons, cubes, cones, cylinders, and spheres).	K.G.2
7	Learning about Set Size	Counting and Cardinality	Know number names and the count sequence.	K.CC.3
8	More Sets	Operations and Algebraic Thinking	Understand addition as putting together and adding to, and understand subtraction as taking apart and taking from.	K.OA.1
9	Number Sequence	Counting and Cardinality	Know number names and the count sequence.	K.CC.1
10	Identifying Pattern and Quantity	Counting and Cardinality	Compare numbers.	K.CC.6
11	Matching Patterns and Quantities	Counting and Cardinality	Know number names and the count sequence.	K.CC.3
12	Introduction to the Number Line	Counting and Cardinality	Know number names and the count sequence.	K.CC.2
			Count to tell the number of objects.	K.CC.4.c
13	Position on the Number Line	Counting and Cardinality	Count to tell the number of objects.	K.CC.4.c
14	Sequence on the Number Line	Counting and Cardinality	Count to tell the number of objects.	K.CC.4.a
15	Using the Number Line	Counting and Cardinality	Count to tell the number of objects.	K.CC.4.a
16	Moving Up and Down	Counting and Cardinality	Compare numbers.	K.CC.7
17	Counting Rotations	Counting and Cardinality	Know number names and the count sequence.	K.CC.2
18	Adding to a Set	Counting and Cardinality	Know number names and the count sequence.	K.CC.1
		Operations and Algebraic Thinking	Understand addition as putting together and adding to, and understand subtraction as taking apart and taking from.	K.OA.1
19	Subtracting from a Set	Operations and Algebraic Thinking	Understand addition as putting together and adding to, and understand subtraction as taking apart and taking from.	K.OA.1
20	Comparing Sets	Counting and Cardinality	Compare numbers.	K.CC.6
21	Using Graphs	Measurement and Data	Represent and interpret data.	1.MD.4
22	More Subtraction	Operations and Algebraic Thinking	Understand addition as putting together and adding to, and understand subtraction as taking apart and taking from.	K.OA.1
23	Depicting Numbers	Counting and Cardinality	Compare numbers.	K.CC.6
24	Comparing Quantities	Counting and Cardinality	Compare numbers.	K.CC.6
25	Making Comparisons	Counting and Cardinality	Compare numbers.	K.CC.6
26	Adding and Subtracting	Counting and Cardinality	Compare numbers.	K.CC.6
27	More Graphs	Counting and Cardinality	Compare numbers.	K.CC.6
		Measurement and Data	Represent and interpret data.	1.MD.4
28	Higher and Lower	Counting and Cardinality	Know number names and the count sequence.	K.CC.2
29	More Rotations	Counting and Cardinality	Compare numbers.	K.CC.6
30	Sequencing Numbers	Counting and Cardinality	Count to tell the number of objects.	K.CC.4.c
31	Understanding the Analog Clock	Counting and Cardinality	Know number names and the count sequence.	K.CC.2
32	The Dollar Bill and the Penny	Counting and Cardinality	Count to tell the number of objects.	K.CC.4.a
		Operations and Algebraic Thinking	Understand addition as putting together and adding to, and understand subtraction as taking apart and taking from.	K.OA.1

Key Standards by Lesson Week

Key Common Core State Standards

Level B

Week	Lesson	Domain	Cluster	Key Standard
1	Counting Basics	Counting and Cardinality	Know number names and the count sequence.	K.CC.1
2	Counting Objects	Counting and Cardinality	Know number names and the count sequence.	K.CC.1
3	Adding One	Operations and Algebraic Thinking	Understand addition as putting together and adding to, and understand subtraction as taking apart and taking from.	K.OA.1
4	Subtracting One	Operations and Algebraic Thinking	Understand addition as putting together and adding to, and understand subtraction as taking apart and taking from.	K.OA.1
5	Counting and Comparing	Counting and Cardinality	Compare numbers.	K.CC.6
		Operations and Algebraic Thinking	Understand addition as putting together and adding to, and understand subtraction as taking apart and taking from.	K.OA.1
6	More Counting and Comparing	Counting and Cardinality	Compare numbers.	K.CC.6
		Operations and Algebraic Thinking	Understand addition as putting together and adding to, and understand subtraction as taking apart and taking from.	K.OA.1
7	Comparing Sets	Counting and Cardinality	Comparing numbers.	K.CC.6
8	Making Comparisons	Measurement and Data	Classify objects and count the number of objects in each category.	K.MD.3
9	Matching Sets	Counting and Cardinality	Know number names and the count sequence.	K.CC.3
10	Matching Quantities and Numbers	Counting and Cardinality	Know number names and the count sequence.	K.CC.3
11	Comparing Quantities	Counting and Cardinality	Compare numbers.	K.CC.6
				K.CC.7
12	Numbers on a Vertical Scale	Counting and Cardinality	Count to tell the number of objects.	K.CC.4.c
13	Numbers on a Line	Counting and Cardinality	Count to tell the number of objects.	K.CC.4.c
14	Adding and Subtracting on a Line	Counting and Cardinality	Compare numbers.	K.CC.6
		Operations and Algebraic Thinking	Understand addition as putting together and adding to, and understand subtraction as taking apart and taking from.	K.OA.1
15	Comparing Positions on a Line	Counting and Cardinality	Count to tell the number of objects.	K.CC.4.c
16	Moving on a Vertical Scale	Counting and Cardinality	Count to tell the number of objects.	K.CC.4.c
17	Predicting Quantities	Operations and Algebraic Thinking	Understand addition as putting together and adding to, and understand subtraction as taking apart and taking from.	K.OA.1
18	Numeral Magnitude	Counting and Cardinality	Know number names and the count sequence.	K.CC.3
			Count to tell the number of objects.	K.CC.4.c
		Operations and Algebraic Thinking	Understand addition as putting together and adding to, and understand subtraction as taking apart and taking from.	K.OA.4
19	Moving and Comparing on a Number Line	Counting and Cardinality	Count to tell the number of objects.	K.CC.4.c
20	Comparing Representations	Counting and Cardinality	Know number names and the count sequence.	K.CC.3
		Measurement and Data	Classify objects and count the number of objects in each category.	K.MD.3
21	Numbers on a Dial	Counting and Cardinality	Count to tell the number of objects.	K.CC.4.b
			Compare numbers.	K.CC.6
22	Adding and Subtracting on a Dial	Operations and Algebraic Thinking	Understand addition as putting together and adding to, and understand subtraction as taking apart and taking from.	K.OA.1
23	Understanding Numerals	Counting and Cardinality	Count to tell the number of objects.	K.CC.5
24	Adding and Subtracting on a Vertical Scale	Operations and Algebraic Thinking	Understand addition as putting together and adding to, and understand subtraction as taking apart and taking from.	K.OA.1
25	More Operations on a Vertical Scale	Operations and Algebraic Thinking	Understand addition as putting together and adding to, and understand subtraction as taking apart and taking from.	K.OA.1
26	Making Bar Graphs	Measurement and Data	Represent and interpret data.	1.MD.4
27	Recording Data on a Number Line	Measurement and Data	Classify objects and count the number of objects in each category.	K.MD.3
28	Problem Solving	Counting and Cardinality	Compare numbers.	K.CC.6
29	More Vertical Bar Graphs	Measurement and Data	Represent and interpret data.	1.MD.4
30	Creating Sets	Counting and Cardinality	Know number names and the count sequence.	K.CC.3
31	Understanding the Analog Clock	Counting and Cardinality	Know number names and the count sequence.	K.CC.2
32	The Five-Dollar Bill and the Nickel	Operations and Algebraic Thinking	Understand addition as putting together and adding to, and understand subtraction as taking apart and taking from.	K.OA.2

Key Standards by Lesson Week

Key Common Core State Standards

Level C

Week	Lesson	Domain	Cluster	Key Standard
1	Counting	Counting and Cardinality	Know number names and the count sequence.	K.CC.1
2	Counting and Comparing	Counting and Cardinality	Know number names and the count sequence.	K.CC.3
3	More Counting and Comparing	Counting and Cardinality	Know number names and the count sequence.	K.CC.3
4	Matching Dot Sets to Numerals	Counting and Cardinality	Know number names and the count sequence.	K.CC.1
5	Number Sequence and Number Lines	Counting and Cardinality	Know number names and the count sequence.	K.CC.1
6	More Number Sequence and Number Lines	Counting and Cardinality	Know number names and the count sequence.	K.CC.1
6		Operations and Algebraic Thinking	Understand addition as putting together and adding to, and understand subtraction as taking apart and taking from.	K.OA.1
7	Number Neighborhoods	Counting and Cardinality	Know number names and the count sequence.	K.CC.1
8	More Number Neighborhoods	Counting and Cardinality	Know number names and the count sequence.	K.CC.1
9	Adding Numbers	Operations and Algebraic Thinking	Understand addition as putting together and adding to, and understand subtraction as taking apart and taking from.	K.OA.1
10	More Adding	Operations and Algebraic Thinking	Understand addition as putting together and adding to, and understand subtraction as taking apart and taking from.	K.OA.1
11	Sequencing Numbers	Operations and Algebraic Thinking	Work with addition and subtraction equations.	1.OA.8
12	Writing Equations	Operations and Algebraic Thinking	Understand addition as putting together and adding to, and understand subtraction as taking apart and taking from.	K.OA.1
13	Counting and Adding	Operations and Algebraic Thinking	Add and subtract within 20.	1.OA.6
14	Making Equations	Operations and Algebraic Thinking	Understand addition as putting together and adding to, and understand subtraction as taking apart and taking from.	K.OA.1
15	Graphing and Comparing Numbers	Operations and Algebraic Thinking	Understand addition as putting together and adding to, and understand subtraction as taking apart and taking from.	K.OA.1
16	More Counting and Adding	Operations and Algebraic Thinking	Understand addition as putting together and adding to, and understand subtraction as taking apart and taking from.	K.OA.1
17	Solving Equations	Operations and Algebraic Thinking	Understand addition as putting together and adding to, and understand subtraction as taking apart and taking from.	K.OA.1
18	Adding and Subtracting	Operations and Algebraic Thinking	Understand addition as putting together and adding to, and understand subtraction as taking apart and taking from.	K.OA.1
19	Subtracting	Operations and Algebraic Thinking	Understand addition as putting together and adding to, and understand subtraction as taking apart and taking from.	K.OA.1
20	Subtracting and Predicting	Operations and Algebraic Thinking	Work with addition and subtraction equations.	1.OA.8
21	Adding and Comparing	Counting and Cardinality	Count to tell the number of objects.	K.CC.4.c
22	Subtracting to Zero	Operations and Algebraic Thinking	Add and subtract within 20.	1.OA.6
23	More Adding and Subtracting	Operations and Algebraic Thinking	Add and subtract within 20.	1.OA.6
24	Numbers to 100	Counting and Cardinality	Know number names and the count sequence.	K.CC.1
24		Operations and Algebraic Thinking	Understand addition as putting together and adding to, and understand subtraction as taking apart and taking from.	K.OA.1
25	More Numbers to 100	Counting and Cardinality	Know number names and the count sequence.	K.CC.1
26	Addition Stories	Operations and Algebraic Thinking	Understand addition as putting together and adding to, and understand subtraction as taking apart and taking from.	K.OA.2
27	Tens and Ones	Number and Operations in Base Ten	Understand place value.	1.NBT.2.a 1.NBT.2.c
28	Adding and Subtracting Length	Operations and Algebraic Thinking	Understand addition as putting together and adding to, and understand subtraction as taking apart and taking from.	K.OA.1
29	Addition and Subtraction Stories	Operations and Algebraic Thinking	Understand addition as putting together and adding to, and understand subtraction as taking apart and taking from.	K.OA.1
29			Represent and solve problems involving addition and subtraction.	1.OA.1
30	Making a Map	Operations and Algebraic Thinking	Understand addition as putting together and adding to, and understand subtraction as taking apart and taking from.	K.OA.1
30			Represent and solve problems involving addition and subtraction.	1.OA.1
31	Understanding the Analog Clock	Operations and Algebraic Thinking	Add and subtract within 20.	1.OA.5
32	The Ten-Dollar Bill and the Dime	Operations and Algebraic Thinking	Represent and solve problems involving addition and subtraction.	1.OA.1

Key Standards by Lesson Week

Key Common Core State Standards

Level D

Week	Lesson	Domain	Cluster	Key Standard
\multicolumn{5}{c}{**Unit 1 Number Sense within 100**}				
1	Constructing Whole Numbers	Number and Operations in Base Ten	Understand place value.	1.NBT.2a
2	Numbers on a Line	Number and Operations in Base Ten	Understand place value.	2.NBT.1
3	Tens and Ones	Number and Operations in Base Ten	Understand place value.	1.NBT.2
4	Visualizing and Constructing Whole Numbers	Number and Operations in Base Ten	Understand place value.	2.NBT.1
5	Number Patterns	Number and Operations in Base Ten	Understand place value.	2.NBT.2
6	Whole Number Relationships	Number and Operations in Base Ten	Understand place value.	2.NBT.4
\multicolumn{5}{c}{**Unit 2 Number Sense to 1,000**}				
1	Understanding the Base-Ten Number System	Number and Operations in Base Ten	Understand place value.	2.NBT.2
2	Constructing Whole Numbers to 999	Number and Operations in Base Ten	Understand place value.	2.NBT.3
3	Representing Number Systems	Number and Operations in Base Ten	Understand place value.	2.NBT.1
4	Place Value to 1,000	Number and Operations in Base Ten	Understand place value.	2.NBT.3
5	Skip Counting within 1,000	Number and Operations in Base Ten	Understand place value.	2.NBT.2
6	Comparing Whole Numbers	Number and Operations in Base Ten	Understand place value.	2.NBT.4
\multicolumn{5}{c}{**Unit 3 Addition**}				
1	Addition Fundamentals	Operations and Algebraic Thinking	Add and subtract within 20.	2.OA.2
2	Mastering the Basic Facts	Operations and Algebraic Thinking	Add and subtract within 20.	2.OA.2
3	Solving Addition Problems	Number and Operations in Base Ten	Use place value understanding and properties of operations to add and subtract.	2.NBT.5
4	Addition Tools and Strategies	Number and Operations in Base Ten	Use place value understanding and properties of operations to add and subtract.	2.NBT.5
5	Addition Word Problems within 100	Operations and Algebraic Thinking	Represent and solve problems involving addition and subtraction.	2.OA.1
6	Solving Addition Word Problems within 1,000	Number and Operations in Base Ten	Use place value understanding and properties of operations to add and subtract.	2.NBT.7
\multicolumn{5}{c}{**Unit 4 Subtraction**}				
1	Subtraction Fundamentals	Operations and Algebraic Thinking	Add and subtract within 20.	2.OA.2
2	Mastering Basic Subtraction Facts	Operations and Algebraic Thinking	Add and subtract within 20.	2.OA.2
3	Solving Subtraction Problems	Number and Operations in Base Ten	Use place value understanding and properties of operations to add and subtract.	2.NBT.5
4	Subtraction Tools and Strategies	Number and Operations in Base Ten	Use place value understanding and properties of operations to add and subtract.	2.NBT.5
5	Subtraction Word Problems within 100	Operations and Algebraic Thinking	Represent and solve problems involving addition and subtraction.	2.OA.1
6	Solving Subtraction Word Problems within 1,000	Number and Operations in Base Ten	Use place value understanding and properties of operations to add and subtract.	2.NBT.7
\multicolumn{5}{c}{**Unit 5 Geometry and Measurement**}				
1	Linear Measurement	Measurement and Data	Measure and estimate lengths in standard units.	2.MD.3
2	Measurement Tools	Measurement and Data	Measure and estimate lengths in standard units.	2.MD.1
3	Time Measurement to the Half Hour	Measurement and Data	Work with time and money.	2.MD.7
4	Time Measurement to the Nearest Five Minutes	Measurement and Data	Work with time and money.	2.MD.7
5	Attributes of Shapes	Geometry	Reason with shapes and their attributes.	2.G.1
6	Graphs	Measurement and Data	Represent and interpret data.	2.MD.10

Key Standards by Lesson Week

Key Common Core State Standards

Level E

Week	Lesson	Domain	Cluster	Key Standard
Unit 1	**Number Sense**			
1	Place Values to 999	Number and Operations in Base Ten	Use place value understanding and properties of operations to perform multi-digit arithmetic.	3.NBT.1
2	Round to the Nearest 10 or Hundred	Number and Operations in Base Ten	Use place value understanding and properties of operations to perform multi-digit arithmetic.	3.NBT.1
3	Skip Counting	Number and Operations in Base Ten	Use place value understanding and properties of operations to perform multi-digit arithmetic.	3.NBT.3
4	Introduction to Fractions	Number and Operations- Fractions	Develop understanding of fractions as numbers.	3.NF.1
5	Representing Fractions	Number and Operations- Fractions	Develop understanding of fractions as numbers.	3.NF.1 3.NF.3d
6	Comparing Fractions	Number and Operations- Fractions	Develop understanding of fractions as numbers.	3.NF.3a 3.NF.3b 3.NF.3d
Unit 2	**Addition**			
1	Mental Addition Strategies	Number and Operations in Base Ten	Use place value understanding and properties of operations to perform multi-digit arithmetic.	3.NBT.2
2	Grouping Strategies	Number and Operations in Base Ten	Use place value understanding and properties of operations to perform multi-digit arithmetic.	3.NBT.2
3	Regrouping Strategies	Number and Operations in Base Ten	Use place value understanding and properties of operations to perform multi-digit arithmetic.	3.NBT.2
4	Solving Addition Problems to 1,000	Number and Operations in Base Ten	Use place value understanding and properties of operations to perform multi-digit arithmetic.	3.NBT.2
5	Computational Estimation	Operations and Algebraic Thinking	Solve problems involving the four operations, and identify and explain patterns in arithmetic.	3.OA.8
6	Variables and Equality	Operations and Algebraic Thinking	Solve problems involving the four operations, and identify and explain patterns in arithmetic.	3.OA.8
Unit 3	**Subtraction**			
1	Subtraction Strategies	Number and Operations in Base Ten	Use place value understanding and properties of operations to perform multi-digit arithmetic.	3.NBT.2
2	More Subtraction Strategies	Number and Operations in Base Ten	Use place value understanding and properties of operations to perform multi-digit arithmetic.	3.NBT.2
3	Subtraction with Regrouping	Number and Operations in Base Ten	Use place value understanding and properties of operations to perform multi-digit arithmetic.	3.NBT.2
4	More Subtraction with Regrouping	Number and Operations in Base Ten	Use place value understanding and properties of operations to perform multi-digit arithmetic.	3.NBT.2
5	Solving Subtraction Problems within 100	Operations and Algebraic Thinking	Solve problems involving the four operations, and identify and explain patterns in arithmetic.	3.OA.8
6	Solving Word Problems within 1,000	Operations and Algebraic Thinking	Solve problems involving the four operations, and identify and explain patterns in arithmetic.	3.OA.8
Unit 4	**Multiplication and Division**			
1	Models for Multiplication	Operations and Algebraic Thinking	Represent and solve problems involving multiplication and division	3.OA.1
2	Number Lines and Arrays	Operations and Algebraic Thinking	Understand properties of multiplication and the relationship between multiplication and division.	3.OA.5
3	Building Multiplication Facts	Operations and Algebraic Thinking	Multiply and divide within 100	3.OA.7
4	Beyond the Basic Facts	Operations and Algebraic Thinking	Understand properties of multiplication and division and the relationship between multiplication and division.	3.OA.5
5	Constructing Division	Operations and Algebraic Thinking	Represent and solve problems involving multiplication and division	3.OA.2
6	Solving Word Problems	Operations and Algebraic Thinking	Represent and solve problems involving multiplication and division	3.OA.3
Unit 5	**Geometry and Measurement**			
1	Measuring Weight	Measurement and Data	Solve problems involving measurement and estimation of intervals of time, liquid volume, and masses of objects.	3.MD.2
2	Finding Area	Measurement and Data	Geometric measurement: understand concepts of area and relate area to multiplication and to addition.	3.MD.6
3	Solving Problems Involving Area	Measurement and Data	Geometric measurement: understand concepts of area and relate area to multiplication and to addition.	3.MD.7
4	Finding Perimeter	Measurement and Data	Geometric measurement: recognize perimeter as an attribute of place figures and distinguish between linear and area measures.	3.MD.8
5	Understanding Shapes	Geometry	Reason with shapes and their attributes.	3.G.1
6	Creating and Interpreting Graphs	Measurement and Data	Represent and interpret data.	3.MD.3

Key Standards by Lesson Week

Key Common Core State Standards

Level F

Week	Lesson	Domain	Cluster	Key Standard
Unit 1	**Number Sense**			
1	Place Value	Number and Operations in Base Ten	Generalize place value understanding for multi-digit whole numbers.	4.NBT.2
2	Rounding	Number and Operations in Base Ten	Generalize place value understanding for multi-digit whole numbers.	4.NBT.3
3	Prime Numbers and Factors	Operations and Algebraic Thinking	Gain familiarity with factors and multiples.	4.OA.4
4	Equivalent Fractions	Number and Operations- Fractions	Extend understanding of fraction equivalence and ordering.	4.NF.1
5	Working with Fractions	Number and Operations- Fractions	Extend understanding of fraction equivalence and ordering.	4.NF.2
6	Number Patterns	Operations and Algebraic Thinking	Generate and analyze patterns.	4.OA.5
Unit 2	**Addition and Subtraction**			
1	Adding and Subtracting One- and Two-Digit Numbers	Number and Operations in Base Ten	Use place value understanding and properties of operations to perform multi-digit arithmetic.	4.NBT.4
2	Addition and Subtraction with Regrouping	Operations and Algebraic Thinking	Use the four operations with whole numbers to solve problems.	4.OA.3
3	Adding and Subtracting Fractions	Number and Operations- Fractions	Build fractions from unit fractions by applying and extending previous understandings of operations on whole numbers.	4.NF.3d
4	Adding Mixed Numbers	Number and Operations- Fractions	Build fractions from unit fractions by applying and extending previous understandings of operations on whole numbers.	4.NF.3c
5	Subtracting Mixed Numbers without Regrouping	Number and Operations- Fractions	Build fractions from unit fractions by applying and extending previous understandings of operations on whole numbers.	4.NF.3d
6	Subtracting Mixed Numbers with Regrouping	Number and Operations- Fractions	Build fractions from unit fractions by applying and extending previous understandings of operations on whole numbers.	4.NF.3c
Unit 3	**Multiplication**			
1	Products of Whole Numbers	Number and Operations in Base Ten	Use place value understanding and properties of operations to perform multi-digit arithmetic.	4.NBT.5
2	Multiplication Strategies	Number and Operations in Base Ten	Use place value understanding and properties of operations to perform multi-digit arithmetic.	4.NBT.5
3	Solving Multiplication Problems	Number and Operations in Base Ten	Use place value understanding and properties of operations to perform multi-digit arithmetic.	4.NBT.5
4	Multiplying Multi-Digit Numbers	Number and Operations in Base Ten	Use place value understanding and properties of operations to perform multi-digit arithmetic.	4.NBT.5
5	Multistep Word Problems with Multiplication	Operations and Algebraic Thinking	Use the four operations with whole numbers to solve problems.	4.OA.3
6	Multiplying Fractions by Whole Numbers	Number and Operations- Fractions	Build fractions from unit fractions by applying and extending previous understandings of operations on whole numbers.	4.NF.4, 4.NF.4a, 4.NF.4b
Unit 4	**Division**			
1	Relating Multiplication to Division	Number and Operations in Base Ten	Use place value understanding and properties of operations to perform multi-digit arithmetic.	4.NBT.6
2	Dividing 2-digit Numbers by 1-digit Numbers	Number and Operations in Base Ten	Use place value understanding and properties of operations to perform multi-digit arithmetic.	4.NBT.6
3	Dividing 3-digit Numbers by 1-digit Numbers	Number and Operations in Base Ten	Use place value understanding and properties of operations to perform multi-digit arithmetic.	4.NBT.6
4	Dividing 2- and 3-digit Numbers by 1-digit Numbers	Number and Operations in Base Ten	Use place value understanding and properties of operations to perform multi-digit arithmetic.	4.NBT.6
5	Dividing 4-digit by 1-digit Numbers (with Remainders)	Number and Operations in Base Ten	Use place value understanding and properties of operations to perform multi-digit arithmetic.	4.NBT.6
6	Multistep Word Problems with Division	Operations and Algebraic Thinking	Use the four operations with whole numbers to solve problems.	4.OA.3
Unit 5	**Geometry and Measurement**			
1	Area and Perimeter of a Rectangle	Measurement and Data	Solve problems involving measurement and conversion of measurements from a larger unit to a smaller unit.	4.MD.3
2	Measuring Angles Using a Protractor	Measurement and Data	Geometric measurement: understand concepts of angle and measure angles.	4.MD.6
3	Points, Lines, Line Segments, Angles, and Rays	Geometry	Draw and identify lines and angles, and classify shapes by properties of their lines and angles.	4.G.1
4	Classifying Polygons	Geometry	Draw and identify lines and angles, and classify shapes by properties of their lines and angles.	4.G.2
5	Converting Units of Time and Length within the Same System	Measurement and Data	Solve problems involving measurement and conversion of measurements from a larger unit to a smaller unit.	4.MD.1
6	Converting Units within the Same System	Measurement and Data	Solve problems involving measurement and conversion of measurements from a larger unit to a smaller unit.	4.MD.1

Key Standards by Lesson Week

Key Common Core State Standards

Level G

Week	Lesson	Domain	Cluster	Key Standard
Unit 1 Number Sense				
1	Working with Decimals	Number and Operations in Base Ten	Understand the place value system.	5.NBT.3
2	Read and Write Decimals Using Expanded Form	Number and Operations in Base Ten	Understand the place value system.	5.NBT.3a
3	Place Value	Number and Operations in Base Ten	Understand the place value system.	5.NBT.3b
4	Multiplying and Dividing by Powers of Ten	Number and Operations in Base Ten	Understand the place value system.	5.NBT.3b
5	Converting Words to Mathematical Symbols	Operations and Algebraic Thinking	Write and interpret numerical expressions.	5.OA.2
6	Order of Operations	Operations and Algebraic Thinking	Write and interpret numerical expressions.	5.OA.1
Unit 2 Multiplication and Division				
1	Prime Numbers and Factorization	Number and Operations in Base Ten	Perform operations with multi-digit whole numbers and with decimals to hundredths.	5.NBT.5
2	Multiplication Properties	Number and Operations in Base Ten	Perform operations with multi-digit whole numbers and with decimals to hundredths.	5.NBT.5
3	Multiplication Strategies	Number and Operations in Base Ten	Perform operations with multi-digit whole numbers and with decimals to hundredths.	5.NBT.5
4	Division Basics	Number and Operations in Base Ten	Perform operations with multi-digit whole numbers and with decimals to hundredths.	5.NBT.6
5	Division Using Whole Numbers	Number and Operations in Base Ten	Perform operations with multi-digit whole numbers and with decimals to hundredths.	5.NBT.6
6	Tables and Graphs	Number and Operations in Base Ten	Perform operations with multi-digit whole numbers and with decimals to hundredths.	5.NBT.5 5.NBT.6
Unit 3 Operations with Decimals				
1	Adding Decimals	Number and Operations in Base Ten	Perform operations with multi-digit whole numbers and with decimals to hundredths.	5.NBT.7
2	Subtracting Decimals	Number and Operations in Base Ten	Perform operations with multi-digit whole numbers and with decimals to hundredths.	5.NBT.7
3	Multiplication with Models	Number and Operations in Base Ten	Perform operations with multi-digit whole numbers and with decimals to hundredths.	5.NBT.7
4	Multiplying Decimals	Number and Operations in Base Ten	Perform operations with multi-digit whole numbers and with decimals to hundredths.	5.NBT.7
5	Division with Models	Number and Operations in Base Ten	Perform operations with multi-digit whole numbers and with decimals to hundredths.	5.NBT.7
6	Solving Problems with Division	Number and Operations in Base Ten	Perform operations with multi-digit whole numbers and with decimals to hundredths.	5.NBT.7
Unit 4 Operations with Fractions				
1	Understanding Fractions	Number and Operations- Fractions	Use equivalent fractions as a strategy to add and subtract fractions.	5.NF.1
2	Adding/Subtracting Fractions with Unlike Denominators	Number and Operations- Fractions	Use equivalent fractions as a strategy to add and subtract fractions.	5.NF.1
3	Adding and Subtracting Mixed Numbers	Number and Operations- Fractions	Use equivalent fractions as a strategy to add and subtract fractions.	5.NF.1
4	Multiplying Whole Numbers by Fractions	Number and Operations- Fractions	Apply and extend previous understandings of multiplication and division to multiply and divide fractions.	5.NF.4
5	Multiplying Fractions by Fractions	Number and Operations- Fractions	Apply and extend previous understandings of multiplication and division to multiply and divide fractions.	5.NF.4
6	Division with Whole Numbers and Fractions	Number and Operations- Fractions	Apply and extend previous understandings of multiplication and division to multiply and divide fractions.	5.NF.7
Unit 5 Geometry and Measurement				
1	Converting Measurements within the Same System	Measurement and Data	Convert like measurement units within a given measurement system.	5.MD.1
2	Volume of Rectangular Prisms	Measurement and Data	Geometric measurement: understand concepts of volume and relate volume to multiplication and to addition.	5.MD.4
3	Coordinate Grids, Quadrant I	Geometry	Graph points on the coordinate place to solve real-world and mathematical problems.	5.G.1
4	Coordinate Grids	Geometry	Graph points on the coordinate place to solve real-world and mathematical problems.	5.G.1
5	Graphing Ordered Pairs	Operations and Algebraic Thinking	Analyze patterns and relationships.	5.OA.3
6	Classifying Polygons	Geometry	Classify two-dimensional figures into categories based on their properties.	5.G.4

Key Standards by Lesson Week

Key Common Core State Standards

Level H

Week	Lesson	Domain	Cluster	Key Standard
Unit 1 Number Sense				
1	Comparing Fractions	Ratios and Proportional Relationships	Understand ratio concepts and use ratio reasoning to solve problems.	6.RP.1
2	Comparing Decimals	The Number System	Compute fluently with multi-digit numbers and find common factors and multiples.	6.NS.3
3	Ratios and Rates	Ratios and Proportional Relationships	Understand ratio concepts and use ratio reasoning to solve problems.	6.RP.2
4	Using Ratios and Rates	Ratios and Proportional Relationships	Understand ratio concepts and use ratio reasoning to solve problems.	6.RP.3b
5	Integers	The Number System	Apply and extend previous understandings of numbers to the system of rational numbers.	6.NS.6
6	Using Integers	The Number System	Apply and extend previous understandings of numbers to the system of rational numbers.	6.NS.6
Unit 2 Operations Sense				
1	Operations with Whole Numbers	The Number System	Compute fluently with multi-digit numbers and find common factors and multiples.	6.NS.2
2	Decimals	The Number System	Compute fluently with multi-digit numbers and find common factors and multiples.	6.NS.3
3	Fractions	The Number System	Apply and extend previous understandings of multiplication and division to divide fractions by fractions.	6.NS.1
4	Order of Operations	Expressions and Equations	Apply and extend previous understandings of multiplication and division to divide fractions by fractions.	6.EE.3
5	Creating Function Tables	Expressions and Equations	Reason about and solve one-variable equations and inequalities.	6.EE.7
6	Functions and Graphs	Expressions and Equations	Reason about and solve one-variable equations and inequalities.	6.EE.7
Unit 3 Algebra				
1	Exponents	Expressions and Equations	Apply and extend previous understandings of multiplication and division to divide fractions by fractions.	6.EE.1
2	Expressions	Expressions and Equations	Apply and extend previous understandings of multiplication and division to divide fractions by fractions.	6.EE.2
3	Equivalent Expressions	Expressions and Equations	Apply and extend previous understandings of multiplication and division to divide fractions by fractions.	6.EE.3
4	Creating Equations	Expressions and Equations	Reason about and solve one-variable equations and inequalities.	6.EE.7
5	Solving Linear Equations	Expressions and Equations	Reason about and solve one-variable equations and inequalities.	6.EE.7
6	Inequalities	Expressions and Equations	Reason about and solve one-variable equations and inequalities.	6.EE.8
Unit 4 Statistical Analysis				
1	Showing Data in Line Plots	Statistics and Probability	Summarize and describe distributions.	6.SP.4
2	Data Displays 1	Statistics and Probability	Summarize and describe distributions.	6.SP.4
3	Measures of Center	Statistics and Probability	Summarize and describe distributions.	6.SP.4
4	Data Displays 2	Statistics and Probability	Summarize and describe distributions.	6.SP.4
5	Describing and Comparing Data	Statistics and Probability	Summarize and describe distributions.	6.SP.5 6.SP.5a 6.SP.5b 6.SP.5c 6.SP.5d
6	Statistical Measurements	Statistics and Probability	Summarize and describe distributions.	6.SP.5 6.SP.5a 6.SP.5b 6.SP.5c 6.SP.5d
Unit 5 Geometry and Measurement				
1	Areas of Triangles	Geometry	Solve real-world and mathematical problems involving area, surface area, and volume.	6.G.1
2	Areas of Quadrilaterals and Polygons	Geometry	Solve real-world and mathematical problems involving area, surface area, and volume.	6.G.1
3	Volume of Right Rectangle Prisms	Geometry	Solve real-world and mathematical problems involving area, surface area, and volume.	6.G.2
4	Volume of Prisms	Geometry	Solve real-world and mathematical problems involving area, surface area, and volume.	6.G.2
5	Nets	Geometry	Solve real-world and mathematical problems involving area, surface area, and volume.	6.G.4
6	Graphing Polygons	Geometry	Solve real-world and mathematical problems involving area, surface area, and volume.	6.G.3

Key Standards by Lesson Week

Key Common Core State Standards

Level I

Week	Lesson	Domain	Cluster	Key Standard
Unit 1 Number Sense				
1	Rational Numbers	The Number System	Apply and extend previous understandings of operations with fractions to add, subtract, multiply, and divide rational numbers.	7.NS.3
2	Fractions, Decimals, and Percents	The Number System	Apply and extend previous understandings of operations with fractions to add, subtract, multiply, and divide rational numbers.	7.NS.2d
3	Ratios and Rates	Ratios and Proportional Relationships	Analyze proportional relationships and use them to solve real-world and mathematical problems.	7.RP.1
4	Unit Rates	Ratios and Proportional Relationships	Analyze proportional relationships and use them to solve real-world and mathematical problems.	7.RP.1
5	Percents, Discounts, and Commissions	Ratios and Proportional Relationships	Analyze proportional relationships and use them to solve real-world and mathematical problems.	7.RP.3
6	Markups, Markdowns, and Simple Interest	Ratios and Proportional Relationships	Analyze proportional relationships and use them to solve real-world and mathematical problems.	7.RP.3
Unit 2 Operations Sense				
1	Adding and Subtracting Integers	The Number System	Apply and extend previous understandings of operations with fractions to add, subtract, multiply, and divide rational numbers.	7.NS.1
2	Properties of Addition and Subtraction	The Number System	Apply and extend previous understandings of operations with fractions to add, subtract, multiply, and divide rational numbers.	7.NS.1d
3	Applications Using Addition and Subtraction	The Number System	Apply and extend previous understandings of operations with fractions to add, subtract, multiply, and divide rational numbers.	7.NS.3
4	Multiplying and Dividing Integers	The Number System	Apply and extend previous understandings of operations with fractions to add, subtract, multiply, and divide rational numbers.	7.NS.2
5	Properties of Operations for Rational Numbers	The Number System	Apply and extend previous understandings of operations with fractions to add, subtract, multiply, and divide rational numbers.	7.NS.2c
6	Applications with Rational Numbers	The Number System	Apply and extend previous understandings of operations with fractions to add, subtract, multiply, and divide rational numbers.	7.NS.3
Unit 3 Algebra				
1	Algebraic Expressions	Expressions and Equations	Solve real-life and mathematical problems using numerical and algebraic expressions and equations.	7.EE.4
2	Solving Simple Equations	Expressions and Equations	Solve real-life and mathematical problems using numerical and algebraic expressions and equations.	7.EE.4
3	Solving Two-Step Equations	Expressions and Equations	Solve real-life and mathematical problems using numerical and algebraic expressions and equations.	7.EE.4
4	Equations for Real-World Problems	Expressions and Equations	Solve real-life and mathematical problems using numerical and algebraic expressions and equations.	7.EE.4a
5	Solving Inequalities	Expressions and Equations	Solve real-life and mathematical problems using numerical and algebraic expressions and equations.	7.EE.4b
6	Inequalities for Real-World Problems	Expressions and Equations	Solve real-life and mathematical problems using numerical and algebraic expressions and equations.	7.EE.4b
Unit 4 Statistical Analysis				
1	Introduction to Probability	Statistics and Probability	Investigate chance processes and develop, use, and evaluate probability models.	7.SP.7
2	Data and Probability	Statistics and Probability	Investigate chance processes and develop, use, and evaluate probability models.	7.SP.7
3	Theoretical and Experimental Probability	Statistics and Probability	Investigate chance processes and develop, use, and evaluate probability models.	7.SP.8
4	Applications Using Probability	Statistics and Probability	Investigate chance processes and develop, use, and evaluate probability models.	7.SP.8
5	Box Plots	Statistics and Probability	Draw informal comparative inferences about two populations.	7.SP.4
6	Interpreting Box Plots	Statistics and Probability	Draw informal comparative inferences about two populations.	7.SP.4
Unit 5 Geometry and Measurement				
1	Angles and Shapes	Geometry	Solve real-life and mathematical problems involving angle measure, area, surface area, and volume.	7.G.5
2	Parallel and Perpendicular Lines	Geometry	Solve real-life and mathematical problems involving angle measure, area, surface area, and volume.	7.G.5
3	Pi, Circumference, and Diameter	Geometry	Solve real-life and mathematical problems involving angle measure, area, surface area, and volume.	7.G.4
4	Area	Geometry	Solve real-life and mathematical problems involving angle measure, area, surface area, and volume.	7.G.6
5	Surface Area	Geometry	Solve real-life and mathematical problems involving angle measure, area, surface area, and volume.	7.G.6
6	Space Figures	Geometry	Solve real-life and mathematical problems involving angle measure, area, surface area, and volume.	7.G.6

Key Standards by Lesson Week

Key Common Core State Standards

Level J

Week	Lesson	Domain	Cluster	Key Standard
Unit 1 Number Sense				
1	Rational Numbers	The Number System	Know that there are numbers that are not rational, and approximate them by rational numbers.	8.NS.1
2	Exponents	Expressions and Equations	Work with radicals and integer exponents.	8.EE.1
3	Operations with Exponents	Expressions and Equations	Work with radicals and integer exponents.	8.EE.1
4	Scientific Notation, Greater Than 1	Expressions and Equations	Work with radicals and integer exponents.	8.EE.4
5	Scientific Notation, Between 1 and 0	Expressions and Equations	Work with radicals and integer exponents.	8.EE.4
6	Real Numbers	The Number System	Know that there are numbers that are not rational, and approximate them by rational numbers.	8.NS.1
Unit 2 Operations Sense				
1	Squares	Expressions and Equations	Work with radicals and integer exponents.	8.EE.2
2	Cubes	Expressions and Equations	Work with radicals and integer exponents.	8.EE.2
3	Adding and Subtracting Integers	Expressions and Equations	Analyze and solve linear equations and pairs of simultaneous linear equations.	8.EE.7
4	Multiplying and Dividing Integers	Expressions and Equations	Analyze and solve linear equations and pairs of simultaneous linear equations.	8.EE.7
5	Solving Equations	Expressions and Equations	Analyze and solve linear equations and pairs of simultaneous linear equations.	8.EE.7
6	Word Problems with Equations	Expressions and Equations	Analyze and solve linear equations and pairs of simultaneous linear equations.	8.EE.7
Unit 3 Algebra				
1	Solving Multistep Equations	Expressions and Equations	Analyze and solve linear equations and pairs of simultaneous linear equations.	8.EE.7b
2	Investigating Slope	Functions	Define, evaluate, and compare functions.	8.F.3
3	Functions and Graphs	Functions	Define, evaluate, and compare functions.	8.F.1
4	Linear Functions and Graphs	Functions	Define, evaluate, and compare functions.	8.F.3
5	Comparing Lines	Functions	Use functions to model relationships between quantities.	8.F.5
6	Nonlinear Functions and Graphs	Functions	Define, evaluate, and compare functions.	8.F.3
Unit 4 Statistical Analysis				
1	Box Plots	Statistics and Probability	Investigate patterns of association in bivariate data.	8.SP.1
2	Scatter Plots	Statistics and Probability	Investigate patterns of association in bivariate data.	8.SP.1
3	Scatter Plot Relationships	Statistics and Probability	Investigate patterns of association in bivariate data.	8.SP.1
4	Line of Best Fit	Statistics and Probability	Investigate patterns of association in bivariate data.	8.SP.2
5	Using One-Way Tables	Statistics and Probability	Investigate patterns of association in bivariate data.	8.SP.4
6	Using Two-Way Tables	Statistics and Probability	Investigate patterns of association in bivariate data.	8.SP.4
Unit 5 Geometry and Measurement				
1	Pythagorean Theorem	Geometry	Understand and apply the Pythagorean Theorem.	8.G.7
2	Symmetry	Geometry	Understand congruence and similarity using physical models, transparencies, or geometry software.	8.G.3
3	Translations	Geometry	Understand congruence and similarity using physical models, transparencies, or geometry software.	8.G.3
4	Rotations and Dilations	Geometry	Understand congruence and similarity using physical models, transparencies, or geometry software.	8.G.3
5	Circles and Cylinders	Geometry	Solve real-world and mathematical problems involving volume of cylinders, cones, and spheres.	8.G.9
6	Cones, Cylinders, and Spheres	Geometry	Solve real-world and mathematical problems involving volume of cylinders, cones, and spheres.	8.G.9

Scope and Sequence

	A	B	C	D	E	F	G	H	I	J
Algebra										
Expressions										
Writing expressions					•		•		•	
Simplifying and evaluating expressions							•	•	•	
Equations										
Patterns	•	•	•	•	•	•	•			
Writing equations				•	•		•	•	•	
Solving one-step equations	•	•	•	•	•		•	•	•	•
Solving two-step equations									•	•
Graphing linear equations							•	•	•	•
Inequalities										
Writing inequalities								•	•	
Solving inequalities									•	
Graphing inequalities									•	
Functions										
Using functions							•	•	•	•
Coordinate Graphing										
Graphing on the coordinate grid							•	•	•	•
Slope										•
Decimal Numbers										
Comparing and Ordering										
Comparing decimal numbers							•	•		
Ordering decimal numbers							•			
Relationships										
Decimal numbers and fractions							•	•	•	
Decimal numbers and percents									•	
Addition										
Whole number and decimal number addends							•	•		
Decimal number addends							•	•		
Subtraction										
Whole number/decimal number subtrahends							•	•		
Decimal number subtrahends							•	•		
Multiplication										
Whole number/decimal number factors							•	•		
Decimal number factors							•	•		
Division										
Whole number divisor							•	•		
Decimal number divisor							•	•		

Scope and Sequence

	A	B	C	D	E	F	G	H	I	J
Decimal Numbers										
Money										
Counting currency		•	•	•				•		
Operations with currency	•	•	•					•		
Fractions										
Place Value										
Fractions of a whole					•	•		•		
Fractions of a group					•	•		•		
Comparing and Ordering										
Comparing fractions					•	•		•		
Ordering fractions					•	•		•		
Written Fractions										
Equivalent fractions					•	•	•	•		
Simplest form						•	•	•		
Mixed numbers							•	•		
Improper fractions							•			
Relationships										
Fractions and decimal numbers							•	•	•	
Fractions and percents									•	
Addition										
Like denominators						•	•	•		
Unlike denominators							•	•		
Mixed numbers							•	•		
Subtraction										
Like denominators						•	•	•		
Unlike denominators							•	•		
Mixed numbers							•	•		
Multiplication										
Whole number and fraction factors						•	•	•		
Fraction factors							•	•		
Division										
Whole number divisors							•	•		
Fraction divisors							•	•		
Geometry										
Plane Figures										
Classifying	•	•	•	•	•	•	•	•		
Symmetry										•
Transformations										
Angles				•	•	•		•	•	

Scope and Sequence

	A	B	C	D	E	F	G	H	I	J
Geometry										
Triangles				•		•	•	•	•	
Quadrilaterals				•	•	•	•	•		
Other polygons				•	•	•	•	•		
Parallel and perpendicular lines				•	•	•	•	•		
Perimeter					•	•				
Circumference									•	•
Area					•	•		•	•	•
Pythagorean Theorem										•
Solid Figures										
Classifying	•	•	•						•	•
Rotational symmetry										•
Surface area								•	•	
Volume							•	•	•	•
Integers										
Place Value										
Meaning of a negative integer								•	•	•
Properties of integers								•		•
Comparing and Ordering										
Comparing integers							•		•	
Ordering integers							•			
Addition										
Adding integers									•	•
Subtraction										
Subtracting integers									•	•
Multiplication										
Multiplying integers									•	•
Division										
Dividing integers									•	•
Measurement										
Length										
Customary units				•		•	•		•	
Metric units				•		•	•			
Mass/Weight										
Customary units					•	•	•			
Metric units						•	•	•		
Capacity/Volume										
Customary units						•	•			
Metric units						•	•			

Scope and Sequence

	A	B	C	D	E	F	G	H	I	J
Measurement										
Telling Time										
To the hour		•	•	•						
To the half hour			•	•						
To the quarter hour				•						
To the nearest five minutes				•						
Percents										
Place Value										
Meaning of percent									•	•
Percent increase/decrease									•	•
Sales tax, commission, tips									•	•
Other								•	•	
Relationships										
Percents and decimal numbers									•	
Percents and fractions									•	
Probability										
Probability										
Meaning of									•	
Likely/unlikely									•	
Theoretical probability									•	
Experimental probability									•	
Ratios and Proportions										
Place Value										
Meaning of ratio								•	•	
Meaning of proportion								•	•	
Rates							•	•	•	
Statistics and Graphing										
Graphs										
Picture graphs	•	•	•	•	•					
Bar graphs	•	•	•	•	•			•		
Line graphs			•					•		•
Histograms								•		
Box and whisker plots								•	•	•
Scatter Plots										
Analyzing graphs	•	•	•	•	•			•		•
Statistics										
Finding the mean								•	•	•
Finding the median								•	•	•

Scope and Sequence

	A	B	C	D	E	F	G	H	I	J
Statistics and Graphing										
Finding the mode								•	•	•
Finding the range								•	•	•
Whole Numbers										
Numbers and Numeration										
Reading and writing numbers	•	•	•	•						
Counting	•	•	•	•						
Skip counting			•	•	•	•				
Even and odd numbers				•						
Prime and composite numbers					•					
Factors and prime factorization							•	•		•
Common factors and GCF							•	•		
Common multiples and LCM							•			
Place Value										
Place Value			•	•	•	•				
Decomposing numbers			•	•	•					
Expanded form					•	•				
Absolute value									•	•
Comparing and Ordering										
Comparing whole numbers	•	•	•	•	•	•				
Ordering whole numbers	•	•	•	•	•	•				
Rounding										
To nearest ten				•	•	•				
To nearest hundred					•	•				
To nearest thousand						•				
Beyond thousands				•		•				
Order of Operations										
Order of Operations							•	•		
Exponents										
Meaning of										•
Squares and square roots										•
Scientific notation										•
Addition										
Basic Facts	•	•	•	•	•					
Properties		•	•	•	•	•		•	•	
Three or more addends			•	•	•					
Near doubles/Doubles plus one				•	•					
Two-digit addends				•	•	•	•			

Scope and Sequence

	A	B	C	D	E	F	G	H	I	J
Whole Numbers										
Three-digit or greater addends				•	•	•		•		
Estimating sums				•	•					
Subtraction										
Relationship to addition				•	•					
Fact families	•	•	•	•						
Two-digit subtrahends				•	•	•		•		
Three-digit or greater subtrahends				•	•	•		•		
Estimating differences	•	•	•	•						
Multiplication										
Basic Facts					•	•		•		
Properties					•	•	•	•	•	
One-digit multipliers					•	•	•	•		
Two-digit or greater multipliers					•	•	•			
Estimating products					•	•	•			
Division										
Relationship to multiplication					•	•	•			
Fact families					•					
Remainders					•	•	•			
One-digit divisors					•	•	•	•		
Two-digit or greater divisors					•	•	•			
Estimating quotients				•	•	•	•			

Glossary

A

absolute value the distance of a number from 0 (e.g., the absolute value of 7, written as |7|, is 7; the absolute value of −4, written as |−4|, is 4)

acute angle an angle with a measure greater than 0 degrees and less than 90 degrees

acute triangle a triangle with three acute angles

add to perform a mathematical operation based on "putting things together"

addend numbers being added (e.g., in the number sentence 15 + 63 = 78, the numbers *15* and *63* are addends)

addition a mathematical operation based on "putting things together"

additive inverses two numbers whose sum is 0 (e.g., 9 + (−9) = 0; the additive inverse of 9 is −9, and the additive inverse of −9 is 9)

adjacent angles two angles with a common side that do not otherwise overlap

algebraic expression a combination of variables, numbers, and at least one operation

algebraic sentence an equation or inequality that compares two algebraic expressions

algorithm a step-by-step procedure for carrying out a computation or for solving a problem

angle two rays with a common endpoint; the common endpoint is called the *vertex* of the angle

area the measure of the interior, or inside, of a figure

array a group of objects arranged in an orderly way in rows and columns

Associative Property the property that states that the sum or product of three or more quantities will be the same regardless of the way in which they are grouped

average See *mean*.

axis (plural **axes**) a number line used in a coordinate grid

B

bar graph a graph in which the lengths of horizontal or vertical bars represent the magnitude of the data represented

base a two-dimensional figure upon which a three-dimensional figure sits

bilateral symmetry having the same size and shape across a dividing line (the line of symmetry)

bisect to divide a segment, angle, or figure into two parts of equal measure

box and whisker plot a diagram that summarizes data using the median, the upper and lower quartiles, and the extreme values

broken line graph a graph in which points are connected by line segments to represent data

C

capacity the amount a container can hold

centi- a prefix for units in the metric system meaning one hundredth

centimeter (cm) a metric unit of length defined as $\frac{1}{100}$ of a meter; equal to 10 millimeters or $\frac{1}{10}$ of a decimeter

circle the set of all points in a plane that are a given distance (the radius) from a given point (the center of the circle)

circumference the distance around a circle

closed figure a figure that divides the plane into two regions, inside and outside the figure; a closed space figure divides space into two regions in the same way

coefficient a number in front of a variable (e.g., in the expression 3x + 2, 3 is a coefficient)

common denominator any nonzero number that is a multiple of the denominators of two or more fractions

common factor a number that is a factor of two or more given numbers

Commutative Property the property that states that the sum or product of two or more quantities will be the same regardless of the order in which they appear

compare to decide whether two numbers or expressions are equal or not equal, and, if not equal, to decide which is larger or smaller than the other

complementary angles two angles whose measures total 90 degrees

Glossary

composite number a whole number that has more than two whole number factors; See also *prime number*.

cone a space figure having a circular base, a curved surface, and one vertex

congruent having identical sizes and shapes

constant a value that remains unchanged

coordinate grid a grid formed by two perpendicular number lines that intersect; the point of intersection is called the *origin*

coordinate one of two numbers used to locate a point on a coordinate grid; See also *ordered pair*.

core pattern the pattern from which the rest of the pattern grows; part of the pattern that repeats

corresponding angles two angles in the same relative position in two figures, or in similar locations in relation to a transversal intersecting two lines

corresponding sides two sides in the same relative position in two figures

cube a space figure whose six faces are congruent squares that meet at right angles

cubic centimeter (cm³) a metric unit of volume; the volume of a cube 1 centimeter on an edge; one cubic centimeter is equal to 1 milliliter.

cubic unit a unit that describes volume and capacity

customary system of measurement the measuring system used most often in the United States; units for linear measure (length, distance) include inch, foot, yard, and mile; units for weight include ounce and pound; units for capacity (amount of liquid or other substance a container can hold) include fluid ounce, cup, pint, quart, and gallon.

cylinder a space figure bound by two congruent, parallel circles and a curved surface that connects around the circumferences of both circles

D

data facts, figures, or other items of information

decimal equivalent a decimal that names the same number as a fraction (e.g., the decimal equivalent of $\frac{3}{4}$ is 0.75)

decimal number a number written in standard notation, usually containing a decimal point, such as 3.78

decimal point a dot used in separating the ones digit from the tenths digit

degree (°) a unit of measure for angles based on dividing a circle into 360 equal parts; a unit of measure for temperature

degree Celsius (°C) the metric system's unit for measuring temperature; water freezes at 0°C and boils at 100°C

degree Fahrenheit (°F) the customary system's unit for measuring temperature; water freezes at 32°F and boils at 212°F

denominator the denominator of a fraction indicates the number of equal parts into which the whole or a set is divided; See also *numerator*.

diameter a line segment containing the center, whose endpoints are points of a circle or sphere

difference the answer to a subtraction problem

digit any of the Arabic numerals from 0 to 9

Distributive Property the property that states that the sum of two numbers multiplied by a third number is equal to the sum of the product of each number and the third number [e.g., $2 \times (3 + 5) = (2 \times 3) + (2 \times 5)$]

divide to separate into parts or pieces; to show how many times one number contains the other number

dividend See *division*.

division a mathematical operation based on "equal sharing" or "separating into equal parts." The *dividend* is the total before sharing. The *divisor* is the number of equal parts or the number in each equal part. The *quotient* is the result of division. The number left over when a set of objects is shared equally or separated into equal groups is called the *remainder*.

divisor See *division*.

E

edge the line segment where two faces of a space figure meet

eight the Arabic numeral 8

endpoint the point at either end of a line segment; the point at the end of a ray

Glossary

equal identical in value or notation

equation a mathematical sentence that states the equality of two expressions (e.g., $3 + 7 = 10$, $y = x + 7$, and $4 + 7 = 8 + 3$)

equilateral polygon a polygon in which all sides are the same length

equivalent equal in value, but in a different form (e.g., $\frac{1}{2}$, $\frac{2}{4}$, 0.5, and 50% are equivalent forms of the same number)

equivalent fractions fractions that have different numerators and denominators but name the same number (e.g., $\frac{2}{3}$ and $\frac{6}{9}$ are equivalent fractions)

estimate to calculate an answer that is close to but not exactly the correct answer

even number a whole number that can be divided by 2 with no remainder; See also *odd number*.

event a happening or an occurrence (e.g., the tossing of a coin)

expanded form a number written as the sum of the values of the digits (e.g., 462.38 in expanded form is $400 + 60 + 2 + 0.3 + 0.08$)

exponent a numeral or symbol placed at the upper right side of another numeral or symbol to indicate the number of times it is to be used as a factor (e.g., in 3^2, 2 is the exponent and indicates multiplying 3 times itself two times; $3^2 = 3 \times 3$)

expression a group of mathematical symbols (numbers, operation signs, variables, grouping symbols) that represents a number (or can represent a number if values are assigned to any variables it contains)

F

face a flat surface of a space figure

fact family a group of addition or multiplication facts grouped together with the related subtraction or division facts (e.g., $4 + 8 = 12$, $8 + 4 = 12$, $12 - 4 = 8$, and $12 - 8 = 4$ form an addition fact family; the facts $4 \times 3 = 12$, $3 \times 4 = 12$, $12 \div 3 = 4$, and $12 \div 4 = 3$ form a multiplication fact family)

factor *v.* to represent a quantity as a product of factors (e.g., 20 factors to 4×5 and 2×10, or $2 \times 2 \times 5$)

factors *n.* the numbers being multiplied

first preceding all others in time or order

five the Arabic numeral 5

four the Arabic numeral 4

fraction a number in the form $\frac{a}{b}$, where a and b are integers and b is not 0

fraction in simplest form a fraction in which the greatest common factor of the numerator and denominator is 1

frequency the number of times an event or a value occurs in a set of data

function a relationship that pairs every element of one set with an element of a second set

function machine an imaginary machine that processes numbers according to a certain rule; a number (input) is put into the machine and is transformed into a second number (output) by application of the rule

G

graph a diagram showing the relationship between two or more sets of data

greater than larger in number or measure

greatest common divisor the greatest whole number by which each number in a set is divisible

greatest common factor the largest factor that two or more numbers have in common

H

height (of a parallelogram) the length of the line segment between the base of the parallelogram and the opposite side (or an extension of the opposite side), running perpendicular to the base

height (of a rectangle) the length of the side perpendicular to the side considered the base of the rectangle

height (of a triangle) the length of the line segment perpendicular to the base of the triangle (or an extension of the base) from the opposite vertex

hexagon a polygon with six sides

histogram a bar graph in which the labels for the bars are numerical intervals

hour a measure of time equal to 60 minutes

Glossary

hundreds the place denoting groups of 10 tens; the third place to the left of the decimal point

hypotenuse in a right triangle, the side opposite the right angle

I

Identity Property of Addition the property that states that the sum of zero and any number (or quantity) equals itself (e.g., 5 + 0 = 5)

Identity Property of Multiplication the property that states that the product of one and any number (or quantity) equals itself (e.g., 2 × 1 = 2)

improper fraction a fraction that names a number greater than or equal to 1; a fraction whose numerator is equal to or greater than its denominator

inch (in.) the customary system's unit of length equal to $\frac{1}{12}$ of a foot

inequality a number sentence stating that two quantities are not equal; relation symbols for inequalities include < (less than), > (greater than), and ≠ (not equal to)

input/output table a table that represents the relationship between two quantities; the second quantity (the output) is generated by applying a rule to the first quantity (the input)

integer a set of integers consists of whole numbers and their opposites; the set of integers is {... −4, −3, −2, −1, 0, 1, 2, 3, 4...}

interior the set of all points in a plane "inside" a closed plane figure, such as a polygon or circle; the set of all points in space "inside" a closed space figure, such as a polyhedron or sphere

intersection the points or sets of points that are common to two or more geometric figures

inverse operation an operation that undoes the results of another operation (e.g., addition and subtraction are inverse operations)

irregular polygon a polygon with sides of different lengths or angles of different measures

isosceles trapezoid a trapezoid that has two non-parallel sides of the same length

isosceles triangle a triangle that has two sides of the same length

K

kilo- a prefix for units in the metric system meaning "one thousand"

L

least common denominator the least common multiple of the denominators of every fraction in a given set of fractions (e.g., 12 is the least common denominator of $\frac{2}{3}$, $\frac{1}{4}$, and $\frac{5}{6}$) See also *least common multiple*.

least common multiple the smallest number that is a multiple of two or more numbers (e.g., some common multiples of 6 and 8 are 24, 48, and 72; twenty-four is the least common multiple of 6 and 8)

leg of a right triangle a side of a right triangle that is not the hypotenuse

less than smaller in number or measure

line a straight path that extends infinitely in opposite directions; thought of as having length, but no thickness

line of symmetry a line that can be drawn through a figure dividing it into two congruent parts

line plot a data display that uses a number line and *X*s to display a set of numerical data

line segment a straight path joining two points, called *endpoints* of the line segment

line symmetry See *bilateral symmetry*.

liter (L) a metric unit of capacity, equal to the volume of a cube 10 centimeters on an edge; 1 L = 1,000 mL = 1,000 cm³ See also *milliliter (mL)*.

M

map scale a ratio that compares the distance between two locations shown on a map with the actual distance between them

math sentence two statements containing numbers, possibly variables, and operation signs linked by a sign of equality or inequality (e.g., 2 + 2 = 4; 3x = 12; 5 < 17)

mean a typical or central value that may be used to describe a set of numbers; it can be found by adding the numbers in the set and dividing the sum by the number of numbers; the mean is often referred to as the *average*

Glossary

measure to find the size or amount of something

median the middle value in a set of data when the data are listed in order from least to greatest (or greatest to least); if the number of values in the set is even (so that there is no "middle" value), the median is the mean of the two middle values

meter (m) the basic unit of length in the metric system, equal to 10 decimeters, 100 centimeters, and 1,000 millimeters

metric system of measurement a measurement system based on the base-ten numeration system and used in most countries in the world; units for linear measure (length, distance) include millimeter, centimeter, meter, kilometer; units for mass (weight) include gram and kilogram; units for capacity (amount of liquid or other substance a container can hold) include milliliter and liter.

midpoint a point halfway between two points

milli- a prefix for units in the metric system meaning "one thousandth"

milliliter (mL) a metric unit of capacity, equal to $\frac{1}{1,000}$ of a liter and 1 cubic centimeter

millimeter (mm) a metric unit of length equal to $\frac{1}{10}$ of a centimeter and $\frac{1}{1,000}$ of a meter

millions the place denoting groups of 1,000 thousands; the seventh place to the left of the decimal point

minuend See *subtraction*.

minute a measure of time equal to 60 seconds

mixed number a number greater than 1, written as a whole number and fraction less than 1 (e.g., $5\frac{1}{2}$ is equal to $5 + \frac{1}{2}$)

mode the value or values that occur most often in a set of data

multiple of a number (*n*) The product of a whole number and the number *n* (e.g., the numbers 0, 4, 8, 12, and 16 are all multiples of 4 because $4 \times 0 = 0, 4 \times 1 = 4, 4 \times 2 = 8, 4 \times 3 = 12$, and $4 \times 4 = 16$)

multiplication the process of adding a number or quantity to itself a certain number of times (e.g., 2×3 is 2 groups of 3 or adding 3 two times: $3 + 3$); the numbers or quantities being multiplied are called *factors;* the result of the multiplication is called the *product* (e.g., $2 \times 3 = 6$: 2 and 3 are the factors and 6 is the product)

multiplicative inverses two numbers whose product is 1 (e.g., the multiplicative inverse of $\frac{2}{5}$ is $\frac{5}{2}$, and the multiplicative inverse of 8 is $\frac{1}{8}$); multiplicative inverses are also called *reciprocals* of each other

multiply to find the total number of things in several equal groups, or to find a quantity that is a certain number of times as much or as many as another number

N

negative number a number less than 0

net a two-dimensional figure that can be folded to form a three-dimensional shape

nine the Arabic numeral 9

number sentence a sentence that is made up of numerals and a relation symbol (<, >, or =); most number sentences also contain at least one operation symbol, and they may also have grouping symbols, such as parentheses

numeral a symbol that represents a number (e.g., *7* or *VII* are symbols representing seven)

numerator the numerator of a fractions indicates how many parts are being referred to; See also *denominator*.

O

obtuse angle an angle with a measure greater than 90 degrees and less than 180 degrees

obtuse triangle a triangle with an obtuse angle

octagon an eight-sided polygon

odd number a whole number that is not divisible by 2, such as 1, 3, 5, and so on

one the Arabic numeral 1

ones the place denoting single units; the place that is one place to the left of the decimal point

opposite of a number a number that is the same distance from 0 on the number line as the given number, but on the opposite side of 0; if *a* is a negative number, the opposite of *a* will be a positive number (e.g., if $a = -5$, then $-a$ is 5) See also *additive inverses*.

ordered pair two numbers or objects for which order is important; two numbers in a specific order used to locate a point on a coordinate grid; they are usually written inside parentheses [e.g., (2, 3)]; See also *coordinate*.

Glossary

ordinal number a number used to express position or order in a series, such as first, third, tenth

origin the point where the *x*- and *y*-axes intersect on a coordinate grid; the coordinates of the origin are (0, 0)

outcome the result of an event (e.g., heads and tails are the two outcomes of the event of tossing a coin)

P

parallel lines lines that are the same distance apart and that extend in the same directions but never meet

parallel rays rays that are part of two different parallel lines

parallel segments line segments that are part of two parallel lines

parallelogram a quadrilateral that has two pairs of parallel sides; pairs of opposite sides and opposite angles of a parallelogram are *congruent*.

parentheses a pair of symbols, ( ), surrounding parts of an equation and used to show in which order operations should be done

pattern a model, plan, or rule that uses words or variables to describe a set of shapes or numbers that repeat in a predictable way

pentagon a polygon with five sides

percent a rational number that can be written as a fraction with a denominator of 100; the symbol % is used to represent percent; 1% means "$\frac{1}{100}$" or "0.01"

perimeter the distance around the boundary of a closed plane figure

perpendicular lines lines that intersect in right angles

pi the ratio of the circumference of a circle to its diameter; it is the same for every circle, approximately 3.14 or $\frac{22}{7}$; pi is also written as the Greek letter π

picture graph a graph that uses pictures or icons to represent data; a picture graph may also be called a *pictograph* or *pictogram*

place value the value of a digit determined by its position within a number

plane figure See *two-dimensional*.

polygon a closed plane figure consisting of line segments (sides) connected endpoint to endpoint

positive number a number greater than 0

power a product of factors that are all the same (e.g., 6 × 6 × 6 (or 216) is called 6 to the third power, or the third power of 6, because 6 is a factor three times); the expression 6 × 6 × 6 can also be written as 6^3

prime factorization a whole number expressed as a product of prime factors (e.g., the prime factorization of 18 is 2 × 3 × 3); a number has only one prime factorization, except for the order in which the factors are written

prime number a whole number greater than 1 that has exactly two whole number factors: 1 and itself; See also *composite number*.

prism a space figure with two congruent, parallel faces (bases) connected by parallelograms; prisms are classified according to the shape of the bases

probability a number between 0 and 1 that indicates the likelihood that something (an event) will happen; the closer a probability is to 1, the more likely it is that an event will happen

product See *multiplication*.

proportion an equation stating that two ratios are equal

pyramid a space figure in which one face (the base) is a polygon and the other faces are triangles with a common vertex (the apex); a pyramid is classified according to the shape of its base

Pythagorean Theorem a mathematical theorem, proven by the Greek mathematician Pythagoras, that states if the legs of a right triangle have lengths *a* and *b*, and the hypotenuse has length *c*, then $a^2 + b^2 = c^2$

Q

quadrilateral a polygon with four sides

quotient See *division*.

R

radius a line segment that goes from the center to any point on the circle

random sample a sample taken from a population in a way that gives all members of the population the same chance of being selected

Glossary

range the difference between the maximum and minimum values in a set of data

rate a ratio comparing two quantities with unlike units

ratio a comparison of two quantities using division

rational number any number that can be written as a fraction

ray a straight path that extends infinitely in one direction from a point, which is called its *endpoint*

reciprocal two numbers whose product is 1

rectangle a parallelogram with four right angles

reflection a transformation in which a figure "flips" so that its image is the reverse of the original

regular polygon a polygon in which all the sides are the same length and all the angles have the same measure

relation symbol a symbol used to express the relationship between two numbers or expressions; symbols used in number sentences are = for "equal to," < for "less than," > for "greater than," and ≠ for "not equal to."

remainder the number left over when a set of objects is shared equally or separated into equal groups

rhombus a parallelogram whose sides are all the same length

right angle an angle with a measure of 90 degrees

right triangle a triangle that has one right angle

rotation a transformation in which a figure "turns" around a center point or axis

rotational symmetry property of a figure that can be rotated around a point (less than a full, 360-degree turn) in such a way that the resulting figure exactly matches the original figure; if a figure has rotational symmetry, its order of rotational symmetry is the number of different ways it can be rotated to match itself exactly; "no rotation" is counted as one of the ways

S

sample a subset of a group used to represent the whole group

scale drawing an accurate picture of an object in which all parts are drawn to the same scale (e.g., if an actual object measures 32 by 48 meters, a scale drawing of it might measure 32 by 48 millimeters)

scale the ratio of the distance on a map or drawing to the actual distance

scalene triangle a triangle in which all three sides have different lengths

[1]**second** the position next to the first in a sequence

[2]**second** a measure of time equal to one-sixtieth of a minute

set a collection of numbers, points, objects, or other things that are grouped together

seven the Arabic numeral 7

similar figures figures that are exactly the same shape but not necessarily the same size

six the Arabic numeral 6

slope a ratio that indicates the steepness of a line

space figure See *three-dimensional.*

sphere the set of all points in space that are a given distance (the radius) from a given point (the center)

square a plane figure that has four sides of equal length and four right angles

square of a number the product of a number multiplied by itself

square root the square root of a number *n* is a number which, when multiplied by itself, results in the number *n*

square unit a unit used to measure area

standard form a number written so that the location of the digit indicates its value (e.g., in 538, the *3* has a value of 30 because the *3* is in the tens place)

statistics the science of collecting, classifying, and interpreting numerical data as it is related to a particular subject

straight angle an angle of 180 degrees; a line with one point identified as the vertex of the angle

subtract to perform a mathematical operation based on "taking away" or comparing "How much more?"

subtraction a mathematical operation based on "taking away" or comparing "How much more?"; the number being subtracted is called the *subtrahend;* the number it is subtracted from is called the *minuend*

subtrahend See *subtraction.*

sum the result of an addition problem

Glossary

supplementary angles two angles whose measures total 180 degrees

surface area the sum of the areas of all the faces of a space figure

symmetry an arrangement of parts that are alike on either side of a central line or when rotated around a central point

T

table an organized display of information

tally chart a chart that uses tally marks to display data

ten the Arabic numeral 10

tens the place denoting the groups of 10 single units; the second place to the left of the decimal point

tessellation an arrangement of closed shapes that covers a surface completely without overlaps or gaps

tetrahedron a space figure with four faces, each formed by an equilateral triangle

thousands the place denoting groups of 10 hundreds; the fourth place to the left of the decimal point

three the Arabic numeral 3

three-dimensional having length, width, and height

transformation an operation that moves or changes a geometric figure in a specified way; rotations, reflections, and translations are all types of transformations

translation a transformation in which a figure "slides" along a line

transversal a line which intersects two or more other lines

trapezoid a quadrilateral with exactly one pair of parallel sides

triangle a polygon with three sides

two the Arabic numeral 2

two-dimensional having length and width

U

unit (of measure) an agreed-upon standard with which measurements are compared

unit cost the cost of one item or one specified amount of an item

unit fraction a fraction whose numerator is 1

unit rate a rate with a denominator of 1

unit square a square whose sides all have a length of 1

V

variable a letter or another symbol that represents a number, one specific number, or many different values

Venn diagram a picture that uses circles to show relationships between sets; elements that belong to more than one set are placed in the overlap between the circles

vertex the point at which the rays of an angle, two sides of a polygon, or the edges of a polyhedron meet

vertical angles two intersecting lines forming four adjacent angles

volume a measure of the amount of space occupied

W

whole number a number that tells how many complete things there are; the set of whole numbers is {0, 1, 2, 3, 4, …}

word form a number written using words instead of numbers (e.g., 2,446 in word form is "two thousand, four hundred forty-six")

Z

zero the number that, when used as an addend, leaves any number unchanged

Zero Property of Multiplication the property that states any number or quantity multiplied by zero equals zero (e.g., $6 \times 0 = 0$)